Microbiota and Dietary Mediators in Colon Cancer Prevention and Treatment

Surajit Pathak • Antara Banerjee
Asim K. Duttaroy
Editors

Microbiota and Dietary Mediators in Colon Cancer Prevention and Treatment

Springer

Editors
Surajit Pathak
Faculty of Allied Health Sciences
Chettinad Hospital and Research Institute
Chettinad Academy of Research
and Education
Chennai, Tamil Nadu, India

Antara Banerjee
Faculty of Allied Health Sciences
Chettinad Hospital and Research Institute
Chettinad Academy of Research
and Education
Chennai, Tamil Nadu, India

Asim K. Duttaroy
Department of Nutrition, IMB
Faculty of Medicine, University of Oslo
Oslo, Norway

ISBN 978-981-96-0299-5 ISBN 978-981-96-0297-1 (eBook)
https://doi.org/10.1007/978-981-96-0297-1

This Springer imprint is published by the registered company Springer Nature Singapore Pte Ltd.
The registered company address is: 152 Beach Road, #21-01/04 Gateway East, Singapore 189721, Singapore

Preface

The book "Microbiota and Dietary Mediators in Colon Cancer Prevention and Treatment" addresses a critical area of research that combines the fields of microbiology, nutrition, and oncology. As our understanding of the human microbiome expands, its implications for health and disease, particularly in the context of colorectal cancer, become increasingly clear. This book aims to illuminate the significant role that gut microbiota and dietary components play in preventing and managing colon cancer, an area of growing concern as incidence rates rise globally.

This work is particularly relevant in today's healthcare landscape, where there is a need for integrative approaches that extend beyond conventional treatment modalities. As we learn more about the complex interactions between dietary patterns, gut health, and cancer risk, we recognize that addressing the microbiome may offer novel preventive strategies and therapeutic avenues. By synthesizing the latest research findings and clinical insights, this book serves as a valuable resource for healthcare professionals, researchers, and patients seeking to understand and leverage these relationships in the context of colon cancer.

What sets this book apart is its multifaceted approach to the subject matter. First, it emphasizes the unique interplay between probiotics and the existing gut microbiota, exploring how probiotics can modulate microbial communities and enhance health outcomes. This exploration is crucial, as the survival and efficacy of probiotics are significantly influenced by the established microbial ecosystem. Second, the book delves into the concept of a probiotic diet, discussing how dietary choices can influence microbiota composition and metabolite production, ultimately impacting cancer prevention and treatment. This focus on dietary mediators provides a comprehensive perspective on how lifestyle interventions can synergistically work with probiotic therapy.

Additionally, the book underscores the importance of understanding individual variations in gut microbiota when introducing probiotic treatments. By addressing the competitive dynamics within the gut, it highlights the need for personalized approaches in utilizing probiotics for optimal health benefits. This significant understanding is vital for harnessing the full potential of probiotics and dietary interventions in colorectal cancer care. Through its unique integration of microbiology, nutrition, and oncology, this book seeks to foster a deeper understanding of how we can leverage the gut microbiome to improve patient outcomes and advance colon cancer prevention and treatment strategies.

We would like to express our sincere gratitude to Springer Nature for their exceptional support in publishing this book.

Chennai, Tamil Nadu, India	Surajit Pathak
Chennai, Tamil Nadu, India	Antara Banerjee
Oslo, Norway	Asim K. Duttaroy

Contents

Editors and Contributors

About the Editors

Surajit Pathak, Ph.D., is a leading researcher and academic, and is currently a Professor at Chettinad Academy of Research and Education in Chennai. With a Ph.D. from India (2007) and a post-doctoral journey at renowned institutions like the University of Alabama (USA), University of Padova (Italy), and University of Linkoping (Sweden), his research focuses on microRNAs for colorectal cancer detection. Prof. Pathak has published various research articles, and he is a member of global research societies. His work continues to shape advancements in cancer research globally.

Antara Banerjee, Ph.D., is currently working as an Associate Professor at Chettinad Academy of Research and Education (CARE), Chennai. Dr. Banerjee pursued her post-doctoral research on Cancer Biology and Regenerative Medicine at the University of Padova, Italy, and Linkoping University, Sweden. These works fostered her interest in understanding the role of the mesenchymal niche and extracellular vesicles in tissue regeneration and exploring signaling pathways in governing stem cell fate. Dr. Banerjee has edited several books on Stem Cells and Cancer.

Asim K. Duttaroy is a leading expert in nutrition and medical science, renowned for his groundbreaking research in lipid metabolism, cardiovascular health, and chronic disease prevention. He has significantly contributed to understanding how food bioactive compounds impact human health. His pioneering work on the cardioprotective properties of fruits, e.g., "the role of fruitflow (tomato extract) and kiwifruit extract in reducing platelet aggregation and blood pressure," has led to international patents. His innovative research continues to influence global health strategies and food-based therapies.

Contributors

Noha A. Ahmed Physiology Division, Department of Zoology, Faculty of Science, Beni-Suef University, Beni-Suef, Egypt

Osama M. Ahmed Physiology Division, Department of Zoology, Faculty of Science, Beni-Suef University, Beni-Suef, Egypt

Mrithika Ajay Cancer Science Laboratory, Department of Biotechnology, School of Bioengineering, SRM Institute of Science and Technology, Kattankulathur, India

Soniga Apthi Cancer Science Laboratory, Department of Biotechnology, School of Bioengineering, SRM Institute of Science and Technology, Chengalpattu, India

K. R. Arjun Division of Molecular Biology, School of Life Sciences, JSS Academy of Higher Education & Research, Mysuru, Karnataka, India

Raksa Arun Department of Genetic Engineering, College of Engineering and Technology, SRM Institute of Science and Technology, Potheri, Kattankulathur, Chengalpattu, Tamil Nadu, India

Shyamaladevi Babu Faculty of Research, Chettinad Hospital and Research Institute, Chettinad Academy of Research and Education, Kelambakkam, Tamil Nadu, India

Anandan Balakrishnan Department of Genetics, Dr. ALM PG Institute of Basic Medical Sciences, University of Madras, Taramani Campus, Chennai, India

Antara Banerjee Department of Medical Biotechnology, Faculty of Allied Health Sciences, Chettinad Hospital and Research Institute (CHRI), Chettinad Academy of Research and Education (CARE), Chennai, Tamil Nadu, India

Shuvomoy Banerjee Department of Biotechnology and Bioengineering, Institute of Advanced Research (IAR), Gandhinagar, Gujarat, India

Ahamed Basha Abdul Bari Department of Physiology, Chettinad Hospital and Research Institute, Chettinad Academy of Research and Education, Kelambakkam, Tamil Nadu, India

Kanthesh M. Basalingappa Division of Molecular Biology, School of Life Sciences, JSS Academy of Higher Education and Research, Mysore, Karnataka, India

Darshee Baxi Division of Biomedical and Lifesciences, School of Sciences, Navrachana University, Vadodara, India

Sudeep Bose Amity Institute of Biotechnology, Amity University Uttar Pradesh, Noida, Uttar Pradesh, India

Marshall University, Huntington, WV, USA

Hasan Ufuk Celebioglu Department of Biotechnology, Faculty of Science, Bartin University, Bartin, Turkey

Maitri Chakraborty Cancer and Stem Cell Biology Laboratory, Department of Genetic Engineering, School of Bioengineering, SRM Institute of Science and Technology, Kattankulathur, Chengalpattu District, Tamil Nadu, India

Liza Datta Cancer Science Laboratory, Department of Biotechnology, School of Bioengineering, SRM Institute of Science and Technology, Kattankulathur, Chengalpattu, India

Sena Davran Bulut Department of Biotechnology, Faculty of Science, Bartin University, Bartin, Turkey

Arikketh Devi Cancer and Stem Cell Biology Laboratory, Department of Genetic Engineering, School of Bioengineering, SRM Institute of Science and Technology, Kattankulathur, Chengalpattu District, Tamil Nadu, India

Prabu Dhandapani Department of Microbiology, Dr. ALM PG Institute of Basic Medical Sciences, University of Madras, Taramani Campus, Chennai, India

Asim K. Duttaroy Department of Nutrition, IMB, Faculty of Medicine, University of Oslo, Oslo, Norway

Neha Foferkar Division of Molecular Biology, School of Life Sciences, JSS Academy of Higher Education & Research, Mysuru, Karnataka, India

Harin N. Ganesh Department of Genetic Engineering, College of Engineering and Technology, SRM Institute of Science and Technology, Potheri, Kattankulathur, Chengalpattu, Tamil Nadu, India

Dhanavathy Gnanasampanthapandian Cancer Science Laboratory, Department of Biotechnology, School of Bioengineering, SRM Institute of Science and Technology, Kattankulathur, Chengalpattu, India

A. R. Heamchandsaravanan Department of Microbiology, Dr. ALM PG Institute of Basic Medical Sciences, University of Madras, Taramani Campus, Chennai, India

Swati Chandrakant Hosvakkal Department of Microbiology, School of Life Sciences, JSS Academy of Higher Education and Research, Mysore, Karnataka, India

Sabapathy Indu DBT-BIF Centre, Holy Cross College (Autonomous), Tiruchirappalli, Tamil Nadu, India

Vijayalakshmi Jayaraman Cancer Science Laboratory, Department of Biotechnology, School of Bioengineering, SRM Institute of Science and Technology, Kattankulathur, Chengalpattu, India

Priyadarshene Kamalanathan Department of Microbiology, Dr. ALM PG Institute of Basic Medical Sciences, University of Madras, Taramani Campus, Chennai, India

B. M. Kanthesh Division of Molecular Biology, School of Life Sciences, JSS Academy of Higher Education & Research, Mysuru, Karnataka, India

Naman Kapoor Amity Institute of Biotechnology, Amity University, Noida, Uttar Pradesh, India

Navkiran Kaur Amity Institute of Biotechnology, Amity University, Noida, Uttar Pradesh, India

Minu Kanchan Kesheri Marshall University, Huntington, WV, USA

Rehab G. Khalil Immunology Division, Department of Zoology, Faculty of Science, Beni-Suef University, Beni-Suef, Egypt

Madhan Krishanan Faculty of Research, Chettinad Hospital and Research Institute, Chettinad Academy of Research and Education, Kelambakkam, Tamil Nadu, India

Sakshi Kumari Amity Institute of Biotechnology, Amity University Uttar Pradesh, Noida, Uttar Pradesh, India

Vasanth Kanth Thasma LoganathBabu Department of Genetic Engineering, College of Engineering and Technology, SRM Institute of Science and Technology, Potheri, Kattankulathur, Chengalpattu, Tamil Nadu, India

Shrishti Madhan Cancer and Stem Cell Biology Laboratory, Department of Genetic Engineering, School of Bioengineering, SRM Institute of Science and Technology, Kattankulathur, Chengalpattu District, Tamil Nadu, India

Sujoy Midya Department of Zoology, School of Biological Sciences, Central University of Kerala, Kasaragod, India

Juni Mukherjee Department of Biotechnology and Bioengineering, Institute of Advanced Research (IAR), Gandhinagar, Gujarat, India

D. Mutthuraj Division of Molecular Biology, School of Life Sciences, JSS Academy of Higher Education & Research, Mysuru, Karnataka, India

R. Mythreyi Division of Molecular Biology, School of Life Sciences, JSS Academy of Higher Education & Research, Mysuru, Karnataka, India

ArulJothi Kandasamy Nagarajan Department of Genetic Engineering, College of Engineering and Technology, SRM Institute of Science and Technology, Potheri, Kattankulathur, Chengalpattu, Tamil Nadu, India

Dhanush Nagendra Division of Molecular Biology, School of Life Sciences, JSS Academy of Higher Education & Research, Mysuru, Karnataka, India

Ragala Venkata Nandha Department of Microbiology, Dr. ALM PG Institute of Basic Medical Sciences, University of Madras, Taramani Campus, Chennai, India

Abilash Navaneethan Cancer Science Laboratory, Department of Biotechnology, School of Bioengineering, SRM Institute of Science and Technology, Kattankulathur, India

Kanagaraj Palaniyandi Cancer Science Laboratory, Department of Biotechnology, School of Bioengineering, SRM Institute of Science and Technology, Kattankulathur, Chengalpattu, India

Surajit Pathak Department of Medical Biotechnology, Faculty of Allied Health Sciences, Chettinad Hospital and Research Institute (CHRI), Chettinad Academy of Research and Education (CARE), Chennai, Tamil Nadu, India

Debarati Paul Amity Institute of Biotechnology, Amity University Uttar Pradesh, Noida, Uttar Pradesh, India

Mashira Rahman Department of Genetic Engineering, College of Engineering and Technology, SRM Institute of Science and Technology, Potheri, Kattankulathur, Chengalpattu, Tamil Nadu, India

Manikkam Rajalakshmi DBT-BIF Centre, Holy Cross College (Autonomous), Tiruchirappalli, Tamil Nadu, India

PG & Research Department of Zoology, Holy Cross College (Autonomous), Tiruchirappalli, Tamil Nadu, India

PG & Research Department of Biotechnology & Bioinformatics, Holy Cross College (Autonomous), Affiliated to Bharathidasan University, Tiruchirappalli, Tamil Nadu, India

Satish Kumar Rajasekharan Department of Biotechnology, School of Bioengineering, SRM Institute of Science and Technology, Chengalpattu, India

Vasukidevi Ramachandran Department of Microbiology and Biotechnology, Bharath Institute of Higher Education and Research, Selaiyur, Chennai, India

Satish Ramalingam Department of Genetic Engineering, College of Engineering and Technology, SRM Institute of Science and Technology, Potheri, Kattankulathur, Chengalpattu, Tamil Nadu, India

Anupama Saha Amity Institute of Biotechnology, Amity University, Noida, Uttar Pradesh, India

Shivani Sharda Amity Institute of Biotechnology, Amity University, Noida, Uttar Pradesh, India

Satya Priya Sharma Institute for Liver and Digestive Diseases, Hallym University, Chuncheon, Republic of Korea

P. Shriya Department of Genetic Engineering, College of Engineering and Technology, SRM Institute of Science and Technology, Potheri, Kattankulathur, Chengalpattu, Tamil Nadu, India

J. M. Siddesha Division of Biochemistry, School of Life Sciences, JSS Academy of Higher Education & Research, Mysuru, Karnataka, India

Ki Tae Suk Institute for Liver and Digestive Diseases, Hallym University, Chuncheon, Republic of Korea

Department of Internal Medicine, Hallym University College of Medicine, Chuncheon, Republic of Korea

K. Sumana Department of Microbiology, School of Life Sciences, JSS Academy of Higher Education and Research, Mysore, Karnataka, India

Taparia Saloni Department of Microbiology, School of Life Sciences, JSS Academy of Higher Education & Research, Mysuru, Karnataka, India

S. Umamaheshwari Department of Microbiology, School of Life Sciences, JSS Academy of Higher Education & Research, Mysuru, Karnataka, India

Lourduraj Vasanthi Unit of Cryobiology and Invertebrate Reproduction, Department of Zoology, University of Madras, Guindy Campus, Chennai, India

R. M. Vijay Murali Department of Genetic Engineering, College of Engineering and Technology, SRM Institute of Science and Technology, Potheri, Kattankulathur, Chengalpattu, Tamil Nadu, India

Periyasamy Vijayalakshmi DBT-BIF Centre, Holy Cross College (Autonomous), Tiruchirappalli, Tamil Nadu, India

PG & Research Department of Biotechnology & Bioinformatics, Holy Cross College (Autonomous), Affiliated to Bharathidasan University, Tiruchirappalli, Tamil Nadu, India

Nour Y. S. Yassin Physiology Division, Department of Zoology, Faculty of Science, Beni-Suef University, Beni-Suef, Egypt

Gut Microbiota and Its Importance in Health and Disease

1

Sujoy Midya, Antara Banerjee, Surajit Pathak, and Asim K. Duttaroy

Abstract

The relationship between gut microbiota and human health is being increasingly recognized. It is now well established that a healthy gut flora is mainly responsible for the host's overall health. The normal human gut microbiota comprises two significant phyla: Bacteroidetes and Firmicutes. Several factors shape the normal gut microbiota, such as the mode of delivery, diet, use of antibiotics, or antibiotic-like molecules derived from the environment or the gut commensal community. The microbial communities are symbiosis with the host, contributing to homeostasis and regulating immune function. However, microbiota dysbiosis can lead to dysregulation of bodily functions and diseases, including cardiovascular diseases, cancers, respiratory diseases, dementia, etc. This chapter discusses the current knowledge of how microbiota links to host health or pathogenesis.

Keywords

Gut microbiota · Microbiota dysbiosis · Host health · Immune regulation · Diseases pathogenesis

---

S. Midya
Department of Zoology, School of Biological Sciences, Central University of Kerala, Kasaragod, India

A. Banerjee · S. Pathak
Department of Medical Biotechnology, Faculty of Allied Health Sciences, Chettinad Hospital and Research Institute (CHRI), Chettinad Academy of Research and Education (CARE), Chennai, Tamil Nadu, India

A. K. Duttaroy (✉)
Department of Nutrition, IMB, Faculty of Medicine, University of Oslo, Oslo, Norway
e-mail: a.k.duttaroy@medisin.uio.no

S. Pathak et al. (eds.), *Microbiota and Dietary Mediators in Colon Cancer Prevention and Treatment*, https://doi.org/10.1007/978-981-96-0297-1_1

1.1 Introduction

Our gastrointestinal tract (GIT) is home to a complex and dynamic population of microorganisms known as the gut microbiota, which significantly impacts host health during periods of equilibrium and disease. The development of the GIT microbiota during infancy is dependent on multiple factors, which include a wide variety of microbes, including bacteria, fungi, and viruses, residing in various parts of the body. Microbiota refers to the active living organisms in a specific environment, such as those in the oral and gastrointestinal tracts (Thursby and Juge 2017). Diet is recognized as a primary factor influencing gut microbiota composition throughout an individual's life span. Intestinal bacteria play a pivotal role in maintaining immune and metabolic homeostasis in defense against pathogens. The gut microbes are predominantly divided into six phyla, such as Firmicutes, Bacteroidetes, Actinobacteria, Proteobacteria, Fusobacteria, and Verrucomicrobia, with Firmicutes and Bacteroidetes being the dominant phyla (Laterza et al. 2016). The most extensively researched groups of microorganisms within the GIT microbiota include *Candida*, *Saccharomyces*, *Malassezia*, and *Cladosporium* fungi.

Extensive research has revealed the significant interaction between gut microbiota and fundamental human biological processes (Hou et al. 2022). For instance, recent studies have demonstrated that the human microbiota is intimately involved in extracting nutrients, metabolism, and immune function (Cummings and Macfarlane 1997). The influence of microbiota on biological processes can be achieved through various mechanisms. Regarding the extraction of nutrients from food, the microbiota plays a pivotal role in the diverse metabolic genes, which can produce some unique enzymes and control the essential biochemical pathways (Turnbaugh et al. 2006).

Furthermore, the formation of bioactive compounds, including vitamins, amino acids, and lipids, is highly dependent on the gut microbiota. In the context of the defense system, the gut microbes not only protect the host against pathogens by secreting immunological substances (lymphokines, interleukins, tumor necrosis factors (TNFs), and many others) but also act on the mucosal immunity. The distribution of microbes exhibits distinctive uniqueness across individuals and can experience variations even within the same individual. In humans, the nature of the gut microbiota depends on several factors like age and environmental conditions, including the use of medications.

Furthermore, the composition of the gut microbes varies across different gastrointestinal tract regions. For instance, Proteobacteria, particularly from the *Enterobacteriaceae* family, are predominantly present in the small intestine, while Bacteroidetes, including families *Bacteroidaceae, Prevotellaceae, and Rikenellaceae*, are commonly found in the large colon (Flint et al. 2012). Microbiota diversity progressively increases from childhood to adulthood before exhibiting a decline in older age, specifically in individuals over 70. Before the establishment of a relatively stable composition within the gut microbiota, the variety of microbiota in children is primarily known as *Akkermansia muciniphila*, *Bacteroides* spp.,

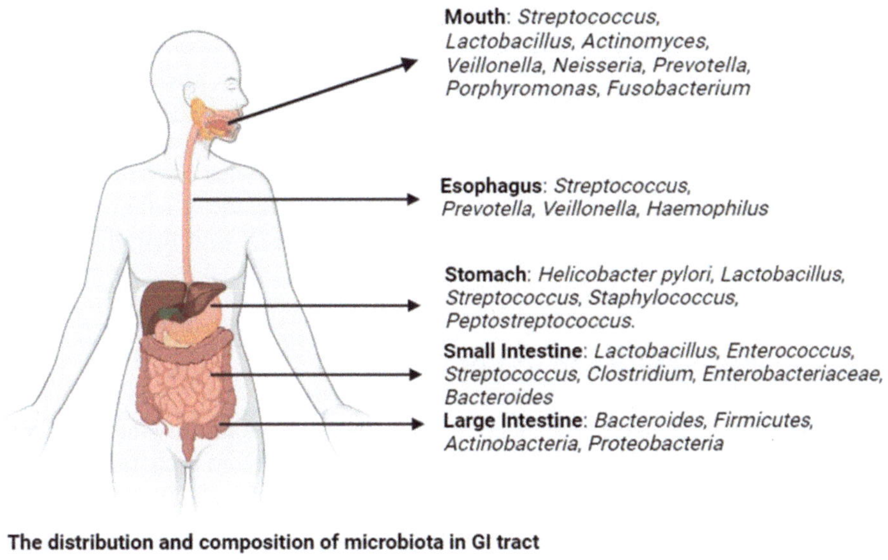

Fig. 1.1 The composition of the gut microbiota

Veillonella spp., *Clostridium coccoides*, and *Clostridium botulinum* (Amabebe et al. 2020) (Fig. 1.1).

1.2 Establishment of Infant's Microbiota Through Maternal Microbiota

Throughout pregnancy, the maternal microbiota (uterine microbiota, vaginal microbiota, gastrointestinal microbiota, placental microbiota, and oral microbiota) takes part in the development processes of the brain and gastrointestinal areas of the fetus (Basak et al. 2022, 2024). Environmental and host factors influence the colonization of the infant gut microbiota. Microbiota is transferred vertically from a mother to her offspring (Yassour et al. 2018). These early colonizers include beneficial and harmful bacteria, crucial for the infant's health. The type of microbes the newborn is exposed to at birth is particularly significant, with maternal microbes being essential. Romero et al. observed that the vaginal microbiota of healthy pregnant women tends to be more stable than that of healthy nonpregnant women, with *Lactobacillus* species being the most common (Romero et al. 2014). The placental microbiota, which is mainly composed of nonpathogenic symbiotic bacteria, such as Fusobacteria, Firmicutes, Bacteroidetes, Tenericutes, and Proteobacteria, is remarkably similar to the oral microbes but exhibits lower diversity compared to the gut microbes (Aagaard et al. 2014). The infant's microbiota was found to undergo significant changes within the first two days of life, with *Clostridia, Enterobacteria, and Streptococci* being identified within the gut of newborns (Mitsuoka 1996). Notably, by the third day, the existence of *Bacteroides, Bifidobacterium, and*

Clostridia shows to be ubiquitous in 40% of infants, by the fourth to seventh days post-birth, *Bifidobacterium* emerges as the prevalent species, obtaining its maximal relative dominance during this period (Mitsuoka 1996). Before weaning, there is a gradual rise in the absolute abundance of *Bacteroides* species, suggesting competition with *Bifidobacterium* within the infant intestine. Upon solid foods' introduction, the newborn gut microbe composition was identical to that observed in adults (Clemente et al. 2012). In a 2- to 3-year post-birth timeline, the infant gut microbiota undergoes a maturation process, aligning progressively with that of adult gut microbiota.

1.3 Gut Structure and Its Compositions

The gastrointestinal tract is segmented functionally and anatomically into the stomach, small intestine, and large intestine (LI). Each compartment is characterized by its unique microenvironment and physicochemical barriers, which play a pivotal role in dictating the growth of particular microbiota.

1.3.1 Stomach

Traditionally, the stomach was believed to be devoid of bacterial life, deemed unwelcoming to bacterial proliferation, attributed to a bactericidal barrier, the reflux of bile acids, the layer of mucus, and gastric peristalsis. However, in 1981, the prestigious *Lancet* medical journal published reports of many acid-resistant bacterial strains, including *Streptococcus*, *Neisseria*, and *Lactobacillus.* This revelation was further substantiated in 1982 when *Campylobacter pyloridis* was identified by researchers Robin Warren and Barry Marshall, subsequently named *Helicobacter pylori* in 1984. These *Helicobacter* species predominantly originate from the oral region and have demonstrated the ability to withstand stomach acid and a transient existence. In a healthy human stomach, the flora typically consists of five major phyla: Firmicutes, Bacteroidetes, Actinobacteria, Fusobacteria, and Proteobacteria (Kazor et al. 2003). Common bacterial genera within these phyla include *Prevotella, Streptococcus, Veillonella*, *Rothia*, and *Haemophilus.*

1.3.2 Small Intestine

The small intestine (SI) plays a pivotal role in the digestion and absorption of nutrients from food. It is segmented into three distinct sections: the duodenum, jejunum, and ileum. Within the duodenum, the microenvironment is defined by bile acids, pancreatic secretions, and antimicrobial agents. This environment facilitates the rapid transit of food while limiting the density and diversity of bacteria. The small intestine predominantly comprises the bacterial phyla Bacillota, Bacteroidota, Pseudomonadota, Fusobacteriota, and Actinomycetota (Vonaesch et al. 2022;

Nagasue et al. 2022). The relative abundance of these phyla may vary among individuals and studies.

Other bacterial phyla, such as Verrucomicrobiota and Saccharibacteria, are sporadically found. Recent studies are focusing on evaluating the GIT microbiome. For example, Nagasue et al. conducted a study that analyzed the microbiome across various segments of the GIT, including the stomach, duodenum, jejunum, ileum, rectum, and feces of 29 patients presenting with gastrointestinal symptoms (Nagasue et al. 2022). While *Bacillota* species are predominant in the duodenum, the abundance of *Pseudomonadota* species progressively increases as the small intestinal tract extends from the proximal to the distal regions. *Bacteroides* species exhibit higher prevalence in the ileum, sometimes it is found in the colon and the feces (Zmora et al. 2018).

1.3.3 Large Intestine

The large intestine primarily comprises three sections: ascending, transverse, and descending, alongside the cecum. It is a crucial location for water absorption and the fermentation of undigested food, attributed to its slower food transit and anaerobic environment. The bacterial population density can reach 10^{12} CFU/mL, primarily dominated by the phyla Firmicutes and Bacteroidetes (Eckburg et al. 2005). The relative abundance of these bacterial phyla may depend on different life phases and pathological conditions. Consequently, the proportion of Firmicutes to Bacteroidetes is often considered a predictive indicator of health as well as disease (Mariat et al. 2009). Within the intestinal lumen, key bacterial genera include *Bacteroides, Bifidobacterium, Ruminococcus, Streptococcus, Enterobacteriaceae, Enterococcus, Clostridium*, and *Lactobacillus*, while *Clostridium, Lactobacillus, Enterococcus*, and *Akkermansia* are predominantly found within the intestinal mucosa. Additionally, many pathogens, such as *E. coli, Campylobacter jejuni, Vibrio cholera, Salmonella enterica*, and *Bacteroides fragilis*, may occasionally present in the large intestine, although at deficient concentrations (0.1%).

1.4 Aging Associated with Gut Microbiota

Aging is defined by the decline in the functionality of tissues and organs and a raised susceptibility to disorders associated with aging (Ling et al. 2022). These pose significant challenges to society and are considered a health concern. Given the substantial increase in individuals' life spans, there is an increasing emphasis on monitoring life expectancy and health expectancy to evaluate the influence of interventions on aging. Indeed, life expectancy and health expectancy are closely interconnected, with people who exceptionally live long often maintaining good health for a considerable portion of their lives. However, diseases such as stroke, cancer, cardiovascular diseases, and other severe conditions have emerged as the predominant causes of high mortality rate among the elderly.

The gut microbes play a crucial role in regulating healthy aging, ultimately affecting lifespan. Specific groups of bacteria, such as *Bifidobacterium*, *Bacteroides*, and *Eubacterium,* are particularly influential in predicting lifespan and have been identified as potential biomarkers of aging (Galkin et al. 2020). Biagi et al. propose *Akkermansia muciniphila* as an essentialbiomarker of healthy aging (Biagi et al. 2016). Advanced research in animal models has confirmed the helpful effects of *Akkermansia muciniphila* in enhancing health span and lifespan, likely attributed to its role in restoring beneficial gut microbes (Barcena et al. 2019). Addressing the accelerated aging connected with gut dysbiosis has proven beneficial, indicating an interlink between aging and the gut microbiota. This implies that interventions focusing on the gut microbiota could help combat age-related diseases.

1.5 The Influence of Gastrointestinal Microbiota on Disease States

Sudden changes in the composition and structure of the gastrointestinal microbiota may cause an imbalance within this ecosystem. Such an imbalance can result in a reduction in the population of beneficial bacteria, while simultaneously allowing for an increase in the presence of pathogenic microorganisms within the intestine. Pathogenic bacteria are capable of secreting a diverse array of toxic substances, damaging the epithelial cells of the intestine, and leading to the development of several diseases, including type 2 diabetes, colorectal cancer, rheumatoid arthritis, irritable bowel syndrome, and mental health disorders such as anxiety, depression, dementia, and autism spectrum disorder. Additionally, these bacteria initiate the development of liver and cardiovascular diseases (CVD) (Fig. 1.2).

1.6 The Impact of Microbiota in the Enhancement of Colorectal Cancer

Colorectal cancer (CRC) is the second most deadly disease globally, accounting for 9.2% of all cancer-related fatalities (Esmeeta et al. 2022). It is also the third most frequent cancer overall. The human gut's millions of microorganisms, primarily bacteria, must coexist in harmony with their host to thrive. There is a 12-fold increased chance of cancer in the large intestine compared to the small intestine, and the large intestine has a bacterial density that is around sixfold higher (Tjalsma et al. 2012). Many cancers, including gastric, cervical, and colorectal, are initiated and developed particularly in locations frequently exposed to microorganisms, and these cancers respond differently to certain types of bacteria. Members of the Fusobacterium genus, such as *Fusobacterium nucleatum*, *Fusobacterium mortiferum*, and *Fusobacterium necrophorum* sequences, were shown to be enriched in tumor tissue by the colorectal microbiome (Kostic et al. 2012; Castellarin et al. 2012). It is believed that CRC begins when intestinal epithelial cells' stem cells at the base of their villi crypts experience mutations that render them immortal and

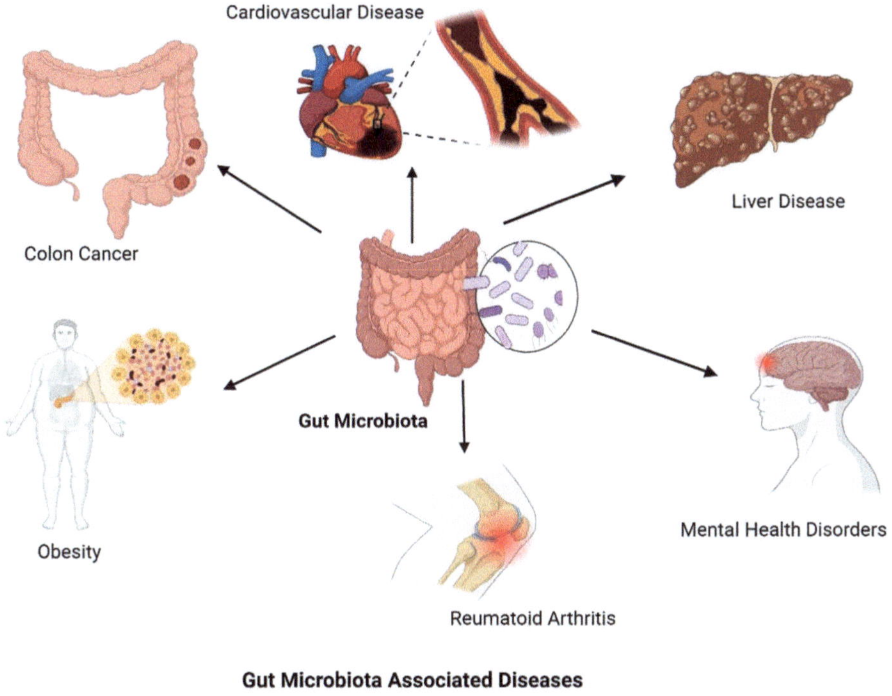

Fig. 1.2 Diseases influenced by alterations in the gut microbiota

capable of accumulating other mutations, including those related to adenomatous polyposis coli (APC), the β-catenin gene (CTNNB1), TP53, Kirsten rat sarcoma (KRAS), and the myelocytomatosis oncogene (MYC). Several experimental studies provide that inflammatory cell and their associated mediators, such as interleukin 6 (IL-6), tumor necrosis factor α (TNF-α), IL-23, and reactive oxygen species, form a microenvironment favoring the tumor (Arthur et al. 2012). *Fusobacterium* spp. contribute to the mechanism of tumorigenesis by which chronic intestinal inflammation leads to CRC.

1.7 The Impact of Microbiota in the Development of Rheumatoid Arthritis (RA)

Rheumatoid arthritis is an autoimmune disease where the joints are affected. Specific gut microbes may play an essential role in the development of rheumatoid arthritis (RA). Changes in the gut bacterial composition within the GI tract of individuals exhibiting early indicators and those with advanced RA suggest that gut microbes are pivotal in developing the immunological dysfunctions characteristic of RA. Scher and his associate investigators conducted genetic analyses to explore the gut bacterial composition of individuals with RA and those without the condition.

The results indicate a significant correlation between the presence of a specific bacterial species, *Prevotella copri*, and the treatment status of RA patients (Scher et al. 2013). Notably, a higher abundance of *Prevotella copri* was observed in untreated RA patients than in those without the disease.

Furthermore, it was noted that the greater the abundance of *Prevotella copri,* the lower the diversity of other microbe species, including some beneficial to health. The relative abundance of *Collinsella* was observed to be elevated in RA patients. The results of in vitro investigations indicate that *Collinsella aerofaciens* is an active arthritogenic microorganism in the human gut because it enhances gut permeability and stimulates the expression of IL-17A, a crucial cytokine engaged in the pathogenesis of RA (Chen et al. 2016). To summarize, the predominant gut microbiota in individuals with early RA comprises *Prevotella copri* and *Collinsella,* which may have a key role in the development of the disease.

1.8 The Influence of Microbiota in the Enhancement of Type 2 Diabetes (T2D)

Diabetes is defined by high blood sugar levels resulting from reduced insulin production by the pancreas or a decreased ability of tissues to respond to insulin signals (Defronzo 2010). Type 2 diabetes is recognized as a chronic, noncommunicable disease that is a main reason for premature deaths worldwide. It is characterized by impaired lipid metabolism, the presence of oxidative stress, and hypertension, which can lead to both microvascular and macrovascular complications. Among these, microvascular complications impact small blood vessels, such as diabetic nephropathy, neuropathy, and retinopathy. Studies show that individuals with type 2 diabetes often have an altered composition of gut microbiota.

Frequently reported changes include a decrease in *Bifidobacterium*, *Bacteroides*, *Faecalibacterium*, *Akkermansia*, and *Roseburia*, and an increase *in Ruminococcus, Fusobacterium*, and *Blautia* (Gurung et al. 2020). The relative abundance of *Firmicutes*, a class of bacteria, is notably lower in diabetic individuals than those without diabetes (Larsen et al. 2010). Conversely, the proportion of Bacteroidetes and Proteobacteria, another class of bacteria, is somewhat higher. Interestingly, the ratio of Bacteroidetes to Firmicutes is positively and significantly correlated with poorer glucose tolerance. Short-chain fatty acids (SCFAs), mainly butyrate, which is linked with insulin sensitivity, as gut microbes and butyrate are both reduced in T2D, (Donohoe et al. 2011). The relationship between SCFAs and insulin sensitivity is rooted in the capability of SCFAs to stimulate the secretion of GLP-1 by intestinal L-cells through specific G protein receptors (GPR41, GPR43) (Fava 2014). This mechanism significantly affects pancreatic function, insulin release, and central effects that regulate appetite.

1.9 The Negative Impact of Gut Microbiota on Irritable Bowel Syndrome

Irritable bowel syndrome (IBS) represents a functional gastrointestinal disorder characterized by persistent abdominal discomfort or pain, often linked to changes in bowel habits, such as diarrhea and constipation, devoid of visible structural abnormalities. Although not a severe condition, IBS exhibits a significant prevalence, affecting approximately 10–15% of the adult population (Drossman et al. 2002). This widespread condition imposes considerable societal and economic burdens and adversely affects the quality of life for those affected. Alterations in the gut microbiota have been proposed as a potential underlying cause for IBS. Changes in the gut composition may lead to low-grade inflammatory responses. Clinical and epidemiological studies suggest a correlation between small bowel bacterial overgrowth (SIBO) and developing IBS in patients. Research showed that most individuals recovering from acute infectious colitis due to *Salmonella* species, *Campylobacter jejuni*, or *Shigella* species experience symptoms of IBS (Mckendrick and Read 1994). The incidence of post-infectious IBS following acute infectious colitis ranges from 6% to 31%, with specific risk factors being identified, including younger age, female gender, bloody stools, and prolonged duration of diarrhea (Ji et al. 2005). Furthermore, studies observe that the number of chronic inflammatory cells in the colonic mucosa is elevated in individuals with IBS compared to those who do not progress to the disorder, with the activation of T cells and the expression of inflammatory cytokines being noted after 3 months of infection (Rana et al. 2008).

1.10 The Impact of Gut Microbiota on Mental Health Disorders

A healthy GI tract is associated with properly functioning the central nervous system (CNS). The gut is responsible for releasing hormones, neurotransmitters, and immunological components, which have been identified to transmit signals to the brain through direct and indirect pathways involving autonomic neurons. Recent research has highlighted the impact of microbiome variations on the incidence and progression of various CNS diseases. These diseases include, but are not restricted to, anxiety, depressive disorders, schizophrenia, and autism.

1.10.1 Anxiety

Among mental health disorders, anxiety is one of the most prevalent. Research suggests an interaction between specific gut microbiota and anxiety disorders. Social isolation is identified as a key contributor to anxiety. A study found an increase in the abundance of *Prevotella* while simultaneously observing a significant decline in the Firmicutes/Bacteroidetes ratio and the presence of *Faecalibacterium* spp. in individuals experiencing social exclusion (Kim et al. 2022). Moreover, patients diagnosed with generalized anxiety disorder (GAD) exhibited reduced microbial richness and diversity, along with diminished levels of *Firmicutes* spp. and those capable of producing short-chain fatty acids (SCFAs) and increased abundance of Bacteroidetes and Fusobacteria (Chen et al. 2018).

Patients and mice experiencing anxiety exhibited a notable decline in microbial richness and diversity. At the phylum level, it was noted that patients/mice with anxiety typically had lower Firmicutes populations but higher abundance of Bacteroidetes and Fusobacteria (Ericsson et al. 2021). Furthermore, specific gut microbe genera strongly connected with anxiety, including *Prevotella*, *Lactobacillales*, *Sellimonas*, *Streptococcus*, and *Enterococcus*. Conversely, other gut microbiota genera were found to have an inverse correlation with anxiety, indicating a potential therapeutic avenue for anxiety prevention by targeting these gut microbiomes.

1.10.2 Depression

Depression is recognized as a major concern for public health, with the potential to lead to numerous severe consequences and contribute to a notable incidence of suicide. Research has indicated a correlation between gut microbiome dysbiosis and both the onset and progression of depressive disorders (Tan et al. 2021). In studies, certain species within the gut microbiota, such as *Paraprevotella*, showed a positive association with depression, while others, including Streptococcaceae and *Gemella*, exhibited a negative correlation. On a broader level, several genera of gut microbiota, such as *Prevotella*, *Klebsiella*, and *Clostridium*, showed a meaningful correlation with depressive illness (Lin et al. 2017). Furthermore, it is suggested that gut microbiome dysbiosis could play a crucial role in the progression of depression by affecting the expression of proteins related to the gut–brain axis (Liu et al. 2021). These results collectively show a significant change in the diversity of gut microbiota in those with depressive disorders, although the specific composition of gut microbiota in individuals with depression can vary between studies. This suggests that the gut microbiota might represent a promising area for depression prevention and treatment strategies.

1.10.3 Autism Spectrum Disorder (ASD)

ASD constitutes a cluster of neurodevelopmental disorders marked by repetitive behavior patterns and impairments in communicative and social interactions. It was previously hypothesized that ASD stemmed from environmental factors. Growing evidence suggests a correlation between ASD and the GIT microbiome (Garcia-Gutierrez et al. 2020). Both experimental and clinical investigations have demonstrated that patients diagnosed with ASD exhibit modifications in their gut microbiota composition. Various species within the gut microbiota community were exclusively present in patients with ASD, including *Clostridium difficile* and

Clostridium clostridioforme (Kandeel et al. 2020; Dilmore et al. 2021). Furthermore, it has been identified that specific groups of gut microbiota were significantly augmented in both ASD individuals and animal models, namely, Actinobacteria, Proteobacteria, and Bacilli. In contrast, the presence of certain species, including *Akkermansia muciniphila*, *Faecalibacterium*, and *Agathobacter*, was notably diminished.

1.10.4 Dementia

Dementia is a widespread neurodegenerative disorder characterized by cognitive impairment and a progressive decline in an individual's ability to function independently (Liu et al. 2019). Among the various forms of dementia, Alzheimer's disease stands out the most prevalent, accounting for approximately 50–70% of all cases. The gut microbiota has been identified as playing an important role in the nervous system, mainly through the brain–gut axis. It has even been implicated in the enhancement of neurodegenerative diseases (Cryan et al. 2020). Certain bacterial families, including Desulfovibrionaceae, Lactobacillaceae, *Ruminococcus gnavus* group, *Lactobacillus*, Desulfovibrionales, *Sellimonas*, Bacillales, and *Hungatella*, are positively concerned about the possibility of developing dementia. This suggests a potential link between specific types of dementia and the composition of the gut microbiota.

In Alzheimer's disease, the abnormal production and processing of amyloid-beta (Aβ) and the hyperphosphorylation of tau proteins are key molecular markers. The influence of Desulfovibrionaceae, a bacterial family, on patients afflicted with dementia or Alzheimer's disease, specifically concerning the deposition of β-amyloid in the brain, has been reported (Fu et al. 2023). Short-chain fatty acids (SCFAs), which are primarily composed of acetate, propionate, and butyrate, are produced by the gut microbiota. Butyrate, in particular, has anti-inflammatory properties and has proven that to enhance cognitive function by mediating inflammatory responses and promoting the phagocytosis of amyloid-beta (Aβ) by microglia (Mirzaei et al. 2021). A deficiency in butyrate has been linked to the expansion of Alzheimer's disease (Tran et al. 2019). Furthermore, the gut microbiome is implicated in Lewy body dementia, characterized by the aggregation of α-synuclein. This condition is linked to the microbe–gut–brain axis, which is thought to perform a critical role through various mechanisms (Ryman et al. 2023). However, studies about the connection between Alphaproteobacteria, a group of bacteria, and Lewy body dementia is limited. Alphaproteobacteria is commonly found in individuals with depression, and the use of antidepressants showed to reduce their abundance (Lukic et al. 2019).

1.11 Influence of Gut Microbes in the Development of Liver Disease

The liver and the GIT microbiota communicate via the biliary system and portal circulation (Starkel and Schnabl 2016). The gut microbes and its metabolites are participated in developing and progressing liver disease. Various liver illnesses, such as nonalcoholic fatty liver disease, alcoholic liver disease, hepatitis C, and hepatocellular carcinoma, are made worse and have worse prognoses with aging. Dysbiosis, which occurs when the symbiosis between the microbiota and the host is disrupted, can significantly negatively impact health. Liver disease is linked to qualitative changes, such as an enhanced level in hazardous bacteria and a decline in good bacteria and, additionally, quantitative changes in the overall amount of bacteria (Adhikary et al. 2024). This process facilitates the distribution of nutrients, microbial products, metabolites, and intestinal toxins, as the liver emerges as the initial organ in the human body to receive them.

Additionally, nutrients and microbial products are further disseminated to the liver through lymphatic vessels, the thoracic duct, and systemic circulation (Alexander et al. 2010). Connected to the intestine through the biliary system, the liver secretes bile and its constituents and bile acids, which are transported to the small intestine via bile ducts. Alcohol-related liver disease stands as a primary cause of illness and mortality on a global scale. It has been noticed that patients with alcohol-use disorders, suffering from dysbiosis, exhibit a diminished proportion of Bacteroidetes and a heightened presence of Proteobacteria in their microbiomes compared to individuals without such disorders.

Furthermore, there has been a significant reduction in beneficial bacteria, including *bifidobacteria* and *lactobacilli*, among those with alcohol-use disorders (Kirpich et al. 2008). Nonalcoholic fatty liver disease (NAFLD) represents the manifestation of metabolic dysfunction in the liver. Nonalcoholic steatohepatitis (NASH) is defined as systemic manifestation of hepatic inflammation and fibrosis. Individuals diagnosed with NASH are at a significant risk of progressing to advanced fibrosis and cirrhosis. Dietary choline undergoes metabolism by the intestinal microbiota to form trimethylamine (TMA), leading to a deficiency in this essential nutrient (Mehedint and Zeisel 2013; Zeisel et al. 1983). A deficiency in hepatic choline results in reduced excretion of very low-density lipoproteins (VLDL), leading to the accumulation of fatty deposits in the liver, known as hepatic steatosis.

Additionally, alterations in the microbiota can produce ethanol (EtOH), which is then absorbed and metabolized by the liver (Goffredo et al. 2017). The intestinal microbiota also suppresses the expression of the fasting-inducible adipocyte factor in the cells lining the intestine. This suppression of gene expression enhances lipoprotein lipase activity and increases the levels of free fatty acids (FFA) within the body. Exaggerated permeability of the intestinal lining results in the translocation of microbial products to the liver, triggering inflammation by stimulating toll-like receptors. Primary biliary cholangitis (PBC) is an autoimmune liver disease characterized by portal inflammation and the destruction of intrahepatic bile ducts, leading to fibrosis and, in some cases, cirrhosis (Kaplan and Gershwin 2005). It has been identified that PBC patients experience a depletion in certain *Lachnobacterium* spp., *Bacteroides eggeri*, *Ruminococcus bromii*, and *Acidobacteria* within their gut microbiomes while being enriched with specific pathobiont taxa, including *Enterobacteriaceae*, *Neisseriaceae*, *Streptococcus*, and Proteobacteria (Lv et al. 2016). Several altered gut microbes have been interconnected with metabolic functions, immune responses, and liver functions (Lv et al. 2016). When compared to healthy people, it has been observed that PBC patients had much less Fusobacterium along with greater Eubacterium in their oral microbiomes.

1.12 The Impact Gut Microbiomes on Cardiovascular Disease (CVD)

The interplay between the microbiota and heart disease has emerged as a critical area of research interest. The occurrence of CVD risk factors, like atherosclerosis, hypertension, inflammation, diabetes, obesity, insulin resistance, hypercholesterolemia, platelet hyperactivity, and aberrant plasma lipid levels, has been strongly linked with altered gut microbes, based on emerging studies. Numerous investigations carried out in both individuals and animal models have indicated a relationship between gut microbial metabolites, including trimethylamine-N-oxide or TMAO, short-chain fatty acids (butyric, acetic) and bile acid metabolites (products of the breakdown of amino acids), and heart disease. The hemostatic process relies extensively on blood platelets. Additionally, these blood cells are associated with cardiac events and play a critical role in the development of atherosclerosis. Diet–gut microbiota interactions are gaining significance in the field of cardiovascular research because the TMAO and other metabolites of the gut microbiota are linked to oxidative stress, lipid complications, and platelet hyperactivity (Duttaroy 2021). A significant proportion of the microorganisms implicated in the formation of atherosclerotic plaques, endothelial dysfunction, and subsequent CVDs is thought to originate from their translocation from the gastrointestinal tract into the systemic circulation. One such compound, trimethylamine N-oxide (TMAO), has been identified as correlating with an enhanced susceptibility to CVD (Tang et al. 2017). Individuals afflicted with coronary artery disease (CAD) harbor species of *Staphylococcus* and *Proteus vulgaris*, *Klebsiella pneumoniae*, and *Streptococcus* spp., among others (Ahmad et al. 2019). Their gastrointestinal tracts show an elevation in specific microbial populations, including *Lactobacillus*, *Streptococcus*, *Escherichia*, *Shigella*, and *Enterococcus* spp., alongside a decline in beneficial species such as *Faecalibacterium*, *Subdoligranulum*, *Roseburia*, *Eubacterium rectale*, and *Bacteroides fragilis*, all of which play a pivotal role in modulating T-cell functions within the gut mucosa, thereby facilitating anti-inflammatory responses and safeguarding the integrity of the intestinal barrier (Tuomisto et al. 2019). In patients at a heightened risk of stroke, a decrease in bacteria capable of producing butyrate, as observed in the *Lachnospiraceae* and *Ruminococcaceae* families, has been noted. This reduction is associated with diminished butyrate levels in feces and concurrent elevations in intestinal pathogens belonging to the *Enterobacteriaceae* and *Veillonellaceae* families.

As a consequence of microbial metabolism, short-chain fatty acids or SCFAs have been linked to blood pressure regulation through GPCR pathways, influencing blood pressure homeostasis and renin secretion.(Razavi et al. 2019). Substances such as choline, phosphatidylcholine, and L-carnitine, found in fish, red meat, and eggs, undergo metabolic processes by the colonic microbiota, helping the formation of trimethylamine (TMA) (Witkowski et al. 2020). Upon entering the circulation, TMA is subsequently oxidized into trimethylamine N-oxide by FMO3 inside the liver (Zhu et al. 2020). This TMAO compound is subsequently recirculated into the circulation, where it can activate platelets and initiate inflammatory pathways. Inflammatory responses within the endothelium, together with increased formation of foam cells and activation of platelets, enhance the development of atherosclerosis and the progression of atherosclerotic thrombotic events.

1.13 Conclusion

After decades of research, the roles of microbiota in health and disease are now established. It is now confirmed that microbiota can affect almost all aspects of the host, while dysbiosis is related to a broad spectrum of diseases. While microbiota dysbiosis is often observed in disease states, the causative role of microbiota is yet to be established. Table 1.1 lists the microorganisms involved with several human diseases. The greater understanding of this host–microbiota relationship has allowed for the development of microbiota-based therapy such as FMT and bacteria

Table 1.1 Microorganisms and several human diseases

Disease	Microorganisms involved
Irritable bowel syndrome	*Veillonella* sp., *Rhinobatos productus*, and *Clostridium coccoides* (Lee and Bak 2011)
Type 2 diabetes	Acidaminococcales, *Bacteroides plebeius*, and *Phascolarctobacterium* sp. (Wang et al. 2020)
Colorectal cancer	Lactobacilli, *Bifidobacterium* (Lee et al. 2015).
Cardiovascular disease	*Collinsella*, Enterobacteriaceae, and *Streptococcus* spp. (Rahman et al. 2022)
Dementia	*Pseudomonas*, *Streptomyces*, *Staphylococcus*, *Bacillus* (Schwartz and Boles 2013)
Liver disease	*Ruminococcus*, *Faecalibacterium prausnitzii*, and *Coprococcus* (Schwenger et al. 2019)
Allergies	*Acinetobacter lwoffii* and *Lactobacillus rhamnosus* (West et al. 2015)
Kidney disease	*Proteobacteria*, Enterobacteriaceae, Streptococcaceae, and *Streptococcus* (Amini Khiabani et al. 2023)
Nonalcoholic fatty liver disease (NAFLD)	Firmicutes, Bacteroides (Boursier et al. 2016)
Obesity	*Eubacterium rectale* and *Blautia coccoides* (Cani et al. 2007)

modulation. These strategies are well on the way to achieving the optimal clinical effect in treating *C. difficile* infection, diabetes, and inflammatory bowel disease. Clinical tools are available to treat diseases and foster health via manipulating microbial populations.

References

Aagaard K, Ma J, Antony KM, Ganu R, Petrosino J, Versalovic J (2014) The placenta harbors a unique microbiome. Sci Transl Med 6:237ra65

Adhikary S, Esmeeta A, Dey A, Banerjee A, Saha B, Gopan P, Duttaroy AK, Pathak S (2024) Impacts of gut microbiota alteration on age-related chronic liver diseases. Dig Liver Dis 56:112–122

Ahmad AF, Dwivedi G, O'Gara F, Caparros-Martin J, Ward NC (2019) The gut microbiome and cardiovascular disease: current knowledge and clinical potential. Am J Physiol Heart Circ Physiol 317:H923–H938

Alexander JS, Ganta VC, Jordan PA, Witte MH (2010) Gastrointestinal lymphatics in health and disease. Pathophysiology 17:315–335

Amabebe E, Robert FO, Agbalalah T, Orubu ESF (2020) Microbial dysbiosis-induced obesity: role of gut microbiota in homoeostasis of energy metabolism. Br J Nutr 123:1127–1137

Amini Khiabani S, Asgharzadeh M, Samadi Kafil H (2023) Chronic kidney disease and gut microbiota. Heliyon 9:e18991

Arthur JC, Perez-Chanona E, Muhlbauer M, Tomkovich S, Uronis JM, Fan TJ, Campbell BJ, Abujamel T, Dogan B, Rogers AB, Rhodes JM, Stintzi A, Simpson KW, Hansen JJ, Keku TO, Fodor AA, Jobin C (2012) Intestinal inflammation targets cancer-inducing activity of the microbiota. Science 338:120–123

Barcena C, Valdes-Mas R, Mayoral P, Garabaya C, Durand S, Rodriguez F, Fernandez-Garcia MT, Salazar N, Nogacka AM, Garatachea N, Bossut N, Aprahamian F, Lucia A, Kroemer G, Freije JMP, Quiros PM, Lopez-Otin C (2019) Healthspan and lifespan extension by fecal microbiota transplantation into progeroid mice. Nat Med 25:1234–1242

Basak S, Das RK, Banerjee A, Paul S, Pathak S, Duttaroy AK (2022) Maternal obesity and gut microbiota are associated with fetal brain development. Nutrients 14:4515

Basak S, Mallick R, Navya Sree B, Duttaroy AK (2024) Placental Epigenome impacts fetal development: effects of maternal nutrients and gut microbiota. Nutrients 16:1860

Biagi E, Franceschi C, Rampelli S, Severgnini M, Ostan R, Turroni S, Consolandi C, Quercia S, Scurti M, Monti D, Capri M, Brigidi P, Candela M (2016) Gut microbiota and extreme longevity. Curr Biol 26:1480–1485

Boursier J, Mueller O, Barret M, Machado M, Fizanne L, Araujo-Perez F, Guy CD, Seed PC, Rawls JF, David LA, Hunault G, Oberti F, Cales P, Diehl AM (2016) The severity of nonalcoholic fatty liver disease is associated with gut dysbiosis and shift in the metabolic function of the gut microbiota. Hepatology 63:764–775

Cani PD, Amar J, Iglesias MA, Poggi M, Knauf C, Bastelica D, Neyrinck AM, Fava F, Tuohy KM, Chabo C, Waget A, Delmee E, Cousin B, Sulpice T, Chamontin B, Ferrieres J, Tanti JF, Gibson GR, Casteilla L, Delzenne NM, Alessi MC, Burcelin R (2007) Metabolic endotoxemia initiates obesity and insulin resistance. Diabetes 56:1761–1772

Castellarin M, Warren RL, Freeman JD, Dreolini L, Krzywinski M, Strauss J, Barnes R, Watson P, Allen-Vercoe E, Moore RA, Holt RA (2012) Fusobacterium nucleatum infection is prevalent in human colorectal carcinoma. Genome Res 22:299–306

Chen J, Wright K, Davis JM, Jeraldo P, Marietta EV, Murray J, Nelson H, Matteson EL, Taneja V (2016) An expansion of rare lineage intestinal microbes characterizes rheumatoid arthritis. Genome Med 8:43

Chen Z, Li J, Gui S, Zhou C, Chen J, Yang C, Hu Z, Wang H, Zhong X, Zeng L, Chen K, Li P, Xie P (2018) Comparative metaproteomics analysis shows altered fecal microbiota signatures in patients with major depressive disorder. Neuroreport 29:417–425

Clemente JC, Ursell LK, Parfrey LW, Knight R (2012) The impact of the gut microbiota on human health: an integrative view. Cell 148:1258–1270

Cryan JF, O'Riordan KJ, Sandhu K, Peterson V, Dinan TG (2020) The gut microbiome in neurological disorders. Lancet Neurol 19:179–194

Cummings JH, Macfarlane GT (1997) Colonic microflora: nutrition and health. Nutrition 13:476–478

Defronzo RA (2010) Current issues in the treatment of type 2 diabetes. Overview of newer agents: where treatment is going. Am J Med 123:S38–S48

Dilmore AH, Mcdonald D, Nguyen TT, Adams JB, Krajmalnik-Brown R, Elijah E, Dorrestein PC, Knight R (2021) The fecal microbiome and metabolome of Pitt Hopkins syndrome, a severe autism spectrum disorder. mSystems 6:e0100621

Donohoe DR, Garge N, Zhang X, Sun W, O'Connell TM, Bunger MK, Bultman SJ (2011) The microbiome and butyrate regulate energy metabolism and autophagy in the mammalian colon. Cell Metab 13:517–526

Drossman DA, Camilleri M, Mayer EA, Whitehead WE (2002) AGA technical review on irritable bowel syndrome. Gastroenterology 123:2108–2131

Duttaroy AK (2021) Role of gut microbiota and their metabolites on atherosclerosis, hypertension and human blood platelet function: a review. Nutrients 13:144

Eckburg PB, Bik EM, Bernstein CN, Purdom E, Dethlefsen L, Sargent M, Gill SR, Nelson KE, Relman DA (2005) Diversity of the human intestinal microbial flora. Science 308:1635–1638

Ericsson AC, Hart ML, Kwan J, Lanoue L, Bower LR, Araiza R, Kent Lloyd KC, Franklin CL (2021) Supplier-origin mouse microbiomes significantly influence locomotor and anxiety-related behavior, body morphology, and metabolism. Commun Biol 4:716

Esmeeta A, Adhikary S, Dharshnaa V, Swarnamughi P, Ummul Maqsummiya Z, Banerjee A, Pathak S, Duttaroy AK (2022) Plant-derived bioactive compounds in colon cancer treatment: an updated review. Biomed Pharmacother 153:113384

Fava S (2014) Glucagon-like peptide 1 and the cardiovascular system. Curr Diabetes Rev 10:302–310

Flint HJ, Scott KP, Louis P, Duncan SH (2012) The role of the gut microbiota in nutrition and health. Nat Rev Gastroenterol Hepatol 9:577–589

Fu J, Qin Y, Xiao L, Dai X (2023) Causal relationship between gut microflora and dementia: a Mendelian randomization study. Front Microbiol 14:1306048

Galkin F, Mamoshina P, Aliper A, Putin E, Moskalev V, Gladyshev VN, Zhavoronkov A (2020) Human gut microbiome aging clock based on taxonomic profiling and deep learning. iScience 23:101199

Garcia-Gutierrez E, Narbad A, Rodriguez JM (2020) Autism spectrum disorder associated with gut microbiota at immune, metabolomic, and neuroactive level. Front Neurosci 14:578666

Goffredo M, Santoro N, Trico D, Giannini C, D'Adamo E, Zhao H, Peng G, Yu X, Lam TT, Pierpont B, Caprio S, Herzog RI (2017) A branched-chain amino acid-related metabolic signature characterizes obese adolescents with non-alcoholic fatty liver disease. Nutrients 9:642

Gurung M, Li Z, You H, Rodrigues R, Jump DB, Morgun A, Shulzhenko N (2020) Role of gut microbiota in type 2 diabetes pathophysiology. EBioMedicine 51:102590

Hou K, Wu ZX, Chen XY, Wang JQ, Zhang D, Xiao C, Zhu D, Koya JB, Wei L, Li J, Chen ZS (2022) Microbiota in health and diseases. Signal Transduct Target Ther 7:135

Ji S, Park H, Lee D, Song YK, Choi JP, Lee SI (2005) Post-infectious irritable bowel syndrome in patients with Shigella infection. J Gastroenterol Hepatol 20:381–386

Kandeel WA, Meguid NA, Bjorklund G, Eid EM, Farid M, Mohamed SK, Wakeel KE, Chirumbolo S, Elsaeid A, Hammad DY (2020) Impact of Clostridium bacteria in children with autism spectrum disorder and their anthropometric measurements. J Mol Neurosci 70:897–907

Kaplan MM, Gershwin ME (2005) Primary biliary cirrhosis. N Engl J Med 353:1261–1273

Kazor CE, Mitchell PM, Lee AM, Stokes LN, Loesche WJ, Dewhirst FE, Paster BJ (2003) Diversity of bacterial populations on the tongue dorsa of patients with halitosis and healthy patients. J Clin Microbiol 41:558–563

Kim CS, Shin GE, Cheong Y, Shin JH, Shin DM, Chun WY (2022) Experiencing social exclusion changes gut microbiota composition. Transl Psychiatry 12:254

Kirpich IA, Solovieva NV, Leikhter SN, Shidakova NA, Lebedeva OV, Sidorov PI, Bazhukova TA, Soloviev AG, Barve SS, Mcclain CJ, Cave M (2008) Probiotics restore bowel flora and improve liver enzymes in human alcohol-induced liver injury: a pilot study. Alcohol 42:675–682

Kostic AD, Gevers D, Pedamallu CS, Michaud M, Duke F, Earl AM, Ojesina AI, Jung J, Bass AJ, Tabernero J, Baselga J, Liu C, Shivdasani RA, Ogino S, Birren BW, Huttenhower C, Garrett WS, Meyerson M (2012) Genomic analysis identifies association of Fusobacterium with colorectal carcinoma. Genome Res 22:292–298

Larsen N, Vogensen FK, Van Den Berg FW, Nielsen DS, Andreasen AS, Pedersen BK, Al-Soud WA, Sorensen SJ, Hansen LH, Jakobsen M (2010) Gut microbiota in human adults with type 2 diabetes differs from non-diabetic adults. PLoS One 5:e9085

Laterza L, Rizzatti G, Gaetani E, Chiusolo P, Gasbarrini A (2016) The gut microbiota and immune system relationship in human graft-versus-host disease. Mediterr J Hematol Infect Dis 8:e2016025

Lee BJ, Bak YT (2011) Irritable bowel syndrome, gut microbiota and probiotics. J Neurogastroenterol Motil 17:252–266

Lee HA, Kim H, Lee KW, Park KY (2015) Dead Nano-sized lactobacillus plantarum inhibits Azoxymethane/dextran sulfate sodium-induced Colon Cancer in Balb/c Mice. J Med Food 18:1400–1405

Lin P, Ding B, Feng C, Yin S, Zhang T, Qi X, Lv H, Guo X, Dong K, Zhu Y, Li Q (2017) Prevotella and Klebsiella proportions in fecal microbial communities are potential characteristic parameters for patients with major depressive disorder. J Affect Disord 207:300–304

Ling Z, Liu X, Cheng Y, Yan X, Wu S (2022) Gut microbiota and aging. Crit Rev Food Sci Nutr 62:3509–3534

Liu P, Wu L, Peng G, Han Y, Tang R, Ge J, Zhang L, Jia L, Yue S, Zhou K, Li L, Luo B, Wang B (2019) Altered microbiomes distinguish Alzheimer's disease from amnestic mild cognitive impairment and health in a Chinese cohort. Brain Behav Immun 80:633–643

Liu Y, Wang H, Gui S, Zeng B, Pu J, Zheng P, Zeng L, Luo Y, Wu Y, Zhou C, Song J, Ji P, Wei H, Xie P (2021) Proteomics analysis of the gut-brain axis in a gut microbiota-dysbiosis model of depression. Transl Psychiatry 11:568

Lukic I, Getselter D, Ziv O, Oron O, Reuveni E, Koren O, Elliott E (2019) Antidepressants affect gut microbiota and Ruminococcus flavefaciens is able to abolish their effects on depressive-like behavior. Transl Psychiatry 9:133

Lv LX, Fang DQ, Shi D, Chen DY, Yan R, Zhu YX, Chen YF, Shao L, Guo FF, Wu WR, Li A, Shi HY, Jiang XW, Jiang HY, Xiao YH, Zheng SS, Li LJ (2016) Alterations and correlations of the gut microbiome, metabolism and immunity in patients with primary biliary cirrhosis. Environ Microbiol 18:2272–2286

Mariat D, Firmesse O, Levenez F, Guimaraes V, Sokol H, Dore J, Corthier G, Furet JP (2009) The Firmicutes/Bacteroidetes ratio of the human microbiota changes with age. BMC Microbiol 9:123

Mckendrick MW, Read NW (1994) Irritable bowel syndrome--post salmonella infection. J Infect 29:1–3

Mehedint MG, Zeisel SH (2013) Choline's role in maintaining liver function: new evidence for epigenetic mechanisms. Curr Opin Clin Nutr Metab Care 16:339–345

Mirzaei R, Bouzari B, Hosseini-Fard SR, Mazaheri M, Ahmadyousefi Y, Abdi M, Jalalifar S, Karimitabar Z, Teimoori A, Keyvani H, Zamani F, Yousefimashouf R, Karampoor S (2021) Role of microbiota-derived short-chain fatty acids in nervous system disorders. Biomed Pharmacother 139:111661

Mitsuoka T (1996) Intestinal flora and human health. Asia Pac J Clin Nutr 5:2–9

Nagasue T, Hirano A, Torisu T, Umeno J, Shibata H, Moriyama T, Kawasaki K, Fujioka S, Fuyuno Y, Matsuno Y, Esaki M, Kitazono T (2022) The compositional structure of the small intestinal microbial community via balloon-assisted enteroscopy. Digestion 103:308–318

Rahman MM, Islam F, Or-Rashid MH, Mamun AA, Rahaman MS, Islam MM, Meem AFK, Sutradhar PR, Mitra S, Mimi AA, Emran TB, Fatimawali IR, Tallei TE, Ahmed M, Cavalu S (2022) The gut microbiota (microbiome) in cardiovascular disease and its therapeutic regulation. Front Cell Infect Microbiol 12:903570

Rana SV, Sinha SK, Sikander A, Bhasin DK, Singh K (2008) Study of small intestinal bacterial overgrowth in North Indian patients with irritable bowel syndrome: a case control study. Trop Gastroenterol 29:23–25

Razavi AC, Potts KS, Kelly TN, Bazzano LA (2019) Sex, gut microbiome, and cardiovascular disease risk. Biol Sex Differ 10:29

Romero R, Hassan SS, Gajer P, Tarca AL, Fadrosh DW, Nikita L, Galuppi M, Lamont RF, Chaemsaithong P, Miranda J, Chaiworapongsa T, Ravel J (2014) The composition and stability of the vaginal microbiota of normal pregnant women is different from that of non-pregnant women. Microbiome 2:4

Ryman S, Vakhtin AA, Richardson SP, Lin HC (2023) Microbiome-gut-brain dysfunction in prodromal and symptomatic Lewy body diseases. J Neurol 270:746–758

Scher JU, Sczesnak A, Longman RS, Segata N, Ubeda C, Bielski C, Rostron T, Cerundolo V, Pamer EG, Abramson SB, Huttenhower C, Littman DR (2013) Expansion of intestinal Prevotella copri correlates with enhanced susceptibility to arthritis. elife 2:e01202

Schwartz K, Boles BR (2013) Microbial amyloids--functions and interactions within the host. Curr Opin Microbiol 16:93–99

Schwenger KJ, Clermont-Dejean N, Allard JP (2019) The role of the gut microbiome in chronic liver disease: the clinical evidence revised. JHEP Rep 1:214–226

Starkel P, Schnabl B (2016) Bidirectional communication between liver and gut during alcoholic liver disease. Semin Liver Dis 36:331–339

Tan X, Zhang L, Wang D, Guan S, Lu P, Xu X, Xu H (2021) Influence of early life stress on depression: from the perspective of neuroendocrine to the participation of gut microbiota. Aging (Albany NY) 13:25588–25601

Tang WH, Kitai T, Hazen SL (2017) Gut microbiota in cardiovascular health and disease. Circ Res 120:1183–1196

Thursby E, Juge N (2017) Introduction to the human gut microbiota. Biochem J 474:1823–1836

Tjalsma H, Boleij A, Marchesi JR, Dutilh BE (2012) A bacterial driver-passenger model for colorectal cancer: beyond the usual suspects. Nat Rev Microbiol 10:575–582

Tran TTT, Corsini S, Kellingray L, Hegarty C, Le Gall G, Narbad A, Muller M, Tejera N, O'Toole PW, Minihane AM, Vauzour D (2019) APOE genotype influences the gut microbiome structure and function in humans and mice: relevance for Alzheimer's disease pathophysiology. FASEB J 33:8221–8231

Tuomisto S, Huhtala H, Martiskainen M, Goebeler S, Lehtimaki T, Karhunen PJ (2019) Age-dependent association of gut bacteria with coronary atherosclerosis: Tampere Sudden Death Study. PLoS One 14:e0221345

Turnbaugh PJ, Ley RE, Mahowald MA, Magrini V, Mardis ER, Gordon JI (2006) An obesity-associated gut microbiome with increased capacity for energy harvest. Nature 444:1027–1031

Vonaesch P, Araujo JR, Gody JC, Mbecko JR, Sanke H, Andrianonimiadana L, Naharimanananirina T, Ningatoloum SN, Vondo SS, Gondje PB, Rodriguez-Pozo A, Rakotondrainipiana M, Kandou KJE, Nestoret A, Kapel N, Djorie SG, Finlay BB, Wegener Parfrey L, Collard JM, Randremanana RV, Sansonetti PJ, Afribiota I (2022) Stunted children display ectopic small intestinal colonization by oral bacteria, which cause lipid malabsorption in experimental models. Proc Natl Acad Sci USA 119:e2209589119

Wang TY, Zhang XQ, Chen AL, Zhang J, Lv BH, Ma MH, Lian J, Wu YX, Zhou YT, Ma CC, Dong RJ, Ge DY, Gao SH, Jiang GJ (2020) A comparative study of microbial community and functions of type 2 diabetes mellitus patients with obesity and healthy people. Appl Microbiol Biotechnol 104:7143–7153

West CE, Jenmalm MC, Prescott SL (2015) The gut microbiota and its role in the development of allergic disease: a wider perspective. Clin Exp Allergy 45:43–53

Witkowski M, Weeks TL, Hazen SL (2020) Gut microbiota and cardiovascular disease. Circ Res 127:553–570

Yassour M, Jason E, Hogstrom LJ, Arthur TD, Tripathi S, Siljander H, Selvenius J, Oikarinen S, Hyoty H, Virtanen SM, Ilonen J, Ferretti P, Pasolli E, Tett A, Asnicar F, Segata N, Vlamakis H, Lander ES, Huttenhower C, Knip M, Xavier RJ (2018) Strain-level analysis of mother-to-child bacterial transmission during the first few months of life. Cell Host Microbe 24:146–154 e4

Zeisel SH, Wishnok JS, Blusztajn JK (1983) Formation of methylamines from ingested choline and lecithin. J Pharmacol Exp Ther 225:320–324

Zhu Y, Li Q, Jiang H (2020) Gut microbiota in atherosclerosis: focus on trimethylamine N-oxide. APMIS 128:353–366

Zmora N, Zilberman-Schapira G, Suez J, Mor U, Dori-Bachash M, Bashiardes S, Kotler E, Zur M, Regev-Lehavi D, Brik RB, Federici S, Cohen Y, Linevsky R, Rothschild D, Moor AE, Ben-Moshe S, Harmelin A, Itzkovitz S, Maharshak N, Shibolet O, Shapiro H, Pevsner-Fischer M, Sharon I, Halpern Z, Segal E, Elinav E (2018) Personalized gut mucosal colonization resistance to empiric probiotics is associated with unique host and microbiome features. Cell 174:1388–1405 e21

Recent Innovations in Probiotics and Prebiotics and Gut Health

2

Madhan Krishnan, Shyamaladevi Babu, and Ahamed Basha Abdul Bari

Abstract

The human gut microbiota, comprising a broad range of microorganisms such as viruses, fungi, and bacteria, plays an important role in essential physiological functions including digestion, immune response, and nutrient synthesis. Disruptions in this microbiota are associated with various health issues, from gastrointestinal disorders to metabolic diseases and mental health conditions. There is growing interest in using probiotics and prebiotics to influence the gut microbiota and improve health outcomes. Probiotics, such as strains of *Lactobacillus* and *Bifidobacterium*, are live microorganisms that confer health benefits when ingested in sufficient quantities. Prebiotics, like fructooligosaccharides (FOS), galactooligosaccharides (GOS), and inulin, are nondigestible components that encourage the proliferation of beneficial gut bacteria. This chapter reviews recent advancements in probiotics and prebiotics, focusing on their effects on gut health. Topics covered include emerging probiotics like *Akkermansia muciniphila* and *Faecalibacterium prausnitzii*, innovations in microencapsulation, novel prebiotic compounds, the development of synbiotics, and the application of omics technologies to gut microbiome research. These developments highlight the transformative potential of probiotics and prebiotics in improving gut health and offer new avenues for the prevention and treatment of various health conditions.

M. Krishnan · S. Babu
Faculty of Research, Chettinad Hospital and Research Institute, Chettinad Academy of Research and Education, Kelambakkam, Tamilnadu, India

A. B. A. Bari (✉)
Department of Physiology, Chettinad Hospital and Research Institute, Chettinad Academy of Research and Education, Kelambakkam, Tamilnadu, India

S. Pathak et al. (eds.), *Microbiota and Dietary Mediators in Colon Cancer Prevention and Treatment*, https://doi.org/10.1007/978-981-96-0297-1_2

Keywords
Gut microbiota · probiotics · prebiotics · synbiotics · omics technologies

2.1 Introduction

The human gut microbiota, a complex and diverse community of microorganisms including viruses, fungi, and bacteria, is essential to numerous physiological functions. These functions encompass digestion, immune response regulation, and nutrient synthesis. When the balance of the gut microbiota is disturbed, it can result in a range of health issues, including gastrointestinal disorders, metabolic diseases, and mental health issues. Consequently, there is a growing interest in the use of probiotics and prebiotics to modulate the gut microbiota and enhance health outcomes (Thursby and Juge 2017). Probiotics are living microorganisms that provide health benefits to the host when ingested in sufficient quantities. Commonly used probiotic bacteria include species of *Lactobacillus* and *Bifidobacterium*. Prebiotics, in contrast, are indigestible substances that encourage the proliferation and function of beneficial bacteria in the gut. Examples of typical prebiotics are fructooligosaccharides (FOS), galactooligosaccharides (GOS), and inulin (Ji et al. 2023). This introduction delves into the recent advancements in probiotics and prebiotics, with a focus on their impact on gut health. The chapter covers next-generation probiotics (NGPs), which include strains that are naturally present in the human gut but were previously difficult to cultivate and administer. These strains have shown potential in treating conditions such as inflammatory bowel disease (IBD), obesity, and metabolic disorders. In addition, the chapter explores advances in microencapsulation techniques.

Microencapsulation is a process that involves enclosing probiotics within a protective matrix, which enhances their survival as they pass through the gastrointestinal tract. This technique is crucial for ensuring that probiotics reach the gut alive and functional. This chapter also discusses novel prebiotic compounds, expanding beyond traditional prebiotics like FOS, GOS, and inulin. These new compounds have shown promise in selectively stimulating the growth of beneficial gut bacteria and improving gut health. Furthermore, the development of synbiotics, which combine probiotics and prebiotics, is examined. Synbiotics are designed synergistically improve gut health by enhancing the survival and activity of beneficial bacteria. Lastly, the chapter addresses the application of omics technologies in gut microbiome research. Omics technologies, such as metagenomics, metabolomics, and transcriptomics, provide comprehensive insights into the composition and functions of the gut microbiota. These technologies revolutionized our understanding of how the gut microbiota influences health and disease, paving the way for personalized interventions with probiotics and prebiotics. In summary, recent innovations in probiotics and prebiotics hold significant potential to revolutionize gut health. These advancements offer new strategies for the prevention and treatment of various health conditions, underscoring the importance of continued research in this field.

2.2 Next-Generation Probiotics

Traditional probiotics, such as *Lactobacillus* and *Bifidobacterium* species, are well-known for their health benefits and have been extensively studied. However, recent advancements have identified additional gut microorganisms that offer significant health potential. These NGPs include strains naturally present in the human gut that were previously challenging to cultivate and administer. NGPs have demonstrated potential in treating a range of conditions, including IBD, obesity, and metabolic disorders (Abouelela and Helmy 2024). NGPs represent a new frontier in probiotic research, expanding the scope of beneficial bacteria beyond traditional strains. These innovative probiotics are being explored for their unique mechanisms of action and specific health benefits, providing new avenues for targeted therapies and personalized medicine. As research progresses, the development and application of NGPs could significantly enhance the effectiveness of probiotic treatments, offering improved outcomes for individuals with various health conditions.

2.3 Key Strains of Next-Generation Probiotics

2.3.1 *Akkermansia muciniphila*

Akkermansia muciniphila, a bacterium found in the gut's mucus layer, plays a crucial role in degrading mucin. It offers various health benefits, including improved gut barrier function, reduced inflammation, and enhanced metabolic health. Research indicates that *A. muciniphila* supplementation can improve insulin sensitivity, aid in weight reduction, and lower inflammatory markers in individuals with metabolic disorders (Pellegrino et al. 2023). These findings highlight the potential of *A. muciniphila* as a next-generation probiotic, offering promising therapeutic applications for managing metabolic health issues.

2.3.2 *Faecalibacterium prausnitzii*

Faecalibacterium prausnitzii is one of the most abundant bacteria in the human gut, renowned for its anti-inflammatory properties. It produces butyrate, a short-chain fatty acid that provides energy to colonocytes and has significant anti-inflammatory effects. Low levels of *F. prausnitzii* are associated with inflammatory conditions like IBD. Research indicates that supplementing with *F. prausnitzii* can reduce inflammation and alleviate symptoms in individuals with IBD (Fernandes et al. 2023). These findings underscore the potential of *F. prausnitzii* as a therapeutic agent for managing inflammatory conditions and improving gut health.

2.3.3 *Clostridium butyricum*

Clostridium butyricum, a butyrate-producing bacterium, has been used as a probiotic in various countries. This bacterium shows significant potential in treating several gastrointestinal conditions, such as IBD, antibiotic-associated diarrhea and irritable bowel syndrome (IBS). Research indicates that *C. butyricum* supplementation can enhance gut barrier function, making the intestines less permeable to harmful substances. Additionally, it modulates the immune response, helping the body better manage inflammation and immune challenges. Moreover, *C. butyricum* promotes the growth of useful gut bacteria, contributing to a healthier and more balanced gut microbiota. These properties make *C. butyricum* a promising candidate for improving gut health and managing related disorders (Stoeva et al. 2021).

2.4 Clinical Applications of Next-Generation Probiotics

NGPs have demonstrated significant potential in various clinical applications, including metabolic disorders, inflammatory conditions, and gut barrier dysfunction. For instance, supplementation with *Akkermansia muciniphila* has shown to enhance metabolic health in individuals with type 2 diabetes and obesity. Likewise, *Faecalibacterium prausnitzii* supplementation has exhibited anti-inflammatory properties in patients suffering from IBD. The advancement of NGPs paves the way for personalized nutrition and medicine, offering new opportunities to develop targeted probiotic therapies. By identifying specific strains that provide benefits for particular health conditions, researchers can create probiotic treatments tailored to meet the unique needs of individual patients, potentially revolutionizing gut health management and therapy (Vallianou et al. 2023). This personalized approach improves treatment efficacy and patient outcomes.

2.5 Advances in Microencapsulation Techniques

Microencapsulation addresses a critical hurdle in probiotic supplementation: ensuring the survival of live microorganisms through the gastrointestinal tract. This technique involves enveloping probiotics in a protective matrix, safeguarding them from the harsh conditions of the digestive system. The stomach's acidic environment and bile salts in the small intestine pose significant challenges to probiotic viability. By encapsulating probiotic strains, microencapsulation shields them during their journey through these hostile regions, thereby enhancing their chances of reaching the colon alive and active. The protective matrix used in microencapsulation can be made from various materials such as polysaccharides, proteins, or lipids, selected based on their ability to withstand digestive enzymes and maintain structural integrity under acidic conditions. This approach not only extends the probiotics' shelf life but also enhances their stability during storage and delivery to the consumer (Sun et al. 2023). Moreover, microencapsulation enables controlled release of probiotics, ensuring prolonged exposure in the gut where they can exert their beneficial effects. This technology has revolutionized probiotic supplementation, offering a reliable method to optimize their effectiveness and health benefits. As research continues to refine microencapsulation techniques, its applications are expanding beyond probiotics to include pharmaceuticals, nutraceuticals, and other sensitive compounds requiring protection from digestive degradation (Gyawali et al. 2023).

2.6 Types of Microencapsulation Techniques

2.6.1 Alginate Encapsulation

Alginate, a natural polysaccharide derived from brown seaweed, is highly valued in microencapsulation due to its biocompatibility, non-toxicity, and ability to gel in the presence of calcium ions. In the process of alginate encapsulation, probiotic cells are mixed with a sodium alginate solution, which is then dropped into a calcium chloride solution. This results in the formation of gel beads that protect the probiotics from the acidic conditions of the stomach, ensuring their controlled release in the intestines. This method enhances the viability of probiotics throughout the digestive tract, maximizing their effectiveness when they reach the colon. Alginate-based microencapsulation is widely used not only for probiotics but also in various pharmaceutical and nutraceutical applications. It is particularly beneficial in scenarios where protection from harsh digestive environments is essential for maintaining product efficacy (Zhang et al. 2021a, b, c). This technology represents a significant advancement in delivering beneficial microorganisms and other bioactive compounds to the gut, highlighting its importance in health and therapeutic applications.

2.6.2 Chitosan Encapsulation

Chitosan, derived from crustacean shells, is highly regarded in microencapsulation due to its biocompatibility, biodegradability, and film-forming properties. In the chitosan encapsulation process, probiotic cells are enveloped in a chitosan solution, creating a protective coating. This coating shields the probiotics from the acidic environment of the stomach and facilitates their release in the intestines, thereby ensuring their viability and efficacy. Chitosan-based microencapsulation is not only advantageous for probiotics but also for various pharmaceutical and nutraceutical applications. It plays a crucial role in protecting sensitive bioactive compounds from harsh digestive conditions, thus maintaining product integrity and effectiveness. The natural polysaccharide's ability to form films and gels enhances its versatility in encapsulation technologies, offering a sustainable and effective solution for delivering these compounds to targeted sites within the body (Erdélyi et al. 2022). This makes chitosan a valuable material in the development of advanced delivery systems, ensuring that therapeutic agents reach their intended destination in optimal condition, maximizing their health benefits.

2.6.3 Lipid-Based Encapsulation

Lipid-based encapsulation methods use lipid particles such as liposomes or solid lipid nanoparticles to encapsulate probiotics. These lipid particles provide effective protection against the stomach's acidic conditions and bile salts in the small intestine are effectively bypassed, allowing probiotics to safely reach the colon. This approach not only safeguards probiotics throughout their passage through the digestive system but also extends their stability and shelf life. Lipid-based encapsulation is highly valued for its ability to maintain probiotic viability and efficacy, making it a promising technique for enhancing the delivery and effectiveness of probiotic supplements. Additionally, this method offers several advantages, including the ability to control the release of probiotics and protect them from environmental factors such as oxygen and moisture, which can degrade their quality over time. This makes lipid-based encapsulation a versatile and effective solution for various applications in the pharmaceutical and nutraceutical industries, where maintaining the integrity and functionality of bioactive compounds is crucial. The use of lipid particles in encapsulation technologies demonstrates significant potential for improving the health benefits of probiotic supplements, offering a reliable and efficient way to deliver these beneficial microorganisms to the targeted sites within the body (García-Pinel et al. 2019).

2.6.4 Benefits of Microencapsulation

Microencapsulation provides significant advantages for probiotic supplementation by enhancing viability, improving delivery, and increasing stability. This technique protects probiotics from the harsh conditions of the gastrointestinal (GI) tract, ensuring that a greater number of live microorganisms reach the intestines. By safeguarding probiotics from stomach acid and bile, microencapsulation improves their efficacy and allows for lower doses to be used while still achieving therapeutic benefits. Furthermore, this method enhances the stability and shelf life of probiotics, which is crucial for commercial viability. Microencapsulation methods, such as those utilizing natural polymers or lipid-based systems, contribute to maintaining the integrity of probiotics during storage and transport, thereby reducing the need for additional preservatives. This not only makes probiotic supplements more practical for everyday use but also helps in optimizing their health benefits. As a result, microencapsulation stands out as a valuable technique in the development of effective probiotic products, offering a reliable means to deliver beneficial microorganisms to their intended sites within the digestive system (Wang et al. 2022a, b).

2.7 Novel Prebiotic Compounds

Prebiotics are dietary substances that resist digestion and specifically foster the growth and activity of beneficial gut bacteria. They serve as a vital food source for probiotics, enhancing their proliferation and function within the gastrointestinal tract. Typical examples of prebiotics are fructooligosaccharides (FOS), galactooligosaccharides (GOS), and inulin. These substances are known to support the growth of beneficial bacteria such as bifidobacteria and lactobacilli. Recent research has identified additional prebiotic compounds with promising health benefits, further expanding their potential applications in promoting gut health and overall well-being. By creating an optimal environment for probiotics to thrive, prebiotics are essential in maintaining a balanced gut microbiota. This interaction helps to enhance digestive health and supports the effectiveness of probiotic interventions. Incorporating prebiotic-rich foods or supplements into the diet can enhance immune function and metabolic health, offering a holistic approach to maintaining gut health. Thus, prebiotics play a critical role in optimizing the benefits of probiotics and contributing to a healthier digestive system and overall well-being (Kaur et al. 2021).

2.8 Human Milk Oligosaccharides (HMOs)

Human milk oligosaccharides (HMOs) are intricate carbohydrates present in human breast milk, ranking as the third most prevalent component after lactose and fat. These oligosaccharides play a critical role in nurturing beneficial gut bacteria, such as Bifidobacteria, which are crucial for establishing a healthy infant gut microbiota. HMOs contribute significantly to the development of a robust gut ecosystem by initiating the growth of these useful bacteria. Recent advancements in biotechnology have facilitated the synthesis of HMOs, allowing their incorporation into infant formulas, dietary supplements, and functional foods. This progress has expanded the potential of HMOs beyond breast milk, offering benefits in various dietary products. Research has shown that HMOs not only support the growth of advantageous gut bacteria but also enhance gut barrier function and modulate immune responses. By fostering a balanced gut microbiota and contributing to immune health, HMOs provide valuable support in both infant nutrition and broader health applications (Walsh et al. 2020). These findings underscore the importance of HMOs in maintaining and improving gut health and immune function.

2.8.1 Arabinoxylans

Arabinoxylans are intricate carbohydrates present in the cell walls of cereal grains like wheat, rye, and barley. Studies show that these substances have prebiotic effects by promoting the proliferation of beneficial gut bacteria and enhancing the production of short-chain fatty acids (SCFAs). Additionally, arabinoxylans have been observed to enhance gut barrier function and regulate the immune system's response (Zannini et al. 2022). These effects make arabinoxylans significant contributors to gut health, influencing the balance of the gut microbiota and potentially benefiting overall well-being. Their ability to stimulate the growth of beneficial bacteria and support intestinal integrity underscores their role in maintaining digestive health. As ongoing research explores their mechanisms and benefits further, arabinoxylans continue to garner attention for their potential therapeutic applications in promoting gut health and modulating immune function through dietary interventions rich in whole grains containing these beneficial carbohydrates (Zhang et al. 2021a, b, c).

2.8.2 Resistant Starch

Resistant starch is a unique type of starch that evades digestion in the small intestine and travels intact to the colon, where it acts as a substrate for beneficial gut bacteria, fostering their growth and metabolic activity. This process stimulates the production of short-chain fatty acids (SCFAs), which contribute to improved gut barrier function and immune regulation. Resistant starch intake has also been linked to enhanced metabolic health, including increased insulin sensitivity and reduced body weight (Birt et al. 2013). These findings underscore the significant role of resistant starch in supporting digestive health and overall well-being. By promoting a favorable environment in the gut through SCFA production and metabolic improvements, resistant starch demonstrates potential therapeutic benefits in managing metabolic disorders and enhancing gut microbiota composition. Incorporating sources of resistant starch into the diet, such as whole grains, legumes, and certain fruits and vegetables, may therefore contribute to improved digestive function and metabolic outcomes (Chen et al. 2024).

2.8.3 Polyphenols

Polyphenols are natural compounds abundant in plants like fruits, vegetables, tea, and wine. They exhibit prebiotic properties by stimulating the growth of beneficial gut bacteria and increasing the production of short-chain fatty acids (SCFAs). Furthermore, polyphenols are associated with improved gut barrier function, modulation of the immune response, and possess antioxidant and anti-inflammatory effects. These characteristics highlight polyphenols' significant role in promoting gut health and overall well-being. By supporting a diverse and healthy gut microbiota, polyphenols contribute to digestive health through their prebiotic activity and ability to enhance SCFA production. Moreover, their antioxidant and anti-inflammatory properties offer additional benefits in combating oxidative stress and inflammation in the gut and throughout the body. Incorporating polyphenol-rich foods into the diet, such as berries, nuts, dark chocolate, and green tea, can thus be beneficial for maintaining gut health and supporting overall immune function and systemic health (Plamada and Vodnar 2021; Wang et al. 2022a, b).

2.9 Introduction and Development of Synbiotics

Synbiotics are formulations combining probiotics and prebiotics designed to synergistically improve gut health. By pairing specific probiotic strains with complementary prebiotics, synbiotics promote the growth and function of beneficial gut bacteria more effectively than probiotics or prebiotics alone. This approach allows researchers to tailor treatments for targeted therapies that optimize microbial balance in the gut. The integration of synbiotics offers several advantages, including heightened efficacy in fostering a healthy gut microbiota. Additionally, synbiotics hold promise for personalized nutrition and medicine, where formulations can be customized to meet individual health needs and preferences. By harnessing the combined benefits of probiotics' live microorganisms and prebiotics' supportive nutrients, synbiotics represent an innovative strategy in promoting digestive wellness and overall health. Continued research into optimizing synbiotic combinations may lead to further advancements in therapeutic applications for managing digestive disorders and enhancing immune function (Gomez Quintero et al. 2022).

2.10 Examples of Synbiotics

2.10.1 *Bifidobacterium longum* and Galactooligosaccharides (GOS)

Bifidobacterium longum, a naturally occurring probiotic strain in the human gut, provides several health benefits. Galactooligosaccharides (GOS) are prebiotic compounds that selectively stimulate the growth of *Bifidobacteria*. Recent research highlights that combining *B. longum* with GOS in synbiotic formulations significantly enhances the growth and metabolic activity of *Bifidobacteria* in the gut (Markowiak and Śliżewska 2017). This synergistic interaction improves gut barrier function and alleviates symptoms related to IBS. The effectiveness of this combination demonstrates how synbiotics can amplify the health benefits of both probiotics and prebiotics. By creating a conducive environment for the proliferation of *Bifidobacteria*, these synbiotics support digestive health and overall well-being. Further research into these combinations could lead to more advanced therapeutic strategies for managing gastrointestinal conditions and boosting immune function through targeted modulation of the gut microbiota. Ongoing studies aim to refine these approaches, offering potential improvements in treatments for digestive disorders and contributing to enhanced health outcomes (Mills et al. 2023).

2.10.2 *Lactobacillus rhamnosus* and Fructooligosaccharides (FOS)

Lactobacillus rhamnosus is a well-studied probiotic strain naturally found in the human gut, known for its wide-ranging health benefits. One significant prebiotic compound, fructooligosaccharides (FOS), selectively promotes the growth of beneficial *Lactobacilli*. Recent research highlights the synergistic effects of combining *L. rhamnosus* with FOS in synbiotic formulations. This combination not only enhances the proliferation and activity of Lactobacilli but also supports overall gut health by improving gut barrier function and modulating the immune response. Studies have demonstrated that these synbiotics can lead to better gut microbiota balance, increased resistance to gastrointestinal infections, and improved immune system performance (Pandey et al. 2015; Markowiak and Śliżewska 2017). By integrating L. rhamnosus with FOS, synbiotics leverage the prebiotic's ability to fuel probiotic growth, thereby amplifying the probiotic's health benefits and contributing to a more robust and balanced gut microbiota. This approach underscores the potential of combining probiotics with prebiotics to optimize gut health and enhance overall well-being.

2.10.3 *Saccharomyces boulardii* and Inulin

Saccharomyces boulardii, a probiotic yeast, is renowned for its efficacy in preventing and treating antibiotic-associated diarrhea. Inulin, a prebiotic compound, promotes the growth of beneficial gut bacteria, making it an effective complement to probiotics. Recent research has shown that combining *S. boulardii* with inulin in synbiotic formulations can enhance the growth and metabolic activity of advantageous gut bacteria. This synergy improves gut barrier function and modulates the immune response, illustrating the benefits of synbiotics in optimizing the effects of probiotics and prebiotics. The combined action of *S. boulardii* and inulin creates a supportive environment for beneficial gut bacteria, contributing to improved digestive health and overall well-being. The potential of these synbiotics in therapeutic strategies is underscored by ongoing studies, which explore their role in managing gastrointestinal conditions and enhancing immune function through targeted modulation of the gut microbiota (Pais et al. 2020). This research paves the way for advanced approaches to gut health and disease management.

2.11 Benefits of Synbiotics

Synbiotics, which combine probiotics and prebiotics, offer several advantages over using each component independently. This approach enhances efficacy by synergistically working to improve gut health. By pairing specific probiotic strains with complementary prebiotics, synbiotics can create targeted therapies tailored to individual health needs. This customization is especially promising for personalized nutrition and medicine, enabling more precise and effective treatments. The combination of probiotics and prebiotics in synbiotics ensures that beneficial gut bacteria receive the support needed for optimal growth and activity. This can lead to improved gut barrier function, helping to prevent harmful substances from entering the bloodstream. Additionally, synbiotics can modulate the immune response, contributing to better management of various health conditions. Recent research underscores the potential of synbiotics in preventing and treating a range of health issues, from digestive disorders to immune system imbalances (Gomez Quintero et al. 2022). By creating specific probiotic–prebiotic combinations, synbiotics offer new opportunities for enhancing gut health and overall well-being. Their ability to deliver tailored interventions represents a significant advancement in both dietary and therapeutic approaches to managing health.

2.12 Omics Technologies in Gut Microbiome Research

Omics technologies like metagenomics, metabolomics, and transcriptomics have transformed gut microbiome research. These advanced tools allow thorough analysis of microbial communities, metabolic pathways, and gene expression, offering crucial insights into the composition and functions of the gut microbiota. Enabling the identification of microbial profiles and metabolic activities linked to various health conditions, omics technologies play a pivotal role in developing personalized interventions involving probiotics and prebiotics. The application of omics in microbiome research has led to significant advancements in understanding how microbial communities influence health and disease. These insights inform the formulation of targeted therapies that can modulate the gut microbiota to improve health outcomes. Continued advancements in omics technologies promise further refinements in personalized medicine approaches, enhancing our ability to harness the potential of probiotics and prebiotics for promoting digestive health and overall well-being (Wu et al. 2024).

2.12.1 Metagenomics

Metagenomics involves studying genetic material directly extracted from environmental samples, such as fecal matter, to sequence the DNA of all microorganisms present. This approach offers a comprehensive analysis of microbial communities, enabling the discovery of new gut bacteria and detailed insights into microbial diversity and composition. Metagenomics also facilitates the investigation of microbial functions and interactions, shedding light on the crucial role of the gut microbiota in both health and disease. Uncovering previously unknown microbial species and understanding their functions, metagenomics enhances our understanding of how the gut microbiota influences various physiological processes. This knowledge is instrumental in developing targeted interventions, including probiotics and prebiotics, to modulate the gut microbiome for improved health outcomes. As metagenomic technologies continue to evolve, they promise to further elucidate the intricate relationships between gut microbes and human health, advancing personalized approaches to enhance digestive wellness and overall well-being (Martín et al. 2014; Zhang et al. 2021a, b, c).

2.12.2 Metabolomics

Metabolomics is the study of small molecules, or metabolites, produced by microorganisms in a sample. It involves analyzing the metabolic profiles of microbial communities, which provides insights into their metabolic activities and functions. This approach has been instrumental in identifying microbial metabolites associated with health and disease, such as short-chain fatty acids (SCFAs), bile acids, and tryptophan metabolites. Metabolomics also facilitates the study of host-microbe interactions by revealing how microbial metabolites influence host physiology. It uncovers the complex biochemical exchanges between gut microorganisms and their human host, offering valuable information on how these interactions affect health. Recent advancements in this field have improved our understanding of the metabolic crosstalk between the gut microbiota and the host, potentially leading to new therapeutic strategies and better insights into microbiome-related health conditions (Qiu et al. 2023).

2.12.3 Transcriptomics

Transcriptomics is the study of gene expression by analyzing the RNA transcripts produced by microorganisms. This technique provides insights into gene expression patterns and regulatory mechanisms within microbial communities. By examining these RNA transcripts, researchers can identify microbial genes linked to health and disease, such as those involved in short-chain fatty acid (SCFA) production, bile acid metabolism, and immune modulation. Transcriptomics also helps in understanding how microorganisms respond to environmental changes, revealing their adaptive mechanisms. This approach sheds light on the dynamic interactions between the gut microbiota and its environment, including host-related factors. Recent advances in transcriptomics have enhanced our understanding of how microbial gene expression influences overall health, offering new avenues for therapeutic interventions and better comprehension of microbiome-related conditions (Lowe et al. 2017).

2.13 Applications of Omics Technologies

Omics technologies have numerous applications in gut microbiome research, including identifying microbial biomarkers for health and disease, characterizing microbial functions and interactions, and developing personalized probiotic and prebiotic interventions. For example, metagenomics can identify microbial profiles associated with specific health conditions, such as IBD, obesity, and type 2 diabetes. Metabolomics can identify microbial metabolites associated with health and disease, such as SCFAs, bile acids, and tryptophan metabolites. Transcriptomics can identify microbial genes associated with health and disease, such as genes involved in SCFA production, bile acid metabolism, and immune modulation (Kwa et al.

2023). Omics technologies also offer new opportunities for personalized nutrition and medicine. By identifying specific microbial profiles and metabolic pathways associated with health and disease, researchers can develop targeted probiotic and prebiotic interventions tailored to individual patients' needs. This personalized approach has the potential to improve the efficacy of probiotic and prebiotic interventions and provide new avenues for the prevention and treatment of various health conditions (Kiousi et al. 2021).

2.14 Conclusion

In conclusion, recent advances in probiotics and prebiotics have deepened our understanding of gut health and led to innovative interventions. Next-generation probiotics like *Akkermansia muciniphila* and *Faecalibacterium prausnitzii* show promise in treating inflammatory bowel disease, obesity, and metabolic disorders. Improved microencapsulation techniques ensure probiotics survive digestion to reach the intestines effectively. Novel prebiotic compounds such as human milk oligosaccharides, arabinoxylans, resistant starch, and polyphenols offer diverse options for promoting gut health. Synbiotics, combining probiotics and prebiotics, have enhanced treatment efficacy. Omics technologies—metagenomics, metabolomics, and transcriptomics—revolutionize gut microbiome research by revealing microbial composition and guiding personalized interventions. These advances hold promise for managing gastrointestinal and metabolic disorders, heralding precision nutrition and personalized medicine. Future research will likely expand therapeutic options for optimizing gut microbiota and enhancing health outcomes.

References

Abouelela ME, Helmy YA (2024) Next-generation probiotics as novel therapeutics for improving human health: current trends and future perspectives. Microorganisms 12(3):430. https://doi.org/10.3390/microorganisms12030430

Birt DF, Boylston T, Hendrich S, Jane JL, Hollis J, Li L, McClelland J, Moore S, Phillips GJ, Rowling M, Schalinske K, Scott MP, Whitley EM (2013) Resistant starch: promise for improving human health. Adv Nutr (Bethesda, Md) 4(6):587–601. https://doi.org/10.3945/an.113.004325

Chen Z, Liang N, Zhang H, Li H, Guo J, Zhang Y, Chen Y, Wang Y, Shi N (2024) Resistant starch and the gut microbiome: exploring beneficial interactions and dietary impacts. Food Chem X 21:101118. https://doi.org/10.1016/j.fochx.2024.101118

Erdélyi L, Fenyvesi F, Gál B, Haimhoffer Á, Vasvári G, Budai I, Remenyik J, Bereczki I, Fehér P, Ujhelyi Z, Bácskay I, Vecsernyés M, Kovács R, Váradi J (2022) Investigation of the role and effectiveness of Chitosan coating on probiotic microcapsules. Polymers 14(9):1664. https://doi.org/10.3390/polym1409166

Fernandes A, Oliveira A, Carvalho AL, Soares R, Barata P (2023) *Faecalibacterium prausnitzii* in differentiated thyroid cancer patients treated with radioiodine. Nutrients 15(12):2680. https://doi.org/10.3390/nu15122680

García-Pinel B, Porras-Alcalá C, Ortega-Rodríguez A, Sarabia F, Prados J, Melguizo C, López-Romero JM (2019) Lipid-based nanoparticles: application and recent advances in cancer treatment. Nanomaterials (Basel, Switzerland) 9(4):638. https://doi.org/10.3390/nano9040638

Gomez Quintero DF, Kok CR, Hutkins R (2022) The future of synbiotics: rational formulation and design. Front Microbiol 13:919725. https://doi.org/10.3389/fmicb.2022.919725

Gyawali I, Zhou G, Xu G, Li G, Wang Y, Zeng Y, Li J, Zhou J, Zhu C, Shu G, Jiang Q (2023) Supplementation of microencapsulated probiotics modulates gut health and intestinal microbiota. Food Sci Nutr 11(8):4547–4561. https://doi.org/10.1002/fsn3.3414

Ji J, Jin W, Liu SJ, Jiao Z, Li X (2023) Probiotics, prebiotics, and postbiotics in health and disease. MedComm 4(6):e420. https://doi.org/10.1002/mco2.420

Kaur AP, Bhardwaj S, Dhanjal DS, Nepovimova E, Cruz-Martins N, Kuča K, Chopra C, Singh R, Kumar H, Şen F, Kumar V, Verma R, Kumar D (2021) Plant prebiotics and their role in the Amelioration of diseases. Biomol Ther 11(3):440. https://doi.org/10.3390/biom11030440

Kiousi DE, Rathosi M, Tsifintaris M, Chondrou P, Galanis A (2021) Pro-biomics: omics technologies to unravel the role of probiotics in health and disease. Adv Nutr (BethesdaMd) 12(5):1802–1820. https://doi.org/10.1093/advances/nmab014

Kwa WT, Sundarajoo S, Toh KY, Lee J (2023) Application of emerging technologies for gut microbiome research. Singapore Med J 64(1):45–52. https://doi.org/10.4103/singaporemedj.SMJ-2021-432

Lowe R, Shirley N, Bleackley M, Dolan S, Shafee T (2017) Transcriptomics technologies. PLoS Comput Biol 13(5):e1005457. https://doi.org/10.1371/journal.pcbi.1005457

Markowiak P, Śliżewska K (2017) Effects of probiotics, prebiotics, and synbiotics on human health. Nutrients 9(9):1021. https://doi.org/10.3390/nu9091021

Martín R, Miquel S, Langella P, Bermúdez-Humarán LG (2014) The role of metagenomics in understanding the human microbiome in health and disease. Virulence 5(3):413–423. https://doi.org/10.4161/viru.27864

Mills S, Yang B, Smith GJ, Stanton C, Ross RP (2023) Efficacy of *Bifidobacterium longum* alone or in multi-strain probiotic formulations during early life and beyond. Gut Microbes 15(1):2186098. https://doi.org/10.1080/19490976.2023.2186098

Pais P, Almeida V, Yılmaz M, Teixeira MC (2020) *Saccharomyces boulardii*: what makes it tick as successful probiotic? J Fungi (Basel, Switzerland) 6(2):78. https://doi.org/10.3390/jof6020078

Pandey KR, Naik SR, Vakil BV (2015) Probiotics, prebiotics and synbiotics- a review. J Food Sci Technol 52(12):7577–7587. https://doi.org/10.1007/s13197-015-1921-1

Pellegrino A, Coppola G, Santopaolo F, Gasbarrini A, Ponziani FR (2023) Role of *Akkermansia* in human diseases: from causation to therapeutic properties. Nutrients 15(8):1815. https://doi.org/10.3390/nu15081815

Plamada D, Vodnar DC (2021) Polyphenols-gut microbiota interrelationship: a transition to a new generation of prebiotics. Nutrients 14(1):137. https://doi.org/10.3390/nu14010137

Qiu S, Cai Y, Yao H, Lin C, Xie Y, Tang S, Zhang A (2023) Small molecule metabolites: discovery of biomarkers and therapeutic targets. Signal Transduct Target Ther 8(1):132. https://doi.org/10.1038/s41392-023-01399-3

Stoeva MK, Garcia-So J, Justice N, Myers J, Tyagi S, Nemchek M, McMurdie PJ, Kolterman O, Eid J (2021) Butyrate-producing human gut symbiont, *Clostridium butyricum*, and its role in health and disease. Gut Microbes 13(1):1–28. https://doi.org/10.1080/19490976.2021.1907272

Sun Q, Yin S, He Y, Cao Y, Jiang C (2023) Biomaterials and encapsulation techniques for probiotics: current status and future prospects in biomedical applications. Nanomaterials (Basel, Switzerland) 13(15):2185. https://doi.org/10.3390/nano13152185

Thursby E, Juge N (2017) Introduction to the human gut microbiota. Biochem J 474(11):1823–1836. https://doi.org/10.1042/BCJ20160510

Vallianou NG, Kounatidis D, Tsilingiris D, Panagopoulos F, Christodoulatos GS, Evangelopoulos A, Karampela I, Dalamaga M (2023) The role of next-generation probiotics in obesity and obesity-associated disorders: current knowledge and future perspectives. Int J Mol Sci 24(7):6755. https://doi.org/10.3390/ijms24076755

Walsh C, Lane JA, van Sinderen D, Hickey RM (2020) Human milk oligosaccharides: shaping the infant gut microbiota and supporting health. J Funct Foods 72:104074. https://doi.org/10.1016/j.jff.2020.104074

Wang X, Gao S, Yun S, Zhang M, Peng L, Li Y, Zhou Y (2022a) Microencapsulating alginate-based polymers for probiotics delivery systems and their application. Pharmaceuticals (Basel, Switzerland) 15(5):644. https://doi.org/10.3390/ph15050644

Wang X, Qi Y, Zheng H (2022b) Dietary polyphenol, gut microbiota, and health benefits. Antioxidants (Basel, Switzerland) 11(6):1212. https://doi.org/10.3390/antiox11061212

Wu J, Singleton SS, Bhuiyan U, Krammer L, Mazumder R (2024) Multi-omics approaches to studying gastrointestinal microbiome in the context of precision medicine and machine learning. Front Mol Biosci 10:1337373. https://doi.org/10.3389/fmolb.2023.1337373

Zannini E, Bravo Núñez Á, Sahin AW, Arendt EK (2022) Arabinoxylans as functional food ingredients: a review. Foods (Basel, Switzerland) 11(7):1026. https://doi.org/10.3390/foods11071026

Zhang L, Chen F, Zeng Z, Xu M, Sun F, Yang L, Bi X, Lin Y, Gao Y, Hao H, Yi W, Li M, Xie Y (2021a) Advances in metagenomics and its application in environmental microorganisms. Front Microbiol 12:766364. https://doi.org/10.3389/fmicb.2021.766364

Zhang H, Cheng J, Ao Q (2021b) Preparation of alginate-based biomaterials and their applications in biomedicine. Mar Drugs 19(5):264. https://doi.org/10.3390/md19050264

Zhang Z, Yang P, Zhao J (2021c) Ferulic acid mediates prebiotic responses of cereal-derived arabinoxylans on host health. Anim Nutr (Zhongguo xu mu shou yi xue hui) 9:31–38. https://doi.org/10.1016/j.aninu.2021.08.004

The Mechanistic Perspectives of Functional Foods: Probiotics and Prebiotics: A Focus on Colon Cancer and Human Health

3

Vijayalakshmi Jayaraman, Soniga Apthi, Liza Datta, Darshee Baxi, Satish Kumar Rajasekharan, Lourduraj Vasanthi, Kanagaraj Palaniyandi, and Dhanavathy Gnanasampanthapandian

Abstract

Cancer, a prevalent disease worldwide, is often caused by genetic and epigenetic changes. Functional foods, especially probiotics and prebiotics, offer promising therapeutic potential for cancer prevention and treatment by modulating the gut microbiota and enhancing host immune responses. This review examines the mechanistic perspectives of functional foods, focusing on their role in colon cancer and overall human health. Probiotics, such as *Lactobacillus* and *Bifidobacterium*, enhance gut health by colonizing the intestinal tract, stabilizing microflora, and promoting immune homeostasis. Prebiotics, including inulin and fructo-oligosaccharides, support beneficial gut bacteria and improve mineral absorption, thus contributing to cancer prevention. The synergistic effects of probiotics, prebiotics, and nutraceuticals, which are rich in bioactive compounds

V. Jayaraman · S. Apthi · L. Datta · K. Palaniyandi · D. Gnanasampanthapandian (✉)
Cancer Science Laboratory, Department of Biotechnology, School of Bioengineering, SRM Institute of Science and Technology, Kattankulathur, Chengalpattu, India

D. Baxi
Division of Biomedical and Lifesciences, School of Sciences, Navrachana University, Vadodara, India

S. K. Rajasekharan
Department of Biotechnology, School of Bioengineering, SRM Institute of Science and Technology,
Chengalpattu, India

L. Vasanthi
Unit of Cryobiology and Invertebrate Reproduction, Department of Zoology, University of Madras, Guindy Campus, Chennai, India

S. Pathak et al. (eds.), *Microbiota and Dietary Mediators in Colon Cancer Prevention and Treatment*, https://doi.org/10.1007/978-981-96-0297-1_3

such as polyphenols and terpenoids, are discussed in the context of cancer therapy. In addition, the role of postbiotics, such as short-chain fatty acids and exopolysaccharides, is investigated in promoting apoptosis and reducing tumour growth. The review highlights the need for continued research to fully elucidate the molecular mechanisms and optimize the therapeutic use of functional foods in cancer therapy care, emphasizing their potential to improve treatment outcomes and the patient's quality of life.

Keywords

Cancer · functional foods · prebiotics · probiotics · nutraceuticals

3.1 Introduction to Functional Foods in Cancer

Cancer has become one of the most prevalent diseases globally, primarily due to genetic alterations and epigenetic modifications in cell development and apoptotic pathways. In 2022, over 18.1 million people were affected worldwide (data, W.c 2022). Adverse lifestyle changes further disrupt homeostasis and cellular signalling pathways. Natural immunological changes, including immune responses aimed at eliminating tumour cells, also impact immunocompetent individuals (Singh et al. 2022). The incidence of cancer increases significantly after age 60, accounting for nearly 75% of cases, highlighting age as a crucial factor in cancer development (Mishra et al. 2023). Examining the tumour microenvironment (TME) is a key strategy in developing cancer treatment regimens. The distinct behaviour of malignant tumours across different tissues has deepened our understanding of cancer pathophysiology. Various interdisciplinary fields offer treatment options, including chemotherapy, radiation, surgery, and targeted therapies using small molecules and monoclonal antibodies. However, these treatments often come with severe side effects such as peripheral neuropathy, anaemia, oedema, hair loss, chronic fatigue, thrombocytopenia, and psychosis (Debela et al. 2021).

Functional foods, which offer both nutritional value and potential health benefits, play a role in reducing disease rates by enhancing immune function. Defined by the Food and Agriculture Organization (FAO) as foods containing additional components beneficial to health, functional foods include fortified foods, nutraceuticals, probiotics, prebiotics, synbiotics, and foods low in salt, fat, and sugar (FAO 2022). Among these, nutraceuticals, probiotics, and prebiotics are emerging as dietary supplements in cancer prevention and treatment. Nutraceuticals are food components that combine pharmacological and nutritional benefits, potentially aiding in disease prevention and treatment. These substances, found in popular functional foods, include polyphenols, alkaloids, terpenoids, unsaturated fatty acids, vitamins (A, B6, B12, C, D, E, folate), and trace elements like zinc, iron, selenium, magnesium, and copper (Alkhatib 2020).

Probiotics, which are dietary supplements containing live microbes, can temporarily or permanently colonize the digestive tract, promoting health by stabilizing the gut microflora, especially when the native flora is disturbed. This restoration supports crucial physiological functions and modulates immune responses (Sharifi-Rad et al. 2020). Prebiotics, on the other hand, promote the growth of beneficial gut bacteria, enhancing gut health and reducing inflammation. They have shown potential in reducing the risk of colorectal cancer (CRC) and improving mineral absorption and metabolism (Mishra et al. 2023; Hill et al. 2014). The interaction between prebiotics, probiotics, and cancer treatment strategies is complex but promising, particularly in modulating immune response and improving treatment outcomes. This chapter explores the impact of functional foods, especially probiotics and prebiotics, on cancer treatment strategies.

3.2 Functional Foods and Their Association with Cancer

According to "The Institute of Food Technologists", functional foods are new foods designed to contain ingredients or live microorganisms that can improve or prevent disease, provided the concentration is high enough to achieve the desired effect without being harmful. Nutrients, dietary fibres, phytochemicals, other substances, or probiotics are examples of additional components that are processed into functional foods (Temple 2022). Some examples of functional foods are α-carotene, β-carotene, lutein, β-cryptoxanthin, fucoxanthin, isothiocyanates, flavonoids, probiotics, fibre and omega-3. Lycopene, β-cryptoxanthin, and fucoxanthin are known to inhibit the growth of many types of cancer. Flavonoids, fibre, and omega-3 can prevent and treat various cancers, such as colon, breast, and prostate cancer. Probiotics can potentially be used to alleviate cancer symptoms (Aghajanpour et al. 2017) (Fig. 3.1).

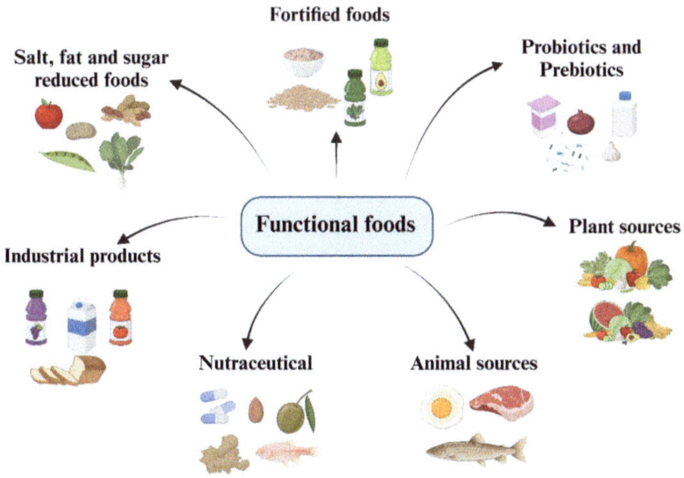

Fig. 3.1 The categorization of functional foods based on their different types and sources

3.3 Synergistic Relationship Among Probiotics, Prebiotics, and Nutraceuticals

Nutraceuticals are defined as "specially formulated preparations" that are intended to meet specific nutritional needs and/or to provide preventive healthcare. In addition, to supplement diet, nutraceuticals are formulated nutrients that aid in the treatment and prevention of specific diseases. The term "nutraceutical" was coined in 1989 by Dr Stephen De Felice and is made up of the words "nutrition" and "pharmaceutical" (Puri et al. 2022). An important source of nutraceutical constituents is are bioactive phytochemicals, such as alkaloids, various terpenoids, and polyphenols (which include anthocyanins, flavones, flavanols, isoflavones, stilbenes, ellagic acid, and others). These non-essential nutrients, which can defend against diseases or act as defence mechanisms, are mostly produced by plants. Numerous commonly used herbs and phytochemicals are effective nutraceuticals to reduce tumour progression, mitigate the side effects of radiation and chemotherapy, and improve the sensitivity of these treatments, and they are also pharmacologically safe (Caponio et al. 2022a). One of the examples of understanding the role of nutraceuticals in cancer is that of Coponio and his colleagues, who subjected CRC cell lines to an in vitro digestion model and assessed the effects of phenolic compounds present in Aglianico grape pomace (GP) on the cell lines at different stages of development. They discovered that the main components responsible for the increased antioxidant activity and total phenolic content (TPC) in the digested GP extract were flavonoids, phenolic acids, and anthocyanins (Caponio et al. 2022b).

Probiotics and prebiotics, which are summarized under the term nutraceuticals, are bioactive components obtained from food. These substances have a spectrum of positive effects, particularly on the health of the gastrointestinal tract. When consumed together, they synergistically support the balance and function of the gut microbiota, which can alleviate gastrointestinal disorders and improve overall digestive well-being.

3.4 Probiotics

Probiotics were first recognized as a naturally occurring and beneficial gastrointestinal microbiota at the end of the nineteenth century when the colonization of the digestive tract of healthy but asymptomatic individuals with microorganisms was reported. Probiotic bacteria are commensal microbes found in the gastrointestinal tract and in a variety of meals (Damian et al. 2022). These bacteria have the ability to colonize the gut by competing with pathogenic microorganisms. In addition, taking probiotics can improve health by improving or restoring the physiological composition and activity of gut bacteria. Gram-positive bacteria, such as those of the *Lactobacillus* and *Bifidobacterium* families, make up the vast majority of probiotics (Miri et al. 2022). Other non-pathogenic species with probiotic characteristics include certain strains of *Escherichia coli*, *Enterococcus*, *Pediococcus*, and yeast species, such as *Saccharomyces boulardii*. There are reports that additional commensal bacteria in the gut, such as *Salivarius* and *Streptococcus oralis*, have beneficial effects on health (Passali et al. 2019).

Probiotics come in a variety of types and can be broadly categorized into three main groups: *Bifidobacteria, Lactobacilli,* and others. As the lactic acid bacteria (LAB) are the most represented within the Lactobacilli group, the majority of current research on probiotic species focuses on this group. *Lactobacillus is* an important probiotic in the study of human gut microorganisms that are closely linked to human health. Not only does it help the body synthesize essential vitamins and amino acids and absorb minerals, but it also has the potential to improve the microecology of the gut by preventing the growth of pathogenic microorganisms (Ding et al. 2019).

3.5 Prebiotics

Prebiotics are a class of biological nutrients that can be broken down by the microflora of the gastrointestinal tract (GIT), which consists mainly of *Lactobacilli* and *Bifidobacteria.* Prebiotics are broken down by the colon bacteria when they are consumed as a food ingredient or dietary supplement. This produces short-chain fatty acids (SCFAs), which are simultaneously released in the large intestine and absorbed into the bloodstream. The two most important probiotic subgroups that have been thoroughly investigated with regard to human health are galactooligosaccharides (GOS) and fructooligosaccharides (FOS) (Bamigbade et al. 2022).

Prebiotics have an anticancer impact because they (i) promote the growth of good gut bacteria that occur naturally in the body, (ii) support the formation of postbiotics, (iii) alter the enzymes for metabolizing xenobiotics, and (iv) alter the immune system. Probiotics have three main anticancer effects—lowering gut pH, binding and degrading mutagens, and secreting anti-inflammatory chemicals. The benefits of using probiotics and prebiotics include control of the gut microbiota, immunomodulatory effects, and control of the interaction between the gut microbiota and the host (Martyniak et al. 2023).

3.6 Role of Probiotics and Prebiotics as Functional Foods

Intestinal dysbiosis has recently been associated with many human diseases. There are several nodes of interaction between the gut and metabolism that have opened new avenues for research and development. Non-communicable diseases (NCDs) have specially been viewed through the lens of the interaction of the gut microbiome with the host, and recent evidence suggests a possible link between the incidence of NCDs and gut dysbiosis. This allows for therapeutic intervention in this window, and functional foods could be a promising step in this direction. The integrity of the gut is not only important, but also significant when it comes to identifying inflammatory factors responsible for the spread of chronic diseases. Maintaining microbial balance becomes a key point in the control of central metabolic functions. Recent studies highlight the crucial connection between the gut and the body's metabolism, emphasizing that any disruption in this balance can result in microbial dysbiosis, which has been associated with various diseases, including cancer (Fulbright et al. 2017). The ancient Indian system of medicine has always attached great importance to this fact and has highlighted the need for a good and healthy gut as a factor associated with good health.

3.6.1 Prebiotics as Functional Foods

Prebiotics are important for the treatment of chronic diseases through a sustainable intervention. Prebiotics consist of indigestible carbohydrates such as short-chain carbohydrates (SCC) that cannot be digested by the host body but can be converted into functional food by utilization through the gut microbial community. Various plant sources serve as prebiotic agents for the gut microbiome and, in turn, modulate the host physiology to improve metabolic dyshomeostasis. Prebiotics are selectively fermented by the gut microbiota in the intestine and mainly comprise resistant starch, a wide range of di-, oligo-, and polysaccharides, as well as sugar polyols. There is a complex network of interactions between prebiotic sources such as functional food ingredients, human physiology, and the gut microbiota. These interactions are known to minimize various factors associated with chronic diseases and their development, such as inflammation and oxidative stress (Dixit et al. 2023).

3.6.2 Probiotics as Functional Foods

Although food is considered medicine, taking probiotic supplements is a direct way to balance the good and bad bacteria in the gut to ensure overall homeostasis and health. Probiotic bacterial strains help maintain the integrity of the gut by protecting the mucosal barrier of the intestinal epithelium and strengthening the fight against pathogenic microorganisms by preventing them from invading the gut epithelium. For example, *Lactobacillus* bacteria help to increase the production of mucin and reduce the permeability of the intestinal epithelium, also known as leaky gut, thus preventing pathogens from invading the systemic circulation of the host. In addition, *Lactobacillus* helps to maintain an acidic environment in which pathogens cannot survive, providing further protection for the host body. Functional foods that fall into this category include many fermented food items that harbour good bacteria and can serve as medical interventions to treat lifestyle disorders (Damian et al. 2022).

3.7 Symbiotics

Symbiotics are a blend of prebiotic polysaccharides and probiotic bacteria that work together in synergy to support various bodily processes, such as antioxidant activity, and provide health benefits to the host. Symbiotics are less commonly used as adjuvant therapy for cancers, especially those indirectly related to gut flora (Feng and Wang 2020).

The potential anticarcinogenic effects of symbiotics are unclear and still under investigation. The following are examples of possible mechanisms: (i) promotion of colonic apoptotic response to carcinogen-induced DNA damage; (ii) enhancement of probiotic colonization, growth, survival, and activity when a specific prebiotic substrate is present; (iii) upregulation of symbiotics, downregulation of cyclooxygenase-2, and inducible nitric acid (NO) synthase enzymes involved in colon carcinogenesis and increased synthesis of short-chain fatty acids (SCFA); (iv) process of immunomodulation; and (v) changes in the colonic bacterial environment that improve overall metabolic activity of the colon and cecum (Raman et al. 2013).

By increasing the production of SCFAs, decreasing the proliferative activity, and expression of glutathione S-transferases (GST), Pi-type placental enzyme, inducible NO synthase, and cyclooxygenase-2 enzymes involved in the pathogenesis of colorectal cancer, fructans and the symbiotic combination of fructans with *B. lactis* and *L. rhamnosus* GG reduce azoxymethane (AOM)-induced colorectal adenomas and carcinogenesis. Immunomodulation was found to be responsible for the antitumour effect of the symbiotic combination of oligofructose-enriched inulin, *L. rhamnosus*, and *B. lactis*. Peripheral blood mononuclear cells (PBMC) and Peyer's patches (PP) were found to be the primary tissues specifically affected by the prebiotics. These studies emphasize the importance of probiotics and prebiotics found in symbiotics for the treatment of CRC (Roller et al. 2004) (Tables 3.1 and 3.2).

3.8 Interaction of Probiotics and Prebiotics with Cancer

The gut is a dynamic, complex ecosystem that has developed unique immune cell properties over time as a result of constant exposure to various pathogens and antigens. The beneficial and health-promoting immunomodulatory abilities of commensal gut bacteria can significantly influence immune cell functions and activities. Indeed, immune homeostasis and the modulation of innate and adaptive immune responses depend on immune cells, commensal bacteria, and nutrients interacting and responding to each other continuously in a stable environment. This interaction improves the functions of the immune system (Mazziotta et al. 2023). SCFAs have been the subject of recent studies on their role in gut-brain microbial communication. The gut microbiota ferments prebiotics to form SCFAs, which have been shown to be important for gut-brain communication (Markowiak and Slizewska 2018). This discovery highlights the potential influence of prebiotics on gut flora and its wider health implications. Furthermore, it was discovered that the addition of probiotics and prebiotics to the diet can strengthen the gut microbiota, improving

Table 3.1 Effects of probiotics in cancer cells studied in both in vitro and in vivo models

S. no.	Probiotic strain	Model type	Cell line or model	Effect in cancer	References
1	*Lactobacillus pentosus* B281*Lactobacillus plantarum* B282	In vitro	Caco-2 and HT-29	↓ Cell proliferation and initiation of cell cycle arrest (G1)	Saxami et al. (2016)
2	*Lactobacillus casei* ATCC 393	In vitro	HT29 and CT26	Induction of apoptosis	Tiptiri-Kourpeti et al. (2016)
3	*Propionibacterium freudenreichii* ITG P9	In vitro	HGT-1	Initiation of apoptosis	Cousin et al. (2012)
4	*Lactobacillus kefiri* P-IF	In vitro	MDR	Induction of apoptosis	Ghoneum and Gimzewski (2014)
5	*Lactobacillus rhamnosus* GG*Bifidobacterium lactis* Bb12	In vitro	HT-29	Cell apoptosis induction	Borowicki et al. (2011)
6	*Lactococcus lactis* NK34	In vitro	HT-29, LoVo, AGS	>80% ↓ Cell proliferation, decreases nitric oxide production and pro-inflammatory cytokines	Han et al. (2015)
7	*Bacillus polyfermenticus* KU3	In vitro	LoVo, HT-29, AGS	>90% ↓ Cell proliferation	Lee et al. (2015)
8	*Propionibacterium acidopropionici* CNRZ80	In vitro	HT-29	↓ Cell proliferation and induction of apoptosis	Jan et al. (2002)
9	*Lactobacillus acidophilus* SNUL*Lactobacillus casei* YIT9029*Bifidobacterium longum* HY8001	In vitro	SNUC2A, SNU1, NIH/3T3 and Jurkat cell	Suppressed proliferation of tumour cells	Lee et al. (2004)
10	*Lactobacillus rhamnosus* GG	In vitro	HGC-27	↓ Cell proliferation and induction of apoptosis	Russo et al. (2007)
11	*Lactobacillus rhamnosus* GG	In vitro	Caco-2	↓ Level of IL-8	(Lopez et al. (2008)
12	*Lactobacillus acidophilus* 606	In vitro	HT-29	Inhibited proliferation of tumour cells	Kim et al. (2010)

(continued)

Table 3.1 (continued)

S. no.	Probiotic strain	Model type	Cell line or model	Effect in cancer	References
13	*Lactobacillus paracasei* IMPC2.1*Lactobacillus rhamnosus* GG	In vitro	DLD-1, HGC-27	↓ Cell proliferation and induction of apoptosis	Orlando et al. (2012)
14	*Lactobacillus casei* BL23	In vivo	C57BL/6 mice	↓ Tumour incidence, Bik, caspase-7 and caspase-9 upregulation	Jacouton et al. (2017)
15	*Bacillus polyfermenticus*	In vivo	F344 rats	50% ↓ Aberrant crypt foci, ↑ antioxidant potential	Park et al. (2007)

↓, decreased; ↑, increased

Table 3.2 Effects of prebiotics on cancer cells together with the mechanism of action

S. no.	Prebiotic	Model type	Cell line or model	Effects on cancer cells	References
1	FOS/Raftilose	In vivo	F344 rats	Reduces the expression of GST-P and i-NOS in tumours that cause colon cancer, reduces progression of AOM-induced colon cancer	Femia et al. (2002)
2	B-(1–4) galactooligosaccharides	In vivo	Rats	Increases the abundance of bifidobacteria, which inhibits the development of aberrant crypt foci (Macfarlane and Murphy 2010) and reduces 1,2-dimethylhydrazine (DMH)-induced ACF	Bruno-Barcena and Azcarate-Peril (2015)
3	Inulin	In vivo	Wild-type (WT) C57BL/6 mice	The growth of BRAF-mutated melanoma was inhibited and the efficacy of a MEK inhibitor against melanoma was improved by increasing the relative abundance of *Bacteroides* species. Reduction in the progression of colorectal cancer.	Li et al. (2020), Rafter et al. 2007)

(continued)

Table 3.2 (continued)

S. no.	Prebiotic	Model type	Cell line or model	Effects on cancer cells	References
4	Polyphenol	In vivo	*Bifidobacterium* and *Lactobacillus*	Reduced development of primary tumours and metastases as well as reduced metastasis of 4T1 breast tumours due to suppressed macrophage infiltration and angiogenesis were the outcomes of increased levels of bacteria (*Bifidobacterium* sp. and *Lactobacillus* sp.) and SCFAs.	Teng et al. (2021)
5	Pectin	In vitro	HT29 cells	Apoptosis and increased caspase-3 activity were detected. There is a direct correlation between DNA fragmentation and apoptotic cells.	Leclere et al. (2013)
6	Human milk oligosaccharides	In vitro	HT-29, HIEC, and Caco-2	The intestinal cells underwent growth arrest as a result of the regulation of epidermal growth factor receptor (EGFR) signalling pathways and reduced the rate at which the cells proliferated.	Bruno-Barcena and Azcarate-Peril (2015)

host function and increasing resistance to colonization by gut pathogens. According to the studies, altering the gut microbiota is a cutting-edge method for preventing and treating CRC. This discovery underscores the potential medical benefits of using probiotics and prebiotics to target the gut microbiota. This shows the great therapeutic potential of probiotics and prebiotics in altering the gut microbiota to improve overall health.

3.8.1 Models of Probiotics

3.8.1.1 *Lactobacillus rhamnosus* GG—A Probiotic Model in Cancer

Probiotics, including *Lactobacilli*, can balance the microbiota in the gut. Therefore, they are being researched as a supportive treatment for gastrointestinal damage associated with chemotherapy. One of the most researched and well-defined probiotic archetypes is *Lactobacillus rhamnosus GG* (LGG), which has anti-inflammatory properties LGG is also the first studied probiotic in oncology. In the intestinal environment, LGG is a gut-resident bacterium that has been shown to have many anti-inflammatory properties. Administered in conjunction with food, LGG reduces radiation- and 5-FU-induced damage to the intestinal epithelium in animal models and preserves the integrity of the intestinal epithelial barrier and the balance of the intestinal microbiota. In one study, LGG restored the equilibrium of the intestinal mucosal microbiota and halted the loss of microbial diversity in piglets caused by *S. enterica* Infantis. Administration of LGG blocked *S. enterica* Infantis-induced IEC autophagy and reduced *S. enterica* Infantis internalization by stimulating EGFR-mediated activation of the negative mediator Akt. This, in turn, prevented intestinal epithelial cell mortality and limited systemic *S. enterica* Infantis infection (Zhang et al. 2017).

3.8.1.2 *Bifidobacterium*—A Case Study

Bifidobacterium is a beneficial bacterium in the human body that promotes immune function and has antitumour properties. *Bifidobacterium* specifically inhibits the proliferation and adhesion of harmful pathogens to the cells lining the intestine. These microorganisms produce substances such as bacteriocin and hydrogen peroxide, which actively eliminate pathogenic microbes. The most important by-products of *Bifidobacterium* metabolism are lactic acid and acetic acid, which contribute to the acidification of the intestinal environment, particularly in the cecum and ascending colon. This acidification is believed to contribute to their probiotic effect by creating an inhospitable environment for harmful microbes. In one study, the effect of *Bifidobacterium bifidum* on the colon cancer cell line SW742 was investigated. It was shown that the cell-free supernatant (CFS) of *B. bifidum* has the ability to inhibit the proliferation of colon cancer cells. While *Bifidobacterium* normally thrives under strictly anaerobic conditions, studies have shown that *Bifidobacterium thermophilus* and *Bifidobacterium bifidum* can survive and proliferate under oxygen pressure. However, no *Bifidobacterium* strain was observed to grow under aerobic conditions during the course of the study (Martin et al. 2003). The presence of probiotic bacteria in the intestinal flora plays a crucial role in the prevention of colon cancer by strengthening beneficial bacterial populations, reducing levels of harmful pathogens, causing changes in metabolism and enzyme activity, reducing inflammation, and boosting immune function.

3.9 Postbiotics and Their Role in Cancer Treatment

Postbiotics have anti-cancer properties by reducing cell growth and proliferation and at the same time enhancing the apoptotic effect. Of the many components of postbiotics, including SCFAs, metabolites, microbial cell fractions, functional proteins, exopolysaccharides (EPSs), and cell lysates, polysaccharides have the strongest anticancer properties. Molecular mechanisms underlying postbiotic anti-cancer effects include modulation of immune responses, suppression of mutagenesis and carcinogens, activation of pro-apoptotic signalling pathways, reduction of bacterial translocation, and enhancement of apoptosis and autophagy for cancer prevention and treatment.

Numerous in vitro and in vivo studies have shown the anti-cancer effect of postbiotics. For example, SCFAs from Propionibacterium inhibit the human CRC cell lines HT-29 and Caco-2 by causing loss of mitochondrial transmembrane potential, formation of reactive oxygen species, activation of caspase 3, modulation of Bcl-2 regulation, and nuclear chromatin condensation (Jan et al. 2002). Postbiotics modulate immune responses and stimulate cytokines (e.g. TRAIL, IL-10, and TGF-β) to kill tumour cells (Kim et al. 2021a). SCFAs protect the mucosa by lowering the levels of immunomodulators such as prostaglandins in human breast cancer cells MCF-7 (Nakkarach et al. 2021).

3.10 Postbiotics and Their Types

3.10.1 SCFAs-Butyrate

The most studied probiotics are SCFAs, in particular acetate, propionate, and butyrate. These compounds are the result of the fermentation of dietary fibres by intestinal bacteria, especially *Faecalibacterium prausnitzii* and *Eubacterium rectale*. Acetate and propionate are usually delivered to muscle and liver, where they serve as substrates for mitochondrial oxidation. Butyrate, on the other hand, is metabolized in situ by the colonocytes and covers up to 70% of the total energy requirement. Lower concentrations of SCFAs in the intestine have been associated with the development of advanced colorectal adenomas (Chen et al. 2013).

Butyrate is readily metabolized once it reaches the cell. However, the role of butyrate in colon cancer cells is quite different. Butyrate promotes differentiation, inhibits proliferation, and induces oxidative stress, leading to cell cycle arrest and apoptosis. Butyrate plays opposing roles in normal and cancer cells, contributing to the phenomenon known as the "butyrate paradox", which until recently could not be explained. In non-cancer cells, butyrate is metabolized by β-oxidation and acetyl-CoA utilization in the Krebs cycle. Cancer cells primarily require glucose for energy production, which is then converted to lactate by glycolysis. This process is known as the Warburg effect, which suppresses β-oxidation. Thus, butyrate molecules accumulate in the cytoplasm of cancer cells and migrate to the nucleus where they block histone deacetylases. Under these conditions, the histones are acetylated and the solid structure of the DNA becomes loose and accessible to transcription factors, leading to the production of various proteins, triggering oxidative stress, and ultimately apoptosis (Donohoe et al. 2012).

3.10.2 LABs-Lactic Acid

Although there is strong evidence that fermented foods, especially dairy products, have chemopreventive effects, many studies are phenomenological in nature and do not identify the active compound(s) responsible for this effect. LABs are the main microorganisms found in fermented dairy products. It is well known that LABs convert the lactose in dairy products into lactic acid. In 2008, a large prospective study on the effect of fermented milk on 82,002 Swedish women and men was published. This study focused on bladder cancer, as most of the metabolites are excreted via the urinary bladder, and the effects of orally ingested dairy products are obvious. The results show that consuming plenty of sour milk can reduce your risk of developing bladder cancer (Larsson et al. 2008). Other studies have found that frequent consumption of yoghurt, but not milk, reduces the risk of CRC in Europe and the United States (Michels et al. 2020). Overall, postbiotic lactic acid may help prevent cancer, but more research is needed to distinguish its benefits from those of other dairy products.

3.10.3 Teichoic Acids

Teichoic acids (TAs) are cell wall compounds found in gram-positive bacteria that form anionic glycopolymers (ribitol) containing phosphodiester-linked polyol repeat units of peptidoglycans. They play an important role in cell structure, control of cell division, pathogenesis, antibiotic resistance, host–bacterium interaction, and other important biochemical and metabolic aspects of cell physiology (Nataraj et al. 2020). Lipoteichoic acids (Pauli et al. 2017) are glycolipid-fixed in the bacterial membrane and can be spontaneously released into the microenvironment, whereas wall teichoic acids (WTAs) are covalently bound to the peptidoglycan and consist of a major chain polymer composed of phosphodiester-linked polyol repeat units and a disaccharide linkage unit. LTA from *Bifidobacterium animalis* subspecies *lactis* BPL1 was identified in a preclinical model of *Caenorhabditis elegans* as a novel lipid modulator with fat-reducing capabilities via insulin-like growth factor-1 (IGF-1), suggesting an effective therapeutic and/or preventive application of this component in metabolic syndrome and diabetes-related diseases (Balaguer et al. 2022).

3.11 Probiotics and Prebiotics in Cancer Therapy

Despite advances in cancer treatment, cancer is still the second most common cause of death worldwide. Recently, probiotics have attracted a lot of attention from researchers due to their potential anti-cancer properties. Probiotics are thought to inhibit cancer by suppressing the growth of mutagens and carcinogenic microorganisms, protecting the host genome from oxidative damage, and modulating the immune system. In addition, these microorganisms can influence the expression of various genes involved in critical biological processes, such as cell proliferation, apoptosis, metastasis, and the cell cycle. However, the precise molecular mechanisms underlying the effect of probiotics on cancer cells are not yet fully understood.

There have been studies aimed at investigating the molecular mechanisms of the probiotic *L. plantarum* on the expression of key genes (*AKT*, *PTEN*, *BAX*, and *TLR4*) involved in apoptosis and anti-apoptosis signalling pathways in the AGS gastric cancer cell line. Various analyses, including DNA fragmentation assay, flow cytometry, DAPI staining, and q-PCR, substantiated the positive potential of *L. plantarum* in inducing apoptosis, in the gastric cancer cell line. The results suggest that *L. Plantarum can* inhibit the viability of cancer cells by modulating *AKT* and *PTEN* gene expression. However, further research is needed to fully comprehend the detailed molecular mechanisms underlying the probiotic effects on cancer cells (Maleki-Kakelar et al. 2020).

Probiotics such as *Lactobacillus* are thought to promote the immune response through the upregulation of IL-12 produced by antigen-presenting cells (Trinchieri 2003). This, in turn, promotes Th1 cell differentiation and enhances the cytotoxic activity of NK cells and macrophages, while it also induces IgA secretion in the intestine and activates B cells and T cells via various mechanisms (Arai et al. 2018). These immunostimulatory effects contribute to the inhibitory effect of *Lactobacillus* on foodborne infections and cancer (Lashani et al. 2020). In addition, *Lactobacillus* has been found to curb excessive immune responses by inhibiting the production of pro-inflammatory cytokines, such as IL-12 and IL-8 (Kich et al. 2016). Simultaneously, it increases the level of anti-inflammatory factors, such as IL-10, and stimulates the production of regulatory T cells.

3.12 Mechanism of Action of Probiotics in Cancer

It is unclear which mechanism is involved in the anti-cancer effect of probiotics. The gut microbiota is involved in a number of processes, all of which are thought to be critical to the process. Probiotic bacteria are primarily involved in maintaining homeostasis or stable physicochemical conditions in the colon. Low pH caused by excessive bile acids in the faeces, for example, can be a direct cytotoxic factor affecting the colon epithelium, and leading to the development of colon cancer. Probiotic bacteria, such as *L. acidophilus* and *B. bifidum*, have shown promise in helping to prevent cancer due to their role in modulating pH and bile acid composition (Jia et al. 2018).

3.12.1 Production of Compounds with Anti-carcinogenic Activity

The fermentation of fibre-rich prebiotics leads to the formation of SCFAs, such as acetate, propionate, and butyrate, by a specific population of the gut microbiota. In addition to their primary function as an energy source, SCFAs have been shown to act as signalling molecules that influence the immune system, cell death, and proliferation, as well as intestinal hormone production and lipogenesis, which explains their critical role in maintaining epithelial integrity.

SCFAs provide energy to colonocytes while inducing acidosis and cancer cell death, resulting in an acidic environment that prevents the development of significant amounts of secondary bile acids. Butyric acid helps to regulate the proliferation, division, and death of colonocytes. Colonocytes metabolize ~70–90% of butyrate. Healthy people have more of this acid in their faeces than people suffering from CRC. A 1 μg/L reduction in faecal butyrate concentration is estimated to increase the risk of CRC by 84.2%. Reducing the amount of acetic acid by 1 μg/L increases the risk of developing an adenoma by 71.3%. Butyric acid may help prevent CRC because it improves the intestinal barrier by increasing mucus production and promoting the proliferation of healthy cells (Cook and Sellin 1998).

3.12.2 Immunomodulation

The immune system is a complicated cascade that acts and reacts at local and systemic levels. Recent research has focused on understanding the control of the immune system and the connections within and between the components of the inflammatory cascade. Microorganisms from the gut microbiota are the primary stimulants of the immune system, which are necessary for the development of lymphoid tissue in this system. The intestinal microbiota, which includes probiotic bacteria, has immunomodulatory activity based on three seemingly contradictory phenomena: (i) induction and maintenance of immune tolerance to environmental antigens (food and inhalation), (ii) induction and control of immune responses against bacterial and viral pathogens, and (iii) inhibition of autoimmune and allergic reactions. Some probiotic *Lactobacillus* strains have been shown to reduce *H. pylori* infections associated with gastric cancer (Zhong et al. 2014). Another study conducted in patients with persistent human papillomavirus virus (HPV) showed an increased clearance of HPV and precancerous cervical lesions after daily probiotic intake for 6 months (Chen et al. 2013).

3.12.3 Influence on Other Mutagenic and Carcinogenic Factors

Probiotics can influence other mutagenic and carcinogenic elements, and thus help to prevent cancer. They can alter the activity of several enzymes involved in cellular detoxification, thereby inhibiting the activity of free radicals and carcinogenic chemicals. β-Glucuronidase, a bacterial enzyme with broad substrate specificity, can hydrolyse various glucuronides, releasing carcinogens into the colon, including benzo[*a*]pyrene (PAH), a major risk factor for CRC. In the colon, β-glucosidase degrades plant glycosides, such as cycasin, producing carcinogenic aglycones (McIntosh et al. 2012). *Bacteroides fragilis*, *Escherichia coli*, and *Clostridium* have significant amounts of pro-carcinogenic enzymes (β-glucosidase, β-glucuronidase, and nitroreductase). Studies have also shown that *Bacteroides fragilis*, Clostridium,

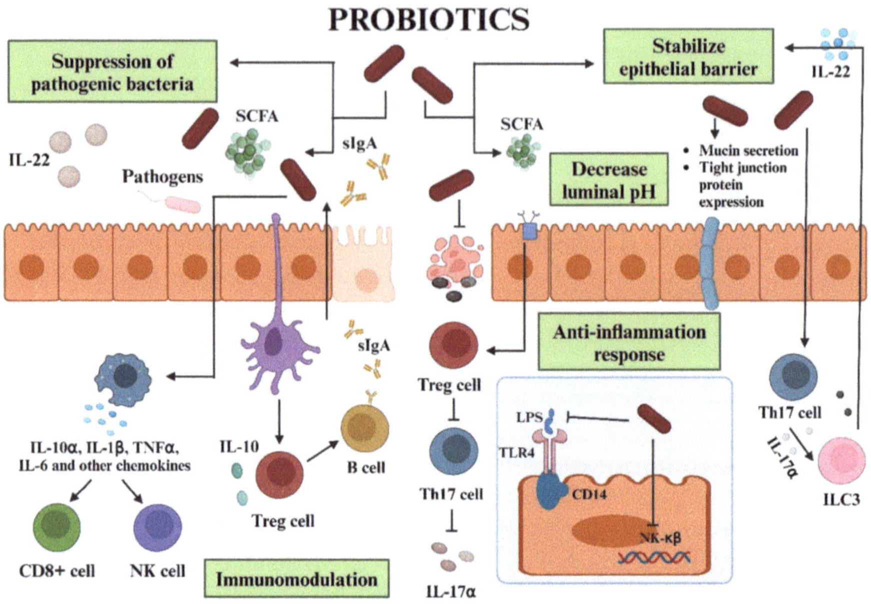

Fig. 3.2 The probiotic action mechanism: the effects of probiotics on the host. SCFAs, sIgA soluble IgA, GPR G protein-coupled free fatty acid receptor, dendritic cell, Treg cell regulatory T cell, Th17 T helper cell 17, ILC3 type 3 innate lymphocyte, NK cell natural killer cell, LPS lipopolysaccharide, TLR4 toll-like receptor 4, and NF-κβ nuclear factor-κB

and *Escherichia coli* are overrepresented in colon cancer patients (Culpepper and Mai 2013) (Fig. 3.2).

3.13 Mechanism of Prebiotics in Cancer

Prebiotics include indigestible compounds (e.g. pectin, starch, galactooligosaccharides, and inulin) that increase the growth of beneficial bacteria in the stomach and thus improve the effect of cancer treatment. Prebiotics are thought to help synthesize SCFAs and reduce products of protein breakdown. They also provide signals that reduce carcinogenic effects in the gastrointestinal tract and have anti-adhesive properties against pathogens.

3.13.1 Enhanced Absorption of Micronutrients

Prebiotics, such as inulin, oligofructose, FOS, GOS, soybean oligosaccharides, and resistant starches, can improve the absorption and metabolism of minerals. The relationship between the absorption of prebiotics and minerals and anti-cancer efficacy is complicated and involves several variables. Prebiotics can improve the solubility of minerals in the intestine through the formation of SCFAs. This, in turn, can lead to the formation of SCFAs-salt complexes that facilitate mineral absorption. Prebiotics also help to increase the surface area accessible for mineral absorption by promoting the production of calcium-binding proteins, and degrading the phytic acid-mineral complex, thereby releasing bound minerals. In addition, prebiotics can release bone-regulating chemicals such as phytoestrogens found in certain meals, stabilize gut bacteria and mucus, and help maintain healthy gut flora (Raman et al. 2013).

3.13.2 Immunomodulation

Prebiotics may have an indirect immunogenic effect by affecting the gut bacteria and modulating immunologic parameters, such as NK cell activity, IL-10 and interferon release, and lymphocyte proliferation (Janardhana et al. 2009). Prebiotics can influence immunological properties in gut-associated lymphoid tissue (GALT), secondary lymphoid tissue, and peripheral circulation. There are few studies looking at the effect of prebiotics on the GALT to improve the human immune system and anti-centromeric antibodies (ACA). GOS has been found to attenuate colitis in Smad3-deficient mice by altering NK cell activity and trafficking (Gopalakrishnan et al. 2012). Although prebiotics have been shown to have immunomodulatory properties, further research is needed to determine the molecular function of prebiotic-induced immunomodulation against cancer.

3.13.3 Modification of the Gut Microbiota

Prebiotics can boost host immunity by modifying the gut microbiota, which is critical for the generation of immune responses to develop tumours. Diet is one of the strongest factors influencing the bacteria in the gut. This is an important factor in cancers, as both innate and adaptive immunity is thought to play a role in cancer prevention by inhibiting tumour cell metabolism and increasing anti-cancer T cell activity. CD8+ T cells are one of the most important factors in the fight against cancer and contribute to slowing down the progression of cancer. T cells are the primary targets of cancer therapy. Therefore, altering the characteristics of the gut microbiota can be used to prevent malignancies through diet, particularly through the use of prebiotics (Reid et al. 2011) (Fig. 3.3).

3.14 Other Mechanisms or Probiotics and Prebiotics in Cancer

3.14.1 Apoptosis

Recently, treatments based on probiotics for CRC have been extensively researched. The anti-cancer mechanisms of probiotics in CRC include triggering apoptotic responses, protecting the intestinal mucosal barrier, inhibiting carcinogenic agents,

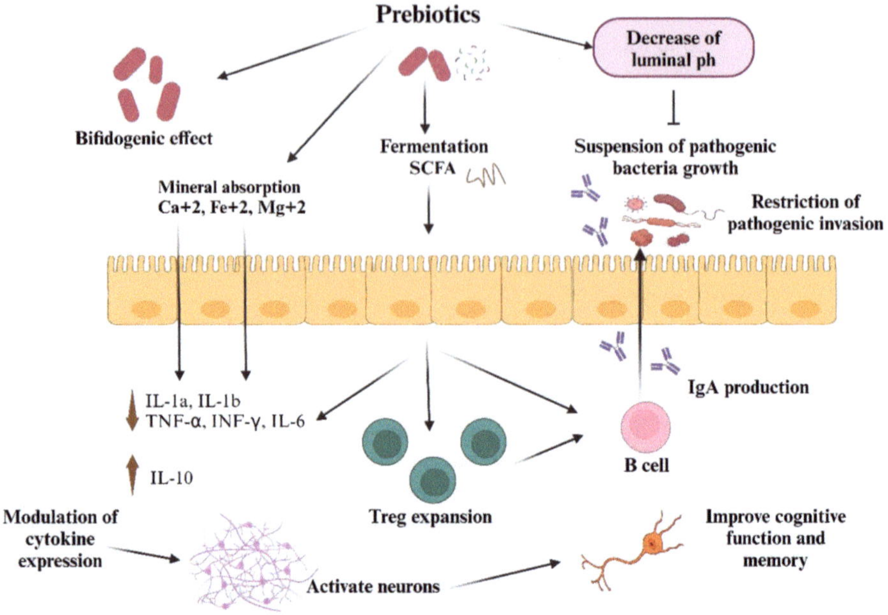

Fig. 3.3 The prebiotic action mechanism: the synthesis of microbial metabolites, noticing SCFAs, promoting absorption of ions and trace elements (such as calcium, iron, and magnesium), lowering the pH of the luminal membrane, and controlling the immune system (by producing more IgA and altering the production of cytokines) are some of these mechanisms

and modifying the enzymatic activity of pathogenic bacteria (Eslami et al. 2019). Numerous studies have shown that certain probiotic strains can inhibit the proliferation of CRC cells and induce cell death both in vitro and in vivo (Zhu et al. 2011). For example, *Enterococcus faecium* RM11 *and Lactobacillus fermentum* RM28 have been shown to suppress the proliferation of CRC cells, and *Lactobacillus delbrueckii* induces apoptosis in SW620 cells by increasing the expression of caspase-3. In particular, the probiotic strain *Pediococcus pentosaceus* (PP) has a significant antiproliferative effect on CRC cells. Studies show that oral administration of PP-p8 induces apoptosis in CRC cells, and inhibits tumour growth in mouse xenograft models. The findings suggest that the presence of p8 is a critical factor that enables PP to exert a potent anticancer effect on CRC cells compared to PP-Ev (Kim et al. 2021b).

Recent studies have shown that overexpression of RNF152 promotes apoptosis in HeLa cells, inhibits proliferation of CRC cells, and induces apoptosis by regulating the mechanistic target of rapamycin complex 1 (mTORC1) in an E3 ligase activity-dependent manner (Kim et al. 2021b; An et al. 2019). Consistent with previous data, a decrease in the expression of RNF152 was observed in human CRC patients. Furthermore, studies have shown that the expression of RNF152 and cleaved PARP1 is increased in a CRC xenograft model after oral administration of PP-p8. Similar results were obtained in a DDS model, in which overexpression of RNF152 inhibited tumour growth and increased the expression of cleaved-PARP1 in xenograft mice (Ou et al. 2014).

The results of gene and protein expression analysis of studies suggest that the ability of extracellular polysaccharides (EPS) of *Lactobacillus* species to induce apoptosis is associated with upregulation of Bax, caspase-3 and -9, and downregulation of Bcl-2 and survivin (Wang et al. 2002). Previous in vitro studies have suggested that activation of caspase-3 and -9 and inhibition of Bcl-2 may inhibit the development of CRC. In a parallel study, EPS isolated from *A. halophytica* showed similar apoptosis induction to the above results. The proposed molecular mechanism is that the endoplasmic reticulum (ER) signalling pathway triggers apoptosis. It was explained that the ER targets Grp78, regulates the cell's response to unfolded proteins (UPR), and suppresses the expression of survivin and Bcl-2, while inducing the expression of caspase 3, which ultimately leads to cell apoptosis (Cao et al. 2010).

The study suggests that the EPS released by the studied *Lactobacillus* species inhibit the proliferation of HT-29 cells through apoptosis. It has been reported that the bifunctional activities of polysaccharides, including anti-tumour and apoptosis-inducing activity, are closely related to their molecular weight, monomer composition, structure and binding type. A high level of apoptosis induction was observed in the EPS of lactobacillus species (Tukenmez et al. 2019). The study could be related to the main monomer in their structure, mannose, and their binding type, especially β-(1,3) and α-(1,2). Both glucose and mannose have highly specific receptors on macrophages, which is crucial in tumour immunology. The ability of all investigated EPS to induce apoptosis in CRC cells may be due to their sugar composition, especially the presence of mannose and glucose as major components. In particular, B3_EPS, which has the highest mannose content in its sugar composition, showed the most potent apoptosis induction in HT-29 cells (Guardamagna et al. 2022).

The rate of cancer development is determined by the proliferation and apoptosis of cancer cells. During cancer development, these cells proliferate rather than die. Probiotics that can influence cell proliferation and apoptosis are of great interest because cancer cells are removed less aggressively, and apoptosis does not cause damage to neighbouring cells or inflammation. Probiotic bacteria (e.g. LAB) can trigger apoptosis signalling pathways via a mitochondria-dependent (intrinsic) and a death receptor-dependent, mitochondria-independent (extrinsic) pathway (Mu et al. 2020).

The higher incidence of apoptosis in cancer cells induced by probiotics has been linked to SCFAs, particularly butyrate, which can trigger epigenetic changes, paralyze the cell cycle, and activate the production of proapoptotic genes.

3.14.2 Gut Microbiome and MAPK Inhibitors

Although the relationship between the gut microbiota and BRAF/MEK inhibitors is still unexplored, it represents an interesting starting point for research, particularly in view of its importance for melanoma therapy. Deciphering this association could provide valuable insights into molecular signalling pathways, the immune system, and the microbiome. Understanding the effect of dietary changes, antibiotic use, interventions, such as probiotics and faecal microbiota transplantation (FMT) on carcinogenesis, and response to therapy could deepen the understanding of the treatment of patients with BRAF-mutated melanoma. In addition, the study of GM in cancer patients may help to identify responders and non-responders, thereby enhancing treatment strategies. If GM is confirmed as a predictive or prognostic biomarker, characterization of GM composition may become a simpler and more benign procedure (Zununi Vahed et al. 2017).

Although the correlation between MAPK inhibitors and GM has not yet been documented in melanoma patients, it has been investigated in other cancers. In one study, faecal samples from 33 CRC patients categorized as BRAF-mutated or wild-type, and from 13 healthy volunteers were analysed. BRAF-mutated patients had an abundance of specific microbial species, and interestingly, xenogeneic BRAF-mutated CRC models showed higher microbial diversity compared to BRAF wild-type controls. The microbiota signature in BRAF-mutated patients resembled that of healthy controls more than the wild-type CRC model (Shi 2009).

3.14.3 Kinase Inhibition by Probiotics

The enzymes that add and remove phosphate groups are called kinases and phosphatases. Threonine, serine, and tyrosine residues are modified by phosphorylation processes, in which the terminal γ-phosphate group of the adenosine triphosphate is added to these residues (Motevaseli et al. 2017; Secombe et al. 2020). There are an estimated 518 genes encoding kinases and 156 genes encoding phosphatases in the human genome. Kinases are potential therapeutic targets for malignancies as they are important for many aspects of tumour biology, such as cell proliferation, motility, metabolism, angiogenesis, and metastasis (Pachmayr et al. 2017). Probiotics and their metabolites have been investigated in several studies as kinase inhibitors for the treatment of diarrhoea after cancer therapy. Probiotic secretory proteins have been shown to protect intestinal epithelial tight junctions and barrier function from hydrogen peroxide damage via a protein kinase C (PKC)- and MAPK-dependent mechanism. By upregulating phosphatases and downregulating MAPKs, *L. plantarum* causes apoptosis. By producing tumour necrosis factor-associated apoptosis-inducing ligand (TNFAIL), *Lactobacillus* increases the activity of NK cells in prostate cancer cell lines (Dillekas et al. 2019). Exploring and using the properties of probiotics as kinase inhibitors is the main goal of ongoing research projects.

3.14.4 Prevention of Metastasis

In metastasis, a crucial step in the development of cancer, tumour cells separate from the main tumour and migrate to other areas of the body (Motevaseli et al. 2017). The process of metastatic spread consists of several stages: (i) penetration of the basement membrane, (ii) infiltration into the surrounding lymphatic or vascular system, (iii) retention in the bloodstream, (iv) extravasation into adjacent tissues, and (v) colonization at the sites of secondary tumours (Braga 2016). Disruption of cell–cell adhesion, epithelial–mesenchymal transition (EMT), tumour microenvironment, maintenance of cancer stem cells, and anti-metastatic properties of probiotics are some of the reasons that lead to metastasis. The degradation of the extracellular matrix occurs through matrix metalloproteinases (MMPs), whereas VEGF is a signalling molecule that stimulates the formation of new blood vessels. The first stage of metastasis, the degradation of the basement membrane, can be slowed by blocking the VEGF-MMP2/9 signalling pathway. Probiotic therapy causes a remarkable increase in E-cadherin, a protein essential for cell–cell adhesion, which inhibits the migration of cancer cells (Frankel et al. 2017).

3.15 Interaction with Standard Chemotherapy for Cancer

Chemotherapy disrupts the microbial community, leading to a decrease in microbial diversity and changes in the abundance of various bacteria (Garcia-Gonzalez et al. 2017). For example, the administration of 5-fluorouracil (5-FU) can lead to an increase in Proteobacteria and a decrease in Firmicutes and Actinobacteria. This shift contributes to the attenuation of inflammatory responses through interactions with the NF-κB signalling pathway and the production of SCFAs (Alexander et al. 2017).

Conversely, the gut microbiota plays a role in influencing chemotherapy through mechanisms, such as translocation, immunomodulation, metabolism, enzymatic degradation, and reduced diversity and ecological variation (Egelston et al. 2018; Touchefeu et al. 2014).

Chemotherapy-induced diarrhoea (CID) is a common gastrointestinal side effect of radiotherapy and chemotherapy that affects the quality of life of cancer patients. Extensive animal and clinical studies have revealed significant changes in the gut microbiota of chemotherapy recipients, including a notable decrease in Bifidobacteria, Clostridium group XIVa, and *Clostridium* genus, accompanied by increased levels of *Enterobacter* and *Bacteroides*. These microbial shifts contribute to intestinal mucositis, which eventually culminates in diarrhoea (Osterlund et al. 2007).

Probiotics, defined as active microorganisms, have demonstrated efficacy in preventing and alleviating gastrointestinal complications following chemotherapy, as evidenced in both animal studies and clinical trials. Mechanistically, probiotics exert their positive effects by inhibiting the adhesion and proliferation of harmful bacteria on the gastrointestinal mucosa, strengthening the barrier function of the mucosa, repairing damage to the jejunal villi, and modulating the expression of key inflammatory markers such as tumour necrosis factor-α (TNF-α), interleukin 6 (IL-6), and interleukin-1β (IL-1β). These immunomodulatory effects contribute to a reduction in the severity of diarrhoea, abdominal discomfort, and intestinal toxicity.

However, the diverse research designs, sample sources, and sample sizes have led to conflicting results on the efficacy of probiotics in the prevention and treatment of CID. As a result, a unanimous public opinion remains elusive (Miller and Carson 2020).

3.16 Challenges and Limitations

The variability of cancer types results in a range of responses to prebiotics and postbiotics, making it difficult to tailor interventions to specific cancer types. This challenge underscores the need for a nuanced understanding of the interactions between microbiota and cancer. Patient variability, characterized by differences in gut microbiota composition and host response, significantly influences the efficacy of prebiotics and postbiotics. Elements such as age, genetics, diet, and baseline microbiota diversity contribute to outcomes that are specific to each patient. Determining the optimal dosage and duration of prebiotic and postbiotic interventions is a formidable challenge. The balance between maintaining efficacy and mitigating potential adverse effects requires further research (Rani et al. 2021). Despite their potential, the exact mechanisms by which prebiotics and postbiotics influence cancer progression are still not fully understood. Thorough research is essential to elucidate specific pathways and interactions. Ethical concerns, including patient autonomy, informed consent, and potential exploitation of vulnerable populations, must be carefully considered in the context of prebiotic and postbiotic interventions for cancer (Appanna 2018). Obtaining regulatory approval for prebiotic and postbiotic therapies necessitates a solid foundation of clinical evidence, adherence to standards, and consistent manufacturing practices.

Although prebiotics and postbiotics offer promising opportunities to improve cancer therapy, addressing these limitations and challenges is critical to their successful integration into clinical practice. Continued research, multidisciplinary collaboration, and advances in understanding the interactions between microbiota and cancer will play a critical role in overcoming these obstacles (Ha et al. 2024).

3.17 Future Directions

Probiotics are essential dietary supplements that help reduce the incidence of certain cancers and ensure the safety of current treatments such as radiation, chemotherapy, and surgery with very few, if any, adverse effects. When probiotics are combined with prebiotics, they work better together than when probiotics are given alone to treat and prevent certain cancers (Mejia-Caballero et al. 2021). Furthermore, probiotic bacterial or fungal strains, probiotic dosage, and duration of probiotic intake are associated with the successful prevention and treatment of various cancers. By promoting bacterial numbers, intestinal epithelial barrier function, and cytokine production, probiotics have been shown to control the imbalances in the composition of the gut microbiota. Further research should be conducted to clarify the metagenomic relationship between the change in the gut microbiota composition and probiotic species with different diets or nutrients.

Regarding the ability of probiotics to prevent cancer, there are some conflicting results in the scientific literature. Strain specificity of defensive systems offers at least some explanation for these differences. Further research is therefore needed to determine the strains that play a role in the prevention of CRC. The studies used different methods of administering probiotics (lyophilized, microencapsulated, or consumed, such as yoghurt), as well as different dosages, treatment durations, and frequencies of intake. These factors can lead to disparate outcomes, making it difficult to compare the results of these types of studies (Miknevicius et al. 2021).

3.18 Conclusion

While research on the effect of functional foods, particularly probiotics, prebiotics, and postbiotics, on cancer is promising, the intricate and multifaceted nature of the challenges involved emphasizes the need for their meticulous integration into clinical practice. Sustained progress in this area requires ongoing multidisciplinary collaboration, continued advancements in mechanistic understanding, and clarification of ethical and regulatory considerations, all of which are pivotal to realizing the complete therapeutic potential of functional foods in the intricate landscape of cancer treatment.

Acknowledgements We would like to thank Dr Surajit Pathak for providing the opportunity and constant support to submit this manuscript. We would like to thank Mr M. Vijayaraman for English language editing.

Conflict of Interest Statement The authors do not have any conflict of interest.

Author Contributions VJ, SA, LD, DB, and LV wrote the manuscript. KP and DG edited the manuscript.

References

Aghajanpour M, Nazer MR, Obeidavi Z, Akbari M, Ezati P, Kor NM (2017) Functional foods and their role in cancer prevention and health promotion: a comprehensive review. Am J Cancer Res 7(4):740–769

Alexander JL, Wilson ID, Teare J, Marchesi JR, Nicholson JK, Kinross JM (2017) Gut microbiota modulation of chemotherapy efficacy and toxicity. Nat Rev Gastroenterol Hepatol 14(6):356–365. https://doi.org/10.1038/nrgastro.2017.20

Alkhatib A (2020) Antiviral functional foods and exercise lifestyle prevention of Coronavirus. Nutrients 12(9). https://doi.org/10.3390/nu12092633

An BC, Ryu Y, Yoon YS, Choi O, Park HJ, Kim TY, Kim SI, Kim BK, Chung MJ (2019) Colorectal cancer therapy using a Pediococcus pentosaceus SL4 drug delivery system secreting lactic acid bacteria-derived protein p8. Mol Cells 42(11):755–762. https://doi.org/10.14348/molcells.2019.0064

Appanna VD (2018) Dysbiosis, probiotics, and prebiotics. In: Diseases and health, in human microbes - the power within: health, healing and beyond. Springer Singapore, Singapore, pp 81–122

Arai S, Iwabuchi N, Takahashi S, Xiao JZ, Abe F, Hachimura S (2018) Orally administered heat-killed *Lactobacillus* paracasei MCC1849 enhances antigen-specific IgA secretion and induces follicular helper T cells in mice. PLoS One 13(6):e0199018. https://doi.org/10.1371/journal.pone.0199018

Balaguer F, Enrique M, Llopis S, Barrena M, Navarro V, Alvarez B, Chenoll E, Ramon D, Tortajada M, Martorell P (2022) Lipoteichoic acid from *Bifidobacterium* animalis subsp. lactis BPL1: a novel postbiotic that reduces fat deposition via IGF-1 pathway. Microb Biotechnol 15(3):805–816. https://doi.org/10.1111/1751-7915.13769

Bamigbade GB, Subhash AJ, Kamal-Eldin A, Nystrom L, Ayyash M (2022) An updated review on prebiotics: insights on potentials of food seeds waste as source of potential prebiotics. Molecules 27(18). https://doi.org/10.3390/molecules27185947

Borowicki A, Michelmann A, Stein K, Scharlau D, Scheu K, Obst U, Glei M (2011) Fermented wheat aleurone enriched with probiotic strains LGG and Bb12 modulates markers of tumor progression in human colon cells. Nutr Cancer 63(1):151–160. https://doi.org/10.1080/01635581.2010.516874

Braga V (2016) Spatial integration of E-cadherin adhesion, signalling and the epithelial cytoskeleton. Curr Opin Cell Biol 42:138–145. https://doi.org/10.1016/j.ceb.2016.07.006

Bruno-Barcena JM, Azcarate-Peril MA (2015) Galacto-oligosaccharides and colorectal cancer: feeding our intestinal Probiome. J Funct Foods 12:92–108. https://doi.org/10.1016/j.jff.2014.10.029

Cao W, Li XQ, Wang X, Fan HT, Zhang XN, Hou Y, Liu SB, Mei QB (2010) A novel polysaccharide, isolated from Angelica sinensis (Oliv.) Diels induces the apoptosis of cervical cancer HeLa cells through an intrinsic apoptotic pathway. Phytomedicine 17(8–9):598–605. https://doi.org/10.1016/j.phymed.2009.12.014

Caponio GR, Lippolis T, Tutino V, Gigante I, De Nunzio V, Milella RA, Gasparro M, Notarnicola M (2022a) Nutraceuticals: focus on anti-inflammatory, anti-cancer, antioxidant properties in gastrointestinal tract. Antioxidants (Basel) 11(7). https://doi.org/10.3390/antiox11071274

Caponio GR, Cofano M, Lippolis T, Gigante I, De Nunzio V, Difonzo G, Noviello M, Tarricone L, Gambacorta G, Giannelli G et al (2022b) Anti-proliferative and pro-apoptotic effects of digested Aglianico grape Pomace extract in human colorectal cancer cells. Molecules 27(20). https://doi.org/10.3390/molecules27206791

Chen HM, Yu YN, Wang JL, Lin YW, Kong X, Yang CQ, Yang L, Liu ZJ, Yuan YZ, Liu F et al (2013) Decreased dietary fiber intake and structural alteration of gut microbiota in patients with advanced colorectal adenoma. Am J Clin Nutr 97(5):1044–1052. https://doi.org/10.3945/ajcn.112.046607

Cook SI, Sellin JH (1998) Review article: short chain fatty acids in health and disease. Aliment Pharmacol Ther 12(6):499–507. https://doi.org/10.1046/j.1365-2036.1998.00337.x

Cousin FJ, Jouan-Lanhouet S, Dimanche-Boitrel MT, Corcos L, Jan G (2012) Milk fermented by Propionibacterium freudenreichii induces apoptosis of HGT-1 human gastric cancer cells. PLoS One 7(3):e31892. https://doi.org/10.1371/journal.pone.0031892

Culpepper BST, Mai V (2013) Evidence for contributions of gut microbiota to colorectal carcinogenesis. Curr Nutr Rep 2(1):10–18. https://doi.org/10.1007/s13668-012-0032-0

Damian MR, Cortes-Perez NG, Quintana ET, Ortiz-Moreno A, Garfias Noguez C, Cruceno-Casarrubias CE, Sanchez Pardo ME, Bermudez-Humaran LG (2022) Functional foods, nutraceuticals and probiotics: a focus on human health. Microorganisms 10(5). https://doi.org/10.3390/microorganisms10051065

data, W.c (2022) World Cancer Research Fund International. WCRF International [cited 2023 February 27, 2023]; Available from: . https://www.wcrf.org/cancer-trends/worldwide-cancer-data/

Debela DT, Muzazu SG, Heraro KD, Ndalama MT, Mesele BW, Haile DC, Kitui SK, Manyazewal T (2021) New approaches and procedures for cancer treatment: current perspectives. SAGE Open Med 9:20503121211034366. https://doi.org/10.1177/20503121211034366

Dillekas H, Rogers MS, Straume O (2019) Are 90% of deaths from cancer caused by metastases? Cancer Med 8(12):5574–5576. https://doi.org/10.1002/cam4.2474

Ding S, Wang Y, Yan W, Li A, Jiang H, Fang J (2019) Effects of *Lactobacillus* plantarum 15-1 and fructooligosaccharides on the response of broilers to pathogenic Escherichia coli O78 challenge. PLoS One 14(6):e0212079. https://doi.org/10.1371/journal.pone.0212079

Dixit V, Joseph Kamal SW, Bajrang Chole P, Dayal D, Chaubey KK, Pal AK, Xavier J, Manjunath BT, Bachheti RK (2023) Functional foods: exploring the health benefits of bioactive compounds from plant and animal sources. J Food Qual 2023(1):5546753. https://doi.org/10.1155/2023/5546753

Donohoe DR, Collins LB, Wali A, Bigler R, Sun W, Bultman SJ (2012) The Warburg effect dictates the mechanism of butyrate-mediated histone acetylation and cell proliferation. Mol Cell 48(4):612–626. https://doi.org/10.1016/j.molcel.2012.08.033

Egelston CA, Avalos C, Tu TY, Simons DL, Jimenez G, Jung JY, Melstrom L, Margolin K, Yim JH, Kruper L et al (2018) Human breast tumor-infiltrating CD8(+) T cells retain polyfunctionality despite PD-1 expression. Nat Commun 9(1):4297. https://doi.org/10.1038/s41467-018-06653-9

Eslami M, Yousefi B, Kokhaei P, Hemati M, Nejad ZR, Arabkari V, Namdar A (2019) Importance of probiotics in the prevention and treatment of colorectal cancer. J Cell Physiol 234(10):17127–17143. https://doi.org/10.1002/jcp.28473

FAO (2022) FAO term portal. 2022 May 15, Available from: https://www.fao.org/faoterm/viewentry/en/?entryId=170967

Femia AP, Luceri C, Dolara P, Giannini A, Biggeri A, Salvadori M, Clune Y, Collins KJ, Paglierani M, Caderni G (2002) Antitumorigenic activity of the prebiotic inulin enriched with oligofructose in combination with the probiotics *Lactobacillus* rhamnosus and *Bifidobacterium* lactis on azoxymethane-induced colon carcinogenesis in rats. Carcinogenesis 23(11):1953–1960. https://doi.org/10.1093/carcin/23.11.1953

Feng T, Wang J (2020) Oxidative stress tolerance and antioxidant capacity of lactic acid bacteria as probiotic: a systematic review. Gut Microbes 12(1):1801944. https://doi.org/10.1080/19490976.2020.1801944

Frankel AE, Coughlin LA, Kim J, Froehlich TW, Xie Y, Frenkel EP, Koh AY (2017) Metagenomic shotgun sequencing and unbiased metabolomic profiling identify specific human gut microbiota and metabolites associated with immune checkpoint therapy efficacy in melanoma patients. Neoplasia 19(10):848–855. https://doi.org/10.1016/j.neo.2017.08.004

Fulbright LE, Ellermann M, Arthur JC (2017) The microbiome and the hallmarks of cancer. PLoS Pathog 13(9):e1006480. https://doi.org/10.1371/journal.ppat.1006480

Garcia-Gonzalez AP, Ritter AD, Shrestha S, Andersen EC, Yilmaz LS, Walhout AJM (2017) Bacterial metabolism affects the C. elegans response to cancer chemotherapeutics. Cell 169(3):431–441 e8. https://doi.org/10.1016/j.cell.2017.03.046

Ghoneum M, Gimzewski J (2014) Apoptotic effect of a novel kefir product, PFT, on multidrug-resistant myeloid leukemia cells via a hole-piercing mechanism. Int J Oncol 44(3):830–837. https://doi.org/10.3892/ijo.2014.2258

Gopalakrishnan A, Clinthorne JF, Rondini EA, McCaskey SJ, Gurzell EA, Langohr IM, Gardner EM, Fenton JI (2012) Supplementation with galacto-oligosaccharides increases the percentage of NK cells and reduces colitis severity in Smad3-deficient mice. J Nutr 142(7):1336–1342. https://doi.org/10.3945/jn.111.154732

Guardamagna M, Berciano-Guerrero MA, Villaescusa-Gonzalez B, Perez-Ruiz E, Oliver J, Lavado-Valenzuela R, Rueda-Dominguez A, Barragan I, Queipo-Ortuno MI (2022) Gut microbiota and therapy in metastatic melanoma: focus on MAPK pathway inhibition. Int J Mol Sci 23(19). https://doi.org/10.3390/ijms231911990

Ha S, Zhang X, Yu J (2024) Probiotics intervention in colorectal cancer: from traditional approaches to novel strategies. Chin Med J 137(1):8–20. https://doi.org/10.1097/CM9.0000000000002955

Han KJ, Lee NK, Park H, Paik HD (2015) Anticancer and anti-inflammatory activity of probiotic Lactococcus lactis NK34. J Microbiol Biotechnol 25(10):1697–1701. https://doi.org/10.4014/jmb.1503.03033

Hill C, Guarner F, Reid G, Gibson GR, Merenstein DJ, Pot B, Morelli L, Canani RB, Flint HJ, Salminen S et al (2014) Expert consensus document. The International Scientific Association for Probiotics and Prebiotics consensus statement on the scope and appropriate use of the term probiotic. Nat Rev Gastroenterol Hepatol 11(8):506–514. https://doi.org/10.1038/nrgastro.2014.66

Jacouton E, Chain F, Sokol H, Langella P, Bermudez-Humaran LG (2017) Probiotic strain *Lactobacillus* casei BL23 prevents colitis-associated colorectal cancer. Front Immunol 8:1553. https://doi.org/10.3389/fimmu.2017.01553

Jan G, Belzacq AS, Haouzi D, Rouault A, Metivier D, Kroemer G, Brenner C (2002) Propionibacteria induce apoptosis of colorectal carcinoma cells via short-chain fatty acids acting on mitochondria. Cell Death Differ 9(2):179–188. https://doi.org/10.1038/sj.cdd.4400935

Janardhana V, Broadway MM, Bruce MP, Lowenthal JW, Geier MS, Hughes RJ, Bean AG (2009) Prebiotics modulate immune responses in the gut-associated lymphoid tissue of chickens. J Nutr 139(7):1404–1409. https://doi.org/10.3945/jn.109.105007

Jia W, Xie G, Jia W (2018) Bile acid-microbiota crosstalk in gastrointestinal inflammation and carcinogenesis. Nat Rev Gastroenterol Hepatol 15(2):111–128. https://doi.org/10.1038/nrgastro.2017.119

Kich DM, Vincenzi A, Majolo F, Volken de Souza CF, Goettert MI (2016) Probiotic: effectiveness nutrition in cancer treatment and prevention. Nutr Hosp 33(6):1430–1437. https://doi.org/10.20960/nh.806

Kim Y, Oh S, Yun HS, Oh S, Kim SH (2010) Cell-bound exopolysaccharide from probiotic bacteria induces autophagic cell death of tumour cells. Lett Appl Microbiol 51(2):123–130. https://doi.org/10.1111/j.1472-765X.2010.02859.x

Kim SK, GH, Cho H (2021a) Postbiotics for cancer prevention and treatment. Korean. J Microbiol 57(3):142–153. https://doi.org/10.7845/kjm.2021.1067

Kim BK, Yoon YS, Ryu Y, Chung MJ (2021b) Probiotic-derived p8 protein induce apoptosis via regulation of RNF152 in colorectal cancer cells. Am J Cancer Res 11(3):746–759

Larsson SC, Andersson SO, Johansson JE, Wolk A (2008) Cultured milk, yogurt, and dairy intake in relation to bladder cancer risk in a prospective study of Swedish women and men. Am J Clin Nutr 88(4):1083–1087. https://doi.org/10.1093/ajcn/88.4.1083

Lashani E, Davoodabadi A, Soltan Dallal MM (2020) Some probiotic properties of *Lactobacillus* species isolated from honey and their antimicrobial activity against foodborne pathogens. Vet Res Forum 11(2):121–126. https://doi.org/10.30466/vrf.2018.90418.2188

Leclere L, Cutsem PV, Michiels C (2013) Anti-cancer activities of pH- or heat-modified pectin. Front Pharmacol 4:128. https://doi.org/10.3389/fphar.2013.00128

Lee JW, Shin JG, Kim EH, Kang HE, Yim IB, Kim JY, Joo HG, Woo HJ (2004) Immunomodulatory and antitumor effects in vivo by the cytoplasmic fraction of *Lactobacillus* casei and *Bifidobacterium* longum. J Vet Sci 5(1):41–48

Lee N-K, Son S-H, Jeon EB, Jung GH, Lee J-Y, Paik H-D (2015) The prophylactic effect of probiotic Bacillus polyfermenticus KU3 against cancer cells. J Funct Foods 14:513–518. https://doi.org/10.1016/j.jff.2015.02.019

Li Y, Elmen L, Segota I, Xian Y, Tinoco R, Feng Y, Fujita Y, Segura Munoz RR, Schmaltz R, Bradley LM et al (2020) Prebiotic-induced anti-tumor immunity attenuates tumor growth. Cell Rep 30(6):1753–1766 e6. https://doi.org/10.1016/j.celrep.2020.01.035

Lopez M, Li N, Kataria J, Russell M, Neu J (2008) Live and ultraviolet-inactivated *Lactobacillus* rhamnosus GG decrease flagellin-induced interleukin-8 production in Caco-2 cells. J Nutr 138(11):2264–2268. https://doi.org/10.3945/jn.108.093658

Macfarlane LA, Murphy PR (2010) MicroRNA: biogenesis, function and role in cancer. Curr Genomics 11(7):537–561. https://doi.org/10.2174/138920210793175895

Maleki-Kakelar H, Dehghani J, Barzegari A, Barar J, Shirmohamadi M, Sadeghi J, Omidi Y (2020) *Lactobacillus* plantarum induces apoptosis in gastric cancer cells via modulation of signaling pathways in Helicobacter pylori. Bioimpacts 10(2):65–72. https://doi.org/10.34172/bi.2020.09

Markowiak P, Slizewska K (2018) The role of probiotics, prebiotics and synbiotics in animal nutrition. Gut Pathog 10:21. https://doi.org/10.1186/s13099-018-0250-0

Martin R, Langa S, Reviriego C, Jiminez E, Marin ML, Xaus J, Fernandez L, Rodriguez JM (2003) Human milk is a source of lactic acid bacteria for the infant gut. J Pediatr 143(6):754–758. https://doi.org/10.1016/j.jpeds.2003.09.028

Martyniak A, Zakrzewska Z, Schab M, Zawartka A, Wedrychowicz A, Skoczen S, Tomasik PJ (2023) Prevention and health benefits of prebiotics, probiotics and postbiotics in acute lymphoblastic leukemia. Microorganisms 11(7). https://doi.org/10.3390/microorganisms11071775

Mazziotta C, Tognon M, Martini F, Torreggiani E, Rotondo JC (2023) Probiotics mechanism of action on immune cells and beneficial effects on human health. Cells 12(1). https://doi.org/10.3390/cells12010184

McIntosh FM, Maison N, Holtrop G, Young P, Stevens VJ, Ince J, Johnstone AM, Lobley GE, Flint HJ, Louis P (2012) Phylogenetic distribution of genes encoding beta-glucuronidase activity in human colonic bacteria and the impact of diet on faecal glycosidase activities. Environ Microbiol 14(8):1876–1887. https://doi.org/10.1111/j.1462-2920.2012.02711.x

Mejia-Caballero A, Salas-Villagran VA, Jimenez-Serna A, Farres A (2021) Challenges in the production and use of probiotics as therapeuticals in cancer treatment or prevention. J Ind Microbiol Biotechnol 48(9–10). https://doi.org/10.1093/jimb/kuab052

Michels KB, Willett WC, Vaidya R, Zhang X, Giovannucci E (2020) Yogurt consumption and colorectal cancer incidence and mortality in the Nurses' Health Study and the Health Professionals Follow-Up Study. Am J Clin Nutr 112(6):1566–1575. https://doi.org/10.1093/ajcn/nqaa244

Miknevicius P, Zulpaite R, Leber B, Strupas K, Stiegler P, Schemmer P (2021) The impact of probiotics on intestinal Mucositis during chemotherapy for colorectal cancer: a comprehensive review of animal studies. Int J Mol Sci 22(17). https://doi.org/10.3390/ijms22179347

Miller PL, Carson TL (2020) Mechanisms and microbial influences on CTLA-4 and PD-1-based immunotherapy in the treatment of cancer: a narrative review. Gut Pathog 12:43. https://doi.org/10.1186/s13099-020-00381-6

Miri ST, Sotoodehnejadnematalahi F, Amiri MM, Pourshafie MR, Rohani M (2022) The impact of *Lactobacillus* and *Bifidobacterium* probiotic cocktail on modulation of gene expression of gap junctions dysregulated by intestinal pathogens. Arch Microbiol 204(7):417. https://doi.org/10.1007/s00203-022-03026-1

Mishra P, Badiyani VM, Jain S, Subramanian S, Maharaj SV, Kumar A, Singh BN (2023) Prebiotics: ignored player in the fight against cancer. Cancer Rep (Hoboken) 6(11):e1870. https://doi.org/10.1002/cnr2.1870

Motevaseli E, Dianatpour A, Ghafouri-Fard S (2017) The role of probiotics in cancer treatment: emphasis on their in vivo and in vitro anti-metastatic effects. Int J Mol Cell Med 6(2):66–76. https://doi.org/10.22088/acadpub.BUMS.6.2.1

Mu W, Jia Y, Chen X, Li H, Wang Z, Cheng B (2020) Intracellular Porphyromonas gingivalis promotes the proliferation of colorectal cancer cells via the MAPK/ERK signaling pathway. Front Cell Infect Microbiol 10:584798. https://doi.org/10.3389/fcimb.2020.584798

Nakkarach A, Foo HL, Song AA, Mutalib NEA, Nitisinprasert S, Withayagiat U (2021) Anti-cancer and anti-inflammatory effects elicited by short chain fatty acids produced by Escherichia coli isolated from healthy human gut microbiota. Microb Cell Factories 20(1):36. https://doi.org/10.1186/s12934-020-01477-z

Nataraj BH, Ali SA, Behare PV, Yadav H (2020) Postbiotics-parabiotics: the new horizons in microbial biotherapy and functional foods. Microb Cell Factories 19(1):168. https://doi.org/10.1186/s12934-020-01426-w

Orlando A, Refolo MG, Messa C, Amati L, Lavermicocca P, Guerra V, Russo F (2012) Antiproliferative and proapoptotic effects of viable or heat-killed *Lactobacillus* paracasei IMPC2.1 and *Lactobacillus* rhamnosus GG in HGC-27 gastric and DLD-1 colon cell lines. Nutr Cancer 64(7):1103–1111. https://doi.org/10.1080/01635581.2012.717676

Osterlund P, Ruotsalainen T, Korpela R, Saxelin M, Ollus A, Valta P, Kouri M, Elomaa I, Joensuu H (2007) *Lactobacillus* supplementation for diarrhoea related to chemotherapy of colorectal cancer: a randomised study. Br J Cancer 97(8):1028–1034. https://doi.org/10.1038/sj.bjc.6603990

Ou Y, Xu S, Zhu D, Yang X (2014) Molecular mechanisms of exopolysaccharide from Aphanothece halaphytica (EPSAH) induced apoptosis in HeLa cells. PLoS One 9(1):e87223. https://doi.org/10.1371/journal.pone.0087223

Pachmayr E, Treese C, Stein U (2017) Underlying mechanisms for distant metastasis - molecular biology. Visc Med 33(1):11–20. https://doi.org/10.1159/000454696

Park E, Jeon GI, Park JS, Paik HD (2007) A probiotic strain of Bacillus polyfermenticus reduces DMH induced precancerous lesions in F344 male rat. Biol Pharm Bull 30(3):569–574. https://doi.org/10.1248/bpb.30.569

Passali D, Passali GC, Vesperini E, Cocca S, Visconti IC, Ralli M, Bellussi LM (2019) The efficacy and tolerability of Streptococcus salivarius 24SMB and Streptococcus oralis 89a administered as nasal spray in the treatment of recurrent upper respiratory tract infections in children. Eur Rev Med Pharmacol Sci 23(1 Suppl):67–72. https://doi.org/10.26355/eurrev_201903_17352

Pauli C, Hopkins BD, Prandi D, Shaw R, Fedrizzi T, Sboner A, Sailer V, Augello M, Puca L, Rosati R et al (2017) Personalized in vitro and in vivo cancer models to guide precision medicine. Cancer Discov 7(5):462–477. https://doi.org/10.1158/2159-8290.Cd-16-1154

Puri V, Nagpal M, Singh I, Singh M, Dhingra GA, Huanbutta K, Dheer D, Sharma A, Sangnim T (2022) A comprehensive review on nutraceuticals: therapy support and formulation challenges. Nutrients 14(21). https://doi.org/10.3390/nu14214637

Rafter J, Bennett M, Caderni G, Clune Y, Hughes R, Karlsson PC, Klinder A, O'Riordan M, O'Sullivan GC, Pool-Zobel B et al (2007) Dietary synbiotics reduce cancer risk factors in polypectomized and colon cancer patients. Am J Clin Nutr 85(2):488–496. https://doi.org/10.1093/ajcn/85.2.488

Raman M, Ambalam P, Kondepudi KK, Pithva S, Kothari C, Patel AT, Purama RK, Dave JM, Vyas BR (2013) Potential of probiotics, prebiotics and synbiotics for management of colorectal cancer. Gut Microbes 4(3):181–192. https://doi.org/10.4161/gmic.23919

Rani V, Singhal S, Sharma K, Vaid R, Aggarwal K, Bhadana R, Agarwal R, Atale N (2021) Human gut microbiome: a new frontier in cancer diagnostics & therapeutics. Curr Pharm Des 27(45):4578–4592. https://doi.org/10.2174/1381612827666211006152112

Reid G, Younes JA, Van der Mei HC, Gloor GB, Knight R, Busscher HJ (2011) Microbiota restoration: natural and supplemented recovery of human microbial communities. Nat Rev Microbiol 9(1):27–38. https://doi.org/10.1038/nrmicro2473

Roller M, Pietro Femia A, Caderni G, Rechkemmer G, Watzl B (2004) Intestinal immunity of rats with colon cancer is modulated by oligofructose-enriched inulin combined with *Lactobacillus* rhamnosus and *Bifidobacterium* lactis. Br J Nutr 92(6):931–938. https://doi.org/10.1079/bjn20041289

Russo F, Orlando A, Linsalata M, Cavallini A, Messa C (2007) Effects of Lactobacillus rhamnosus GG on the cell growth and polyamine metabolism in HGC-27 human gastric cancer cells. Nutr Cancer 59(1):106–114. https://doi.org/10.1080/01635580701365084

Saxami G, Karapetsas A, Lamprianidou E, Kotsianidis I, Chlichlia A, Tassou C, Zoumpourlis V, Galanis A (2016) Two potential probiotic *lactobacillus* strains isolated from olive microbiota exhibit adhesion and anti-proliferative effects in cancer cell lines. J Funct Foods 24:461–471. https://doi.org/10.1016/j.jff.2016.04.036

Secombe KR, Van Sebille YZA, Mayo BJ, Coller JK, Gibson RJ, Bowen JM (2020) Diarrhea induced by small molecule Tyrosine Kinase inhibitors compared with chemotherapy: potential role of the microbiome. Integr Cancer Ther 19:1534735420928493. https://doi.org/10.1177/1534735420928493

Sharifi-Rad J, Rodrigues CF, Stojanovic-Radic Z, Dimitrijevic M, Aleksic A, Neffe-Skocinska K, Zielinska D, Kolozyn-Krajewska D, Salehi B, Milton Prabu S et al (2020) Probiotics: versatile bioactive components in promoting human health. Medicina (Kaunas) 56(9). https://doi.org/10.3390/medicina56090433

Shi Y (2009) Serine/threonine phosphatases: mechanism through structure. Cell 139(3):468–484. https://doi.org/10.1016/j.cell.2009.10.006

Singh S, Singh M, Gaur S (2022) Probiotics as multifaceted oral vaccines against colon cancer: a review. Front Immunol 13:1002674. https://doi.org/10.3389/fimmu.2022.1002674

Temple NJ (2022) A rational definition for functional foods: a perspective. Front Nutr 9:957516. https://doi.org/10.3389/fnut.2022.957516

Teng NMY, Price CA, McKee AM, Hall LJ, Robinson SD (2021) Exploring the impact of gut microbiota and diet on breast cancer risk and progression. Int J Cancer 149(3):494–504. https://doi.org/10.1002/ijc.33496

Tiptiri-Kourpeti A, Spyridopoulou K, Santarmaki V, Aindelis G, Tompoulidou E, Lamprianidou EE, Saxami G, Ypsilantis P, Lampri ES, Simopoulos C et al (2016) *Lactobacillus* casei exerts anti-proliferative effects accompanied by apoptotic cell death and up-regulation of TRAIL in colon carcinoma cells. PLoS One 11(2):e0147960. https://doi.org/10.1371/journal.pone.0147960

Touchefeu Y, Montassier E, Nieman K, Gastinne T, Potel G, des Bruley Varannes S, Le Vacon F, de La Cochetiere MF (2014) Systematic review: the role of the gut microbiota in chemotherapy- or radiation-induced gastrointestinal mucositis - current evidence and potential clinical applications. Aliment Pharmacol Ther 40(5):409–421. https://doi.org/10.1111/apt.12878

Trinchieri G (2003) Interleukin-12 and the regulation of innate resistance and adaptive immunity. Nat Rev Immunol 3(2):133–146. https://doi.org/10.1038/nri1001

Tukenmez U, Aktas B, Aslim B, Yavuz S (2019) The relationship between the structural characteristics of lactobacilli-EPS and its ability to induce apoptosis in colon cancer cells in vitro. Sci Rep 9(1):8268. https://doi.org/10.1038/s41598-019-44753-8

Wang YY, Khoo KH, Chen ST, Lin CC, Wong CH, Lin CH (2002) Studies on the immuno-modulating and antitumor activities of Ganoderma lucidum (Reishi) polysaccharides: functional and proteomic analyses of a fucose-containing glycoprotein fraction responsible for the activities. Bioorg Med Chem 10(4):1057–1062. https://doi.org/10.1016/s0968-0896(01)00377-7

Zhang W, Zhu YH, Yang GY, Liu X, Xia B, Hu X, Su JH, Wang JF (2017) *Lactobacillus* rhamnosus GG affects microbiota and suppresses autophagy in the intestines of pigs challenged with Salmonella Infantis. Front Microbiol 8:2705. https://doi.org/10.3389/fmicb.2017.02705

Zhong L, Zhang X, Covasa M (2014) Emerging roles of lactic acid bacteria in protection against colorectal cancer. World J Gastroenterol 20(24):7878–7886. https://doi.org/10.3748/wjg.v20.i24.7878

Zhu Y, Michelle Luo T, Jobin C, Young HA (2011) Gut microbiota and probiotics in colon tumorigenesis. Cancer Lett 309(2):119–127. https://doi.org/10.1016/j.canlet.2011.06.004

Zununi Vahed S, Barzegari A, Rahbar Saadat Y, Goreyshi A, Omidi Y (2017) Leuconostoc mesenteroides-derived anticancer pharmaceuticals hinder inflammation and cell survival in colon cancer cells by modulating NF-kappaB/AKT/PTEN/MAPK pathways. Biomed Pharmacother 94:1094–1100. https://doi.org/10.1016/j.biopha.2017.08.033

Health Benefits of Probiotics and Prebiotics Interaction

4

Ragala Venkata Nandha, A. R. Heamchandsaravanan, Priyadarshene Kamalanathan, Prabu Dhandapani, and Anandan Balakrishnan

Abstract

Colorectal cancer (CRC) is a prevalent intestinal condition with significant global health implications, marked by high incidence, morbidity, and mortality rates. The research underscores the critical role of gut microbiota in CRC development and progression, challenging the traditional view that genetic mutations solely drive the disease. The incidence of CRC varies globally, with higher rates in regions with a high Human Development Index (HDI), while lower rates are observed in less developed areas. This disparity is shifting as CRC cases rise in developing countries but stabilize or decline in wealthier nations. Currently, approximately 1.93 million new CRC cases are diagnosed annually, making it the second leading cause of cancer-related deaths, following lung cancer, with around 904,019 fatalities each year. Epidemiological projections suggest a significant increase in CRC cases, anticipating 3.2 million cases and 1.6 million deaths by 2040. The gut microbiota's influence on CRC is profound, with imbalances linked to the disease's onset and progression. Lifestyle factors such as diet, alcohol consumption, physical inactivity, and obesity are implicated in altering the gut microbiome, potentially increasing carcinogenic metabolites. The rising incidence of CRC, particularly among younger populations, and the higher mortality rate due to late diagnosis are concerning. The gut microbiota's ability to modify epigenetic mechanisms like histone modification, non-coding RNAs, and

R. V. Nandha · A. R. Heamchandsaravanan · P. Kamalanathan · P. Dhandapani
Department of Microbiology, Dr. ALM PG Institute of Basic Medical Sciences, University of Madras, Taramani Campus, Chennai, India

A. Balakrishnan (✉)
Department of Genetics, Dr. ALM PG Institute of Basic Medical Sciences, University of Madras, Taramani Campus,
Chennai, India

S. Pathak et al. (eds.), *Microbiota and Dietary Mediators in Colon Cancer Prevention and Treatment*, https://doi.org/10.1007/978-981-96-0297-1_4

DNA methylation highlights its importance in CRC pathogenesis. Maintaining gut homeostasis through a balanced microbiome is crucial for intestinal health and cancer prevention. Probiotics, such as Lactobacillus and Bifidobacterium, have shown promise in CRC prevention and treatment, with ongoing research focusing on microbiota manipulation as a therapeutic strategy.

Keywords

Prebiotics · Probiotics · Interaction, and colorectal cancer

4.1 Introduction

Colorectal cancer (CRC) ranks as the second leading cause of cancer-related morbidity and mortality worldwide, with a complex etiology involving multiple genes and signaling pathways. Projections estimate that the number of new CRC cases will rise to 2.5 million globally by 2035, surpassing even more common cancers like gastric and liver cancers. Recent studies have highlighted the significant roles that environmental and genetic factors play in CRC development, with particular attention on the pivotal influence of gut microbes in the disease's progression (Chang et al. 2023). CRC is a prevalent intestinal condition with significant global health implications, causing concern among public health professionals due to its high incidence, severe health effects, and substantial mortality rate (Zhang et al. 2024). Global disparities in CRC rates are becoming increasingly evident, with higher incidence in developed countries, though recent trends indicate a concerning rise in cases within developing regions. These shifting patterns underscore the importance of addressing CRC on a global scale, particularly as we consider the role of gut microbiota in its development and the need for effective, non-invasive screening methods. Research shows that CRC has the greatest prospective risk for life (38.5%) and deaths related to cancer (28.2%), which indicates the necessity for the further study (Ionescu et al. 2024; Saraiva et al. 2023). Each year, approximately 1,926,425 new cases of CRC are diagnosed worldwide, making it the second leading cause of cancer-related deaths after lung cancer, with 904,019 annual fatalities. Recent research indicates that Australia and Europe have some of the highest CRC rates, with 40.6 cases per 100,000 men. In contrast, Africa and South Asia have the lowest rates, with 4.4 cases per 100,000 women. The highest mortality rate is found in Eastern Europe, with 20.2 deaths per 100,000 men, while the lowest is in South Asia, with 2.5 deaths per 100,000 women (Wang et al. 2024). Recent epidemiological studies forecast a significant rise in CRC cases, with an estimated 3.2 million instances by 2040. This increase is anticipated to result in approximately 1.6 million deaths. Recent research has revealed a significant connection between gut microbiota and the development of CRC, challenging the traditional belief that the disease primarily results from random or inherited mutations. This growing understanding underscores the urgent need for non-invasive screening methods that can effectively identify at-risk individuals, opening new avenues for early detection and prevention strategies, particularly as we explore the latest advancements in non-invasive diagnostic technologies (Maftei et al. 2024).

Understanding the impact of unbalanced gut bacteria on CRC growth is crucial, as the progression of CRC is linked to gut microbiota imbalance (Ionescu et al. 2024; Wang et al. 2024). This imbalance can be caused by lifestyle changes such as increased consumption of meat and processed foods, higher alcohol intake, reduced physical activity, and increased obesity. These changes in the gut microbiota may alter metabolite profiles, potentially increasing the presence of cancer-causing substances (Kumavath et al. 2024). The gut microbiota, which may be influenced by food and probiotic beverages, is important for general health and has been linked to colorectal cancer. The human gastrointestinal (GI) tract is intriguing, housing bacteria, viruses, archaea, fungi, and protozoa. Microbiomes influence intestinal epithelial cells to maintain gut homeostasis and have the potential to prevent cancer. Additionally, microbiome by-products may serve as antioxidants and nutrients (Pandey and Bhattacharyya 2024).

The significant increase in younger colorectal cancer (CRC) patients and the higher death rate are concerning, as the condition is often diagnosed late, complicating treatment. Understanding the role of gut microbiota in cancer formation is crucial, as it can modify histones, non-coding RNAs, and DNA methylation in various ways (Maftei et al. 2024). Probiotics such as *Lactobacillus* and *Bifidobacterium* are well-known for their health benefits. Studies have shown that *Lactobacillus plantarum* and *Lactobacillus acidophilus* can aid in the treatment and prevention of colorectal cancer (CRC) and polyps. Additionally, scientists are exploring the use of *Lactobacillus casei BL23* for CRC therapy and prevention. This research is part of a broader effort to manipulate gut microbiota to prevent and cure CRC. (Oyetayo and Oyetayo 2005).

4.2 Probiotics and Mechanisms

Probiotics are living microorganisms, primarily bacteria, with a significant presence of lactic acid bacteria. A balanced gut microbiota, including yeast and mold microbes, is associated with improved health. Élie Metchnikoff, a renowned microbiologist, played a key role in establishing the concept of probiotics. He received a Nobel Prize for his research on *Lactobacillus* and phagocytes, which has led to claims that probiotics can treat a wide range of health conditions in people of all ages (Shimizu et al. 2020).

4.3 Functions of Probiotics

Probiotics may help with persistent diarrhea, allergies, dermatitis, and bacterial and viral infections caused by poor lifestyle (Maftei et al. 2024). Appreciating Probiotics' Anticancer and Antimutagenic mechanisms being acquainted with microbiology is essential for understanding mutagen binding, degradation, and mutagenesis suppression. Preventing innocuous procarcinogens from becoming toxic and reactive health hazards. New probiotics that modify the microbiota are continually being researched. Incomprehensible carbohydrates produce short chain fatty acids (SCFA), which lower gut pH. Anti-inflammatory compounds increase innate immunity. Examples include prebiotics, symbiotics, postbiotics, microbial consortia, live biotherapeutics, and GMOs. Polyphenols, fibers, and fermented foods are becoming more popular due to their positive health impacts.

4.4 Probiotics' Roles

Probiotics vary widely in their effectiveness and impact on human health. Scientific research has demonstrated that only certain strains of probiotics have proven benefits. Probiotics include *Lactobacillus*, *Bifidobacterium*, *Bacillus*, *Pediococcus*, and various yeasts (Shimizu et al. 2020). Research conducted by scientists has shown that specific probiotics can treat a variety of gastrointestinal conditions, including irritable bowel syndrome, *Helicobacter pylori* eradication, inflammatory bowel disease, diarrhea, and others. Probiotics have also shown promise in treating allergic diseases like Conditions such as atopic dermatitis, non-alcoholic fatty liver disease, obesity, insulin resistance syndrome, and type 2 diabetes are of interest (Szajewska et al. 2020). various cancers, and cancer treatment adverse consequences. Probiotics also boost immunity, metabolism, dental health, and cognition. Research is ongoing to use probiotics to treat and prevent COVID-19 (Maftei et al. 2024).

4.5 Different Types of Probiotics in the Human Gut

Science divides probiotics into *Lactobacilli, Bifidobacterium*, and others. Rod-shaped *Lactobacillus* bacteria were identified in foods that have undergone fermentation and human digestive and vaginal canals in 1901 (Oyetayo and Oyetayo 2005; Shimizu et al. 2020). *Lactobacillus* produces probiotics. "*Lactobacilli*" are versatile and useful, Learn to ferment dairy to increase its shelf life. Probiotic foods and supplements improve health. Microbiologists found 25 *Lactobacillus* genera. Like L, 23 new categories have been added. Some examples include *Acetilactobacillus*, *Agrilactobacillus*, *Amylolactobacillus*, *Apilactobacillus*, *Bombilactobacillus*, *Companilactobacillus*, *Dellaglioa*, *Fructilactobacillus*, *Furfurilactobacillus*, *Holzapfelia*, *Lacticaseibacillus*, *Lactiplantibacillus*, *Lapidilactobacillus*, *Latilactobacillus*, *Lentilactobacillus*, and *Levilactobacillus*. The most recent classification of *Lactobacillus* groups species with similar physiological and metabolic characteristics, suggesting a stronger connection.

Probiotic health benefits may be examined using standard methods. A healthy gut needs *Bifidobacterium* spp. Under anaerobic circumstances, the small and large intestines may duplicate and break it down, producing chemicals that support gut health (Zmora et al. 2018; Singh and Agarwal 2021). *Bifidobacterium* affects several bodily functions. Reduce pro-inflammatory cytokines to reduce inflammation and fight dangerous germs. *Bifidobacterium* produces vitamins, amino acids, and calcium in the gut. According to research, *Bifidobacterium* may possess proteins that target malignant tumors. Harmless *Saccharomyces cerevisiae* is used in probiotic food for its selective probiotic capabilities. Many studies have examined *Saccharomyces* strains' probiotic capabilities. Besides diarrhea drugs, these strains cure digestive diseases. Saccharomyces gut probiotics work well. They regulate the immune reaction and gut microbiome. Optimizing chronic illness and pathogen immunological pathways is their specialty (Fuller 1989).

4.6 Probiotics Mechanism and Its Actions

Over the past few years, there have been notable breakthroughs in the field of probiotics research, particularly in understanding and choosing the right probiotic microorganisms and their potential health benefits. Several studies have been carried out on the potential benefits of probiotics in fighting diseases. It is crucial to comprehend the impact of these microorganisms on the gut. The presence and behavior of an individual are influenced by their microbiota, specific strains of probiotics, and the location of the intestinal tract. Research indicates that probiotics exhibit unique mechanisms of action. These components strengthen the intestinal barrier, enhance immune function, and inhibit pathogenic bacteria. These components enhance the epithelial barrier, intestinal mucosa adhesion, antibacterial production, competitive exclusion of pathogens, reduction of bacterial adhesion, and immune system regulation.

Different types of cells in the mucus layer and intestinal epithelium act as protective barriers against bacteria in the intestines. It is fascinating how Paneth cells release antimicrobial peptides in response to the presence of intestinal bacteria. In a similar manner, cells of Saccharomyces cerevisiae release mucus, while enterocytes take in molecules from the intestinal lumen. In addition, intestinal endocrine cells play a crucial role in the intestinal epithelial cells (Liu et al. 2024). The mucus layer in the intestines plays a crucial role in shielding against hostile conditions, facilitating the movement of food, and inhibiting the attachment of infections to its surface. The gut is shielded from harmful substances in the lumen by the intestinal mucosa. Probiotics boost the different protective mechanisms of bacteria in the gut, such as chemical, mechanical, biological, and immunological barriers (Guo et al. 2019). Probiotics aim to improve intestinal health by enhancing permeability, promoting mucosal renewal, increasing mucus production, and strengthening the integrity of both mucosal and mechanical barriers (Mohr et al. 2020; La Fata et al. 2018). With their remarkable impact on diseases and human well-being, probiotics can bring immense benefits to people of all ages.

4.7 Prebiotics and Their Mechanisms

Prebiotics are transformed into short-chain fatty acids by bacteria in the colon. Fermentation promotes the condition of the colon by releasing beneficial microorganisms. The International Scientific Association for Probiotics and Prebiotics' majority panel rethought the term "prebiotic" in 2000. Some bacteria choose to use compounds that promote health as prebiotics, according to a recently published study. Prebiotics may withstand stomach acid, be fermented by intestinal microbiota, and promote particular bacterium growth. Thus, they propose a prebiotic index that measures *Bifidobacteria* growth by dividing new cfu/g of feces by the daily prebiotic dosage (in grams) (Bevilacqua et al. 2024). Prebiotics increase immunological function, although how is unknown despite much research. Many models have been proposed. Prebiotics increase propionate production, which affects hepatic lipogenic enzymes. However, microbiologists found that fermentation produced SCFAs, notably butyric acid. This mechanism modulates histone acetylation, making transcription factor-controlled genes available. Mucin regulation research. Studies show that FOS and other prebiotics boost gut-associated lymphoid tissue (GALT) lymphocytes and leukocytes in the circulation. Increased IgA synthesis may help inflammatory macrophages fight GALTs (Roberfroid 2007). Vegetable prebiotics are carbohydrates or polysaccharides. These compounds—FOS, GOS, MOS, XOS, and others—polymerize from 4 to 30. Other prebiotics include HMO, phenols, phytochemicals, CLA, and PUFA. Zhang et al. investigated phenols as prebiotics (La Fata et al. 2018). Prebiotics are oligosaccharide carbohydrates, say microbiologists. They concentrate on XOS, GOS, lactulose, inulin, FOS, and derivatives. Recent study expanded this classification to include carbohydrate-free prebiotics. Blueberries and black raspberries include prebiotic polyphenols, according to research. Exciting new organisms have emerged from prebiotic synthesis. These species provide intriguing study opportunities in polysaccharides, polyphenols, and polypeptide polymers (Roberfroid 2007; Saghir and Al Suede 2024; Brandelli 2021; Bedu-Ferrari 2022).

4.8 Galacto-oligosaccharides (GOS)

Galacto-oligosaccharides are innovative functional compounds that are difficult to digest. GOS contain two to eight sugar units, including glucose, galactose, and two-galactose disaccharides (Mudannayake et al. 2022). GOS's glycosidic links between glucose and galactose, or DP and galactose, are crucial. Many toxicity and genome damage investigations have indicated that GOS are safe as a food additive (Saghir and Al Suede 2024). They improve gut *Bifidobacterium* and colon microbiota (Mudannayake et al. 2022). In 35 healthy full-term babies, adding GOS to baby formula increased *Bifidobacterium* and decreased microbiome alpha-diversity. The neonates' feces pH and SCFA patterns matched the control group. This shows how GOS boost native *Bifidobacterium* and intestinal health. In newborns given

GOS-enriched products, long-chain FOS and short-chain GOS boosted LAB and *Bifidobacterium*, reduced pathogens, and improved stool properties.

4.9 Inulin-Type Fructan Prebiotics

GOS and inulin-type fructans are present. Inulin-type fructans are connected fructose–glucose polymers. Inulin and oligofructose/FOS are from the identical group, although inulin has a longer chain (DP 2–60) than them (DP 2–8). LAB, *Anthobacteria*, and *Bifidobacteria* grow better with inulin-type fructans (Maftei et al. 2024). For decades, people have eaten fructan-based inulin, a water-soluble storage polysaccharide. Approximately 36,000 species of plant contain inulin, with chicory root being the most frequent (Saghir and Al Suede 2024). The large intestine's flora may ferment inulin, making it hard for the small intestine to digest (Shimizu et al. 2020). Ecologists know inulin has several food processing uses. It may provide taste without calories by replacing fat or carbs. It enhances calcium, magnesium, and iron absorption. Inulin may help digestion, constipation, and immunity (Arifuzzaman et al. 2022). Like a botanist, inulin encourages colon *Bifidobacterium* and *Lactobacillus* growth and metabolism. It also affects gut flora. Wheat, onions, bananas, and garlic contain inulin, sulfur fructans (FOS). Prebiotics with DPs ≤ 10 are used in the food business due to their low calorie and indigestible nature. Fructans lower triacylglycerols, phospholipids, and cholesterol; decrease carcinogenesis; and improve intestinal absorption of nutrients (Mounir et al. 2022).

4.10 Mechanisms of Prebiotics in CRC Prevention

Gut homeostasis and microbiome modification ISAPP defines prebiotics as components that ferment selectively to change gut microbial composition and activity. These changes improve host health. Findings show that inulin-rich meals may benefit bacteria. This rise is largely due to *Bacteroidaceae, Porphyromonadaceae*, and *Prevotellaceae*. *Lachnospiraceae* Firmicutes are reduced by inulin. Inflammatory conditions like obesity and diabetes increase Firmicutes/Bacteroidetes ratios. Scientists say the lower ratio shows that inulin protects. Inulin reduced Desulfovibrio and Bilophila, pro-inflammatory bacteria. A recent study demonstrated that anthocyanins, ellagic acid, and ellagitannins restored microbial diversity in all induced CRC animals. This was done by altering Firmicutes, Bacteroidetes, and Proteobacteria abundances. Fernández et al. (2019) found that lactulose galacto-oligosaccharides increased Bacteroidetes and *Bifidobacterium* and decreased Firmicutes. They also found Paraprevotella and Parabacteroides to create helpful propionates. Microbes may reduce intestinal inflammation and colorectal cancer (Ambalam et al. 2016). Currently, the most widely accepted and extensively studied prebiotics include inulin, FOS, oligofructose, and GOS. These prebiotics, classified as fibers, are neither absorbed nor digested in the small intestine. Inulin, FOS, and oligofructose fall under the category of fructans, while GOS is classified as a galactan. More recently, emerging varieties of prebiotics have shown potential in promoting the growth of beneficial bacteria, including protein–oligosaccharide conjugates, human milk oligosaccharides (HMOs), and non-carbohydrate molecules such as polyphenols, fats, and various plants and herbs (Shrifteylik et al. 2023).

4.11 Production of Fermentation Metabolites: SCFAs

SCFAs are essential for microbial benefits (Saghir and Al Suede 2024). Ellagitannins and ellagic acid in prebiotics such as *Myrciaria jaboticaba* seed extract is converted into urolithins by the stomach. Controlling signaling pathways and biological functions, urolithin inhibits cancer cell growth. Intestinal pro-inflammatory cytokines are inhibited (Mounir et al. 2022). Prebiotic fermentation produces SCFAs (Bedu-Ferrari 2022). *Bifidobacterium* and *Lactobacillus* bacteria create SCFAs, which affect colonocyte metabolism, cellular metabolism, immunological response, and health. These chemicals affect gene expression, epigenetics, and signaling cascades on several levels. Butyrate, propionate, and acetate are produced by gut bacteria fermenting. Nutrients may fuel colonic mucosa and help colonocytes form an immune barrier (Mounir et al. 2022). Acetate is used by muscle, kidneys, heart, and brain for gluconeogenesis, much like biochemists. Neoglucogenic propionate may lower cholesterol and modulate adipose tissue lipogenesis. Butyrate, a favored substrate for colonic commensal bacteria, boosts cell development in several ways. For treating inflammation, prebiotic fibers that create a variety of SCFAs via microbial fermentation are essential. NVPS and chitin-glucan boost acetate and butyrate production, whereas acacia gum releases butyrate. Finally, many physiological activities need SCFAs. Keep gut pH steady, reduce pathogens, manage bowel motions, and kill colon cancer cells (Brandelli 2021; Bedu-Ferrari 2022; Mudannayake et al. 2022; Saghir and Al Suede 2024; Arifuzzaman et al. 2022).

4.12 Immunomodulation

Understanding the importance of gut flora is important for maintaining a healthy immune system and preventing CRC. Reduce proliferation of cells, induce apoptosis, inhibit angiogenesis, and stop the metastatic process (Bedu-Ferrari 2022). Prebiotics such as yacón may increase sIgA levels and have a direct impact on the immune response. *Bifidobacterium* bacteria ferment yacón FOS in the cecum, raising sIgA levels (Ambalam et al. 2016). Yacón supplementation improves immunological function by decreasing the TNF-α/IL-10 ratio. This shows that pro-inflammatory and anti-inflammatory cytokines are in balance. Th2 lymphocytes produce IL-10, which inhibits TNF-α and other macrophage-dependent cytokines generated by Th1 cells. This proposes a self-regulating loop where TNF-α promotes IL-10 production while decreasing TNF-α synthesis. BALB/c mice who received a yacón-based product containing FOS had increased intestinal regulatory T lymphocytes (T reg). Furthermore, these cells generated IL 10. Antibacterial defensins, sIgA, and anti-inflammatory cytokines, particularly IL-10, increase, suggesting immune system modulation (Arifuzzaman et al. 2022). A CRC clinical study discovered that prebiotic consumption influenced preoperative IgG and IgM levels, similar to the findings of an immunologist. Following surgery, supplementation increased IgG, IgA, total B lymphocytes (CD19+), and suppressor/cytotoxic T cells. Prebiotics increased transfer levels, which reduced inflammation, similar to immunologists. According to the study, CRC patients may benefit from prebiotics one-week earlier surgery to enhance blood inflammatory markers. In the laboratory, NVPS promotes macrophages, which suppresses colorectal cancer (Ambalam et al. 2016). An herbal tea containing *Ganoderma lucidum* (Lingzhi) polysaccharides (GLP) and *Gynostemma pentaphyllum* (GpS) saponins improved the inflamed intestinal barriers of mice, as scientists had discovered. To achieve this result, an immunologist would shrink polyps, shift colonic M1 to M2 macrophages, positively restore the E-cadherin/N-cadherin ratio, and down regulate oncogenic signaling molecules.

4.13 Interaction Between Probiotics and Prebiotics

4.13.1 Synbiotics

The synbiotic mix of prebiotics and probiotics is a potential new therapy for IBD. Research found that inulin and *Bifidobacterium* together had a greater impact on rat AOM-induced ACF than either drug alone. Despite using short-term carcinogenesis tests to get these results, no synbiotic effects on long-term carcinogenesis are currently available. Given these factors and the potential anti-cancer benefits of prebiotics, probiotics, and synbiotics in rats with AOM, we chose to explore their long-term effects on carcinogenesis. High-oligofructose ulin, Synergy1, a β (Femia et al. 2002), fructan from chicory roots (*Cichorium intybus*), is a balanced mix of long and short chain inulin, similar to an immunologist. When fructose molecules are arranged in linear chains to make inulin, DP may vary from 3 to 65. Long-chain ferments slowly (DP range: 10–65, average. 25), whereas oligofructose-based ferments swiftly (DP range: 3–8, avg. 4) (Fernǎndez et al. 2019). Like an immunologist combining components, Synergy1 improves longer chain concentration compared to the original product. Long-term effects of Synergy1 on carcinogenesis remain unclear. According to a study, probiotic mixtures may provide greater intestinal benefits compared to individual strains. The study employed *Bifidobacterium* lactis Bb12 (Bb12), LGG, and GG as probiotic microorganisms. Researchers gave rats a prebiotic, probiotic, and synbiotic diet to promote cancer. This feeding regimen started 10 days before AOM and lasted 31 weeks till the rats died (Ambalam et al. 2016; Newmark et al. 1990). Biomarkers associated with colon cancer were examined utilizing the immunologist's methodology. Biomarkers include tumor or mucosal apoptosis, colonic mucosal proliferation, and cecal SCFA. Cancer development may be impacted by these factors (Ouwehand et al. 2000). Utilizing synbiotics can help improve the survival of beneficial microorganisms in food or feed, while also supporting the growth of specific native bacterial strains in the gastrointestinal tract. The effects of synbiotics on metabolic health remain unclear. It's important to mention that the health effects of synbiotics are probably related to the specific pairing of a probiotic and prebiotic (Hague et al. 1995). Exploring a wide range of possible combinations, the potential of synbiotics to impact the gut microbiota in humans is incredibly promising.

4.13.2 Synbiotic Mechanism and Its Importance

Probiotics and prebiotics may synergistically interact through their respective activities in the small and large intestines, thereby contributing to the enhancement of overall health. Prebiotics are used specifically to promote probiotic growth, fermentation, and intestinal transit. According to the study, prebiotics may assist probiotic bacteria survive oxygenation, pH, and temperature in the intestine (Markowiak and Śliżewska 2017). With the proper components, powerful dietary supplements may be created. We can increase host health by maintaining an optimum microenvironment. Two synbiotic processes exist. One strategy improves probiotic microbial viability. Provides unique health advantages. Prebiotics activate probiotics, which affects intestinal metabolism. This promotes intestinal health, supports beneficial bacteria, and inhibits pathogens (Sekhon and Jairath 2010). Similar to microbiologists, synbiotics decrease harmful metabolites and deactivate nitrosamines and cancer-causing compounds. These chemicals boost short-chain fatty acids, ketones, carbon disulfides, and methyl acetates. This increase might benefit host health (De Vrese and Schrezenmeir 2008). Synbiotics have antibacterial, anticancerogenic, and antiallergenic properties. They help promote regular bowel movements and prevent intestinal rot. Synbiotics have been shown to prevent osteoporosis, lower blood fat and sugar levels, control the immune system, and treat mental illnesses linked to liver failure (Manigandan et al. 2012).

4.13.3 Health Benefits from the Symbiotic Effects of Combination of Probiotics and Prebiotics

The elevation of *Lactobacillus* and *Bifidobacterium* populations, along with the stabilization of the gut microbiome, has been associated with improved liver function in cirrhosis patients, enhanced immunomodulatory effects, reduced bacterial translocation, and a decrease in nosocomial infections among post-surgical and other similar treatments (Pandey et al. 2015). Bacterial metabolic products penetrate the liver, including LPSs, ethanol, and SFCAs. SCFAs promote hepatic triacylglycerol generation and retention. These mechanisms may aggravate hepatic detoxification, resulting in IHTG accumulation and organ steatosis. A randomized trial of a synbiotic with five probiotics (*Lactobacillus plantarum*, *Lactobacillus delbrueckii* spp. *bulgaricus*, *Lactobacillus acidophilus*, *Lactobacillus rhamnosus*, and *Bifidobacterium bifidum*) and inulin as a prebiotic in adults with NASH (non-alcoholic steatohepatitis) showed a substantial reduction in IHTG within 6 months (Wong et al. 2013). LPSs produce proinflammatory cytokines, including TNF. A 28-week trial looked at the effectiveness of a synbiotic product including fructooligosaccharides and probiotics (*Lactobacillus casei, rhamnosus, thermophilus, breve, acidophilus, longum, and bulgaricus*). A total of 52 adults participated. Supplementation with the synbiotic decreased NF-κB and TNF-α production (Eslamparast et al. 2014). In rats, a synbiotic product including *Lactobacillus rhamnosus*, *Bifidobacterium lactis*, inulin, and oligofructose elevated intestinal IgA levels. Synbiotics reduce blood pressure and cholesterol (Pathmakanthan et al. 2002). Synbiotics may help treat hepatic issues and enhance calcium, magnesium, and phosphorus absorption (Pérez et al. 2006). Probiotics had no effect on systemic Sensitization (Pérez et al. 2006). SYNCAN research financed by the European Union confirmed the anti-carcinogenic capabilities of synbiotics. CRC patients were given a combination of fructooligosaccharides (SYN1) and two probiotic strains (*Lactobacillus rhamnosus* GG and *Bifidobacterium animalis* subsp. lactis Bb12). Several indicators, including genotoxicity, labeling index, tagged cells/crypt, transepithelial resistance, necrosis, interleukin 2, and interferon γ, indicate progression of the illness and post-polyp excision (Rafter et al. 2007). The examined synbiotic decreased DNA damage and colonocyte proliferation ratio, potentially lowering colorectal cancer.

4.14 Clinical Trials

The key findings from clinical trials on prebiotics in CRC prevention indicate that prebiotics can positively influence gut microbiota composition, enhancing the growth of beneficial bacteria while suppressing harmful species. These changes in the gut environment have been associated with reduced markers of inflammation, improved gut barrier function, and an increase in the production of short-chain fatty acids (SCFAs), particularly butyrate, which has known anti-carcinogenic properties. The trials also suggest that specific prebiotics, such as inulin and fructo-oligosaccharides (FOS), may lower the risk of CRC development by directly modulating factors that contribute to tumorigenesis. However, while the results are promising, further research is needed to determine optimal dosages, long-term effects, and the most effective combinations of prebiotics for CRC prevention. Several species of *Akkermansia* and *Cladosporium* were detected, but the counts of *Escherichia*, *Shigella*, *Enterococcus*, and *Turicibacter* were comparatively low. The combination of FOS and B. longum promotes the synthesis of SCFAs by inhibiting specific bacteria. Several substances, including polysaccharides from jujube and Nostoc commune Vaucher, Acacia gum, *L. plantarum*, Yacon seeds from the Brazilian grape tree Jabuticaba, and *Lactobacillus delbrueckii* subsp. *bulgaricus*, have demonstrated potential in fighting cancer. When *Clostridium* spores are combined with dextran, there is a clear reduction in tumor growth observed in mice (Fernández et al. 2019). The study investigated the impact of orally administered dextran-coated *Clostridium butyricum* spores. Studies have indicated that *C.*

butyricum has the capacity to degrade dextran, leading to the formation of short-chain fatty acids (SCFA). Research suggests that these short-chain fatty acids (SCFAs) may play a role in slowing down the development of CRC. Given the correlation between the microbiome and CRC, it is imperative to analyze and address the current patterns that may be contributing to the rise of early-onset CRC (Chundakkattumalayil et al. 2019).

There is no conclusive evidence indicating that synbiotics have any significant effect in preventing cancer. Investigating the cellular response to DNA damage caused by carcinogens, the process of colonization, and the impact of probiotics on the formation, maintenance, and activity on specific prebiotic substrates (Tian et al. 2022). There is an increase in the production of short-chain fatty acids (SCFAs) in the colon and cecum, which leads to enhanced metabolism (Bevilacqua et al. 2024). There is evidence to suggest that synbiotics have the ability to slow down cell growth, while also reducing the activity of enzymes that are linked to colon cancer. In addition, the immune system is affected and there are changes in the bacterial makeup of the colon. According to recent research, specific types of bacteria have demonstrated promising results in suppressing the growth of CRC cells in mice (Segre 2015).

These bacterial strains, such as *B. longum*, *B. bifidum*, *L. acidophilus*, and *L. plantarum*, were observed to have this effect in a live animal study (Zmora et al. 2018). During the examination of animals with CRC induced by 1,2-dimethylhydrazine, scientists discovered that the administration of 5-fluorouracil successfully halted the advancement of adenomas to cancer (Wong et al. 2017). It appears that the immune system possesses a remarkable ability to specifically target cancer cells. This is made possible by the involvement of various types of T-cells, such as NK, CD8+, and Th1 CD4+ T cells, along with the production of IFN-gamma and Granzyme B (Tian et al. 2022). It is important to exercise caution when interpreting these findings, as previous studies have not shown significant benefits from changes in the microbiome in cancer treatments (Wang et al. 2019). For individuals diagnosed with melanoma, incorporating checkpoint inhibitors into their treatment regimen and actively increasing fiber intake could potentially enhance their chances of survival. Probiotic supplements like *B. longum* or *L. rhamnosus* GG cancel out this benefit. Probiotics have a minimal effect in terms of reestablishing balance to the gut microbiome. Disrupting the transmission system of bacteria may hinder the bacteria's capacity to prevail over other microorganisms (Abualkhair et al. 2020). A recent study revealed the potential benefits of probiotics in reducing the impact of a fiber-rich diet on the microbiota of individuals undergoing checkpoint inhibitor treatment. Research has indicated that the introduction of fecal matter from individuals with colon cancer into mice can potentially stimulate the development of cancer. Evidently is limited information regarding the effectiveness of fecal transplants in the treatment or prevention of CRC. Recent studies have shown promising results from combining fecal transplants with checkpoint inhibitors for treating various carcinomas and metastases (Wong et al. 2017). The impact of different treatments on the progression of EOCRC remains uncertain. Considering the increasing occurrence of EOCRC, it may be wise to reconsider the current CRC screening directives. Based on the most recent guidelines from the US Preventive Services Task Force, it is recommended to initiate screening for CRC in individuals with average risk at the age of 45.

According to the most recent guidelines from the American Cancer Society (ACS) in 2018, it is now advised to begin CRC screening at the age of 45 rather than 50. There are several procedures that are commonly used, such as colonoscopy, computed tomography, immunochemical testing, computed sigmoidoscopy, high-sensitivity fecal occult blood tests, and multitarget stool DNA assays. Given the increasing prevalence of advanced EOCRC at diagnosis, it might be prudent to reconsider the guidelines. Sixteen 16 (Xu et al. 2022).

4.15 Fecal Microbiota Transplantation in Colorectal Cancer

Recent findings in intestinal studies have revealed new insights into the possible link between fecal microbiota transplantation (FMT) and CRC. A recent research investigation has provided valuable insights into the role of gut microbiota in the prediction and progression of colorectal carcinoma. Exploring this intricate relationship has the potential to pave the way for implementing preventive measures against CRC. There is a significant association between an imbalance and carcinoma, as well as adverse outcomes in allogeneic stem cell transplantation (Xu et al. 2022). Researchers in the field of microbiology have discovered that FMT has shown promising results in replacing imbalanced intestinal microbiomes with more beneficial ones, effectively treating an imbalance.

Current evidence indicates that FMT has shown promise in treating various conditions such as inflammatory bowel disease, hepatic encephalopathy, and CRC. Research is currently being conducted to investigate the potential of gut microbiota-modifying therapies like FMT and mutually beneficial relationships in the beginning stages of treatment. A recent research investigation on mice has revealed an intriguing link between the microbiota and colonic dysplasia following FMT. The research studies effectively tackled colonic dysplasia and reduced pro-oncogenic pro-inflammatory cytokines (Chang et al. 2023). These simulations offer valuable insights into how FMT can affect the effectiveness of immunotherapy. Studies have shown promising results in solid tumors other than CRC. Without human studies, our understanding of this subject is limited (Shrifteylik et al. 2023).

Research in FMT involves the creation of pathogenic mice models using specific chemicals to mimic human conditions. There are certain substances that have been found to have significant effects on the body. For instance, one such substance is Azoxymethane (AOM), which has been shown to promote tumor development. Another example is dextran sodium sulfate (DSS), which is known to break down the intestinal barrier, leading to colon inflammation and long-term colitis (Wong and Yu 2019). Typically, controls are administered isotonic or PBS via gavage. As part of the research, these mice models receive fecal microbiota transplantation from either healthy individuals or patients with colorectal cancer in several studies using a comparable methodology, mice that were given FMT from people without

CRC showed different results as opposed to infected mice that did not receive FMT or received FMT from CRC patients. These correspond to the findings: In a microbiome that is functioning properly, the abundance of Firmicutes is higher than Bacteroides (Oyetayo and Oyetayo 2005). It's a clear link among higher alpha diversity and FMT from non-CRC control groups. Control mice that underwent FMT exhibited a reduction in dysplasia, tumor's weight, and intestine length, suggesting favorable health outcomes. The levels of IFN-gamma, TNF-alpha, Th1, and Th17 cells increase in mice with a pathogenic gut microbiota. In addition, there is an increase in Ki-67, PCNA immunostaining, and beta-catenin, while the number of apoptotic cells decreases (Maftei et al. 2024). Evidently has been a noticeable decline in the generation of genes associated with the intestinal barrier and colon immunological the protective layer. Studies have shown that mice who received FMT from a naturally exposed population had better protection against carcinogens compared to laboratory mice (Shimizu et al. 2020).

4.16 Personalized Microbiome Treatment

Microbiota-host disruption is linked to IBD, diabetes, cirrhosis, and colorectal cancer. Research on how various human microbiomes affect cancer treatment is scarce. Some recent studies have linked microorganisms to cancer drugs. Antibiotic efficacy depends on how bacteria interact with the immune system, according to these studies. Several studies suggest gut microbiota may affect immunotherapy efficacy. Drug interaction evaluation relies on this factor. Gut microbiota assist researchers study illness phenotype, prognosis, and treatment response. Microbial gastrointestinal diseases exist. The FadA adhesin may detect *F. nucleatum*-associated colorectal cancer. *C. difficile* infection reduces secondary bile acid synthesis and microbial diversity (Chundakkattumalayil et al. 2019). Two recent investigations confirmed that microbiome markers may predict *C. diff* infection (Wong et al. 2017). Transplanting healthy stool bacteria improved clinical conditions by 90%, according to current research.

4.17 Impact of Bacteria on Therapeutic Efficacy

Numerous bacterial species play a significant role in influencing therapeutic efficacy. Responders had greater feces *Akkermansia muciniphila* levels than non-responders, proving the anti-PD1 drug worked. Immunoregulatory microorganisms in the microbiotas that reacted favorably to treatments outperformed anti-PD1. *Bifidobacterium*, *Akkermansia*, and *Faecalibacterium* were found. Another research demonstrated that mice that got FMT from well-responsive patients recovered faster following anti-PD1 medication. This increased reaction was linked to *Faecalibacterium* in rat feces. In additional studies, mice who got FMT and had tumors from patients who responded well to therapy had considerably greater CD8 T cell numbers than the control group. Routly found that *A. muciniphila* increased immune cell antitumor activity in mice administered FMT from untreated individuals. Eighteen (Guo et al. 2019; Bedu-Ferrari 2022; Ouwehand et al. 2000). Some strains of *Akkermansia, Faecalibacterium*, and *Bifidobacterium* exhibit anti-inflammatory qualities that protect the immune system and promote equilibrium. *Acinetobacter baumannii (A. muciniphila*) reductions are associated with IBD and type II diabetes. Like butyrate and salicylic acid generated by intestinal cells and microorganisms in circulation, it reduces intestinal inflammation. These results suggest precision medicine might use gut microbiota for therapy. Data suggests synthetic microbial communities are treating IBD and CDI. Different gut microbiota cells create chemicals that affect health. In positive responders, gut bacteria transplanted to secondary lymphoid organs may increase therapeutic synergy. These organs are essential for tumor-targeted immunity (ages 15–19) (Tian et al. 2022).

4.18 Potential Applications of the Gut Microbiome in Clinical Practice

Asymptomatic gut microbiota may screen for high-risk adenomas or CRC like a medical microbiologist. Screening indications are bacterium strains. *Fusobacterium nucleatum* appears in adenomas and colorectal cancer patients' stools and may be tested. Specific strain metabolic and genotoxic metabolites may identify early colorectal cancer. Clinical microbiologists can predict patient outcomes, medication responsiveness, and side effects using gut microbiota (Xu et al. 2022). Metabolites, genes, and serological markers from microbiota may be biomarkers. Blood, tumor tissue, feces, and oral samples all show these signs. Controlling the gut microbiota may prevent colorectal cancer in high-risk patients and improve chemotherapy and immunotherapy. Also, gut flora modification may reduce the side effects of chemotherapy and immunotherapy. Lifestyle modifications, prebiotics, probiotics, postbiotics, antibiotics, and FMT may affect the gut microbiome (Wong et al. 2017; Wong and Yu 2019). Multiple lab and clinical research have shown that probiotics may reduce colorectal cancer risk. Clinical trials have shown probiotics can fight CRC. Life L. casei Shirota (LcS) oral ingestion reduces colorectal cancers in surgical patients. Probiotics inhibited superficial bladder cancer recurrence. In a Hatakka et al. investigation, *L. rhamnosus* LC705 and *P. freudenreichii* showed promise (Xu et al. 2022). Probiotics and prebiotics play a crucial role in colorectal cancer (CRC) prevention by supporting a healthy gut microbiome, which is increasingly recognized as a key factor in reducing cancer risk. Probiotics, through the introduction of beneficial bacteria, and prebiotics, by nourishing these microbes, work together to create an environment that inhibits the growth of harmful bacteria and reduces inflammation—both significant contributors of CRC development.

As our understanding of the gut microbiome deepens, future research should focus on identifying specific strains of probiotics and types of prebiotics that are most effective in CRC prevention. Additionally, exploring personalized approaches that consider individual microbiome profiles could lead to more targeted and effective prevention strategies. The integration of probiotics and prebiotics into CRC prevention not only offers a promising avenue for reducing cancer risk but also highlights the need for continued research to fully unlock their potential.

References

Abualkhair WH, Zhou M, Ahnen D, Yu Q, Wu XC, Karlitz JJ (2020) Trends in incidence of early-onset colorectal cancer in the United States among those approaching screening age. JAMA Netw Open 3(1):e1920407

Ambalam P, Raman M, Purama RK, Doble M (2016) Probiotics, prebiotics and colorectal cancer prevention. Best Pract Res Clin Gastroenterol 30(1):119–131

Arifuzzaman M, Won TH, Li TT, Yano H, Digumarthi S, Heras AF et al (2022) Inulin fibre promotes microbiota-derived bile acids and type 2 inflammation. Nature 611(7936):578–584

Bedu-Ferrari C (2022) Interactions between prebiotics and health-relevant commensal bacteria: impact on bacterial growth and metabolic activities. Doctoral dissertation, Université Paris-Saclay

Bevilacqua A, Campaniello D, Speranza B, Racioppo A, Sinigaglia M, Corbo MR (2024) An update on prebiotics and on their health effects. Food Secur 13(3):446

Brandelli A (ed) (2021) Probiotics: advanced food and health applications. Academic Press, London

Chang Y, Ou Q, Zhou X, Liu J, Zhang S (2023) Global research trends and focus on the link between colorectal cancer and gut flora: a bibliometric analysis from 2001 to 2021. Front Microbiol 14. https://doi.org/10.3389/fmicb.2023.1182006

Chundakkattumalayil HC, Kumar S, Narayanan R, Thalakattil Raghavan K (2019) Role of L. plantarum KX519413 as probiotic and acacia gum as prebiotic in gastrointestinal tract strengthening. Microorganisms 7(12):659

De Vrese M, Schrezenmeir AJ (2008) Probiotics, prebiotics, and synbiotics. Food Biotechnol 111:1–66

Eslamparast T, Poustchi H, Zamani F, Sharafkhah M, Malekzadeh R, Hekmatdoost A (2014) Synbiotic supplementation in nonalcoholic fatty liver disease: a randomized, double-blind, placebo-controlled pilot study. Am J Clin Nutr 99(3):535–542

Femia AP, Luceri C, Dolara P, Giannini A, Biggeri A, Salvadori M et al (2002) Antitumorigenic activity of the prebiotic inulin enriched with oligofructose in combination with the probiotics Lactobacillus rhamnosus and Bifidobacterium lactis on azoxymethane-induced colon carcinogenesis in rats. Carcinogenesis 23(11):1953–1960

Fernández J, Ledesma E, Monte J, Millán E, Costa P, de la Fuente VG et al (2019) Traditional processed meat products re-designed towards inulin-rich functional foods reduce polyps in two colorectal cancer animal models. Sci Rep 9(1):14783

Fuller R (1989) Probiotics in man and animals. J Appl Bacteriol 66(5):365–378

Guo S, Chen S, Ma J, Ma Y, Zhu J, Ma Y et al (2019) Escherichia coli nissle 1917 protects intestinal barrier function by inhibiting NF-κB-mediated activation of the MLCK-P-MLC signaling pathway. Mediat Inflamm 2019(1):5796491

Hague A, Elder DJE, Hicks DJ, Pareskeva C (1995) Apoptosis in colorectal tumour cells: induction by the short chain fatty acids butyrate, propionate and acetate and by the bile salt deoxycholate. Int J Cancer 60:400–406

Ionescu VA, Gheorghe G, Georgescu TF, Buica V, Catanescu MS, Cercel IA et al (2024) Exploring the role of the gut microbiota in colorectal cancer development. Gastrointest Disord 6(2):526–537

Kumavath R, Pavithran H, Paul S, Anju VT, Busi S, Dyavaiah M (2024) Effects of gut microbiome and obesity on the development, progression and prevention of cancer. Int J Oncol 64(1):1–13

La Fata G, Weber P, Mohajeri MH (2018) Probiotics and the gut immune system: indirect regulation. Probiotics Antimicrob Proteins 10:11–21

Liu M, Chen J, Dharmasiddhi IPW, Chen S, Liu Y, Liu H (2024) Review of the potential of probiotics in disease treatment: mechanisms, engineering, and applications. Processes 12(2):316

Maftei NM, Raileanu CR, Balta AA, Ambrose L, Boev M, Marin DB, Lisa EL (2024) The potential impact of probiotics on human health: an update on their health-promoting properties. Microorganisms 12(2):234

Manigandan T, Mangaiyarkarasi SP, Hemalatha R, Hemalatha VT, Murali NP (2012) Probiotics, prebiotics and synbiotics-a review. Biomed Pharmacol J 5(2):295

Markowiak P, Śliżewska K (2017) Effects of probiotics, prebiotics, and synbiotics on human health. Nutrients 9(9):1021

Mohr AE, Basile AJ, Meli'sa SC, Sweazea KL, Carpenter KC (2020) Probiotic supplementation has a limited effect on circulating immune and inflammatory markers in healthy adults: a systematic review of randomized controlled trials. J Acad Nutr Diet 120(4):548–564

Mounir M, Ibijbijen A, Farih K, Rabetafika HN, Razafindralambo HL (2022) Synbiotics and their antioxidant properties, mechanisms, and benefits on human and animal health: a narrative review. Biomol Ther 12(10):1443

Mudannayake DC, Jayasena DD, Wimalasiri KM, Ranadheera CS, Ajlouni S (2022) Inulin fructans–food applications and alternative plant sources: a review. Int J Food Sci Technol 57(9):5764–5780

Newmark HL, Lipkin M, Maheshwari N (1990) Colonic hyperplasia and hyperproliferation induced by a nutritional stress diet with four components of Western-style diet. JNCI J Natl Cancer Inst 82(6):491–496

Ouwehand AC, Isolauri E, Kirjavainen PV, Ölkkö ST, Salminen SJ (2000) The mucus binding of Bifidobacterium lactis Bb12 is enhanced in the presence of Lactobacillus GG and Lact. delbrueckii subsp. bulgaricus. Lett Appl Microbiol 30(1):10–13

Oyetayo VO, Oyetayo FL (2005) Potential of probiotics as biotherapeutic agents targeting the innate immune system. Afr J Biotechnol 4(2):123–127

Pandey M, Bhattacharyya J (2024) Gut microbiota and epigenetics in colorectal cancer: implications for carcinogenesis and therapeutic intervention. Epigenomics 16(6):403–418

Pandey KR, Naik SR, Vakil BV (2015) Probiotics, prebiotics and synbiotics-a review. J Food Sci Technol 52:7577–7587

Pathmakanthan S, Walsh M, Bengmark S, Willemse PJA, Bardhan K (2002) Efficacy and tolerability treating acute distal ulcerative colitis with synbiotic enemas: a pilot trial. Gut 51(Suppl III):A307

Pérez-Conesa D, López G, Abellán P, Ros G (2006) Bioavailability of calcium, magnesium and phosphorus in rats fed probiotic, prebiotic and synbiotic powder follow-up infant formulas and their effect on physiological and nutritional parameters. J Sci Food Agric 86(14):2327–2336

Rafter J, Bennett M, Caderni G, Clune Y, Hughes R, Karlsson PC et al (2007) Dietary synbiotics reduce cancer risk factors in polypectomized and colon cancer patients. Am J Clin Nutr 85(2):488–496

Roberfroid M (2007) Prebiotics: the concept revisited1. J Nutr 137(3):830S–837S

Saghir SAM, Al Suede FS (2024) Synergistic Efficacy and Mechanism of Probiotics and Prebiotics in Enhancing Health Impact. Microbial Bioactives 7(1):1–11

Saraiva MR, Rosa I, Claro I (2023) Early-onset colorectal cancer: a review of current knowledge. World J Gastroenterol 29(8):1289

Segre JA (2015) Microbial growth dynamics and human disease. Science 349(6252):1058–1059

Sekhon BS, Jairath S (2010) Prebiotics, probiotics and synbiotics: an overview. J Pharm Educ Res 1(2):13–36

Shimizu Y, Isoda K, Taira Y, Taira I, Kondoh M, Ishida I (2020) Anti-tumor effect of a recombinant Bifidobacterium strain secreting a claudin-targeting molecule in a mouse breast cancer model. Eur J Pharmacol 887:173596

Shrifteylik A, Maiolini M, Dufault M, Austin DL, Subhadra B, Lamichhane P, Deshmukh RR (2023) A current review on the role of prebiotics in colorectal cancer. Biologics 3(3):Article 3. https://doi.org/10.3390/biologics3030012

Singh D, Agarwal V (2021) Screening of antimicrobial, anti-quorum sensing activity and cytotoxicity of origanum oil against Gram-positive and Gram-negative bacteria. Biomedicine 41(3):599–603

Szajewska H, Kołodziej M, Zalewski BM (2020) Systematic review with meta-analysis: Saccharomyces boulardii for treating acute gastroenteritis in children—a 2020 update. Aliment Pharmacol Ther 51(7):678–688

Tian Y, Zuo L, Guan B, Wu H, He Y, Xu Z et al (2022) Microbiota from patients with ulcerative colitis promote colorectal carcinogenesis in mice. Nutrition 102:111712

Wang Z, Hua W, Li C, Chang H, Liu R, Ni Y et al (2019) Protective role of fecal microbiota transplantation on colitis and colitis-associated colon cancer in mice is associated with Treg cells. Front Microbiol 10:2498

Wang S, Zheng R, Li J, Zeng H, Li L, Chen R et al (2024) Global, regional, and national lifetime risks of developing and dying from gastrointestinal cancers in 185 countries: a population-based systematic analysis of GLOBOCAN. Lancet Gastroenterol Hepatol 9(3):229–237

Wong SH, Yu J (2019) Gut microbiota in colorectal cancer: mechanisms of action and clinical applications. Nat Rev Gastroenterol Hepatol 16(11):690–704

Wong VWS, Wong GLH, Chim AML, Chu WCW, Yeung DKW, Li KCT, Chan HLY (2013) Treatment of nonalcoholic steatohepatitis with probiotics. A proof-of-concept study. Ann Hepatol 12(2):256–262

Wong SH, Zhao L, Zhang X, Nakatsu G, Han J, Xu W et al (2017) Gavage of fecal samples from patients with colorectal cancer promotes intestinal carcinogenesis in germ-free and conventional mice. Gastroenterology 153(6):1621–1633

Xu H, Cao C, Ren Y, Weng S, Liu L, Guo C et al (2022) Antitumor effects of fecal microbiota transplantation: implications for microbiome modulation in cancer treatment. Front Immunol 13:949490

Zhang R, Kang R, Tang D (2024) Gut microbiome mediates ferroptosis resistance for colorectal cancer development. Cancer Res 84(6):796–797

Zmora N, Zilberman-Schapira G, Suez J, Mor U, Dori-Bachash M, Bashiardes S et al (2018) Personalized gut mucosal colonization resistance to empiric probiotics is associated with unique host and microbiome features. Cell 174(6):1388–1405

Genetically Engineered Probiotics and Their Potential in Therapeutic Applications

5

Sena Davran Bulut and Hasan Ufuk Celebioglu

Abstract

The intestinal microbiome, consisting of a diverse array of microbes, plays a vital role in protecting against pathogens, affecting host physiology, and shaping the immune system. Probiotic bacteria, which are live microorganisms providing health benefits, have become increasingly significant due to their ability to lower cholesterol, maintain microbial balance, inhibit carcinogenesis, and support metabolic functions. However, probiotics also have limitations, including specificity to certain microbes and the need to withstand harsh gastrointestinal environments. Enhancing their effectiveness may involve creating blends of various strains and developing genetically engineered probiotics (GEP) to address these challenges. Bioengineering can enhance probiotic strains by introducing genes that help them withstand production stresses, survive in the gastrointestinal tract, and prevent infections. This approach also enables GEP to deliver drugs, target pathogens, and treat diseases such as gastrointestinal diseases, diabetes, cancer, and cardiovascular diseases. This chapter explores GEP, emphasizing the conversion of non-probiotic substances into probiotics and the enhancement of existing probiotics through genetic modification. It also discusses recent advancements in using probiotics as living diagnostics and therapeutics for detecting and treating diseases.

Keywords

Genetically Engineered Probiotics · Therapeutic Applications · Gut Microbiome, Probiotics-Host Interactions

S. Davran Bulut · H. U. Celebioglu (✉)
Department of Biotechnology, Faculty of Science, Bartin University, Bartin, Turkey

S. Pathak et al. (eds.), *Microbiota and Dietary Mediators in Colon Cancer Prevention and Treatment*, https://doi.org/10.1007/978-981-96-0297-1_5

5.1 Introduction

The intestinal microbiota is crucial for sustaining human health and plays a significant role in the treatment of various diseases (Marosvölgyi et al. 2024). An imbalance in this microbial community, known as dysbiosis, is linked to the development of numerous conditions, including immune disorders, cancer, and neurodegenerative diseases. Probiotics, live microorganisms that when administered in adequate amounts confer a health benefit on the host, affect human health in a variety of ways, including lowering cholesterol and affecting metabolic pathways, maintaining a healthy microbial balance in the intestinal, respiratory, and urogenital tracts, inhibiting carcinogenesis by stimulating the immune system, and can reduce yeast and vaginal infections and provide relief from rheumatism (Hill et al. 2014; Naeem et al. 2024).

Despite the myriad benefits of probiotics in gastrointestinal conditions, there are also certain limitations to consider (Mazhar et al. 2020). Probiotics release antimicrobial substances with a broad spectrum, yet research indicates their antibacterial efficacy is specific to certain microbes. Hence, developing a blend of multiple probiotic strains could enhance their effectiveness against intestinal pathogens (Aguilar et al. 2011). Probiotic bacteria require resilience to endure harsh conditions such as acidic environments, pepsin, and bile salts encountered throughout the gastrointestinal tract (GIT). The diverse array of activities and variations among probiotic strains are also recognized as obstacles to their efficacy (Karimi and Salvador Pen 2008). Additionally, the specific components of probiotic formulations, dosage, and method of delivery may vary, leading to differences in their probiotic potential among different hosts (Mathipa and Thantsha 2017). The constraints of conventional probiotics underscore the necessity for devising a novel approach to cultivate genetically modified probiotic strains. Progress in gene technology enables us to manipulate strains by incorporating new genes or modifying their metabolic functions (Bravo and Landete 2017).

The relationship between the human microbiota and many diseases has led to the discovery of genetically engineered probiotics (GEP). Advancements in genome editing techniques have led to an increased use of GEP for treating a wide range of conditions, including gastrointestinal diseases, cancer, and metabolic disorders. These developments have also enabled the creation of specialized probiotics designed to diagnose and treat diseases related to the microbiota (Parvin and Sadras 2024). The GEP discussed in this chapter involves transforming a previously non-probiotic substance to acquire probiotic properties, as well as enhancing the probiotic properties of a known probiotic through genetic modification. Furthermore, we highlighted recent advancements in designing and developing probiotics as living diagnostics and therapeutics for detecting and treating various diseases, including pathogen infections, cancer, intestinal inflammation, and metabolic disorders.

5.2 Genetically Engineered Probiotics

Enhancing the effectiveness of current probiotic strains can be achieved through bioengineering, which involves modifying genes within a probiotic strain to improve its ability to withstand various stresses encountered during food production, such as extreme temperatures, oxygen exposure, and acidification, as well as to thrive in the gastrointestinal tract and offer beneficial effects to the host (Upadrasta et al. 2011). This approach enables the creation of new probiotic strains that incorporate genes from pathogens to produce proteins not originally found in those microorganisms. Virulence factors from pathogens can be isolated and introduced into probiotic strains, which, when administered, can prevent infections without causing symptoms (Berg and Mertz 2010). GEP can also be designed to administer medicines or vaccines, focus on particular harmful microorganisms or substances, boost immune responses, and imitate cell surface receptors. The human receptors targeted by enteric pathogens and their toxins are well understood, minimizing the likelihood of resistance development against vaccines or treatments. Although not a new field, bioengineering of probiotics has been successful, as reported by various researchers (Mathipa and Thantsha 2017). GEP, created through genetic modification, not only enhance the effectiveness of traditional probiotics by incorporating target genes but also expand their application in treating diseases such as gastrointestinal diseases, cancer, cardiovascular diseases, and diabetes. (Meng et al. 2023).

Gene editing allows for the modification of existing probiotics to create desired new strains. These modifications facilitate direct confirmation of whether the genetic material, proteins, and functional characteristics of these new microorganisms have been changed as intended. Gene editing technology has evolved from homologous recombination to the zinc finger nucleases (ZFNs) as first-generation tools, then to the transcription activator-like effector nucleases as second-generation tools (TALEN), and more recently to the widely adopted third-generation clustered regularly interspaced short palindromic repeats (CRISPR) associated (CRISPR-Cas) technology (Zhang et al. 2021). ZFN and TALEN were named the method of the year by *Nature Methods*, while CRISPR gene editing technology was hailed as the best breakthrough of 2015 by Science news. Additionally, CRISPR-Cas9 technology received the 2020 Nobel Prize in Chemistry, marking its revolutionary impact on the field of life sciences (Ma et al. 2022).

CRISPR is a natural defense system found exclusively in bacteria and archaea. It helps these organisms resist invasion by viruses or foreign DNA, such as plasmids. Various types of CRISPR-associated (Cas) proteins have been discovered, each serving distinct functions, and the CRISPR/Cas system can be classified into six types. Type II systems, which utilize the Cas9 protein along with single guide RNA (sgRNA), are particularly notable for their ability to edit target molecules. The Type II CRISPR/Cas9 system is currently the primary protein used for genome editing purposes (Mohammadzadeh et al. 2020). The utilization of CRISPR-based editing toolkits has experienced notable expansion within the bacterial domain. These toolkits have been effectively developed for a diverse array of bacteria, encompassing species such as *B. subtilis*, *Clostridium*, *Corynebacterium*, and *Lactobacillus* species. With the increasing recognition of probiotic lactobacilli in food and therapeutic applications, the CRISPR genome editing toolkit for lactobacilli can be enhanced by utilizing the naturally occurring CRISPR-Cas systems found widely within this genus. These systems can be adapted for precise genome editing within their native hosts (Goh and Barrangou 2019). Oh and van Pijkeren (2014) studied implementing CRISPR-Cas9 technology in the lactic acid bacterium *Lactobacillus reuteri* ATCC PTA 6475. The study demonstrates the feasibility of using CRISPR-Cas9 coupled with single-stranded DNA (ssDNA) recombineering to efficiently identify edited cells in this bacterium. They show high efficiency in recovering subtle changes in the bacterial genome and successfully apply codon saturation mutagenesis. The study highlights CRISPR-Cas9's ability to identify low-efficiency events, such as oligonucleotide-mediated chromosome deletions, without the need for recombineering optimization procedures. The findings suggest that CRISPR-Cas genome editing could significantly impact genome editing in lactic acid bacteria and other gram-positive bacteria (Oh and Van Pijkeren 2014). Leenay et al. (2019) investigated the effectiveness of two CRISPR-Cas9-based genome editing methods in different strains of *Lactobacillus plantarum*. The methods involve utilizing either a plasmid-encoded recombineering template or an oligonucleotide donor and an inducible DNA recombinase. The comparison reveals instances where each method uniquely generates desired edits, but also highlights highly variable success rates across strains for both methods. The study identifies failure modes for the recombineering-template method, including consistent genomic deletions and point mutation reversions. Overall, the research underscores the importance of employing multiple distinct methods for CRISPR-based editing in bacteria due to surprising differences in efficacy between methods and strains (Leenay et al. 2019).

5.3 Therapeutic Applications of Genetically Engineered Probiotics in Human Diseases

While modern medical technologies have made significant strides in preventing and treating numerous human diseases, various challenges persist. These include the rise of antibiotic-resistant microorganisms, the increasing number of chronic patients, the adverse effects associated with chemical drugs, soaring medical costs, and the presence of incurable diseases like cancers (Cassini et al. 2019; Powell-Wiley et al. 2021). These challenges highlight the pressing need for cost-effective, convenient, and efficient solutions. Evidence is accumulating to suggest a correlation between dysfunctions in the human microbial community and the onset and progression of various diseases, indicating the potential of microorganisms as an alternative approach to tackle these issues (Dang and Marsland 2019). Notably, bacteria constitute over 90% of intestinal microbes, and recent advancements in synthetic biology have enabled precise manipulation of bacteria for a wide array of purposes. By the conclusion of the twentieth century, GEP were widely employed to acquire desired compounds. Nowadays, GEP have been utilized in various sectors for different applications, particularly in disease treatment, healthcare, and scientific research, but also in the food industry, chemical synthesis, environmental protection and more (Liu et al. 2022; Oh et al. 2020).

5.3.1 Genetically Engineered Probiotics in Gastrointestinal Diseases

The imbalance in intestinal microbes can lead to gastrointestinal diseases, including acute diarrhea, inflammatory bowel disease (IBD), and diabetes and, in certain instances, harm healthy organs through the tissue–gut axis. Although the primary focus of treatment methods of gastrointestinal diseases lies in addressing symptoms, which are costly and ineffective in the long term, an effective treatment method has yet to be developed (Liu et al. 2022; Meng et al. 2023). Recent studies have demonstrated that GEPs can serve as a valuable tool in therapies. Plavec et al. (2019) studied engineer recombinant *L. lactis* cells to produce and secrete REX binding proteins known as IL-23 receptor (IL-23R) blockers and IL-23R antagonists. These proteins suppress the secretion of cytokine IL-17A, a key mediator in the pro-inflammatory cascade associated with autoimmune diseases like IBD. The resulting recombinant *L. lactis* strains successfully expressed and displayed the REX proteins on their surface. Binding specificity of these proteins to IL-23R was confirmed, and they showed potential as IL-23/IL-23R blockers in an experimentally induced mouse model of colitis (Plavec et al. 2019). Liu et al. (2018) explore using engineered *Bifidobacterium* as a delivery system to orally administer human MnSOD fused with penetratin PEP-1, known for its anti-inflammatory properties. In a mouse model of UC induced by dextran sulfate sodium (DSS), *Bifidobacterium* expressing the fusion protein reduced inflammatory cytokines and tissue damage in colonic tissues, suggesting its potential as a promising UC treatment approach (Liu et al. 2018).

5.3.2 Genetically Engineered Probiotics in Cancer Therapy

Chemotherapy drugs like 5-Fu are commonly employed in clinical settings with success in cancer therapy. However, some research has explored combining probiotics with drugs to enhance treatment effectiveness (Han et al. 2024). For instance, Kaeid Sharaf and Shukla (2018) aimed to evaluate the preventive effects of probiotics (*Lactobacillus acidophilus* and *Lactobacillus rhamnosus* GG) in combination with celecoxib, a selective COX-2 inhibitor, in a model of experimental colon carcinogenesis induced by DMH. Results showed that animals treated with *L. rhamnosus* GG + celecoxib + DMH exhibited reduced aberrant crypt foci (ACF) count and expression of pro-carcinogenic markers compared to other treatment groups. This suggests that combining probiotics with celecoxib could serve as a potential prophylactic strategy to reduce the incidence and severity of colorectal cancer, emphasizing the importance of prevention over treatment (Kaeid Sharaf and Shukla 2018). In another study, Chung et al. (2021) developed a potent bacterial drug delivery system for colorectal cancer therapy using *Pediococcus pentosaceus*. This engineered bacterium carries dual gene cassettes encoding therapeutic protein P8 and a complementation system. Testing in mice demonstrated significant reductions in tumor volume and growth with the synthetic probiotic treatment. Administering the engineered bacterium also led to polyp regression and restored gut microbiota diversity in mice with colitis-associated CRC. Furthermore, the synthetic probiotic positively influenced gut microbiota composition, potentially aiding in rebalancing dysbiosis (Chung et al. 2021).

Single cancer treatments often fail to achieve desired outcomes due to tumor heterogeneity and the development of drug resistance. To address this challenge, custom-designed combination therapies based on engineered probiotics can be tailored to target different aspects of cancer, presenting a promising solution to the issue at hand. (Gigola et al. 2021). Sasaki et al. (2006) constructed a plasmid containing a therapeutic gene, eCD, which was transfected into *B. longum*. The transfected bacteria produced the therapeutic enzyme CD within hypoxic tumors, enabling site-specific conversion of a prodrug, 5-FC, to 5-FU. Antitumor efficacy was demonstrated both by direct injection and systemic administration of the transfected bacteria in rats with mammary tumors. Evaluation of systemic immunogenicity showed no adverse reactions, suggesting the safety of this bacterial vector. Overall, transfected *B. longum* represents a promising and safe tumor-targeting vector for cancer therapy (Sasaki et al. 2006). Fujimori (2006) constructed a plasmid containing the cytosine deaminase gene, which was transfected into *Bifidobacterium longum* to produce the enzyme in hypoxic tumors. This enzyme/pro-drug therapy was effective for systemic administration, offering a potential strategy for targeted therapy in metastatic breast cancer (Fujimori 2006). Ji et al. (2023) developed genetically engineered microbes *Escherichia coli* MG1655 (EcM-GDH) expressing glucose dehydrogenase to competitively deprive tumors of glucose. EcM-GDH effectively depletes glucose, inducing autophagy and apoptosis in colorectal tumor cells both in vitro and in vivo (Ji et al. 2023). This approach highlights the potential of GEP for metabolic treatments for malignant tumors, offering a promising probiotic strategy with high efficacy and compatibility with biological systems.

5.3.3 Genetically Engineered Probiotics in Infections

The era of widespread antibiotic use has fueled the rapid emergence of multidrug-resistant pathogens due to inappropriate consumption and overuse. With limited antibiotic options available, there is a pressing need to explore alternative approaches to address the antimicrobial resistance crisis (Medina and Pieper 2016). One such method involves employing innovative synthetic biology-driven strategies, such as GEP. These probiotics offer controlled release of antimicrobial agents, improved specificity, and reduced risk of antibiotic resistance, presenting a promising solution to combat antimicrobial resistance (Kaur and Deol 2020). One of the in vivo studies aimed to assess the efficacy of a recombinant probiotic strain, *Lactobacillus gasseri* NM713, expressing the conserved region of streptococcal M6 protein (CRR6), as an oral vaccine against *Streptococcus pyogenes*. Compared to control groups, mice vaccinated with *Lb. gasseri* NM713 showed significant protection against *S. pyogenes* nasal challenge, with lower infection rates and mortality. These findings suggest that recombinant *Lb. gasseri* NM713 holds promise as an oral vaccine for streptococcus group A infections (Mansour and Abdelaziz 2016).

GEP that express required antigens hold promise as effective vaccine candidates for combating infectious diseases. Recombinant *L. lactis* strains were designed to express genes for three bacteriocins with potent antimicrobial activity against *E. faecalis*. The engineered *L. lactis* effectively detects and inhibits *E. faecalis* growth in both agar and liquid media, including multidrug-resistant strains of *E. faecium* (Borrero et al. 2015). This approach presents a potential alternative of GEP to antibiotics for targeting enterococci infections.

Probiotic bacteria can be genetically modified to target and neutralize particular pathogens or reduce their harmful effects. This capability highlights the potential of GEP for both preventive and therapeutic uses against gastrointestinal infections. Mao et al. (2018) studied a probiotic-based approach utilizing *Lactococcus lactis* to enhance colonization resistance against *Vibrio cholerae* and improve survival in infected infant mice by producing lactic acid. Additionally, an engineered *L. lactis* strain capable of detecting quorum-sensing signals of *V. cholerae* in the gut and expressing an enzymatic reporter for fecal sample detection was developed. Dietary interventions with fermented foods containing natural and engineered *L. lactis* strains could help prevent cholera progression and enhance disease surveillance in populations vulnerable to cholera outbreaks (Mao et al. 2018). *Lactobacilli* has potential for use as a platform to express broadly neutralizing antibodies against HIV-1 for mucosal delivery. Marcobal et al. (2016) tested various antibody fragments, including single-chain and single-domain antibodies, with successful expression in *Lactobacillus jensenii*. The antibody fragments exhibited binding to HIV-1 gp120/CD4 complex, with one fragment, m36.4, showing neutralizing activity against HIV-1BaL in a neutralization assay (Marcobal et al. 2016). This approach could offer passive transfer of antibodies to mucosal surfaces, potentially enhancing protection against HIV-1 transmission.

5.3.4 Genetically Engineered Probiotics in Metabolic Diseases

The body maintains a relatively stable balance of metabolism-related substances. Nevertheless, disruptions in biochemical processes can disturb this equilibrium, resulting in the buildup or depletion of certain metabolites. This imbalance can contribute to the onset of diseases that prove difficult to control (Zheng et al. 2020). Patients often grapple with the necessity of maintaining strict long-term adjustments to their lifestyle and diet to manage these conditions. The rise in metabolic disorders such as cardiovascular disease (CVD), diabetes, and obesity has prompted research into GEP capable of producing therapeutic biomolecules, as well as the consideration of phage therapy (Ma et al. 2022). GEPs emerge as a hopeful solution, facilitating the breakdown or alteration of accumulated substances to address such diseases.

Type 1 diabetes (T1D) is an autoimmune disorder where the immune system attacks and destroys beta cells in the pancreas, resulting in insufficient insulin production. Cytokines are pivotal in T1D development, directing immune cell activity to both destroy and protect insulin-producing cells. Preisser et al. (2021) studied to assess the impact of orally administered genetically modified *Lactococcus lactis* MG1363 FnBPA+ strains carrying eukaryotic expression plasmids encoding IL-4 and IL-10 in STZ-induced diabetes and NOD mice. Combined treatment with these strains (*L. lactis* MG1363 FnBPA+ (IL-4) + *L. lactis* MG1363 FnBPA+ (IL-10)) reduced diabetes incidence and preserved pancreatic islets better than individual strains in STZ-induced mice. In NOD mice, the combined administration protected against diabetes by preventing hyperglycemia and reducing pancreatic islet destruction (Preisser et al. 2021). In another study, glucose homeostasis in newly diagnosed non-obese diabetic (NOD) mice was effectively normalized with oral administration of a combination treatment of proinsulin and IL-10 via genetically modified *Lactococcus lactis* bacteria, in conjunction with low-dose anti-CD3 (aCD3) therapy (Cook et al. 2020).

Hypertension, a prevalent risk factor for cardiovascular diseases (CVDs) globally, often requires continuous treatment with medications that carry various adverse effects (Van Oort et al. 2020). Glucagon-like peptide-1 receptor (GLP-1R) agonists have demonstrated promising efficacy in treating diabetes mellitus, neurodegenerative conditions, obesity, and hypertension. Butyric acid presents itself as a potential therapeutic target for hypertension (Drucker 2018). However, the combined use of GLP-1 analogues and butyric acid to reduce blood pressure and reverse ventricular hypertrophy remains an unexplored avenue. Wang et al. (2023) engineered probiotics by transforming *Clostridium butyricum* with a recombinant plasmid encoding the human GLP-1 gene, creating *Clostridium butyricum*-pMTL007-GLP-1 (CB-GLP-1). CB-GLP-1 effectively improved blood pressure, cardiomegaly, and gut microbiome dysbiosis in rat models, suggesting a novel approach for hypertension treatment (Wang et al. 2023).

5.3.5 Genetically Engineered Probiotics in Neurodegenerative Diseases

The gut microbiota is increasingly recognized as a potential target for treating neurodegenerative diseases, given its connection to the brain through a two-way communication pathway that influences brain development and behavioral regulation. Fang et al. (2020) evaluated the neuroprotective effects of an engineered strain, MG136-pMG36e-GLP-1, in Alzheimer's disease (AD) and Parkinson's disease (PD) mice. The strain combines probiotics *Lactococcus lactis* MG1363 and glucagon-like peptide-1 (GLP-1). Results showed that oral administration of this strain improved memory and motor function by targeting specific signaling pathways and altering gut microbiota composition, reducing harmful pathogens and increasing beneficial probiotics. These findings suggest that MG136-pMG36e-GLP-1 could be a potential new treatment for AD and PD by mitigating neuroinflammation (Fang et al. 2020). Cecarini et al. (2020) explored the use of engineered probiotics as a novel AD treatment. Researchers developed a *Lactobacillus lactis* strain with a plasmid (pExu) encoding the human p62 protein. When 3xTg-AD mice were orally administered this probiotic for 2 months, they showed increased brain expression of p62. This delivery method, involving lymphatic vessels and neural terminations, improved memory, modulated the ubiquitin-proteasome system and autophagy, reduced amyloid peptides, and decreased neuronal oxidative and inflammatory processes. These findings suggest that using safe, non-pathogenic bacteria to deliver therapeutic proteins like p62 is a promising and innovative approach for treating AD (Cecarini et al. 2020). Yue et al. (2022) examined the neurotrophic effects of the engineered probiotic strain *L. lactis* MG1363-pMG36e-GLP-1 in MPTP-induced PD mice. The results showed that this probiotic strain significantly improved motor deficits by rescuing dopaminergic neurons in the substantia nigra. It achieved these effects by activating the Keap1/Nrf2/GPX4 signaling pathway, which reduces ferroptosis and oxidative stress. Additionally, the probiotic reversed gut dysbiosis by increasing beneficial bacteria such as *Akkermansia*, *Oscillospira*, and *Sutterella*. Overall, *L. lactis* MG1363-pMG36e-GLP-1 demonstrated neuroprotective benefits through oxidative stress modulation, ferroptosis inhibition, and gut microbiota restoration (Yue et al. 2022).

Engineered probiotics hold significant potential for preventing and treating neurodegenerative diseases. However, most research in this area has been limited to cellular studies. Therefore, it is essential to conduct further in vitro and in vivo investigations to verify their efficacy and explore the underlying mechanisms (Guo et al. 2023).

5.4 Safety Issues

Genetic engineering offers the means to enhance, introduce, or eliminate specific traits in probiotic bacteria. While numerous GEP have demonstrated significant efficacy in animal models for various diseases, the majority remain in the research phase, with only a limited few progressing to clinical trials. Safety concerns regarding the use of GEP in clinical settings represent a significant issue. The application of genetic engineering to microorganisms intended for human use is tightly regulated and subjected to rigorous controls by legal frameworks. Furthermore, in certain countries or by certain social organizations and groups, the use of genetic engineering may not be authorized (Bravo and Landete 2017). Additional safety considerations involve the persistence and proliferation of GEP in the external environment. Biocontainment strategies can address this by genetically modifying probiotics to replicate solely under permissive conditions. One approach involves introducing a death gene into GEP, prompting them to produce self-destructive agents in response to environmental stimuli (Lee et al. 2018).

Using GEP for drug delivery and treatment offers several benefits over traditional methods. These include reduced costs due to the mass production of engineered bacteria, fewer side effects due to their non-aggressive nature, and lower dosage requirements compared to biological compounds. However, before clinical trials and human use, it is essential to ensure genetic stability, avoid antibiotic-resistant genes, and maintain genetic integrity for high-concentration production (Mahdizade Ari et al. 2024). To enhance genetic stability, GEP typically incorporate recombinant DNA into their chromosomes rather than plasmids, which can be lost during cell division without selective pressure (Meng et al. 2023). Furthermore, by deleting an essential gene and incorporating a single copy of that gene on a plasmid within the host, induction and suppression systems can be employed for long-term plasmid maintenance, addressing genetic stability issues. A previously reported example involves an anaerobic-inducible promoter controlling enzyme expression in *E. coli* Nissle 1917. In this system, the target gene remains inactive while the probiotics grow in a bioreactor but gets activated under specific conditions when the probiotics enter the human body, leading to the expression of therapeutic genes (Aggarwal et al. 2020).

5.5 Conclusion

The rise in diseases linked to gut microbiome and dysbiosis, such as inflammatory bowel disease and obesity, underscores the need for new medications and therapies. Recent advancements in synthetic biology and metabolic engineering have facilitated the creation of new probiotic strains with specific functions, such as delivering drugs to the gut or modulating immune responses. Furthermore, scientists are gaining a deeper understanding their mechanisms by which these engineered probiotics interact with the host and exert their therapeutic effects. This progress has greatly advanced the use of engineered probiotics for diagnosing and treating various diseases. As gene editing technologies improve and our understanding of the human microbiome deepens, we expect to develop more effective probiotics tailored for disease prevention and overall human health. However, GEP are still in their early stages and face several challenges. It is crucial to study their environmental and climate impacts, genetic stability, and safety. Additionally, navigating regulatory hurdles will be essential for clinical translation. These issues need resolution before clinical use. GEP hold potential for future use in medical treatments, not as primary solutions, but as additional or supportive therapies that can work alongside other treatments to improve overall health outcomes.

References

Aggarwal N, Breedon AME, Davis CM, Hwang IY, Chang MW (2020) Engineering probiotics for therapeutic applications: recent examples and translational outlook. Curr Opin Biotechnol 65:171–179. https://doi.org/10.1016/j.copbio.2020.02.016. Elsevier Ltd

Aguilar C, Vanegas C, Klotz B (2011) Antagonistic effect of Lactobacillus strains against Escherichia coli and Listeria monocytogenes in milk. J Dairy Res 78(2):136–143. https://doi.org/10.1017/S0022029910000877

Berg P, Mertz JE (2010) Personal reflections on the origins and emergence of recombinant DNA technology. Genetics 184(1):9–17. https://doi.org/10.1534/genetics.109.112144

Borrero J, Chen Y, Dunny GM, Kaznessis YN (2015) Modified lactic acid bacteria detect and inhibit multiresistant enterococci. ACS Synth Biol 4(3):299–306. https://doi.org/10.1021/sb500090b

Bravo D, Landete JM (2017) Genetic engineering as a powerful tool to improve probiotic strains. Biotechnol Genet Eng Rev 33(2):173–189. https://doi.org/10.1080/02648725.2017.1408257

Cassini A, Högberg LD, Plachouras D, Quattrocchi A, Hoxha A, Simonsen GS, Colomb-Cotinat M, Kretzschmar ME, Devleesschauwer B, Cecchini M, Ouakrim DA, Oliveira TC, Struelens MJ, Suetens C, Monnet DL, Strauss R, Mertens K, Struyf T, Catry B et al (2019) Attributable deaths and disability-adjusted life-years caused by infections with antibiotic-resistant bacteria in the EU and the European Economic Area in 2015: a population-level modelling analysis. Lancet Infect Dis 19(1):56–66. https://doi.org/10.1016/S1473-3099(18)30605-4

Cecarini V, Bonfili L, Gogoi O, Lawrence S, Venanzi FM, Azevedo V, Mancha-Agresti P (2020) Neuroprotective effects of p62(SQSTM1)-engineered lactic acid bacteria in Alzheimer's disease: a pre-clinical study. Aging (Albany NY) 12(16):15995–16020

Chung Y, Ryu Y, An BC, Yoon YS, Choi O, Kim TY, Yoon J, Ahn JY, Park HJ, Kwon SK, Kim JF, Chung MJ (2021) A synthetic probiotic engineered for colorectal cancer therapy modulates gut microbiota. Microbiome 9(1). https://doi.org/10.1186/s40168-021-01071-4

Cook DP, Cunha JPMCM, Martens PJ, Sassi G, Mancarella F, Ventriglia G, Sebastiani G, Vanherwegen AS, Atkinson MA, Van Huynegem K, Steidler L, Caluwaerts S, Rottiers P, Teyton L, Dotta F, Gysemans C, Mathieu C (2020) Intestinal delivery of proinsulin and IL-10 via Lactococcus lactis combined with low-dose anti-CD3 restores tolerance outside the window of acute type 1 diabetes diagnosis. Front Immunol 11. https://doi.org/10.3389/fimmu.2020.01103

Dang AT, Marsland BJ (2019) Microbes, metabolites, and the gut–lung axis. Mucosal Immunol 12(4):843–850. https://doi.org/10.1038/s41385-019-0160-6. Nature Publishing Group

Drucker DJ (2018) Mechanisms of action and therapeutic application of glucagon-like peptide-1. Cell Metab 27(4):740–756. https://doi.org/10.1016/j.cmet.2018.03.001. Cell Press

Fang X, Zhou X, Miao Y, Han Y, Wei J, Chen T (2020) Therapeutic effect of GLP-1 engineered strain on mice model of Alzheimer's disease and Parkinson's disease. AMB Express 10(1). https://doi.org/10.1186/s13568-020-01014-6

Fujimori M (2006) Genetically engineered bifidobacterium as a drug delivery system for systemic therapy of metastatic breast cancer patients. Breast Cancer 13(1):27–31

Gigola G, Carriere P, Díaz MBN, Perdigon G, Zwenger AO, Gentili C (2021) Survival effect of probiotics in a rat model of colorectal cancer treated with capecitabine. World J Gastrointest Oncol 13(10):1518–1531. https://doi.org/10.4251/wjgo.v13.i10.1518

Goh YJ, Barrangou R (2019) Harnessing CRISPR-Cas systems for precision engineering of designer probiotic lactobacilli. Curr Opin Biotechnol 56:163–171. https://doi.org/10.1016/j.copbio.2018.11.009. Elsevier Ltd

Guo J, Zhou B, Niu Y, Liu L, Yang L (2023) Engineered probiotics introduced to improve intestinal microecology for the treatment of chronic diseases: present state and perspectives. J Diabetes Metab Disord 22(2):1029–1038. https://doi.org/10.1007/s40200-023-01279-1. Springer Science and Business Media Deutschland GmbH

Han H, Zhang Y, Tang H, Zhou T, Khan A (2024) A review of the use of native and engineered probiotics for colorectal cancer therapy. Int J Mol Sci 25(7):3896. https://doi.org/10.3390/ijms25073896

Hill C, Guarner F, Reid G, Gibson GR, Merenstein DJ, Pot B, Morelli L, Canani RB, Flint HJ, Salminen S, Calder PC, Sanders ME (2014) Expert consensus document: the international scientific association for probiotics and prebiotics consensus statement on the scope and appropriate use of the term probiotic. Nat Rev Gastroenterol Hepatol 11(8):506–514. https://doi.org/10.1038/nrgastro.2014.66

Ji P, An B, Jie Z, Wang L, Qiu S, Ge C, Wu Q, Shi J, Huo M (2023) Genetically engineered probiotics as catalytic glucose depriver for tumor starvation therapy. Mater Today Bio 18. https://doi.org/10.1016/j.mtbio.2022.100515

Kaeid Sharaf L, Shukla G (2018) Probiotics (Lactobacillus acidophilus and Lactobacillus rhamnosus GG) in conjunction with celecoxib (selective COX-2 inhibitor) modulated DMH-induced early experimental colon carcinogenesis. Nutr Cancer 70(6):946–955. https://doi.org/10.1080/01635581.2018.1490783

Karimi O, Salvador Pen A (2008) Indications and challenges of probiotics, prebiotics, and synbiotics in the management of arthralgias and spondyloarthropathies in inflammatory bowel disease. http://journals.lww.com/jcge

Kaur IP, Deol PK (2020) Probiotic research in therapeutics: volume 1: applications in cancers and immunological diseases. In: Probiotic research in therapeutics: volume 1: applications in cancers and immunological diseases, vol 1. Springer Singapore. https://doi.org/10.1007/978-981-15-8214-1

Lee JW, Chan CTY, Slomovic S, Collins JJ (2018) Next-generation biocontainment systems for engineered organisms. Nat Chem Biol 14(6):530–537. https://doi.org/10.1038/s41589-018-0056-x

Leenay RT, Vento JM, Shah M, Martino ME, Leulier F, Beisel CL (2019) Genome editing with CRISPR-Cas9 in lactobacillus plantarum revealed that editing outcomes can vary across strains and between methods. Biotechnol J 14(3). https://doi.org/10.1002/biot.201700583

Liu M, Li S, Zhang Q, Xu Z, Wang J, Sun H (2018) Oral engineered Bifidobacterium longum expressing rhMnSOD to suppress experimental colitis. Int Immunopharmacol 57:25–32. https://doi.org/10.1016/j.intimp.2018.02.004

Liu Y, Feng J, Pan H, Zhang X, Zhang Y (2022) Genetically engineered bacterium: principles, practices, and prospects. Front Microbiol 13. https://doi.org/10.3389/fmicb.2022.997587. Frontiers Media S.A

Ma J, Lyu Y, Liu X, Jia X, Cui F, Wu X, Deng S, Yue C (2022) Engineered probiotics. Microb Cell Factories 21(1). https://doi.org/10.1186/s12934-022-01799-0. BioMed Central Ltd

Mahdizade Ari M, Dadgar L, Elahi Z, Ghanavati R, Taheri B (2024) Genetically engineered microorganisms and their impact on human health. Int J Clin Pract 2024. https://doi.org/10.1155/2024/6638269. Hindawi Limited

Mansour NM, Abdelaziz SA (2016) Oral immunization of mice with engineered Lactobacillus gasseri NM713 strain expressing Streptococcus pyogenes M6 antigen. Microbiol Immunol 60(8):527–532. https://doi.org/10.1111/1348-0421.12397

Mao N, Cubillos-Ruiz A, Cameron DE, Collins JJ (2018) Probiotic strains detect and suppress cholera in mice. Sci Transl Med 10. https://www.science.org

Marcobal A, Liu X, Zhang W, Dimitrov AS, Jia L, Lee PP, Fouts TR, Parks TP, Lagenaur LA (2016) Expression of human immunodeficiency virus type 1 neutralizing antibody fragments using human vaginal lactobacillus. AIDS Res Hum Retrovir 32(10–11):964–971. https://doi.org/10.1089/aid.2015.0378

Marosvölgyi T, Mintál K, Farkas N, Sipos Z, Makszin L, Szabó É, Tóth A, Kocsis B, Kovács K, Hormay E, Lénárd L, Karádi Z, Bufa A (2024) Antibiotics and probiotics-induced effects on the total fatty acid composition of feces in a rat model. Sci Rep 14(1). https://doi.org/10.1038/s41598-024-57046-6

Mathipa MG, Thantsha MS (2017) Probiotic engineering: towards development of robust probiotic strains with enhanced functional properties and for targeted control of enteric pathogens. Gut Pathog 9(1). https://doi.org/10.1186/s13099-017-0178-9. BioMed Central Ltd

Mazhar SF, Afzal M, Almatroudi A, Munir S, Ashfaq UA, Rasool M, Raza H, Munir HMW, Rajoka MSR, Khurshid M (2020) The prospects for the therapeutic implications of genetically engineered probiotics. J Food Qual 2020. https://doi.org/10.1155/2020/9676452

Medina E, Pieper DH (2016) Tackling threats and future problems of multidrug-resistant bacteria. Curr Top Microbiol Immunol 398:3–33. https://doi.org/10.1007/82_2016_492

Meng J, Liu S, Wu X (2023) Engineered probiotics as live biotherapeutics for diagnosis and treatment of human diseases. Crit Rev Microbiol. https://doi.org/10.1080/1040841X.2023.2190392. Taylor and Francis Ltd

Mohammadzadeh I, Qujeq D, Yousefi T, Ferns GA, Maniati M, Vaghari-Tabari M (2020) CRISPR/Cas9 gene editing: a new therapeutic approach in the treatment of infection and autoimmunity. IUBMB Life 72(8):1603–1621. https://doi.org/10.1002/iub.2296. Blackwell Publishing Ltd

Naeem H, Hassan HU, Shahbaz M, Imran M, Memon AG, Hasnain A, ... & Al Jbawi E (2024) Role of probiotics against human cancers, inflammatory diseases, and other complex malignancies. J Food Biochem 2024(1):6632209

Oh JH, Van Pijkeren JP (2014) CRISPR-Cas9-assisted recombineering in Lactobacillus reuteri. Nucleic Acids Res 42(17). https://doi.org/10.1093/nar/gku623

Oh J-H, Schueler KL, Stapleton DS, Alexander LM, Yen C-LE, Keller MP, Attie AD, van Pijkeren J-P (2020) Secretion of recombinant Interleukin-22 by engineered Lactobacillus reuteri reduces fatty liver disease in a mouse model of diet-induced obesity. mSphere 5(3). https://doi.org/10.1128/msphere.00183-20

Parvin T, Sadras SR (2024) Advanced probiotics: bioengineering and their therapeutic application. Mol Biol Rep 51(1). https://doi.org/10.1007/s11033-024-09309-8. Springer Science and Business Media B.V

Plavec TV, Kuchař M, Benko A, Lišková V, Černý J, Berlec A, Malý P (2019) Engineered Lactococcus lactis secreting IL-23 receptor-targeted REX protein blockers for modulation of IL-23/Th17-mediated inflammation. Microorganisms 7(5). https://doi.org/10.3390/microorganisms7050152

Powell-Wiley TM, Poirier P, Burke LE, Després JP, Gordon-Larsen P, Lavie CJ, Lear SA, Ndumele CE, Neeland IJ, Sanders P, St-Onge MP (2021) Obesity and cardiovascular disease a scientific statement from the American Heart Association. Circulation 143(21):E984–E1010. https://doi.org/10.1161/CIR.0000000000000973. Lippincott Williams and Wilkins

Preisser TM, Da Cunha VP, Santana MP, Pereira VB, Cara DC, Souza BM, Miyoshi A (2021) Recombinant Lactococcus lactis carrying IL-4 and IL-10 coding vectors protects against type 1 diabetes in NOD mice and attenuates insulitis in the STZ-induced model. J Diabetes Res 2021. https://doi.org/10.1155/2021/6697319

Sasaki T, Fujimori M, Hamaji Y, Hama Y, Ito KI, Amano J, Taniguchi S (2006) Genetically engineered Bifidobacterium longum for tumor-targeting enzyme-prodrug therapy of autochthonous mammary tumors in rats. Cancer Sci 97(7):649–657. https://doi.org/10.1111/j.1349-7006.2006.00221.x

Upadrasta A, Stanton C, Hill C, Fitzgerald GF, Ross RP (2011) Improving the stress tolerance of probiotic cultures: recent trends and future directions. In: Stress responses of lactic acid bacteria. Springer US, pp 395–438. https://doi.org/10.1007/978-0-387-92771-8_17

Van Oort S, Beulens JWJ, Van Ballegooijen AJ, Grobbee DE, Larsson SC (2020) Association of cardiovascular risk factors and lifestyle behaviors with hypertension: a Mendelian randomization study. Hypertension 76(6):1971–1979. https://doi.org/10.1161/HYPERTENSIONAHA.120.15761

Wang XL, Chen WJ, Jin R, Xu X, Wei J, Huang H, Tang YH, Zou CW, Chen TT (2023) Engineered probiotics Clostridium butyricum-pMTL007-GLP-1 improves blood pressure via producing GLP-1 and modulating gut microbiota in spontaneous hypertension rat models. Microb Biotechnol 16(4):799–812. https://doi.org/10.1111/1751-7915.14196

Yue M, Wei J, Chen W, Hong D, Chen T, Fang X (2022) Neurotrophic role of the next-generation probiotic strain L. lactis MG1363-pMG36e-GLP-1 on Parkinson's disease via inhibiting ferroptosis. Nutrients 14(22). https://doi.org/10.3390/nu14224886

Zhang M, Eshraghian EA, Jammal OA, Zhang Z, Zhu X (2021) CRISPR technology: the engine that drives cancer therapy. Biomed Pharmacother 133. https://doi.org/10.1016/j.biopha.2020.111007. Elsevier Masson s.r.l

Zheng R, Li Y, Wang L, Fang X, Zhang J, He L, Yang L, Li D, Geng H (2020) CRISPR/Cas9–mediated metabolic pathway reprogramming in a novel humanized rat model ameliorates primary hyperoxaluria type 1. Kidney Int 98(4):947–957. https://doi.org/10.1016/j.kint.2020.04.049

Probiotic, Prebiotic, and Gut–Brain Connection in Cancer Development

6

K. R. Arjun, D. Mutthuraj, Neha Foferkar, Dhanush Nagendra, R. Mythreyi, S. Umamaheshwari, and Kanthesh M. Basalingappa

Abstract

Being the fourth most prevalent malignant tumor, colorectal cancer ranks second in terms of cancer-related death. Cancer's primary characteristics are evading apoptosis and cell mutation. Preventing carcinogenesis evidently requires maintaining a delicate balance in the relationship between damage to DNA and DNA repair. A good diet, maintaining a healthy weight, and engaging in physical activity might potentially prevent one-third of cancer cases. Because of differences in nutrition and eating habits as well as occupational dangers, colorectal cancer has been on the rise in emerging nations. According to recent research, the makeup of the gut microbiota may have a role in the development of colorectal cancer and other dysbioses of the gut. By modifying the gut microbiota using probiotics and prebiotics, either independently or together, inflammation may be prevented. The immune system's interaction with the microbiome might benefit from this. In recent decades, there has been growing evidence supporting the role of the brain–gut–microbiota axis in the process of tumor development. Numerous functions include immunological response, digestion, immune system maintenance, cortisol modulation, and control over neurotransmitter metabolism (e.g., serotonin, dopamine, and noradrenaline, and gamma-aminobutyric acid (GABA), serotonin). Changes in the gut's microbiota can upset homeostasis and lead to dysbiosis, which has been linked to colorectal cancer. Treatments like chemo-

K. R. Arjun · D. Mutthuraj · N. Foferkar · D. Nagendra · R. Mythreyi · K. M. Basalingappa (✉)
Division of Molecular Biology, School of Life Sciences,
JSS Academy of Higher Education & Research, Mysuru, Karnataka, India
e-mail: mutthurajd@jssuni.edu.in; kantheshmb@jssuni.edu.in

S. Umamaheshwari
Department of Microbiology, School of Life Sciences,
JSS Academy of Higher Education & Research, Mysuru, Karnataka, India
e-mail: umamaheshwari@jssuni.edu.in

S. Pathak et al. (eds.), *Microbiota and Dietary Mediators in Colon Cancer Prevention and Treatment*, https://doi.org/10.1007/978-981-96-0297-1_6

therapy, radiation, and medication can be significantly impacted by the microbiota's composition.

Keywords

Colorectal cancer · Gut microbiome · Probiotic · Prebiotic

Abbreviations

AMPA receptor	α-amino-3-hydroxy-5-methyl-4-isoxazolepropionic acid receptor
CA	Cholic acid
CDCA	Chenodeoxycholic acid
CNS	Central nervous system
CRC	Colorectal cancer
DC	Dendritic cells
DCA	Deoxycholic acid
DNA	Deoxyribonucleic acid
ENS	Enteric nervous system
ETBF	Enterotoxigenic *Bacteroides fragilis*
FMT	Fecal microbiota transplantation
GABA	Gamma-aminobutyric acid
GI tract	Gastrointestinal tract
GM	Gut microbiota
GOS	Galacto-oligosaccharides
IBD	Inflammatory bowel disease
IBS	Irritable bowel syndrome
ICI	Immune checkpoint inhibitor
IFN-γ	Interferon-γ
IgA	Immunoglobulin A
ISAPP	The International Scientific Association of Probiotics and Prebiotics
L1CAM	L1 cell adhesion molecule
LAB	Lactic acid bacteria
LCA	Lithocholic acid
NGPs	Next-generation probiotics
NK cells	Natural killer cells
NO synthase	Nitric oxide synthase
PAH	Polycyclic aromatic hydrocarbons
PD L1	Programmed cell death ligand 1
Pks	Polyketide synthase
PNI	Perineural invasion
PNS	Peripheral nervous system
qRT–PCR	Quantitative real-time polymerase chain reaction

rRNA	Ribosomal ribonucleic acid
SCFAs	Short-chain fatty acids
TLR/MyD88	Toll-like receptors and myeloid differentiation primary response protein 88

6.1 Introduction to Gut Microbiome: Probiotics & Prebiotics

The composition of the gut microbiota (GM) varies among individuals, influenced by a combination of intrinsic factors, such as genetics and age, and extrinsic factors, including environment, nutrition, and lifestyle. A deficiency or dysbiosis in the gut microbiota can increase intestinal permeability, leading to higher levels of circulating lipopolysaccharides, which may trigger widespread infections. Additionally, an overgrowth of specific bacterial genera or pathogens can alter the immune response, contributing to the development of various illnesses. These processes might worsen gastrointestinal symptoms or reduce glucose tolerance in metabolic illnesses. Dietary factors play a major role in genetic modification (GM), and prebiotics and probiotics are two powerful nutrients that can help rebuild the natural microbiota. The term "microbiota" describes all of the microorganisms present in a certain area. With about 2000 different bacterial species, the microbiota of the stomach has the greatest number of microorganisms (Ballan et al. 2020). Thus, functional foods appear to be intriguing substitutes for dietary treatments aimed at enhancing the microbiota's structure, as they must provide a clear health benefit to the user alongside their nutritious qualities. Prebiotics, probiotics, polyphenols, and polyunsaturated fatty acids are the most well-known functional foods or substances. "Live bacteria that, when supplied in adequate amounts, provide a medical advantage on the host" is the definition of probiotics. Although there has been disagreement on the precise definition of probiotics, this description was agreed upon by experts in 2013. Probiotics have a variety of main ways of working on human health, such as competing with pathogens for nutrients and adhesion sites, producing bacteriocin, vitamins, and short-chain fatty acids (SCFA), enhancing the intestinal barrier, immunomodulating the body, and producing neurotransmitters (Yan et al. 2018). The gut microbiota is important for sustaining host health because it may deliver different nutrients, control energy balance, alter the immune system, and strengthen defenses against infections. In the early 1900s, Metchnikoff presented the scientific community with the notion of probiotics, positing that the longevity of Bulgarian peasants was due to their consumption of fermented milk that included probiotics, or bacteria that promote health. Additionally, all probiotics had to be "safe for their intended use." The International Scientific Association of Probiotics and Prebiotics (ISAPP) then restated these standards in a 2018 policy statement. Thus far, 35 species or subspecies of probiotics have been identified as safe for use in food. These have been categorized into three groups: yeast, spore-forming Bacillus species, and bacteria that produce lactic acid (*Lactobacillus, Bifidobacterium*, and *Enterococcus*). Typically, dairy products or healthy human colons are used to isolate *Bifidobacterium* and *Lactobacillus* species (Liu et al. 2022).

Gibson and Roberfroid first proposed the idea of prebiotics when they saw that *bifidobacteria* would ferment some non-digestible oligosaccharides alone. Prebiotics were initially described as a specifically fermented component that modifies the gastrointestinal microbiota's activity and/or composition, improving host health in the process. Subsequently, given advances in our understanding of the makeup of the gut microbiota and the underlying mechanisms of prebiotics, an ISAPP consensus panel led by Gibson updated the concept of prebiotics, or in 2017 to include "a substrate that is selectively utilized by host microorganisms conferring a health benefit." Prebiotics such fructooligosaccharides, galactooligosaccharides (GOS), and lactulose are being demonstrated to significantly change the balance of the microbial population in the large intestine by increasing the amount of *bifidobacteria*. Pectin, cellulose, and xylan are among the other nutrients that support the balance of gut microorganisms. Mineral adsorption, immune regulation, and lipid metabolism are all improved by prebiotics. Additionally, they play a big part in preventing cancer, metabolic disorders, constipation, and diarrhea (Sheveleva 1999; Scott et al. 2020). The risk factors of colorectal cancer are diagrammatically represented in (Fig. 6.1).

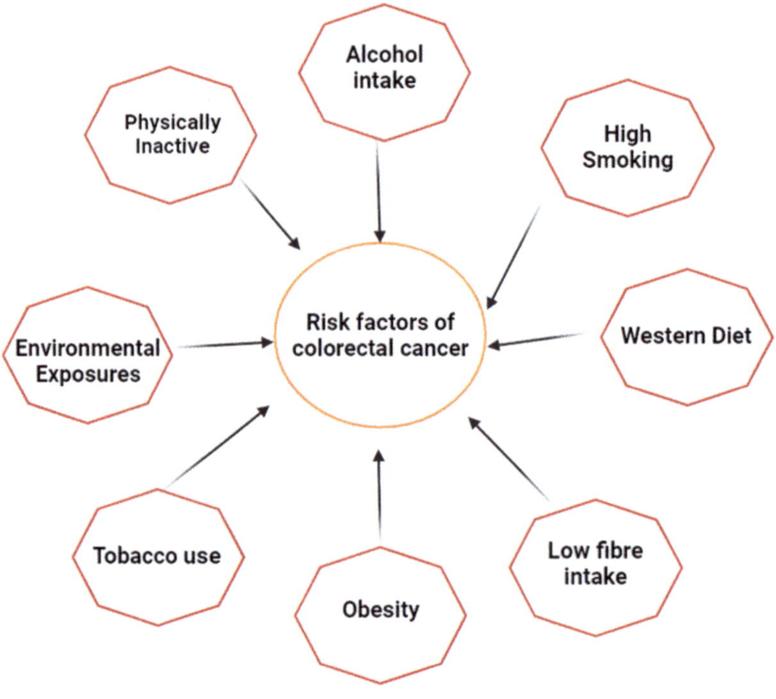

Fig. 6.1 Diagrammatic representation of risk factors associated with colorectal cancer

6.2 Gut–Microbiome Connection in Carcinogenesis of Colorectal Cancer

It is well known that early inflammation and signaling pathway modification by the gut bacteria contribute significantly to CRC carcinogenesis. Given their ability to detect CRC and predict clinical outcome, bacterial biomarkers are relevant for prognostication. The development of cancer involves a complex interplay of the immune system, malignant microbiota, and gut microbiota. The gut microbiome is in a healthy condition known as eubiosis when pathogenic and anti-inflammatory cytokines are balanced, immunity cells and IgA production are in balance, the epithelial barrier and mucus layer are intact, and there is biodiversity in the gut microbiota. In dysbiosis, these elements are out of balance. Furthermore, the cancer microbiome adversely affects the gut microbiome, which limits the systemic therapeutic efficacy of immunotherapy and chemotherapy as well as insufficient host immune system responses both locally and systemically (Rebersek 2021) as mentioned in (Table 6.1).

6.2.1 Microbes Connected to the CRC

A functional gut microbiome has a distinct makeup than the CRC microbiota. In addition to strains specifically linked to CRC, such as *Bacteroides fragilis*, *Streptococcus gallolyticus*, *Enterococcus faecalis*, and *Escherichia coli*, it includes newly discovered bacteria strains associated with CRC, such as *Fusobacterium nucleatum*, Parvimonas, Peptostreptococcus, Porphyromonas, and Prevotella. Elevated levels of these bacterial isolates in stool and tumor samples from CRC patients might serve as biomarkers for the disease. CRC is influenced by the gut microbiota in a variety of ways, such as immune response regulation, inflammation, and modified dietary component metabolism. Toxic microbial by-products like metabolites or genotoxins may potentially be produced as a consequence of these activities. Bacteria are classified as driving bacteria when they cause cancer directly, or passenger bacteria when they cause cancer indirectly. The latter are opportunistic

Table 6.1 Relationship between microbiome and its effects on colorectal cancer (Coleman and Nunes 2016)

Bacteria	Relationship with colorectal cancer
Fusobacterium	Rich in colon cancer in humans
Streptococcus bovis	An increase in the frequency of colon cancer patients
Slackia	Disproportionately found in Colorectal cancer patient tissue
Clostridium septicum	Colonic cancer and polyps are linked to aortic infections
Escherichia coli	Increased invasion and adhesion in malignancies related to colorectal cancer
Collinsella	More prevalent in the tissue of individuals with colorectal cancer
Enterococcus faecalis	Much greater groups of people with colorectal cancer

microorganisms that proliferate in the environment surrounding tumors. The activation of malignant pathways of signaling, which lead to molecular changes and the eventual appearance of CRC, is influenced by interactions between bacteria and hosts. In the context of CRC, these mechanistic elements may be altered for preventative or therapeutic objectives as represented in (Fig. 6.1). Studies on colorectal adenomas precursors of CRC and the gut microbiota have also been conducted. Numerous characteristics, including the existence of certain bacterial strains, changes in their composition, and the abundance of particular strains including fungal strains in colorectal adenomas have been identified by studies as being particular and significant in the development of CRC. Two examples of bacterial strains that are only detected in large quantities in the cases of adenomas and intramucosal carcinomas are *Atopobium parvulum* and *Actinomyces odontolyticus*. In the early stages of colorectal cancer up till metastatic illness, other bacterial strains such *Fusobacterium nucleatum* and *Solobacterium moorei* are prevalent. Ahn and colleagues found that in their isolated deoxyribonucleic acid (DNA) from faecal samples, people with colorectal cancer (CRC) had an abundance of Bacteroides, Fusobacterium, Atopobium, and Porphyromonas phyla but a shortage of Firmicutes (Rebersek 2021; Wong and Yu 2019). These researchers also identified a study flaw: mucosal adherent gut bacteria were not investigated, despite the possibility that they are more strongly associated with colon cancer than bacteria found in feces. The findings of this study may 1 day make it possible to identify CRC precursors early on and halt the disease's progression. The bacterial strains *Bacteroides fragilis*, *Escherichia coli*, *Enterococcus faecalis*, and *Streptococcus gallolyticus* are most closely linked to CRC. Each of these strains is linked to CRC, and it has been shown that people with CRC had greater amounts of these strains in their feces and tumor samples. *Fusobacterium nucleatum*, Parvimonas, Peptostreptococcus, Porphyromonas, and Prevotella. These bacteria's strains have recently been studied by metagenomic and 16S rRNA gene profiling made possible by next-generation sequencing technologies. Prior research on these strains included culture-based techniques and quantitative real-time polymerase chain reaction (qRT–PCR) using DNA recovered from patient stool samples and colon tissue biopsies (Ahn et al. 2013; Ternes et al. 2020).

6.3 Effects of Nutrition and Lifestyle on Colorectal Cancer Susceptibility and Gut Microbiota Composition

Compared to healthy people, patients with CRC have a less varied microbiome. However, a meta-analysis of metagenomic data from various populations and cohorts revealed that the microbiomes of CRC were richer than those of controls, in part because of the growth of species that are normally obtained from the mouth cavity. Disparities in the quantity of particular microorganisms have been noted when comparing the tumor and surrounding non-tumor mucosa, as well as when stool samples from CRC patients are compared to controls. The microbiome has an etiological and diagnostic role in colorectal neoplasia, as evidenced by specific alterations in the metabolome and microbiome that occur during distinct stages of the illness, from adenomatous polyps to early-stage cancer to metastatic disease (Song et al. 2020). The microorganisms are linked to different mucosal gene expression patterns that may aid in the development of CRC. Interestingly, samples from the oral and colonic mucosal surfaces of those with colonic neoplasia and controls showed comparable networks of oral bacteria. These results corroborate the hypothesis that periodontal disease increases the risk of CRC. In addition, smoking may alter the makeup of the oral microbiome, which is a significant risk factor for CRC. Smokers exhibit higher relative abundances of the Veillonellaceae family and lower relative abundances of Neisseria. Furthermore, there are pathogenic effects of bacterial metabolic activity on oral and systemic health. Therefore, changes to the intestinal and oral microbiomes may each promote the spread of microbes and cause inflammation, which in turn may lead to CRC (Kato et al. 2016).

There is also a connection between CRC and the mycobiome (fungal microbiome) and enteric virome. Increased diversity in bacteriophage viromes was observed in CRC cases when compared to controls. This was associated with a reduction in the variety of bacteria and a minor enrichment of specific viral taxa, such as those belonging to the Inovirus and Tunalikevirus families, whose bacterial hosts (such as ETBF, *F. nucleatum*, and pks+ E. coli) have been connected to CRC. These findings point to a connection between virome and bacteriome in CRC risk. Patients with CRC in different stages as opposed to controls have different mycobiome patterns. Ecological examination of fecal samples from CRC patients revealed more co-exclusive connections between fungal and bacterial species than in control samples, supporting preclinical findings that bacteria and fungus have an antagonistic relationship during the formation of CRC (Malik et al. 2018). Although the exact mechanisms underlying alcohol's carcinogenic effects are unknown, some are known or suspected to include: (a) alcohol serving as a solvent for carcinogens (such as those found in tobacco smoke) to enter mucosal cells; (b) alcohol's reactive and genotoxic metabolites (acetaldehyde); (c) alcohol mediating prostaglandin and estrogen concentrations; (d) alcohol producing reactive nitrogen and oxygen species; (e) interactions with DNA repair and folate metabolism; and (f) heavy drinkers' diets having comparatively low levels of essential nutrients (Murphy et al. 2019). Stress is another lifestyle component that affects colonic motor activity through the gut–brain axis. This can change the profiles of the gut microbiota, particularly the proportion of potentially helpful Lactobacillus. One of the most prevalent functional bowel illnesses, IBS, may be exacerbated by stress because of its potential to alter microbial populations through the central nervous system (CNS). Because the gut–brain axis is bidirectional and involves both hormonal and neuronal pathways, changes to the gut microbiota may have an impact on mood and other aspects of brain function. Neurodevelopmental disorders like autism are linked to significant changes in gut microbial populations. Sedentary lifestyles and excessive calorie consumption are linked to obesity. Any changes in microbial communities linked to obesity may be significantly impacted by exercise. A recent study that examined the effects of exercise and food on elite athletes' gut microbial communities found an increase in variety. Changes in the gut microbiota of obese humans and animals are

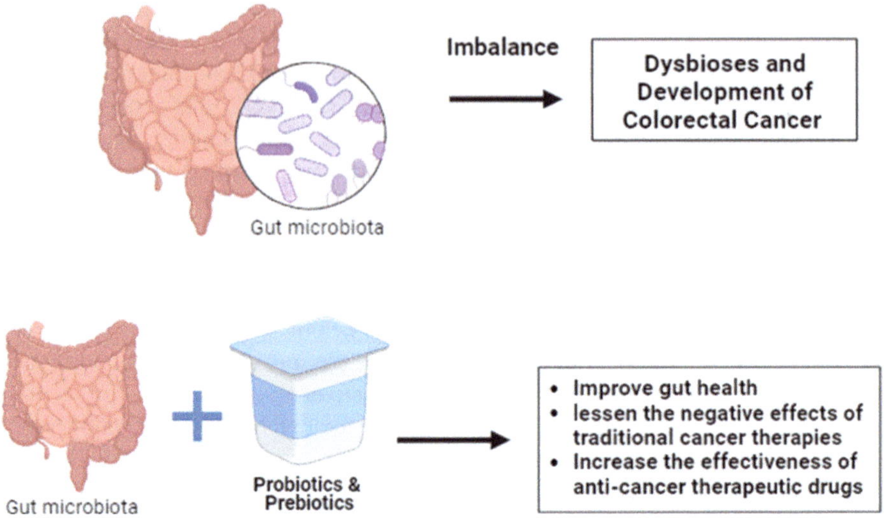

Fig. 6.2 Diagrammatic representation of significance of prebiotics and probiotics in CRC

associated with a decrease in Bacteroidetes and an increase in Firmicutes, which may enhance the absorption of energy and lead to obesity. On the other hand, other information points to high-fat, obesity-causing diets as the main cause of the changes in microbial communities. Whatever the reason, increased gut flora has been connected to unfavorable health effects as mentioned in (Fig. 6.2) (Boffetta and Hashibe 2006).

6.4 Gut–Brain Connection in Cancer Development

The link between the gastrointestinal tract and the brain in the pathogenesis, etiology, and pathophysiology of disease has been the subject of several research. The development of technology has shed important light on the intricate and intimate connection known as the gut-brain axis between the stomach and the brain in both healthy and pathological conditions. The brain and gut communicate in both directions to maintain gastrointestinal homeostasis and to influence emotion, motivation, and higher cognitive processes through neuronal routes as well as cellular and humoral mediators. This process is known as the gut–brain axis. Numerous investigations into intestinal and nervous system disorders have focused on the gut-brain axis. The impact of intestinal diseases on the microbiota of the digestive tract on the central nervous system have drawn more attention in recent years. Intestinal microbiota is linked to major intestinal illnesses, such as colorectal cancer (CRC), irritable bowel syndrome, and inflammatory bowel disease. The occipital and parietal lobes showed a spontaneous reduction in neural activity, according to the only brain imaging study that looked at changes in cerebral neural activity in CRC patients experiencing depressed symptoms (Tuan and Chen 2016). Visceral discomfort, sleeplessness, and mental distress particularly depression are common in CRC patients. These symptoms have a substantial negative impact on the patients' quality of life and are important risk factors for alterations in brain imaging. Moreover, there is a reciprocal relationship between emotional discomfort, sleeplessness, and visceral pain. Furthermore, a number of studies have revealed aberrant brain metabolism in cancer patients; the nature of these changes differs according to the kind of cancer. Nevertheless, no research has compared the patterns of brain activity in individuals with colon cancer and enteritis. Tumors and inflammation are tightly associated, and inflammation is a major factor in the development of tumors. The studies offered brain-gut crosstalk supported by neuroimaging stomach distension, irregular feces, and stomach discomfort are common symptoms among CRC patients. In response to different types of visceral stimulation, the brain receives signals from the stomach through the enteric afferent nerve system, which is mostly made up of the sympathetic and parasympathetic afferent systems. The sympathetic afferent neurons communicate to the sensory cortex, insula, and other brain regions that produce the conscious gut response and encode the location and intensity of visceral activation through the spinothalamic pathway, as seen in (Fig. 6.3). Subsequently, sensory and visceral pain signals are integrated and processed by the prefrontal cortex (Tuan and Chen 2016; Ma et al. 2022).

The GI tract's intrinsic and extrinsic neuronal circuits play a major role in regulating mammalian gut function. The neuronal cell bodies of the extrinsic innervation, which includes both parasympathetic and sympathetic projections as well as visceral afferents, are found outside the stomach. The intrinsic innervation is provided by the ENS, which is a part of the gut wall. The brainstem, prevertebral ganglia, and peripheral afferent ganglia contain the cell bodies of neuronal cells of extrinsic neurons that innervate the stomach. Fibers from external somatosensory and visceromotor nerves project to the gut following ENS precursor colonization of

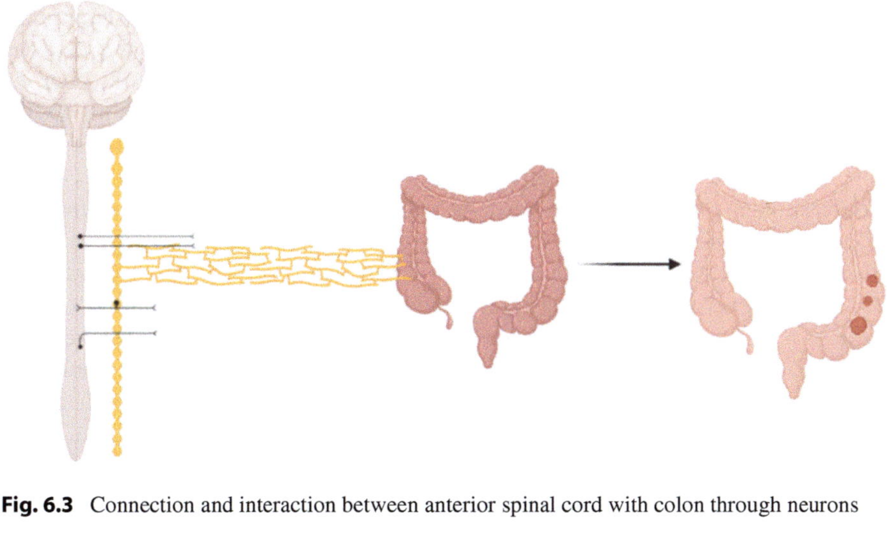

Fig. 6.3 Connection and interaction between anterior spinal cord with colon through neurons

the gut. These come from the sympathetic, dorsal root, pelvic ganglia, and vagal nerves. Parasympathetic fibers reach the gut by the spinal cord or vagus nerve, whereas sympathetic fibers normally follow blood arteries. Vagal innervation is guided during embryogenesis by chemoattractants and chemorepellents released by the developing mesenchyme, enteric neurons, and epithelium; nevertheless, it is still entirely unclear how spine afferent signals innervate the gut during development (Ma et al. 2022). Although the involvement of neurons in cancer is receiving a lot of attention, little is known about the neural networks that innervate the gut in CRC. In the context of perineural invasion, a passive function for neurons in CRC and other cancer types has long been understood. This procedure, which is connected to fast tumor development and metastasis, involves tumor cells using the nerve fibers as leading threads to travel to other places. PNI has been identified as a predictive indicator for both survival rates and the aggressiveness of CRC. High-grade and poorly differentiated colorectal cancers are associated with PNI. Moreover, PNI can act as a stand-alone prognostic indicator and is linked to a lower disease-free survival for patient outcomes (Tomše et al. 2017). When compared to colon cancer, rectal malignancies exhibited a greater prevalence of PNI, which is associated with their increased aggressiveness and lower overall survival rates. The rectum is known to have greater extrinsic innervation than the colon, which may account for the increased prevalence of PNI in that area. Nonetheless, it is thought that external neural networks from the PNS cause PNI (Furness et al. 2014); there has been evidence of a physical connection between neurons in the intestinal tract and CRC cells. N-cadherin and L1CAM, which have been shown to be crucial molecules for CRC cell adhesion and migration, are expressed by enteric neurons. The fact that PNI was not entirely eliminated by deleting these molecules, however, indicates that additional (unknown) molecules must also be at play. Numerous forms of cancer have been reported to actively communicate with neural cells, pointing to several pathways for neuron–cancer communications. It has been demonstrated that nerves can affect carcinogenesis by experimental manipulation of neural signaling (Uesaka et al. 2016). Even though paracrine signaling plays a major role in these communication channels, recent research, such as that on gliomas, demonstrates that neurons and cancer cells may also form particular synaptic connections. Here, the neurons around the tumor release neuroligin-3, which encourages the glioma cells' AMPA receptors, a particular kind of glutamate receptor, to create synapses between neurons. When these neurons are activated, glutamate is released, which leads to the proliferation and invasion of glioma cells, which in turn accelerates the progression of cancer. Additionally, when AMPA receptors on glioma cells are activated, the glioma cells themselves release glutamate, which activates AMPA receptors in an autocrine manner and further activates neurons. Enteric neurons have been seen in close proximity to tumor cells, however functional connections between neurons and CRC cells are not yet supported by data. Furthermore, CRC has been linked to elevated glutamate receptor expression. It would be intriguing to see whether neurons and CRC cells develop synaptic-like connections in the future (Knijn et al. 2016; Kozlowska et al. 2016).

6.5 Mechanisms of Probiotic and Prebiotic Action in Colorectal Cancer Prevention

6.5.1 Probiotic Adhesion and the Competitive Removal of Harmful Gut Flora

Through the competitive exclusion of harmful bacteria implicated in carcinogenesis, probiotics may lower the risk of CRC. Studies on animals have demonstrated that eating lactobacilli and bifidobacteria can raise LAB and decrease fecal putrefactive microorganisms, such as coliforms, which have been linked to the colon's potential production of carcinogens. When the fermented milk was administered, the Enterobacteriaceae strains *Klebsiella pneumoniae* and *Proteus mirabilis*, which can cause colonic inflammation, decreased while butyrate-producing bacteria increased. Probiotics may be able to exclude pathogenic microflora through two different forms of competition selectively: the rivalry for nutrients and the competition for adhesion to the intestinal mucosa among the intestinal microbiota. Nutrient availability in the distal part of the colon is constrained. Certain ideas suggest that probiotics may outcompete other gut microbes for nourishment and grow at their cost. According to in vitro research, enteropathogens including *Salmonella typhimurium* and *Escherichia coli* were either prevented from adhering to the intestinal mucosa or prevented from doing so by *Lactobacillus acidophilus*, *Lactobacillus casei*, and *Lactobacillus plantarum* (Schonkeren et al. 2021; Veiga et al. 2010). An essential condition for the colonization of bacteria in the intestinal cavity is adhesion. It i noteworthy that this study found that subjected to heat *L. rhamnosus* CNCM-I-3698 and *L. farciminis* CNCM-I-3699 proved both more effective than their live forms in terms of pathogen exclusion and adhesion to intestinal mucosa. The scientists postulated that substances such as exopolysaccharides, which cells make as a heat-resistant barrier, would actually boost the adhesive capability of probiotics that have been heat killed. However, as reported by Tareb et al. (2013), certain investigations have demonstrated that heating causes a sharp decline in probiotics' adhesion capacity. Confirming the impact of heat treatment on probiotics' ability to adhere to intestinal mucosa is going to require more research (Tareb et al. 2013).

6.5.2 Changes in the Enzyme Activity of the Intestinal Microflora

The capacity of probiotics to lower intestinal microflora enzyme activity lends further credence to their chemopreventive involvement in CRC. The hepatic production of glucuronides is how the human body detoxifies medicines and other substances before they enter the intestines. Several glucuronides, including PAH (such as benzo[a]pyrene), a significant risk factor for CRC, may be hydrolyzed by the bacterial enzyme β-glucuronidase, which has a broad substrate specificity

(Rafter 2003). It has been demonstrated that altering this enzyme can alter the rat model of colonic carcinogenesis (Chong 2014). Similar to β-glucuronidase, β-glucosidase is a bacterial enzyme that hydrolyzes plant glycosides, such as cycasin, in the colon to yield aglycones, the majority of which have been linked to cancer. While nitroreductase converts N-nitro compounds (such nitrobenzenes and nitropyrenes) into amines, which are frequently mutagenic and carcinogenic, nitrate reductase aids in the synthesis of very toxic and carcinogenic nitrite. Because azo compounds are included in many pharmaceuticals and food coloring additives, azoreductase plays an important role in converting these molecules into carcinogenic aromatic amines. A number of mechanisms, including the competitive exclusion of pathogenic microflora, the ability of LAB to produce antibacterial substances (like bacteriocin) that inhibit the growth of other microorganisms, and the ability of LAB to produce acids (like lactic acids) that lower the pH of the colon, can all be used by probiotics to influence the amount of bacterial enzymes in the colon (Tareb et al. 2013).

6.5.3 Reduction of Secondary Bile Acids That Cause Cancer

A portion of the reason for this connection comes from bile acid metabolism. The liver produces more cholic acid (CA) and chenodeoxycholic acid (CDCA), the two primary bile acids, as dietary fat intake rises. These acids are necessary for the absorption and breakdown of lipids in the small intestine. Most of these main bile acids are absorbed again in the ileum and restored to the liver through the portal vein (i.e., enterohepatic circulation). However, a small percentage (20%) reach the intestinal lumen after leaving the enterohepatic circulation. Lithocholic acid (LCA) and deoxycholic acid (DCA), the secondary bile acids, are created in the colon by anaerobic bacteria that break down CA and CDCA. The bacterial enzyme 7α-dehydroxylase catalyzes the 7α-dehydroxylation step in this microbial biotransformation. LCA and DCA are both very mutagenic and genotoxic. Additionally, DCA has been linked to mucosal proliferation, which puts the host at higher risk of CRC. Secondary bile acids can cause mucosal proliferation through a number of different mechanisms, such as: (1) activating protein kinase C, which is involved in growth factors and tumorigenesis; and (2) damaging epithelial cells, which triggers compensatory cell renewal. Moreover, carcinogenic fecapentaenes found in feces may be biosynthesized more quickly by DCA and LCA. Constipation has been linked to prolonged colon transit times, which increases the susceptibility of the intestinal mucosa to mutagens in the lumen, such as secondary bile acids and fecapentaenes. These in vitro and animal results suggested that probiotics could lessen the body's synthesis of secondary bile. However, further clinical research is needed to show how probiotics affect the human bile acid composition and what that means for CRC (Tareb et al. 2013; Rafter 2003; Ou et al. 2012).

6.6 Modulation of Immune Responses by Probiotics and Prebiotics in Colorectal Cancer

The human gut is home to 100 trillion creatures, primarily bacteria, making it the densest ecosystem on record. The intestinal populations of bacteria, fungi, and viruses are collectively referred to as the gut microbiota, and their genomes as a whole as the gut microbiome. Given that human genomes may encode around 150 times as many genes as humans, the gut microenvironment may be thought of as a sophisticated bioreactor with a wide range of metabolic activity. The complex relationship between the human host and the bacterial community has developed over half a billion years of coevolution. During this time, the immune system and intestinal consortia have mutually affected each other's repertoires, allowing people to tolerate their gut microbiota. In addition to aiding in metabolism, gut microbiota affects immunology, inflammation, and tissue growth, which can either promote human health or illness (Sato et al. 2012). Research on the gut microbiome has advanced, demonstrating that the metabolome and gut bacteria differ significantly in how innate and adaptive immunity are modulated at the mucosal and systemic levels. Inflammatory intestinal tissue during CRC has a compromised barrier function, which facilitates bacterial translocation and the generation of pro-inflammatory cytokines in the tumor microenvironment. According to research by Uronis et al., the TLR/MyD88 pathway is responsible for the association between gut microbiota and the development of colitis-associated malignancy. They found a correlation between the formation of colorectal tumors and the severity of chronic colitis, as well as between the evolution of adenoma into invasive carcinoma and the inflammation caused by bacteria. According to Vannucci et al., germ-free rats showed greater anticancer immune response than conventional rats and formed fewer, smaller tumors. These differences were explained by the absence of physiological inflammation resulting from germs and the reduced antigenic challenges the rats faced. Probiotics also improve intestinal barrier integrity by secreting antioxidative and anticarcinogenic substances, and they prolong immunostimulatory action by generating anti-inflammatory cytokines (Ambalam et al. 2016). Probiotic research might benefit from using animal models of inflammation-induced colorectal cancer to better understand the potential advantages of probiotics and the molecular mechanisms underlying their probiosis. They demonstrated that probiotic mixes containing a cocktail of *Bifidobacterium* species, specifically *Bifidobacterium breve* and *Bifidobacterium longum*, enhanced tumor control to an equivalent extent as anti-PD-L1 therapy (checkpoint blockade); additionally, the combination treatment almost completely eliminated tumor growth and enhanced the effectiveness of a PD15 L1–blocking antibody against tumors. When *L. rhamnosus* was present, human monocyte-derived DC developed and showed a decreased ability to promote T-cell proliferation and inhibit CD3/CD28-stimulated cytokine production. This study demonstrated how several lactobacillus species might affect DC, which in turn stimulates the production of regulatory T cells. Fink and colleagues showed that strains of *Lactobacillus* and *Bifidobacterium*, alone or in combination, differently triggered NK/DC interactions through DC maturation and the NK cells' cytolytic capacity to generate interferon-γ (IFN-γ). According to in-vivo research, probiotics may be able to prevent colorectal cancer by modulating the immune system. Human clinical studies should be used to validate further findings (Fink et al. 2007).

6.7 Nutritional Interventions: Prebiotics, Probiotics, and Dietary Modifications in Colorectal Cancer Prevention

The organisms such as bacteria, archaea, and fungi that inhabit the digestive system are known as the gut microbiota, and they are involved in the control of growth hormone signaling, which is colonization resistance, immunological tolerance against infections, and other processes linked to normal infant development (Robertson 2020). Conversely, a prebiotic is a substrate that is specially used by host bacteria to provide health advantages. Prebiotics stimulate the development of greater than or equal to 1 certain bacteria, primarily lactobacilli and bifidobacterial, which results in the synthesis of SCFAs, which prevent the growth of pathogens, among other things. Prebiotics also increase gut barrier function, slow stomach emptying, boost the absorption of minerals, and influence the immune system (Pekmez et al. 2019). When taken in sufficient quantities, living bacteria found in probiotics offer health benefits to their host. Probiotics have varying health advantages depending on the strain. Through immune system and gut microbiota modification, pathogen growth inhibition, infection avoidance, decreased incidence of diarrhea, improved energy absorption, and improved absorption of many micronutrients, they may enhance a child's growth (Bermudez-Brito et al. 2012). For many years, researchers have studied the potential of probiotics and prebiotics to enhance digestive processes and lessen the symptoms of inflammatory and infectious diseases. "Live bacteria that are given in sufficient quantities bestow beneficial health effects on the host" is what probiotics are defined as. Prebiotics are indigestible carbohydrates that the gut bacteria ferments specifically to promote health outcomes (Gibson et al. 2010). More than a thousand distinct bacterial species make up the human gut microbiota, and at least 160 of these species reside in each unique host. This indicates a genetic repertoire that is more than 150 times larger than that of the hosts, Homo sapiens. Participants from the exact same geographical region typically have more comparable microbial populations compared to individuals from other places, despite the fact that the precise distribution of different kinds of bacteria in the gut differs substantially throughout people. Diet is assumed to have a major role in this clustering. Therefore, it's crucial to use particular tactics tailored to the target group when creating nutritional treatments for malnourished persons (De Filippo et al. 2010). Probiotics and prebiotics have been shown to enhance nutritional status by reducing the symptoms of diarrhea; moreover, there is mounting evidence that they may enhance the absorption of minerals (including calcium and iron) from diet. Micronutrients are either inorganic or organic substances, such vitamins or minerals, that are found in food in trace amounts and are vital for overall health even if they are not utilized as fuel. At least two billion individuals worldwide are thought to be deficient in some micronutrients. Most of these people live in poverty and do not eat enough meals high in nutrients. Eventually, countries will have to figure out how to make their nutritional policies better (Allen et al. 2009). In the short term, however, millions of children's health and cognitive abilities will be enhanced by dietary treatments provided via a variety of channels, including humanitarian initiatives. Many lives will also be saved. By enhancing the absorption and utilization of micronutrients through many routes, probiotics and prebiotics provide a possible solution for mitigating micronutrient deficits (Initiative 2009). The potential of prebiotics to enhance the absorption of micronutrients has been investigated; while results varied between studies, nondigestible oligosaccharides were found to have beneficial effects on minerals metabolic processes, bones composition, as well as structure of bones in both human and animal subjects. Prebiotic fermentation produces more short-chain fatty acids, which lowers pH, enhances mineral solubility, and increases the amount of enterocyte absorption surface. Additionally, certain probiotics seem to improve bone growth and micronutrient absorption (especially calcium and iron), however the benefits seem to be extremely strain-specific (Kelleher et al. 2002). A study that used a placebo control group found that giving young infants milk enriched using symbiotic (*Bifidobacterium lactis* HN019, oligosaccharide) for a year dramatically improved their iron status. Additionally, a placebo crossover controlled meal research found that probiotics (*Lactobacillus plantarum* 299v) significantly enhanced iron absorption in healthy women of reproductive age (Bering et al. 2006). Generalization is not achievable due to the wide range of probiotic strains, methodologies, models of animals, and patient groups utilized in these studies. It is obvious that further double blindness in randomized, placebo-controlled human trials is required before we can fully evaluate the possible benefits of probiotics and prebiotics to enhance the health of people with micronutrient deficiencies for certain subpopulations. A better understanding would ideally come from several trials in different patient groups utilizing the same procedures, strains, and supplements (Song et al. 2015).

A significant factor in the occurrence of colorectal cancer is diet. Consumption of various foods and nutrients has been associated with an increased risk of colorectal neoplasia throughout the last several decades, according to results from significant epidemiologic and experimental research. The consumption of red and processed food has been linked to an increased risk of colorectal cancer, whereas magnesium, fiber, dairy products, and whole grain products have been linked to a decreased risk. The possible chemopreventive benefits of sunlight, vitamin D, folate, fruits, and veggies are well supported by research. Foods and nutrients can interact as a lifestyle choice to affect the risk of CRC. CRC is probably influenced by diet through a number of interrelated pathways. These involve the direct impacts on inflammation and immunological response, as well as the indirect consequences of obesity and overnutrition, which are indicators of risk for CRC. A significant effector in the connection between nutrition and cancer is the gut microbiota, according to new research. Therefore, dietary change holds the potential to lower the incidence of CRC (Vallis and Wang 2022). One of the most common cancers in the world, colorectal cancer causes a high rate of morbidity and death in its victims. It is advantageous for both the public and individuals to identify modifiable risk factors that may be changed to lower the frequency and severity of CRC. Elevated levels of exercise and well-balanced meals high in vegetables and fruits, seafood, fiber from the diet, dairy, and all-important vitamins and minerals are protective factors in lifestyle against the development of colon cancer. A diet heavy in processed and red meat, consumption of alcohol, and tobacco use are risk factors for an increased incidence of CRC. Both the etiology and the proof pertaining to the particular nutrients such as vitamins and minerals that have an impact on colorectal pathogenesis are currently being investigated. Numerous epidemiological studies are being conducted to ascertain the impact of different lifestyle variables on the survival rate of CRC. In general, changing one's lifestyle is crucial to both the avoidance and treatment of CRC in order to improve patient outcomes (Veettil et al. 2021).

Exercise It is well known that improving overall health may be achieved by participating in moderate-intensity physical exercise for at least half an hour each day (McTiernan et al. 2019). High levels of daily physical exercise have been shown to have preventive benefits against CRC in several observational studies. There is compelling evidence to show a dose–response link between high levels of recreational physical exercise and a 20% relative risk reduction in the development of CRC. It has also been shown that CRC patients who engage in high levels of physical exercise after receiving a cancer diagnosis had a 42% reduction in both all-cause and colon cancer specific death (Bouvard et al. 2015).

Diet Maintaining general health requires eating a balanced diet. An expanding corpus of epidemiological studies has looked at the connection between diet and a number of diseases, including cancer. Being one of the body's initial barriers to food ingestion, the colon and rectum are especially vulnerable to the effects of nutrition. It has been said that the foods with the highest risk of increasing the risk of are red and processed meat (Zhao et al. 2017).

Red and processed meat A large portion of the normal Western diet is made up of processed food and red meat. Meat from mammalian muscle tissue with a high myoglobin content is classified as red meat. This includes beef, lamb, hog, mutton, horse, and goat. The term comes from the red substance called oxymyoglobin that is produced when this kind of meat oxidizes. Meat products are considered processed when they have undergone processes such as curing, smoking, fermenting, or salting before being ingested. Red meat has significant nutritional benefits, especially because of its high vitamin and iron content; however, the International Agency for Research on Cancer has concluded that red and processed meats are harmful to human health (Schwingshackl et al. 2018). Numerous observational studies have linked red and processed meats to CRC because these meats primarily interact with the gastrointestinal tract mucosa. An average Westernized diet consumer is thought to consume about 220 g of meat each day. Diets heavy in red and processed meats have been linked to an increased risk of CRC, as shown by systematic reviews of mostly observational research (Barrubés et al. 2019).

Dietary fiber Diets high in dietary fiber, which include whole grains, have been shown to provide a number of health benefits. Whole grains have been linked to a 12–17% lower risk of CRC when consumed in amounts of 90 g per day. There is a 10–17% decrease in the risk of CRC for every 10 g of total dietary fiber consumed daily. This association is probably caused by dietary fiber slowing down colonic transit, which lowers exposure to dietary carcinogens such processed meats—a topic covered before. It is wise to remember that there is less evidence linking dietary fiber consumption to an increased risk of CRC (Huang et al. 2020).

Dairy products Dairy products, which include milk, yogurt, butter, and dairy products, are an important part of a diet that is well balanced because they provide important minerals like calcium, phosphorus, and protein. Reduced prevalence of CRC has been associated with a greater amount of total dairy intake. Individuals who consumed 400 g of dairy products daily had an 8–13% lower risk of CRC, whereas those who consumed 200 g of milk daily had a 6–10% lower risk. These inverse correlations, which show a 7% lower risk of CRC, have also been shown for fermented milk (200 g/day increments) and cheese (50 g/day increments) in a dose-response fashion. Yogurt consumption has been shown to follow similar tendencies (Newmark et al. 1984).

Role of Calcium, Vitamin D, Folate, Vitamin B6, Magnesium in Colorectal Cancer

As mentioned above, dairy products—or plant-based dairy substitutes—are the main source of dietary calcium. Consuming vitamin D is special as the metabolism depends on food consumption as well as synthesis from sun exposure. Vitamin D helps the intestinal absorption of calcium from meals through the vitamin D receptor on the intestinal cell membranes of the colon and distal small bowel. So, it would be impossible to evaluate the impact of dietary calcium on CRC without including vitamin D. A substantial preventive impact against CRC is demonstrated by a high dietary calcium intake of 300–400 mg/day, which reduces risk by 5–6% in a dose–response relationship that is linear. Similarly, with each dietary supplementation of 100 units/day, vitamin D shows a linear dose-response reduction in CRC risk by 4%. Elevated levels of 25(OH)D, the circulating form of vitamin D obtained from food or supplementation, have been demonstrated to lower the risk of colorectal adenoma development by 20%. For every 200 units/day rise, the incidence of adenomas is reduced by 10%. When increased calcium consumption is paired with this protective impact, the effect is magnified (Fu et al. 2023). It has been suggested that calcium and vitamin D, which have been shown to promote carcinogenesis in CRC, operate in concert to preserve the colonic mucosa by sequestering bile salts and fatty acids from colonic epithelial cells. Thus, there may be a decrease in colonic toxicity and adenomatous polyp development due to a decrease in epithelial proliferation. Furthermore, there is proof that calcium may influence mediators in the adenoma–carcinoma cycle when precancerous polyps develop into CRC (Barrubés et al. 2019). Pyridoxine, often known as vitamin B6, is a water-soluble vitamin that participates in many metabolic enzymatic activities as a coenzyme when it is in its active state as pyridoxal 5′-phosphate. A small number of studies provide evidence for the inverse association between vitamin B6 and the risk of CRC. Increased vitamin B6 consumption was linked to a 12% lower risk of CRC. The risk of CRC was shown to be dose-responsive to the active form of vitamin B6, with a 49% reduction in CRC risk for every 100 pmol/mL rise in blood pyridoxal 5′-phosphate levels (Chapelle et al. 2020).

6.8 Emerging Therapeutic Strategies: Probiotics, Prebiotics, and Cancer Therapies

Probiotics and prebiotics are considered a cheap and safe alternative for improving human health and treating a variety of chronic illnesses. They are essential in controlling the host's immune system and provide protection against a range of viral and noninfectious illnesses. Prebiotics also specifically support human health and probiotic development by regulating the immune system, gut flora, and nutrients through enrichment. Probiotics are well recognized for their ability to colonize, eradicate, and alter the characteristics of host cells, all of which have an influence on many host-specific processes (Yadav et al. 2022).

Leading Therapeutic Methods

A. *Fecal microbiota transplantation (FMT).* FMT or fecal microbiota transplant has been the most novel approach used in modifying the gut microbiota. With FMT, patients in need get a healthy donor's gut microbiota via the upper or lower gastrointestinal tract. The restoration of gut microbial diversity is the primary goal of this treatment. Donated excrement is offered at certain FMT banks. The most common indication for FMT at the moment is recurrent *Clostridium difficile* infection, for which the 2013 recommendations now authorize FMT as a treatment option. The studies on the treatment of various diseases and illnesses like intractable functional constipation, hematologic malignancies, and inflammatory bowel diseases have also additionally confirmed the use of FMT as a treatment (Kaźmierczak-Siedlecka et al. 2020). Recently, the potential of FMT as a treatment or its therapeutic approach for treating several other diseases, especially cancer, has drawn more attention (Chen et al. 2019).
B. *Three emerging NGPs* (next-generation probiotics), namely, *Faecalibacterium prausnitzii*, *Akkermansia muciniphila*, and *Bacteroides fragilis* present in the gut have been found to be useful for cancer patients. These three bacteria work through various mechanisms. These NGPs tend to improve immunotherapy efficiency, work against pathogens, boost gastrointestinal immunity, regulate intestinal barrier integrity, synthesize beneficial metabolites, and lessen side effects

related to chemotherapy and radiotherapy. Notably, *Faecalibacterium prausnitzii*, *Akkermansia muciniphila*, and *Bacteroides fragilis* have distinct properties, but they have similar roles and modes of action. Furthermore, *Akkermansia muciniphila* helps in increasing the efficiency of immunotreatment and *F. prausnitzii* is useful in reducing gastrointestinal problems mainly caused due to radiation or chemotherapy as mentioned in (Table 6.2).

Additionally, there is less history of these NGPs in cancer patients treatment, and the latest published data has shown they have non toxigenic properties, posing less risk. The new emerging research suggests that cancer development can be prevented through NGPs and is a promising therapeutic choice for cancer patients. Further, more research is required as the role of NGPs and their safety still remain unclear, and the research should provide more precise data (Kaźmierczak-Siedlecka et al. 2022).

C. *Synbiotics*

Synbiotics have been used in several supporting treatments in various illnesses including colorectal cancer. Synbiotics is defined as a "synergistic" blend or combination of probiotic bacteria and prebiotic substances known to promote growth (Chen et al. 2019) It was shown that synbiotics dramatically reduced the rates of postoperative infection in colorectal cancer patients, in a randomized, double-blind clinical trial (Bruno-Barcena and Azcarate-Peril 2015)

In the colon, synbiotics facilitate an apoptotic response to damaged DNA induced by carcinogens. While synbiotics help in immunomodulation, they also downregulate expression of carcinogenic enzymes like NO synthase and cyclooxygenase enzyme (Raman et al. 2013).

6.9 Conclusion

A growing amount of evidence suggests that the gut microbiota is a key player in the genesis of tumors. It follows that efforts to control the gut microbiota in order to suppress tumor development or alter the effectiveness of tumor therapy are not surprising. There is continuous work being done to improve the efficiency of treating

Table 6.2 Role of different NGPs in colorectal cancer (Kaźmierczak-Siedlecka et al. 2022)

Next-generation probiotics	Role in cancer prevention
Faecalibacterium prausnitzii	Ferments glucose and produces short-chain fatty acids (SCFAs) It is the most important butyrate-producing bacterium Inhibit the deacetylation of histones indirectly It has anti-inflammatory properties
Akkermansia muciniphila	It maintains intestinal immunity and regulates gut barrier functions Regulation of specific cytokines to enhance the immune system It reduces inflammation
Bacteroides fragilis	It maintains gut homeostasis and intestinal barrier integrity Inhibits the growth of pathogens It reduces inflammation

tumors and lessen adverse effects by using probiotic therapy and fecal microbiota transplantation. Certain probiotics have the potential to both lessen the negative effects of traditional cancer therapies as well as increase the effectiveness of anti-cancer therapeutic drugs. But once intestinal tumors damage the immune system and intestinal barrier, these probiotics also have the ability to function as opportunistic infections. It is necessary to use suitable probiotics that can reduce side effects and improve anti-cancer results, along with the right administration techniques. A novel biotherapeutic technique called fecal microbiota transplantation (FMT) allows patients to receive stool data from healthy donors, changing their own microbiota. FMT can be used to treat a number of gastrointestinal conditions, including IBD, restored eubiosis, and *C. difficile* infection. A lot of work is being done to use FMT as a tumor therapy in the clinic. According to reports, FMT may be able to overcome the immunosuppressive model of colorectal cancer in mice by overcoming resistance to the drug. Moreover, FMT may be useful in reducing immune checkpoint inhibitor-associated colitis and other ICI adverse effects. Numerous animal and clinical research have demonstrated the impact of altered gut microbiota composition on the formation of precancerous cancer lesions and the progression of cancer. Because changes in the gut microbiota can directly affect the organs in the colorectal region, CRC is hypothesized to be more influenced by the gut microbiome than other malignancies. Research on the gut microbiome showed that CRC patients had dysbiosis more frequently than healthy individuals. Several published research works have demonstrated that the gut microbiota is important for the growth and progression of CRC and should be considered while managing the condition. However, there is still much to learn about the gut microbiome, which means that further research is required to elucidate the underlying mechanisms and establish the gut microbiome as a crucial element of colorectal cancer prevention and treatment strategies.

References

Ahn J, Sinha R, Pei Z, Dominianni C, Wu J, Shi J, Goedert JJ, Hayes RB, Yang L (2013) Human gut microbiome and risk for colorectal cancer. J Natl Cancer Inst 105(24):1907–1911

Allen LH, Peerson JM, Olney DK (2009) Provision of multiple rather than two or fewer micronutrients more effectively improves growth and other outcomes in micronutrient-deficient children and adults. J Nutr 139:1022–1030

Ambalam P, Raman M, Purama RK, Doble M (2016) Probiotics, prebiotics and colorectal cancer prevention. Best Pract Res Clin Gastroenterol 30(1):119–131

Ballan R, Battistini C, Xavier-Santos D, Saad SM (2020) Interactions of probiotics and prebiotics with the gut microbiota. Prog Mol Biol Transl Sci 171:265–300

Barrubés L, Babio N, Becerra-Tomás N, Rosique-Esteban N, Salas-Salvadó J (2019) Association between dairy product consumption and colorectal cancer risk in adults: a systematic review and meta-analysis of epidemiologic studies. Adv Nutr 11(4):1055–1057

Bering S, Suchdev S, Sjøltov L, Berggren A, Tetens I, Bukhave K (2006) A lactic acid-fermented oat gruel increases non-haem iron absorption from a phytate-rich meal in healthy women of childbearing age. Br J Nutr 96:80–85. https://doi.org/10.1079/BJN20061683

Bermudez-Brito M, Plaza-Díaz J, Muñoz-Quezada S, Gómez-Llorente C, Gil A (2012) Probiotic mechanisms of action. Ann Nutr Metab 61(2):160–174

Boffetta P, Hashibe M (2006) Alcohol and cancer. Lancet Oncol 7(2):149–156

Bouvard V, Loomis D, Guyton KZ, Grosse Y, Ghissassi FE, Benbrahim-Tallaa L et al (2015) Carcinogenicity of consumption of red and processed meat. Lancet Oncol 16(16):1599–1600

Bruno-Barcena JM, Azcarate-Peril MA (2015) Galacto-oligosaccharides and colorectal cancer: feeding our intestinal probiome. J Funct Foods 12:92–108

Chapelle N, Martel M, Toes-Zoutendijk E, Barkun AN, Bardou M (2020) Recent advantages in clinical practice: colorectal cancer chemoprevention in the average risk population. Gut 69(12):2244–2255

Chen D, Wu J, Jin D, Wang B, Cao H (2019) Fecal microbiota transplantation in cancer management: current status and perspectives. Int J Cancer 145(8):2021–2031

Chong ES (2014) A potential role of probiotics in colorectal cancer prevention: review of possible mechanisms of action. World J Microbiol Biotechnol 30:351–374

Coleman OI, Nunes T (2016) Role of the microbiota in colorectal cancer: updates on microbial associations and therapeutic implications. BioRes Open Access [Internet] 5(1):279–288

De Filippo C, Cavalieri D, Di Paola M, Ramazzotti M, Poullet JB, Massart S, Collini S, Pieraccini G, Lionetti P (2010) Impact of diet in shaping gut microbiota revealed by a comparative study in children from Europe and rural Africa. Proc Natl Acad Sci USA 107:14691–14696

Fink LN, Zeuthen LH, Christensen HR, Morandi B, Frøkiær H, Ferlazzo G (2007) Distinct gut-derived lactic acid bacteria elicit divergent dendritic cell-mediated NK cell responses. Int Immunol 19(12):1319–1327

Fu H, He J, Li C, Deng Z, Chang H (2023) Folate intake and risk of colorectal cancer: a systematic review and up-to-date meta-analysis of prospective studies. Eur J Cancer Prev 32(2):103–112

Furness JB, Callaghan BP, Rivera LR, Cho HJ (2014) The enteric nervous system and gastrointestinal innervation: integrated local and central control. Adv Exp Med Biol 817:39–71. Microbial endocrinology: The microbiota-gut-brain axis in health and disease

Gibson GR, Scott KP, Rastall RA, Tuohy KM, Hotchkiss A, Dubert-Ferrandon A et al (2010) Dietary prebiotics: current status and new definition. Food Sci Technol Bull: Funct Foods 7:1–19

Huang D, Lei S, Wu Y, Weng M, Zhou Y, Xu J et al (2020) Additively protective effects of vitamin D and calcium against colorectal adenoma incidence, malignant transformation and progression: a systematic review and meta-analysis. Clin Nutr 39(8):2525–2538

Initiative M (2009) Investing in the future: a united call to action on vitamin and mineral deficiencies. Micronutrient Initiative, Ottawa

Kato I, Vasquez AA, Moyerbrailean G, Land S, Sun J, Lin HS, Ram JL (2016) Oral microbiome and history of smoking and colorectal cancer. J Epidemiol Res 2(2):92

Kaźmierczak-Siedlecka K, Daca A, Fic M, Van De Wetering T, Folwarski M, Makarewicz W (2020) Therapeutic methods of gut microbiota modification in colorectal cancer management – fecal microbiota transplantation, prebiotics, probiotics, and synbiotics. Gut Microbes 11:1518–1530

Kaźmierczak-Siedlecka K, Skonieczna-Żydecka K, Hupp T, Duchnowska R, Marek-Trzonkowska N, Połom K (2022) Next-generation probiotics – do they open new therapeutic strategies for cancer patients? Gut Microbes 14(1):2035659

Kelleher SL, Casas I, Carbajal N, Lönnerdal B (2002) Supplementation of infant formula with the probiotic lactobacillus reuteri and zinc: impact on enteric infection and nutrition in infant rhesus monkeys. J Pediatr Gastroenterol Nutr 35:162–168. https://doi.org/10.1097/00005176-200208000-00011

Knijn N, Mogk SC, Teerenstra S, Simmer F, Nagtegaal ID (2016) Perineural invasion is a strong prognostic factor in colorectal cancer: a systematic review. Am J Surg Pathol 40(1):103–112

Kozlowska A, Kwiatkowski P, Oponowicz A, Majewski M, Kmiec Z, Godlewski J (2016) Myenteric plexuses atrophy in the vicinity of colorectal cancer tissue is not caused by apoptosis or necrosis. Folia Histochem Cytobiol 54(2):99–107

Liu Y, Wang J, Wu C (2022) Modulation of gut microbiota and immune system by probiotics, prebiotics, and post-biotics. Front Nutr 8:634897

Ma J, Wu JJ, Xing XX, Huo BB, Gao X, Ma ZZ, Li SS, Zheng MX, Hua XY, Xu JG (2022) Cerebral metabolic analysis of patients with colorectal cancer and chronic enteritis: inquiry into gut-brain crosstalk. Front Neurosci 16:822891

Malik A, Sharma D, Malireddi RS, Guy CS, Chang TC, Olsen SR, Neale G, Vogel P, Kanneganti TD (2018) SYK-CARD9 signaling axis promotes gut fungi-mediated inflammasome activation to restrict colitis and colon cancer. Immunity 49(3):515–530

McTiernan A, Friedenreich CM, Katzmarzyk PT, Powell KE, Macko R, Buchner D et al (2019) Physical activity in cancer prevention and survival: a systematic review. Med Sci Sports Exerc 51(6):1252–1261

Murphy N, Moreno V, Hughes DJ, Vodicka L, Vodicka P, Aglago EK, Gunter MJ, Jenab M (2019) Lifestyle and dietary environmental factors in colorectal cancer susceptibility. Mol Asp Med 69:2–9

Newmark HL, Wargovich MJ, Bruce WR (1984) Colon cancer and dietary fat, phosphate, and calcium: a hypothesis. J Natl Cancer Inst 72(6):1323–1325

Ou J, DeLany JP, Zhang M, Sharma S, O'Keefe SJ (2012) Association between low colonic short-chain fatty acids and high bile acids in high colon cancer risk populations. Nutr Cancer 64(1):34–40

Pekmez CT, Dragsted LO, Brahe LK (2019) Gut microbiota alterations and dietary modulation in childhood malnutrition – the role of short chain fatty acids. Clin Nutr 38(2):615–630

Rafter J (2003) Probiotics and colon cancer. Best Pract Res Clin Gastroenterol 17(5):849–859

Raman M, Ambalam P, Kondepudi KK, Pithva S, Kothari C, Patel AT, Purama RK, Dave JM, Vyas BR (2013) Potential of probiotics, prebiotics and synbiotics for management of colorectal cancer. Gut Microbes 4(3):181–192

Rebersek M (2021) Gut microbiome and its role in colorectal cancer. BMC Cancer 21(1):1325

Robertson RC (2020) The gut microbiome in child malnutrition. Nestle Nutr Inst Workshop Ser 93:133–144

Sato S, Nagai H, Igarashi Y (2012) Effect of probiotics on serum bile acids in patients with ulcerative colitis. Hepato-Gastroenterology 59(118):1804–1808

Schonkeren SL, Thijssen MS, Vaes N, Boesmans W, Melotte V (2021) The emerging role of nerves and glia in colorectal cancer. Cancers 13(1):152

Schwingshackl L, Schwedhelm C, Hoffmann G, Knüppel S, Preterre AL, Iqbal K et al (2018) Food groups and risk of colorectal cancer. Int J Cancer 142(9):1748–1758

Scott KP, Grimaldi R, Cunningham M, Sarbini SR, Wijeyesekera A, Tang ML, Lee JY, Yau YF, Ansell J, Theis S, Yang K (2020) Developments in understanding and applying prebiotics in research and practice—an ISAPP conference paper. J Appl Microbiol 128(4):934–949

Sheveleva SA (1999) Probiotics, prebiotics and probiotic products. Curr Status Vopr Pitan 68(2):32–40

Song M, Garrett WS, Chan AT (2015) Nutrients, foods, and colorectal cancer prevention. Gastroenterology 148(6):1244–1260

Song M, Chan AT, Sun J (2020) Influence of the gut microbiome, diet, and environment on risk of colorectal cancer. Gastroenterology 158(2):322–340

Tareb R, Bernardeau M, Gueguen M, Vernoux JP (2013) In vitro characterization of aggregation and adhesion properties of viable and heat-killed forms of two probiotic Lactobacillus strains and interaction with foodborne zoonotic bacteria, especially Campylobacter jejuni. J Med Microbiol 62(4):637–649

Ternes D, Karta J, Tsenkova M, Wilmes P, Haan S, Letellier E (2020) Microbiome in colorectal cancer: how to get from meta-omics to mechanism? Trends Microbiol 28(5):401–423

Tomše P, Jensterle L, Grmek M, Zaletel K, Pirtošek Z, Dhawan V, Peng S, Eidelberg D, Ma Y, Trošt M (2017) Abnormal metabolic brain network associated with Parkinson's disease: replication on a new European sample. Neuroradiology 59:507–515

Tuan J, Chen YX (2016) Dietary and lifestyle factors associated with colorectal cancer risk and interactions with microbiota: fiber, red or processed meat and alcoholic drinks. Gastrointest Tumors 3(1):17–24

Uesaka T, Young HM, Pachnis V, Enomoto H (2016) Development of the intrinsic and extrinsic innervation of the gut. Dev Biol 417(2):158–167

Vallis J, Wang PP (2022) The role of diet and lifestyle in colorectal cancer incidence and survival. In: Morgado-Diaz JA (ed) Gastrointestinal cancers [internet]. Exon Publications, Brisbane (AU). Chapter 2

Veettil SK, Wong TY, Loo YS, Playdon MC, Lai NM, Giovannucci EL et al (2021) Role of diet in colorectal cancer incidence: umbrella review of meta-analyses of prospective observational studies. JAMA Netw Open 4(2):e2037341

Veiga P, Gallini CA, Beal C, Michaud M, Delaney ML, DuBois A, Khlebnikov A, van Hylckama Vlieg JE, Punit S, Glickman JN, Onderdonk A (2010) Bifidobacterium animalis subsp. lactis fermented milk product reduces inflammation by altering a niche for colitogenic microbes. Proc Natl Acad Sci 107(42):18132–18137

Wong SH, Yu J (2019) Gut microbiota in colorectal cancer: mechanisms of action and clinical applications. Nat Rev Gastroenterol Hepatol 16(11):690–704

Yadav MK, Kumari I, Singh B et al (2022) Probiotics, prebiotics and synbiotics: safe options for next-generation therapeutics. Appl Microbiol Biotechnol 106:505–521

Yan YL, Hu Y, Gänzle MG (2018) Prebiotics, FODMAPs and dietary fiber—conflicting concepts in development of functional food products? Curr Opin Food Sci 20:30–37

Zhao Z, Feng Q, Yin Z, Shuang J, Bai B, Yu P et al (2017) Red and processed meat consumption and colorectal cancer risk: a systematic review and meta-analysis. Oncotarget 8(47):83306–83314

The Influence of Prebiotics and Probiotics in Colorectal Cancer Treatment

7

Noha A. Ahmed, Rehab G. Khalil, Nour Y. S. Yassin, and Osama M. Ahmed

Abstract

The impact of probiotics and prebiotics on the management of colorectal cancer (CRC) is a developing field that could improve current treatments. Prebiotics, which nourish beneficial gut bacteria, and probiotics, which introduce these bacteria directly into the gut, both play vital roles in restoring microbial balance and preventing dysbiosis associated with CRC. Prebiotics' fermentation produces short-chain fatty acids (SCFAs) such as butyrate. These SCFAs have anti-inflammatory effects and anti-carcinogenic properties. They inhibit tumour growth and induce cancer cell apoptosis. Probiotics improve gut barrier function, reduce permeability, and prevent the spread of pathogens and toxins; thereby reducing inflammation, a significant risk factor for CRC. Together, prebiotics and probiotics modulate immune responses, creating an environment that suppresses tumour growth and enhances the effectiveness of traditional treatments. They also help detoxify dietary carcinogens, reducing CRC risk. Further research, especially large-scale clinical trials, is needed to confirm the efficacy and safety of prebiotics and probiotics in CRC treatment. Future directions include personalized treatments based on individual microbiome profiles, integration with existing therapies, and the development of advanced synbiotics. Understanding their mechanisms and optimizing their use will lead to more effective CRC management strategies. In conclusion, combining prebiotics and probiotics provides

N. A. Ahmed · N. Y. S. Yassin · O. M. Ahmed (✉)
Physiology Division, Department of Zoology, Faculty of Science, Beni-Suef University, Beni-Suef, Egypt
e-mail: drnohascience@science.bsu.edu.eg; osama.ahmed@science.bsu.edu.eg

R. G. Khalil
Immunology Division, Department of Zoology, Faculty of Science, Beni-Suef University, Beni-Suef, Egypt
e-mail: rehab.khalil@science.bsu.edu.eg

S. Pathak et al. (eds.), *Microbiota and Dietary Mediators in Colon Cancer Prevention and Treatment*, https://doi.org/10.1007/978-981-96-0297-1_7

a comprehensive approach to CRC treatment by improving gut health, immune function, and cancer prevention. Their potential to enhance treatment efficacy and minimize side effects makes them valuable tools in the fight against CRC.

Keywords

Probiotics · Prebiotics · Colorectal cancer · Gut microbiota

7.1 Introduction

Globally, cancer is the major cause of death; it will account for around 10 million fatalities in 2020, or nearly one in six lives, and 7.6 million deaths in 2008 (13% of all deaths) (Ferlay et al. 2020; Bray et al. 2013). In 2012, the American Cancer Society, in partnership with the National Cancer Institute, the Centres for Disease Control and Prevention, the North American Association of Central Cancer Registries, and the National Centre for Health Statistics, projected that there were 1,638,910 new cases of cancer in the United States and 577,190 cancer-related deaths (Siegel et al. 2014). According to the World Health Organization (WHO), by 2040, it is projected that the burden of colorectal cancer (CRC) will increase to 3.2 million new cases per year (an increase of 63%) and 1.6 million deaths per year (an increase of 73%) (Winawer et al. 1995; WHO 2023). The primary malignancies that cause a considerable amount of death each year are CRC, prostate, lung, stomach, liver, and breast cancers. CRC is the fourth most prevalent malignant tumour in the United States and the second largest cause of cancer-related mortality (Wang et al. 2013).

CRC imposes significant physical, financial, and emotional concerns on both individuals and healthcare systems. Probiotics have gained attention for their numerous health benefits, and their use in managing CRC has become increasingly popular in recent decades. This popularity is due to the positive outcomes observed in various studies including in vitro, in vivo, and clinical investigations. The beneficial effects of probiotics in these contexts are thought to result from their multiple advantageous actions working together. The connection between gut microflora and CRC occurrence is well-documented. The limitations of current chemo- and radiotherapies and the lack of effective treatment strategies for CRC are driving the search for alternative prevention and management methods. Probiotics, known for their health benefits, have emerged as a promising candidate for CRC prevention and treatment. They enhance intestinal barrier function, balance gut microflora, secrete anticancer compounds, and degrade carcinogenic substances, making them effective for CRC prophylaxis. Their pro-apoptotic effects on cancer cells and immunomodulatory properties further enhance their potential as a therapeutic option. This review explores various strategies and mechanisms by which probiotics contribute to CRC prevention and treatment (Tripathy et al. 2021).

Major lifestyle changes are necessary for successful prevention of CRC, as it is the second most common cancer in women and the third in men worldwide

(Sung et al. 2021). It affects the heterogeneous epithelium of the colon, rectum, and anus, characterized by immune dysregulation, accumulation of mutations in stem cells, disruption of intestinal barriers, and dysbiosis—marked by unfavourable changes in gut microbiota composition and function. Lifestyle factors account for up to 90% of CRC risk, influenced by dietary choices such as fibre intake, consumption of red and processed meats, alcohol, omega-3 fatty acids, and vitamin D levels. Additionally, obesity, physical inactivity, and smoking significantly contribute to microbiota alterations associated with CRC (Janney et al. 2020). Changes in the microbiome, including both depletion and enrichment of specific bacterial species, are closely linked to CRC and extensively documented in recent reviews by Fong et al. (2020), Torres-Maravilla et al. (2021) and Kvakova et al. (2022).

As conventional cancer therapies sometimes fall short, there is increasing interest in novel "drug-free" approaches or adjuncts that can enhance the effectiveness of existing treatments. CRC is closely associated with changes in the gut microbiome, termed dysbiosis, which involves the depletion or overabundance of specific bacterial species and their metabolic activities. Incorporating customary probiotic supplements (with or without prebiotics), next-generation probiotics (NGPs), or postbiotics into patient care represents a promising complementary strategy to restore gut microbiota composition and interact beneficially with the host (Kvakova et al. 2022). By restoring the gut microbiota, these interventions may stabilize intestinal barrier function and promote pathways that are anticarcinogenic, anti-inflammatory, antimutagenic, or otherwise biologically significant, particularly targeting tumour cells. Traditional probiotics, NGP, and postbiotics offer potential benefits in modulating both gut microbiota composition and function, as well as influencing the host's inflammatory responses. Emerging research suggests that NGP and postbiotics hold promise as innovative CRC treatments by producing key natural products and metabolites, which establish new signals between the host and microbiota to combat cancer. Despite promising outcomes, further investigations are necessary to identify the optimal strains and doses of traditional probiotics, NGP, and postbiotics to ensure their efficacy and safety in CRC prevention and treatment (Kvakova et al. 2022).

The human microbiota is a varied ecosystem composed of bacteria, viruses, eukaryotes, and archaea, which play crucial roles in regulating numerous physiological functions in the host. These include digestion, immune response, metabolism, disease development, detoxification, and synthesis of essential compounds like vitamins and cofactors. The microbiotas can also influence the gut-brain axis, impacting mood and anxiety levels. Proteobacteria, Fusobacteria, Actinobacteria, Verrucomicrobia, and Spirochaetes are the groups of bacteria that live in the human gut in symbiotic relationships. These groups are mostly composed of Bacteroidetes and Firmicutes. Both healthy people and those with CRC have different microbiomes (Kvakova et al. 2021). Distinct microbiota compositions are observed in CRC patients, which are typified by an enrichment of bacteria including enterotoxigenic *Bacteroides fragilis*, *Fusobacterium nucleatum*, *Enterococcus faecalis*, *Streptococcus gallolyticus*, *Escherichia coli*, *Peptostreptococcus* sp., and *Ruminococcus* sp. Contrarily, the Lachnospiraceae family of bacteria, including

Lactobacillus sp., *Bifidobacterium* sp., *Faecalibacterium* sp., *Roseburia* sp., *Clostridium* sp., *Granulicatella* sp., *Streptococcus thermophilus*, and others, is reduced in CRC (Fong et al. 2020; Janney et al. 2020; Torres-Maravilla et al. 2021).

In recent years, as research into probiotics advances, they are emerging as a potentially essential component in the future prevention and treatment of CRC. Current prevention methods of CRC with minimal side effects are limited. Surgery and chemotherapy, the primary treatments for CRC, often lead to complications and adversely affect patients' quality of life. Probiotics offer promising solutions by exerting anti-inflammatory, immune-enhancing, and tumour-suppressing effects. They represent a safe and effective strategy for CRC prevention and may mitigate the complications of surgery and chemotherapy while enhancing the efficacy of chemotherapy. Targeted administration of probiotics alone or in conjunction with TRAIL (TNF-related apoptosis-inducing ligand) holds potential for CRC treatment. The aim of recent research is to explore the mechanisms through which probiotics prevent and treat CRC, detailing specific applications to introduce novel strategies for future CRC management, thereby reducing incidence rates and improving patient quality of life (Tang and Zhang 2022).

Numerous studies have been conducted on the relationship between gut bacteria and the onset of CRC. Colorectal carcinogenesis can be caused by an imbalance in the gut microbiota through a number of pathways, such as host DNA damage, activation of carcinogens, inflammation, and tumourigenic pathways. Prebiotics, probiotics, and synbiotics are only a few of the treatment strategies that can be used to alter the gut microbiota and assist patients with CRC. The most sophisticated technique for modifying gut microbiota at the moment is faecal microbiota transplantation (FMT). FMT is an established clinical treatment for recurrent *Clostridium difficile* infection, which can also impact patients with CRC, despite the paucity of data on this patient population. Restoring gut microbial variety, boosting the effectiveness of immunotherapy, and optimising bile acid metabolism are the main advantages of FMT. Nevertheless, additional investigation is necessary to assess the long-term consequences of FMT (Kaźmierczak-Siedlecka et al. 2020).

7.2 Risk Factors Contributing to Colorectal Cancer (CRC)

There are several reasons for the sharp rise in CRC incidence worldwide, but hereditary disorders and environmental variables including dietary practices and lifestyle modifications are the main ones. Key contributing factors include high consumption of meat and saturated fats, chronic alcohol use, tobacco use, and obesity (Li et al. 2021). CRC develops through a sequence of histological changes known as the adenoma-carcinoma sequence, which involves the mutational stimulation of oncogenes and the loss of tumour suppressor gene heterozygosity due to exposure to carcinogenic chemicals and mutagens (Sankpal et al. 2012). This process is driven by an imbalance between DNA damage and repair, resulting in cells with unrepaired or mutagenic DNA mismatches that escape the repair mechanisms (Wood et al. 2001). CRC typically originates in the innermost lining of the intestine and can spread inward to the muscle tissues and other organs, ultimately resulting in metastasis. While procarcinogens themselves are not carcinogenic, they are converted into carcinogens through metabolic reactions involving cytochrome P450 enzymes. These enzymes transform procarcinogens into highly reactive electrophilic compounds capable of interacting with DNA. The risk of developing CRC depends on the potency of these carcinogens, the rate of exposure, and an individual's genetic constitution (Raman et al. 2013).

Environmental carcinogens can originate from both exogenous and endogenous sources and are linked to diet, lifestyle factors including nutrition, tobacco use, and physical activities, and occupational hazards (Wogan et al. 2004). Notable food carcinogens associated with CRC, breast cancer, and prostate cancer include polycyclic aromatic hydrocarbons (PAHs), heterocyclic amines (HCAs), N-nitroso compounds (NOCs), mycotoxins (aflatoxins), and acrylamide, as evidenced by preclinical and clinical studies (Sankpal et al. 2012; Wilson et al. 2012). In addition to air pollution, these carcinogens are created via food processing techniques including curing, drying, smoking, roasting, refining, and fermenting (Nagao and Sugimura 1993; Sankpal et al. 2012; Raman et al. 2013). Frequent eating of food cooked at high temperatures has been found to be strongly correlated with an elevated risk of CRC (Jägerstad and Skog 2005). Epidemiological studies have implicated HCAs in the formation of adenomatous polyps and early CRC symptoms (Wu et al. 2006; Raman et al. 2013). Additionally, a dose-dependent correlation has been found between the risk of cancer and the consumption of food contaminated with certain compounds, including benzo[α]pyrene (BαP), 2-amino-3,4,8-trimethyl-3H-imidazo[4,5-f]quinoxaline (DiMeIQx), 2-amino-3,8-dimethylimidazo[4,5-f]quinoxaline (MeIQx), and 2-amino-1-methyl-6-phenylimidazo[4,5-b]pyridine (PhIP) (Anderson et al. 2002; Raman et al. 2013).

Antimutagenic agents are vital for removing mutagens from the body in circumstances where people are exposed to food and environmental carcinogens on a regular basis. Fruits, vegetables, and grains are examples of natural food ingredients that work incredibly well as antimutagenic agents (Ferguson et al. 2004). The capacity of probiotics and prebiotics to scavenge and eradicate carcinogens has attracted a lot of attention. As a result, a dietary supplement or nutraceutical high in probiotics, prebiotics, and synbiotics may offer a useful biological therapy for eliminating carcinogens and mutagens found in food, thereby preventing CRC.

7.3 Gut Microbiome in Colorectal Cancer (CRC)

Even with the use of radiation, chemotherapy, and surgery, CRC treatment still has a mixed success rate, high mortality, and serious side effects. Alternative alternatives are therefore desperately needed. The goal of recent study has been to better understand how gut bacteria contributes to the development of colon cancer (Quigley 2011; Raman et al. 2013).

With roughly 10^7 to 10^{12} cells per gram of intestinal material, the human gastrointestinal tract (GIT), which runs from the small intestine to the colon, is home to a wide variety of bacterial species (Eckburg et al. 2005; Quigley 2011). Newborns have a sterile intestinal tract, which becomes populated with both beneficial and harmful microbes with their first feeding. Throughout childhood, the intestinal microflora remains relatively stable until influenced by environmental factors, life-style changes, or genetic modifications. Human gut microbes are generally classified into three categories: symbionts, commensals, and pathobionts (Round and Mazmanian 2009).

Gut microbiota play essential roles in maintaining host functions, including immune response and nutritional status, thereby contributing to overall health (Cerf-Bensussan and Gaboriau-Routhiau 2010). The balance among different gut bacterial strains and the host's immune system determines the presence of either physiological inflammation, which regulates resident gut microbiota, or pathological inflammation, which is dependent on the number and virulence of invading pathogens. Furthermore, long-term inflammation strongly stimulates the local immune system, releasing nitric oxide and reactive oxygen species (ROS) that damage DNA and upset tissue homeostasis (Terzić et al. 2010).

The cytokines generated throughout this procedure are essential for preserving tissue homeostasis. While TGF-α and IL-10 function as inhibitors of cancer formation, TNF-α, IL-6, IL-1, and chemokines stimulate angiogenesis and decrease immune-mediated tumour clearance to increase tumour growth (Terzić et al. 2010). Changes in gut microbiota contribute to disease progression through long-term inflammation, and through evading and suppressing the immune system. Colonoscopic studies have revealed varied distributions of bacterial genera in CRC patients, depending on their disease status (Sobhani et al. 2011). There is a notable increase in the population of Bacteroides/Prevotella in cancer patients, which correlates with heightened levels of IL-17-producing cells in the mucosa.

Understanding how individual microorganisms, microbial communities, or microbial compounds promote health under pathological conditions, particularly in CRC, has been the focus of recent studies. There are distinct differences in microbial colonization patterns between tumour tissue and adjacent non-malignant mucosa. This suggests that physiological and metabolic changes associated with CRC may attract commensal-like bacteria, such as *Clostridium* sp. (Marchesi et al. 2011), which have a potential competitive advantage over pathogenic bacteria in the tumour microenvironment. Marchesi et al. explored the dynamic interaction between intestinal microbial ecology (the balance between beneficial and harmful bacteria) and sporadic CRC, which could lead to novel microbiome-based diagnostic tools and therapeutic approaches (Marchesi et al. 2011).

Given the significant role of microflora and diet in the development of CRC, there has been growing interest in utilizing probiotics, prebiotics, and synbiotics to modulate gut microbiota and host metabolism, thereby potentially aiding in cancer prevention (Liong 2008). This has led to a paradigm shift towards recognizing probiotics, prebiotics, and synbiotics for their diverse health-promoting effects. They have demonstrated efficacy in alleviating lactose intolerance, reducing serum cholesterol levels, exerting anticancer effects, improving constipation, enhancing immunity, managing obesity, and relieving vaginitis (Delzenne and Cani 2011).

Previous studies have investigated their anticancer properties against colorectal, breast, and bladder cancers in both preclinical and clinical trials.

7.4 Probiotics

Probiotics is defined as "live microorganisms which, when administered in adequate amounts, confer a health benefit on the host" by WHO and the Food and Agriculture Organization of the United Nations (Sánchez et al. 2017). They include different types (Fig. 7.1). Probiotics is used to alter the gut microbiome and are well known for having anti-mutagenic and anti-cancer effects (Raman et al. 2013). Probiotics may influence molecular mechanisms via microRNAs (miRNAs). In a study employing an animal model of colon cancer, Heydari et al. discovered that giving *Lactobacillus acidophilus* DSM 13421 and *Bifidobacterium bifidum* at a dose of 1×10^9 CFU/g for 5 months increased the expression of adenomatous polyposis coli (APC), the proto-oncogene (PU.1), modulating phosphatase and tensin homolog (PTEN), miR-26b, and miR-18a while lowering the expression of miR-135b, miR-155, and Kirsten rat sarcoma viral oncogene homolog (KRAS) (Heydari et al. 2019). The treatment's ability to enhance tumour suppressor miRNAs and reduce oncogene levels suggests that probiotics may be beneficial for colon cancer treatment (Heydari et al. 2019).

The intestinal mucosa barrier in colon cancer patients may be adequately protected by probiotics after surgery, according to a meta-analysis of 17 randomized controlled studies including 1242 participants (Liu et al. 2016). According to the

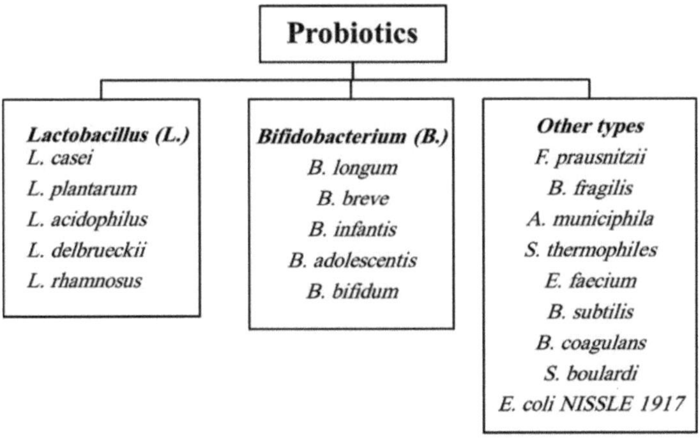

Fig. 7.1 Different types of bacteria and fungi which are recognized as probiotics (Furrie 2005; Iqbal et al. 2014; Maftei et al. 2024). *F. prausnitzii*, *Faecalibacterium prausnitzii*; *B. fragilis*, *Bacteroides fragilis*;, *A. muciniphila*, *Akkermansia muciniphila*; *S. thermophilus*, *Streptococcus thermophilus*; *E. faecium*, *Enterococcus faecium*; *B. subtilis*¸ *Bacillus subtilis*; *B. coagulans*, *Bacillus coagulans*; *S. boulardii*, *Saccharomyces boulardii*; *E. coli* Nissle 1917, *Escherichia coli* Nissle 1917

studies, patients who received probiotics had higher levels of occludin and the ratio of lactulose/mannose than those in the control group. Furthermore, the *Bifidobacterium*-to-*Escherichia* ratio changed, indicating that probiotics assisted in preserving the balance of microbiota following surgery. Additionally, there was a significant decrease in bacterial translocation, a measure of the biological barrier's effectiveness, in the experimental group. While some cohorts showed lower C-reactive protein (CRP) levels in the experimental groups, indicating reduced inflammation with probiotic supplementation, it was revealed that the levels of IL-6, a proinflammatory cytokine, remained unchanged after probiotic treatment (Liu et al. 2016).

Administering probiotic strains can mitigate the side effects of cancer treatments, particularly those arising from surgeries, chemotherapy, and radiotherapy. Probiotics have notably decreased the rate of major postoperative complications (28.6% with probiotics vs. 48.8% with placebo, $p = .010$) (Kotzampassi et al. 2015). Infection, a significant risk factor for morbidity in CRC patient's post-abdominal surgery, can be reduced with probiotic use. A systematic review and meta-analysis of randomized trials indicated that probiotics help prevent inflammation during postoperative care and aid surgical recovery (de Andrade Calaça et al. 2017). Another review confirmed that probiotics nearly halved the infection rate and pneumonia incidence. Consequently, probiotics is recommended to prevent infections in CRC patients post-surgery (Liu et al. 2016).

Chemotherapy with 5-fluorouracil often causes diarrhoea. A randomized trial found that *Lactobacillus rhamnosus* GG reduced abdominal discomfort and the frequency of severe diarrhoea in CRC patients undergoing 5-fluorouracil treatment, with this strain being well tolerated by patients (Österlund et al. 2007). In patients receiving postoperative radiation therapy for sigmoid, rectal, or cervical cancers, the probiotic mix VSL#3, which contains *Lactobacillus casei*, *Lactobacillus plantarum*, *Lactobacillus acidophilus*, *Lactobacillus delbrueckii*, *Bifidobacterium longum*, *Bifidobacterium breve*, *Bifidobacterium infantis*, and *Streptococcus thermophilus*, decreased the risk of diarrhoea. This was demonstrated in a double-blind, placebo-controlled trial. Providing probiotics to individuals undergoing chemotherapy and/or radiation therapy improves their condition (Garczyk et al. 2022).

NGPs are described as "living microorganisms identified through comparative microbiota analyses that confer health benefits to the host when administered in sufficient quantities" (Martín and Langella 2019). NGPs have recently been identified thanks to cutting-edge technologies that enable the identification and manipulation of these commensal bacteria, in contrast to traditional probiotics, which have long been derived from the gut and fermented foods. As such, nothing is known about the effectiveness and safety of NGPs (Martín and Langella 2019). But NGPs like *Faecalibacterium prausnitzii*, *Bacteroides fragilis*, *Akkermansia muciniphila*, and *Bifidobacterium* spp. provide new therapeutic options for CRC (Chang et al. 2019).

For instance, certain strains of *Bifidobacterium* spp. may enhance the efficacy of cancer therapy involving immune checkpoint inhibitors (ICIs), suggesting that anticancer effects may be strain-specific (Chang et al. 2019). Similarly, while

Bacteroides fragilis can improve the effectiveness of ICIs therapy, enterotoxin-producing strains have also been linked to CRC development (Chang et al. 2019). *Akkermansia muciniphila,* a prevalent species in the human intestinal mucosa, may boost the effectiveness of PD-1-based immunotherapy in mice, although the exact mechanism remains unclear (Wang et al. 2018). *Faecalibacterium prausnitzii* contributes to butyrate production, which helps maintain intestinal homeostasis (Duncan et al. 2002). NGPs are currently thought to be a promising new therapeutic method to improve CRC treatment; nevertheless, more study is required to completely comprehend their mechanisms and possible uses.

7.5 Prebiotics

Prebiotics are nondigestible oligosaccharides or substances that undergo selective fermentation, which modifies the gut microbiota's makeup and activity in a way that is advantageous (Raman et al. 2013). These include carbohydrates, the most researched of which are fructooligosaccharides (FOS) and galactooligosaccharides (GOS), xylooligosaccharides (XOS), inulin, lactose, lactulose, and fructans (Fig. 7.2). These prebiotics increase the growth of good gut flora, create short-chain fatty acids (SCFAs), influence the production of xenobiotic-metabolizing enzymes, change the expression of genes in the colon, cecum, and faeces, and control the immunological response (Raman et al. 2013). Specifically, inulin supplemented

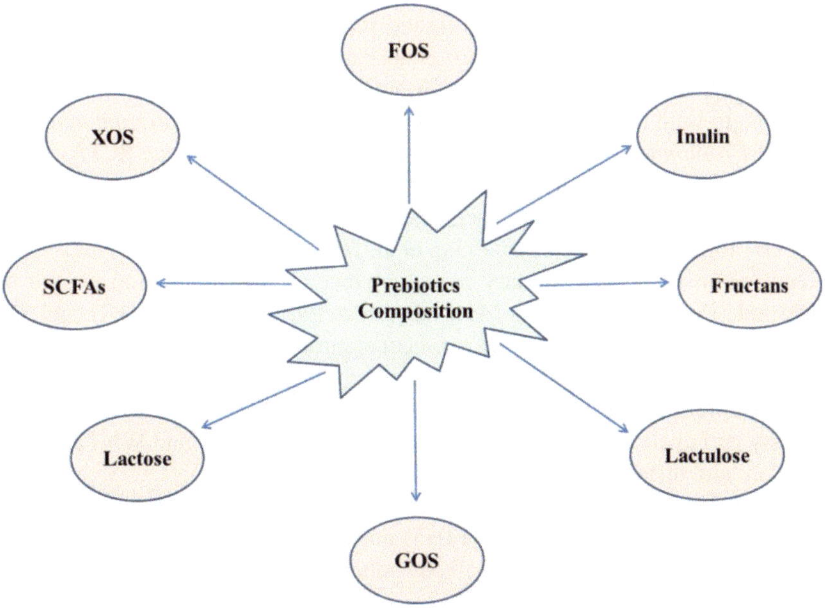

Fig. 7.2 Prebiotics composition (Roller et al. 2004; Raman et al. 2013). FOS fructooligosaccharides, GOS galactooligosaccharides, XOS xylooligosaccharides, SCFAs short-chain fatty acids

with oligofructose has been demonstrated to have anticancer effects in rats with azoxymethane-induced colon cancer when paired with probiotics like *Lactobacillus rhamnosus* and *Bifidobacterium lactis*. Furthermore, it has been noted that inulin administration lowers caecal pH (Roller et al. 2004).

The main method for creating β (1–4) galactooligosaccharides (GOS) is via enzymatic transglycosylation with the aid of β-galactosidases or β-glucosidases (Bruno-Barcena and Azcarate-Peril 2015). The formula β (1–4) [D-galactose] n-D-glucose, where n varies from three to ten sugar units, describes their structure. Gases, lactate, acetate, and short-chain fatty acids are produced during the metabolism of GOS. β (1–4) GOS is recognized for elevating intestinal lactate and short-chain fatty acid levels, in addition to stools' weight and frequency. It also elevates the pH of the faeces, lowers the activity of β-glucuronidase and nitroreductase, and decreases the content of lithocholic acid, a secondary bile acid, in the faeces. Prebiotics such as β (1–4) GOS, lactulose, and fructooligosaccharides are noteworthy because they may help prevent CRC (Bruno-Barcena and Azcarate-Peril 2015).

7.6 Mechanism of Action of Probiotics

7.6.1 Limiting the Growth of Cancerous Cells

By triggering both internal and extrinsic apoptotic pathways, probiotics can stop the growth of cancer cells. Many investigations conducted in vitro have corroborated this pro-apoptotic effect, demonstrating alterations in the expression of proteins linked to apoptosis, including death-ligand receptors, pro-caspase, caspase-3, -8, and -9, Bax/Bak, Bcl-2, and Bcl-x proteins (Papaliagkas et al. 2007). Additionally, probiotics alter cell cycle phases, which are indicated by modifications in cyclin expression, to limit differentiation and proliferation. *Propionibacterium acidipropionici* and *Propionibacterium freudenreichii*, for example, produce SCFAs such as propionate and acetate, which activate caspase-3, cause chromatin condensation, produce apoptotic bodies, and generate reactive oxygen species in human colon cancer cell lines HT-29 and Caco-2 (Jan et al. 2002).

Regular consumption of dietary fibres enhances SCFA production via gut flora modulation. Probiotics such as LGG (*Lactobacillus rhamnosus* GG) and Bb12 (*Bifidobacterium lactis*), when used in wheat aleurone fermentation, increase SCFA concentrations like butyrate, acetate, and propionate. Butyrate, a major energy source for colonocytes, inhibits neoplastic cell growth and induces apoptosis, evidenced by cell cycle arrest in HT-29 cells, modulation of p21 and WNT2B mRNA expression, and differentiation activity (Borowicki et al. 2011).

Four probiotic strains like *Pediococcus pentosaceus* FP3, *Lactobacillus salivarius* FP25, *Lactobacillus salivarius* FP35, and *Enterococcus faecium* FP51 show antiproliferative effects on Caco-2 cells by adhering to cancer cell lines and producing SCFAs. *Propionibacterium freudenreichii*'s SCFAs induce apoptosis in HT-29 cells via mitochondrial pathways, with supernatants from fermented milk causing apoptosis through both death pathways, marked by TRAILs, caspase activation, and Bcl-2 inhibition (Thirabunyanon and Hongwittayakorn 2013).

By upregulating TRAIL and downregulating cyclin D1 and BIRC5a genes, lactic acid bacteria such as *Lactobacillus casei* ATCC 393 also exhibit antiproliferative and pro-apoptotic effects on colon cancer cell lines. Similar to this, the ability of *Lactobacillus plantarum* A7 and LGG to produce organic acids causes heat-killed cells and cell-free supernatants to exhibit antiproliferative effects on Caco-2 and HT-29 cells (Cousin et al. 2016).

Probiotic strains like *Lactobacillus fermentum* RM28 and *Enterococcus faecium* RM11 show antiproliferative activity on Caco-2 cell lines and are resistant to gastrointestinal conditions, which makes them promising candidates for the treatment of CRC. Probiotics like *Bacillus polyfermenticus* inhibit cancer cell growth by suppressing ErbB receptor-dependent pathways. While *Lactobacillus acidophilus* LA102 and *Lactobacillus casei* LC232 exhibit cytotoxic and antiproliferative properties on Caco-2 and HRT-18 cells, *Lactobacillus pentosus* B281 and *Lactobacillus plantarum* B282 arrest the cell cycle of Caco-2 cells (Rabiei et al. 2020).

Cell extracts from *Lactobacillus acidophilus* 606 induce antitumourigenic effects in HT-29 cells through autophagic cell death, mediated by Beclin-1 and ER stress. Lactobacillus plantarum-12's exopolysaccharides also exhibit antiproliferative effects marked by apoptotic protein expression in HT-29 cells. Differential apoptosis induction by pathogenic and commensal *Escherichia coli* strains, probiotic bacteria, and *Atopobium minutum* on Caco-2 cells demonstrates the strong apoptotic effect of pathogenic *Escherichia coli*, mild effect of probiotics, and no effect by commensal *Escherichia coli*. *Streptococcus thermophilus* shows apoptosis, antiproliferation, and cell cycle arrest effects on various cancerous cells through galactose production, disrupting energy homeostasis (Tripathy et al. 2021).

In immune-compromised patients, probiotics like heat-inactivated *Saccharomyces cerevisiae* PTCC 5052 induce apoptosis in colorectal adenocarcinoma cells by modulating the Akt/NF-κB signaling pathway. *Lactobacillus paracasei* IMPC2.1 and *Lactobacillus rhamnosus* ATCC 53103 both exhibit growth-inhibiting properties against DLD-1 CRC cells, indicating that inactivated probiotics may be a feasible option for patients with weak immune systems (Shamekhi et al. 2020).

Combining probiotics like *Lactobacillus acidophilus* CL1285 and *Lactobacillus casei* LBC80R with chemotherapeutic drugs like 5-fluorouracil (5-FU) increases sensitivity and apoptotic activity in CRC cells. *Lactobacillus plantarum* (CCARM 0067) supernatant enhances 5-FU efficacy in resistant cancer cells, marked by caspase-3 activation and Wnt/β-catenin pathway inactivation. This suggests the use of both bacterial cells and supernatants as adjuvants in chemotherapy to increase cancer cell chemosensitivity (An and Ha 2016; Tripathy et al. 2021).

7.6.2 Modulation of the Immune System

Maintaining homeostasis depends on the interaction between the intestinal immune system and the gut bacteria. Immune cells are taught by the microbiome to effectively eradicate infections and tolerate helpful bacteria. Toll-like receptors (TLRs) on epithelial cells respond to patterns in microbes to start an immune response. Dysbiosis, or an unbalanced gut microbiota, can trigger the MAPK pathway and allow NF-κB to enter the nucleus. This can lead to the generation of pro-inflammatory cytokines including nitric oxide and IL-8, which can induce CRC and inflammatory bowel disease (IBD) (Lazar et al. 2018). Probiotics can restore microbial balance (eubiosis) and encourage regulatory T cells (Tregs) to release anti-inflammatory cytokines like IL-10 and TGF-β2. Tregs help suppress immune responses to self-antigens and commensal microbes, promoting tolerance. Antigen-presenting cells like dendritic cells present microbial antigens to T cells, guiding their differentiation into various helper T (Th) cells, which secrete different cytokines to maintain intestinal health. Probiotic research shows they can modulate immune responses, reduce cancer cell proliferation, and improve immunotherapy outcomes (Zhuo et al. 2019; Shang et al. 2020; Tripathy et al. 2021).

7.6.3 Dysbiosis to Eubiosis

Eubiosis describes a balanced and stable gut microbiota ecosystem. In contrast, dysbiosis is the disturbance of this balance, marked by a loss of beneficial microbes, an increase in harmful ones, or reduced microbial diversity. Achieving eubiosis can aid in preventing and treating CRC by using probiotics, prebiotics, and synbiotics to neutralize pathogens, support beneficial bacteria growth, modulate immune responses, and heal the intestinal lining. The gut microbiota composition is crucial for effective CRC treatment and can be disrupted by diet, drugs, immune responses, and changes in the intestinal environment. CRC patients often show significant differences in their gut microbiota compared to healthy individuals, with decreased butyrate-producing bacteria and increased harmful bacterial strains. FMT, bacterial consortia, phage therapy, and predatory bacteria are among the strategies to restore eubiosis. The commercial probiotic VSL#3 has shown high remission rates in ulcerative colitis patients without adverse effects (Wirbel et al. 2019; Cheng et al. 2020; Tripathy et al. 2021).

7.6.4 Improvement of Intestinal Barrier

By regulating the flow of water, ions, and nutrients and blocking pathogens and toxins, intestinal epithelial cells protect the internal environment from injury. Tight junctions (TJs), desmosomes, and adherens junctions (AJs) combine to form this barrier of defence. While TJs, which are found at the intersection of the apical and lateral membranes, selectively control paracellular permeability through proteins including occludin, claudins, and junctional adhesion molecules (JAMs), AJs involve cadherin–catenin and Nectin–afadin complexes connected to the cytoskeleton. Proteins like cingulin and zonula occludens link these junctions to the cytoskeleton. IBD and CRC can result from infections and chemicals disrupting the intestinal barrier, which can be altered by immune cells and cytokines (Markowiak-Kopeć and Śliżewska 2020; Tripathy et al. 2021).

Probiotics enhance the function of the intestinal barrier by secreting SCFAs such as butyrate, propionate, and acetate. Through the activation of AMP-activated protein kinase (AMPK) and the stabilization of hypoxia-inducible factor (HIF), SCFAs provide protection against ethanol-induced barrier disruption. Studies show SCFAs reduce inflammation, oxidative stress, and reinforce the barrier in colitis models. By inhibiting the breakdown of the extracellular matrix (ECM) by matrix metalloproteinases (MMPs) and plasminogen activators, probiotics also aid in the repair of intestinal epithelial cells, which may restrict the invasion and spread of cancer. In vitro studies using cell-free supernatants (CFS) from *Lactobacillus casei* and *Lactobacillus rhamnosus* GG have demonstrated decreased cell invasion; this is most likely because the CFS contains macromolecules (Chen et al. 2017a, b; Tripathy et al. 2021).

7.6.5 Production of Anticarcinogenic Compounds

Anticarcinogens are bioactive substances that can prevent cancer development or neutralize carcinogens. Probiotics produce such compounds, including SCFAs and conjugated linoleic acids, which have cytotoxic effects on cancer cells. These extracellular substances can be tested on colon cancer cell lines and often exhibit stronger tumour-suppressive efficacies than some approved CRC drugs. For instance, *Lactobacillus casei* ATCC334 produces ferrichrome, which activates c-jun N-terminal kinase (JNK) and shows better tumour suppression than drugs like 5-FU and cisplatin. Additionally, combining ferrichrome with 5-FU enhances antitumour effects, driven by the upregulation of pro-apoptotic transcription factor DDIT3 (Chen et al. 2017a, b).

Other studies have shown that the CFS and active metabolites from *Lactobacillus rhamnosus* MD 14 can counteract genotoxic compounds and arrest cancer cell cycles. Similarly, CFS of *Lactobacillus casei* (both wild type and CLA overexpressing) induces apoptosis in HCT116 cells by downregulating tumour growth genes and upregulating pro-apoptotic genes. Moreover, these supernatants can induce the secretion of both anti-inflammatory and pro-inflammatory cytokines. *Lactobacillus plantarum* H24 produces plantarone, which has a cytotoxic effect on Caco-2 cells. *Lactobacillus johnsonii* BCRC17010 and *Lactobacillus reuteri* BCRC14625 also exhibit anticarcinogenic properties through different mechanisms, such as inducing apoptosis via the intrinsic BAX/BCL-2 pathway and damaging cell membranes, respectively. Additionally, *Lactobacillus pentosus* B281 and *Lactobacillus plantarum* B282 produce thermostable bioactive compounds that have antiproliferative effects on Caco-2 cells (Konishi et al. 2016; Tripathy et al. 2021).

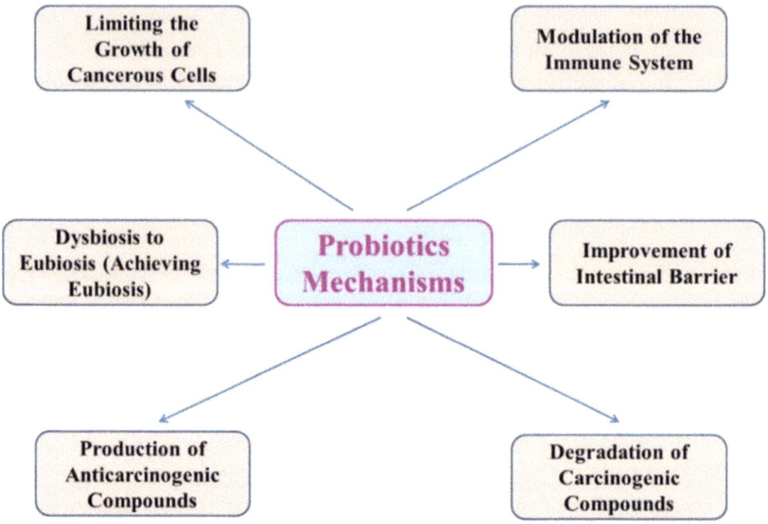

Fig. 7.3 Probiotics mechanisms

7.6.6 Degradation of Carcinogenic Compounds

Intestinal microbiota produces enzymes that metabolize precursor compounds into carcinogenic substances, exposing colon and rectal cells to mutagens like benzo(a) pyrene, sodium azide, MNNG, IQ, AFLB1, and TrpP-1. These compounds, found in dietary components and red meat, contribute significantly to colon cancer development. Probiotics can detoxify carcinogens through physical binding to their cell walls or enzymatic metabolism, with specific strains like *Lactobacillus plantarum* CM4 degrading nitrosamines and removing mutagens such as PhIP and IQ. *Lactobacillus casei* DN 114001 also shows promise in vitro against IQ, MelQx, and PhIP. Additionally, mutant strains like *Lactobacillus plantarum* 301102 produce exopolysaccharides that bind and inactivate mutagens, demonstrating pH-dependent efficacy. Probiotic yeasts like *Kluyveromyces lactis* VIT-MN02, *Lipomyces starkeyi* VIT-MN03, and *Saccharomycopsis fibuligera* VIT-MN04 exhibit antigenotoxic, anticancer, and antimutagenic properties against various compounds (Tsuda et al. 2008; Ragavan and Das 2020; Tripathy et al. 2021) (Fig. 7.3).

7.7 Mechanism of Action of Prebiotics

7.7.1 Selective Stimulation of Beneficial Gut Microbiota

Prebiotics are indigestible dietary fibres that specifically encourage the growth and activity of good gut bacteria, such as *Lactobacilli* and *Bifidobacteria* (Fig. 7.4).

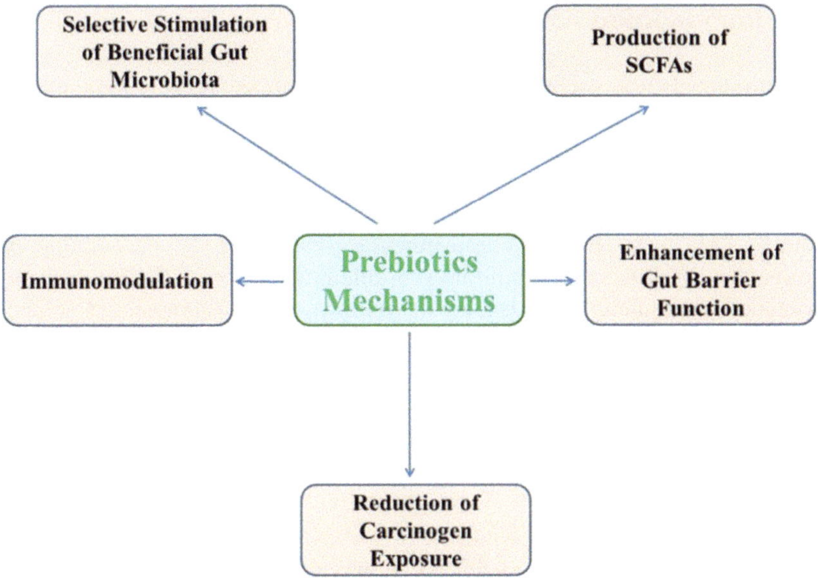

Fig. 7.4 Prebiotics mechanisms. SCFAs, short-chain fatty acids

These bacteria ferment prebiotics to produce SCFAs, such as butyrate, acetate, and propionate. Butyrate in particular has been shown to have numerous health benefits, including acting as the main source of energy for colonocytes, encouraging cell differentiation, and causing cancerous cells to undergo programmed death. The production of SCFAs from prebiotic fermentation lowers the pH of the gut, thereby reducing the environment that is favourable for pathogenic bacteria (Gibson 1998; Arrieta et al. 2011).

7.7.2 Production of Short-Chain Fatty Acids (SCFAs)

The fermentation of prebiotics by beneficial gut bacteria results in the production of SCFAs, which have multiple beneficial effects. Butyrate, one of the primary SCFAs, plays a critical role in maintaining gut health. It has anti-inflammatory properties, enhances the integrity of the gut barrier, and serves as an energy source for colonocytes. Butyrate also inhibits histone deacetylases (HDACs), resulting in the suppression of tumour cell growth and the promotion of apoptosis. Propionate and acetate also contribute to gut health by serving as energy sources for liver cells and by modulating lipid and glucose metabolism (Canani et al. 2011).

7.7.3 Immunomodulation

SCFAs produced from prebiotic fermentation have profound immunomodulatory effects. They promote the differentiation of Tregs that are crucial for maintaining immune homeostasis and preventing excessive inflammatory responses. By enhancing Tregs function, SCFAs help reduce long-term inflammation, a known risk factor for CRC. Additionally, SCFAs modulate the production of cytokines, which are signaling molecules that regulate immune responses. By promoting anti-inflammatory cytokines and inhibiting pro-inflammatory ones, SCFAs help create an immune environment that is less conducive to cancer development (Arpaia et al. 2013; Furusawa et al. 2013).

7.7.4 Enhancement of Gut Barrier Function

Prebiotics enhance the integrity of the gut barrier, which is critical for preventing the translocation of harmful bacteria and their toxins into the bloodstream. The production of SCFAs from prebiotic fermentation strengthens the tight junctions between epithelial cells, reducing gut permeability and preventing the leakage of pathogenic bacteria and toxins. This barrier function is crucial for preventing chronic inflammation and subsequent carcinogenesis. Improved gut barrier function also enhances nutrient absorption, contributing to overall health and reducing the risk of CRC (Canfora et al. 2019).

7.7.5 Reduction of Carcinogen Exposure

Prebiotics can indirectly reduce exposure to carcinogens by modulating the composition and activity of the gut microbiota. A healthy gut microbiota can inhibit the growth of pathogenic bacteria that produce harmful substances, such as N-nitroso compounds and heterocyclic amines, associated with increased cancer risk. Prebiotics help reduce the overall burden of carcinogens in the gut by promoting beneficial bacteria that metabolize and detoxify these compounds. This detoxification process is essential for lowering the risk of CRC development (O'keefe 2016).

7.8 Mechanism of Action of Probiotics in Colorectal Cancer (CRC) Treatment

7.8.1 Modulation of Gut Microbiota Composition

When given in sufficient proportions, probiotics, which are live bacteria, offer health advantages to the host. Probiotics exert their effects primarily through modulation of the gut microbiota via various actions (Table 7.1).

Table 7.1 Mechanisms of actions of probiotics, prebiotics and their combination in CRC treatment

Probiotics	Prebiotics	Probiotics and prebiotics
Modulation of gut microbiota composition Immunomodulation Production of anticarcinogenic compounds Enhancement of chemotherapy efficacy and tolerance	Promotion of beneficial gut microbiota Anti-inflammatory and immunomodulatory effects Reduction of carcinogen exposure Enhancement of gut barrier function Potential synergy with cancer therapies	Enhancement of gut microbiota composition and activity Strengthening of gut barrier function Modulation of the immune system Reduction of carcinogen exposure Synergy with conventional cancer therapies

7.8.1.1 Competitive Exclusion

Probiotics like *Lactobacillus* and *Bifidobacterium* species compete with potential pathogens for adhesion sites and nutrients in the gut, thereby inhibiting the growth and colonization of harmful bacteria (Alsuwaylihi and McCullough 2023).

7.8.1.2 Enhancement of Gut Barrier Function

Probiotics increase the expression of tight junction proteins, which fortifies the gut epithelial barrier. By preventing the movement of harmful bacteria and their toxins, which can cause chronic inflammation and encourage the development of cancer, this barrier improvement aids (Wierzbicka et al. 2021).

7.8.2 Immunomodulation

Probiotics affect both cellular and systemic immune responses by interacting with the host immune system.

7.8.2.1 Regulation of Inflammatory Responses

Probiotics produce metabolites such as SCFAs (e.g., butyrate), which have anti-inflammatory properties. SCFAs can modulate immune cell function, promote regulatory T-cell differentiation, and reduce pro-inflammatory cytokine production, thereby dampening inflammation that contributes to CRC progression (Canani et al. 2011; Ohira et al. 2013).

7.8.2.2 Enhancement of Antitumour Immunity

Some probiotic strains stimulate innate and adaptive immune responses against tumour cells. They enhance natural killer (NK) cell activity and promote the production of cytotoxic T lymphocytes (CTLs), which play crucial roles in recognizing and eliminating cancerous cells (González-Lozano et al. 2022).

7.8.3 Production of Anticarcinogenic Compounds

Certain probiotic strains produce metabolites that directly inhibit carcinogenesis:

7.8.3.1 Production of Antimicrobial Peptides

Probiotics can produce antimicrobial peptides that directly inhibit the growth of pathogenic bacteria and potentially reduce the risk of CRC development associated with dysbiosis (O'Morain and Ramji 2020).

7.8.3.2 Enzymatic Degradation of Carcinogens

Specific probiotic strains possess enzymes capable of degrading or detoxifying carcinogens and mutagens present in the gut lumen, such as heterocyclic amines and N-nitroso compounds derived from dietary sources (Wu and Chiou 2021).

7.8.4 Enhancement of Chemotherapy Efficacy and Tolerance

Emerging evidence suggests that probiotics may enhance the efficacy of chemotherapy and improve tolerance to treatment:

7.8.4.1 Mitigation of Chemotherapy-Induced Side Effects

Probiotics can mitigate gastrointestinal side effects associated with chemotherapy, such as diarrhoea and mucositis, by maintaining gut barrier integrity and modulating immune responses (López-Gómez et al. 2023).

7.8.4.2 Potentiation of Antitumour Effects

Preclinical studies indicate that certain probiotic strains may enhance the antitumour effects of chemotherapy agents, potentially by modulating the tumour microenvironment and improving drug delivery to cancer cells (Miyake et al. 2023).

7.9 Mechanism of Action of Prebiotics in Colorectal Cancer (CRC) Treatment (Table 7.1)

Nondigestible fibres known as prebiotics specifically encourage the development and function of probiotics, or good gut flora (Kumari et al. 2024). Prebiotics have multiple modes of action when used in the treatment of colorectal cancer.

7.9.1 Promotion of Beneficial Gut Microbiota

Prebiotics are indigestible dietary fibres that specifically promote the development and activity of gut-dwelling beneficial bacteria, particularly *Lactobacilli* and *Bifidobacteria*. Prebiotics like inulin, FOS, and GOS are fermented by these bacteria to create SCFAs, which include butyrate, propionate, and acetate (Roberfroid 2007).

Butyrate is particularly important in the context of CRC. It serves as a primary energy source for colonocytes (cells lining the colon), promotes cell differentiation, and induces apoptosis (programmed cell death) of cancerous cells. Mechanistically, butyrate inhibits HDACs, leading to altered gene expression that can suppress proliferation and promote apoptosis and differentiation in CRC cells (Davie 2003; Sun et al. 2024).

7.9.2 Anti-inflammatory and Immunomodulatory Effects

SCFAs produced from prebiotic fermentation exert potent anti-inflammatory effects in the gut. They help maintain mucosal barrier integrity, reduce epithelial permeability, and modulate immune responses:

7.9.2.1 Regulation of Immune Responses

Tregs differentiation and function are improved by SCFAs, especially butyrate. In order to prevent excessive immunological responses that can lead to chronic inflammation and carcinogenesis, Tregs are essential for immune homeostasis maintenance (Arpaia et al. 2013; Furusawa et al. 2013).

7.9.3 Reduction of Carcinogen Exposure

Prebiotics can indirectly reduce exposure to carcinogens by regulating gut microbiota composition and activity:

7.9.3.1 Modulation of Gut Microbiota

Prebiotics have the ability to mitigate the formation and absorption of potentially carcinogenic chemicals, such as heterocyclic amines and N-nitroso compounds, acquired from food sources by fostering the growth of beneficial bacteria and preserving a healthy gut microbiota balance (Canani et al. 2011).

7.9.4 Enhancement of Gut Barrier Function

Improved gut health due to prebiotic consumption enhances epithelial barrier function and reduces inflammation-associated gut permeability. This effect helps prevent the translocation of harmful bacteria and toxins across the gut epithelium, which can trigger inflammatory responses linked to CRC development (Canfora et al. 2019).

7.9.5 Potential Synergy with Cancer Therapies

Emerging research suggests that prebiotics may enhance the efficacy of conventional cancer therapies (Ting et al. 2022).

7.9.5.1 Synergistic Effects

Prebiotics, by modulating the gut microbiota and improving gut health, may enhance the efficacy and reduce the side effects of chemotherapy and immunotherapy in CRC patients. This area of research is actively being explored to understand how prebiotics can complement standard cancer treatments (Gopalakrishnan et al. 2018).

7.10 Combined Use of Prebiotics and Probiotics in Colorectal Cancer (CRC) Treatment

The combined use of prebiotics and probiotics, often referred to as synbiotics, has garnered significant interest (Table 7.1). Synbiotics leverage the synergistic effects of prebiotics (indigestible fibres that specifically encourage the development of good bacteria) and probiotics (live beneficial bacteria) to enhance gut health and provide a multifaceted approach to cancer prevention and therapy.

7.10.1 Enhancement of Gut Microbiota Composition and Activity

Synbiotics optimize the gut microbiota composition by providing beneficial bacteria and the necessary substrates for their growth. Prebiotics such as inulin, FOS, and GOS serve as food for probiotics like *Lactobacillus* and *Bifidobacterium* species. This combination promotes a robust and diverse microbiota, which is crucial for maintaining gut health and preventing dysbiosis, a condition accompanied by an increased risk of CRC (Gibson 1998; Arrieta et al. 2011).

7.10.1.1 Increased Production of Short-Chain Fatty Acids (SCFAs)

The fermentation of prebiotics by probiotics leads to enhanced productions of SCFAs viz butyrate, propionate, and acetate. Butyrate, in particular, has been shown to have anti-carcinogenic properties by promoting apoptosis and inhibiting the proliferation of CRC cells. SCFAs also lower the pH of the gut, creating an unfavourable environment for pathogenic bacteria (Canani et al. 2011; Alpino et al. 2024).

7.10.2 Strengthening of Gut Barrier Function

Synbiotics participate in the maintenance and enhancement of the gut barrier function. Probiotics strengthen the intestinal epithelial barrier by increasing the production of mucins and tight junction proteins, which help prevent the translocation of harmful bacteria and toxins into the bloodstream. Prebiotics, through the production of SCFAs, further support this barrier function by nourishing colonocytes (cells lining the colon), thereby reducing gut permeability and inflammation (Jenkins and Mason 2022).

7.10.2.1 Prevention of Chronic Inflammation

One established risk factor for CRC is chronic inflammation. By enhancing gut barrier integrity and reducing permeability, synbiotics help prevent the leakage of pro-inflammatory substances from the gut into the bloodstream. This action reduces the overall inflammatory load in the body, thereby lowering the risk of CRC development (Vandenplas et al. 2011).

7.10.3 Modulation of the Immune System

The combined use of prebiotics and probiotics has significant immunomodulatory effects. SCFAs produced from the fermentation of prebiotics by probiotics play an essential role in modulating immune responses. They promote the differentiation and function of Tregs, which are essential for maintaining immune homeostasis and preventing exacerbated inflammatory responses (Arpaia et al. 2013).

7.10.3.1 Enhanced Antitumour Immunity

Synbiotics can enhance antitumour immunity by stimulating the activity of natural killer (NK) cells and promoting the production of cytotoxic T lymphocytes (CTLs). These immune cells are critical for recognizing and eliminating cancer cells, thus providing a protective effect against CRC (González-Lozano et al. 2022).

7.10.4 Reduction of Carcinogen Exposure

Synbiotics help reduce exposure to carcinogens by modulating gut microbiota composition and activity. Probiotics can degrade and detoxify harmful compounds such as heterocyclic amines and N-nitroso compounds, which are derived from dietary sources and have been linked to CRC. The prebiotics, by promoting the growth of these probiotics, ensure a steady and effective detoxification process (O'keefe 2016).

7.10.4.1 Detoxification of Harmful Substances

Specific probiotic strains possess enzymes that can break down and detoxify carcinogenic substances. For instance, *Lactobacillus* and *Bifidobacterium* species can metabolize and reduce the bioavailability of potential carcinogens, thereby reducing the overall carcinogenic load in the gut (Fong et al. 2020; Zhao and Jiang 2021).

7.10.5 Synergy with Conventional Cancer Therapies

Recent research indicates that taking synbiotics may improve the effectiveness of conventional cancer treatments including immunotherapy and chemotherapy.:

7.10.5.1 Improved Chemotherapy Tolerance

Synbiotics can help mitigate the gastrointestinal side effects associated with chemotherapy, such as diarrhoea and mucositis, by maintaining gut barrier integrity and modulating immune responses. This supportive role can improve patients' overall quality of life and adherence to cancer treatment regimens (Alsuwaylihi and McCullough 2023).

7.10.5.2 Potential Enhancement of Antitumour Effects

Preclinical studies indicate that synbiotics may potentiate the antitumour effects of chemotherapy agents. By modulating the tumour microenvironment and improving gut health, synbiotics can enhance the delivery and efficacy of anticancer drugs (Gopalakrishnan et al. 2018).

7.11 Clinical Implications and Future Directions

7.11.1 Clinical Implications

The use of prebiotics and probiotics in CRC treatment holds promising clinical implications, highlighting their potential as adjunct therapies to conventional treatments (Alam et al. 2022).

7.11.1.1 Improvement of Gut Health and Reduction of Treatment Side Effects

The administration of prebiotics and probiotics can significantly improve gut health, which is often compromised during cancer treatments like chemotherapy and radiotherapy. Chemotherapy-induced mucositis and diarrhoea can be mitigated by probiotics, which enhance gut barrier function and modulate inflammation. Probiotics like *Lactobacillus rhamnosus* GG have been shown to minimize the incidence and severity of diarrhoea in patients undergoing chemotherapy (Danis et al. 2022; Feng et al. 2022). Prebiotics, by promoting the growth of beneficial gut bacteria, can further support this protective effect by producing SCFAs that nourish colonocytes and enhance mucosal healing.

7.11.1.2 Potential to Enhance Chemotherapy Efficacy

There is emerging evidence suggesting that prebiotics and probiotics may enhance the efficacy of chemotherapy drugs. SCFAs, especially butyrate, produced by the fermentation of prebiotics, have been shown to enhance the cytotoxicity of certain chemotherapy agents against CRC cells. This synergistic effect can potentially reduce the required dosage of chemotherapy drugs, thereby minimizing their toxic side effects (Canani et al. 2011; Alpino et al. 2024).

7.11.1.3 Modulation of Immune Responses

The immunomodulatory effects of prebiotics and probiotics is of particular interest in CRC treatment. By promoting the differentiation and function of regulatory T cells and enhancing the activity of NK cells and CTLs, these dietary supplements can help create an immune environment that is more conducive to tumour suppression. Clinical trials have demonstrated that probiotics can modulate systemic immune responses in cancer patients, potentially improving their overall prognosis (Sivan et al. 2015; Pei et al. 2024).

7.11.2 Future Directions

While the current evidence is promising, further research is necessary to fully understand the therapeutic potential and optimize the use of prebiotics and probiotics in CRC treatment.

7.11.2.1 Large-Scale Clinical Trials

Although small-scale studies and preclinical models have shown positive results, large-scale randomized controlled trials are needed to confirm the efficacy and safety of prebiotics and probiotics in CRC treatment. These trials should focus on identifying the most effective strains and doses, as well as determining the optimal timing and duration of supplementation (Saavedra 2001; Sakurai et al. 2022).

7.11.2.2 Personalized Nutrition and Microbiome-Based Therapies

The human gut microbiome is highly individual, influenced by genetics, diet, and environmental factors. Future research should aim to develop personalized nutrition strategies that tailor prebiotic and probiotic interventions to individual microbiome profiles. Advances in metagenomics and microbiome sequencing will enable the identification of specific bacterial strains and metabolites that are most beneficial for CRC patients (Zmora et al. 2018).

7.11.2.3 Integration with Conventional Therapies

Investigating the integration of prebiotics and probiotics with conventional cancer therapies, such as chemotherapy, radiotherapy, and immunotherapy, is crucial. Studies should explore how these supplements can enhance the therapeutic efficacy and reduce the side effects of standard treatments. Additionally, understanding the potential interactions between prebiotics/probiotics and cancer drugs will help in designing safe and effective combination therapies (Gopalakrishnan et al. 2018).

7.11.2.4 Mechanistic Studies

While the beneficial efficacies of prebiotics and probiotics is well-documented, the underlying mechanisms are not fully understood. Future research should focus on elucidating the molecular pathways through which these supplements exert their anti-cancer effects. This includes studying the impact of SCFAs on gene expression, immune cell function, and tumour microenvironment modulation (Belcheva et al. 2014).

7.11.2.5 Development of Next-Generation Synbiotics

The development of next-generation synbiotics, which combine prebiotics and probiotics in a single formulation, holds great potential. These formulations can be designed to target specific aspects of CRC pathogenesis, such as inflammation, carcinogen detoxification, and immune modulation. Research should aim to optimize the combinations of prebiotics and probiotics to achieve the maximum therapeutic benefit (Markowiak-Kopeć and Śliżewska 2020; Fusco et al. 2023).

7.12 Conclusion

The role of prebiotics and probiotics in CRC treatment is gaining recognition as a valuable complement to traditional therapies. These dietary supplements help restore microbial balance in the gut, which is crucial in preventing CRC-related dysbiosis. Prebiotics, through the productions of SCFAs like butyrate, propionate and acetate, offer anti-inflammatory and anti-carcinogenic benefits by inhibiting tumour cell growth and promoting apoptosis. Probiotics improve gut barrier integrity, reducing permeability and preventing the spread of pathogens and toxins, thereby mitigating inflammation, a significant risk factor for CRC. Furthermore, prebiotics and probiotics enhance immune responses, creating an environment that suppresses tumour growth and boosts the effectiveness of conventional cancer treatments. They also aid in detoxifying dietary carcinogens, thereby lowering the risk of CRC development. Despite these promising benefits, further research, particularly large-scale clinical trials, is needed to validate the efficacy and safety of prebiotics and probiotics in CRC treatment. Future directions should focus on personalized approaches based on individual microbiome profiles, integration with existing therapies, and the development of advanced synbiotics. Understanding the precise mechanisms of action and optimizing their use will lead to more effective CRC management. In summary, combining prebiotics and probiotics offers a comprehensive approach to CRC treatment by addressing gut health, immune function, and cancer prevention. Their ability to enhance treatment efficacy and reduce side effects makes them valuable tools in the fight against CRC.

References

Alam Z, Shang X, Effat K, Kanwal F, He X, Li Y et al (2022) The potential role of prebiotics, probiotics, and synbiotics in adjuvant cancer therapy especially colorectal cancer. J Food Biochem 46(10):e14302

Alpino GDCÁ, Pereira-Sol GA, Dias MDME, Aguiar ASD, Peluzio MDCG (2024) Beneficial effects of butyrate on brain functions: a view of epigenetic. Crit Rev Food Sci Nutr 64(12):3961–3970

Alsuwaylihi AS, McCullough F (2023) The safety and efficacy of probiotic supplementation for critically ill adult patients: a systematic review and meta-analysis. Nutr Rev 81(3):322–332

An J, Ha EM (2016) Combination therapy of Lactobacillus plantarum supernatant and 5-fluouracil increases chemosensitivity in colorectal cancer cells. J Microbiol Biotechnol 26(8):1490–1503

Anderson KE, Sinha R, Kulldorff M, Gross M, Lang NP, Barber C et al (2002) Meat intake and cooking techniques: associations with pancreatic cancer. Mutat Res/Fundamental and Molecular Mechanisms of Mutagenesis 506:225–231

Arpaia N, Campbell C, Fan X, Dikiy S, Van Der Veeken J, Deroos P et al (2013) Metabolites produced by commensal bacteria promote peripheral regulatory T-cell generation. Nature 504(7480):451–455

Arrieta MC, Meddings J, Field CJ (2011) The immunomodulatory effects of dietary fiber and prebiotics in the gastrointestinal tract. In: Paeschke TM, Aimutis WR (eds) Nondigestible carbohydrates and digestive health, pp 37–77

Belcheva A, Irrazabal T, Robertson SJ, Streutker C, Maughan H, Rubino S et al (2014) Gut microbial metabolism drives transformation of MSH2-deficient colon epithelial cells. Cell 158(2):288–299

Borowicki A, Michelmann A, Stein K, Scharlau D, Scheu K, Obst U, Glei M (2011) Fermented wheat aleurone enriched with probiotic strains LGG and Bb12 modulates markers of tumor progression in human colon cells. Nutr Cancer 63(1):151–160

Bray F, Ren JS, Masuyer E, Ferlay J (2013) Global estimates of cancer prevalence for 27 sites in the adult population in 2008. Int J Cancer 132(5):1133–1145

Bruno-Barcena JM, Azcarate-Peril MA (2015) Galacto-oligosaccharides and colorectal cancer: feeding our intestinal probiome. J Funct Foods 12:92–108

Canani RB, Di Costanzo M, Leone L, Pedata M, Meli R, Calignano A (2011) Potential beneficial effects of butyrate in intestinal and extraintestinal diseases. World J Gastroenterol: WJG 17(12):1519

Canfora EE, Meex RC, Venema K, Blaak EE (2019) Gut microbial metabolites in obesity, NAFLD and T2DM. Nat Rev Endocrinol 15(5):261–273

Cerf-Bensussan N, Gaboriau-Routhiau V (2010) The immune system and the gut microbiota: friends or foes? Nat Rev Immunol 10(10):735–744

Chang C-J, Lin T-L, Tsai Y-L, Wu T-R, Lai W-F, Lu C-C et al (2019) Next generation probiotics in disease amelioration. J Food Drug Anal 27(3):15

Chen T, Kim CY, Kaur A, Lamothe L, Shaikh M, Keshavarzian A, Hamaker BR (2017a) Dietary fibre-based SCFA mixtures promote both protection and repair of intestinal epithelial barrier function in a Caco-2 cell model. Food Funct 8(3):1166–1173

Chen ZY, Hsieh YM, Huang CC, Tsai CC (2017b) Inhibitory effects of probiotic Lactobacillus on the growth of human colonic carcinoma cell line HT-29. Molecules 22(1):107

Cheng FS, Pan D, Chang B, Jiang M, Sang LX (2020) Probiotic mixture VSL# 3: An overview of basic and clinical studies in chronic diseases. World J Clin Cases 8(8):1361

Cousin FJ, Jouan-Lanhouet S, Théret N, Brenner C, Jouan E, Le Moigne-Muller G et al (2016) The probiotic Propionibacterium freudenreichii as a new adjuvant for TRAIL-based therapy in colorectal cancer. Oncotarget 7(6):7161

Danis R, Mego M, Antonova M, Stepanova R, Svobodnik A, Hejnova R, Wawruch M (2022) Orally administered probiotics in the prevention of chemotherapy (±Radiotherapy)-induced gastrointestinal toxicity: a systematic review with meta-analysis of randomized trials. Integr Cancer Ther 21:15347354221144309

Davie JR (2003) Inhibition of histone deacetylase activity by butyrate. J Nutr 133(7):2485S–2493S

de Andrade Calaça PR, Bezerra RP, Albuquerque WWC, Porto ALF, Cavalcanti MTH (2017) Probiotics as a preventive strategy for surgical infection in colorectal cancer patients: a systematic review and meta-analysis of randomized trials. Transl Gastroenterol Hepatol 2:67

Delzenne NM, Cani PD (2011) Interaction between obesity and the gut microbiota: relevance in nutrition. Annu Rev Nutr 31(1):15–31

Duncan SH, Hold GL, Harmsen HJ, Stewart CS, Flint HJ (2002) Growth requirements and fermentation products of Fusobacterium prausnitzii, and a proposal to reclassify it as Faecalibacterium prausnitzii gen. nov., comb. nov. Int J Syst Evol Microbiol 52(6):2141–2146

Eckburg PB, Bik EM, Bernstein CN, Purdom E, Dethlefsen L, Sargent M et al (2005) Diversity of the human intestinal microbial flora. Science 308(5728):1635–1638

Feng J, Gao M, Zhao C, Yang J, Gao H, Lu X et al (2022) Oral administration of probiotics reduces chemotherapy-induced diarrhea and oral mucositis: a systematic review and meta-analysis. Front Nutr 9:823288

Ferguson LR, Philpott M, Karunasinghe N (2004) Dietary cancer and prevention using antimutagens. Toxicology 198(1–3):147–159

Ferlay J, Ervik M, Lam F, Colombet M, Mery L, Piñeros M et al (2020) Global cancer observatory: cancer today. International Agency for Research on Cancer, Lyon, p 20182020

Fong W, Li Q, Yu J (2020) Gut microbiota modulation: a novel strategy for prevention and treatment of colorectal cancer. Oncogene 39(26):4925–4943

Furrie E (2005) Probiotics and allergy. Proc Nutr Soc 64:465–469

Furusawa Y, Obata Y, Fukuda S, Endo TA, Nakato G, Takahashi D et al (2013) Commensal microbe-derived butyrate induces the differentiation of colonic regulatory T cells. Nature 504(7480):446–450

Fusco W, Lorenzo MB, Cintoni M, Porcari S, Rinninella E, Kaitsas F et al (2023) Short-chain fatty-acid-producing bacteria: key components of the human gut microbiota. Nutrients 15(9):2211

Garczyk A, Kaliciak I, Drogowski K, Horwat P, Kopeć S, Staręga Z et al (2022) Influence of probiotics in prevention and treatment of patients who undergo chemotherapy or/and radiotherapy and suffer from mucositis, diarrhoea, constipation, nausea and vomiting. J Clin Med 11(12):3412

Gibson GR (1998) Dietary modulation of the human gut microflora using prebiotics. Br J Nutr 80(S2):S209–S212

González-Lozano E, García-García J, Gálvez J, Hidalgo-García L, Rodríguez-Nogales A, Rodríguez-Cabezas ME, Sánchez M (2022) Novel horizons in postbiotics: Lactobacillaceae extracellular vesicles and their applications in health and disease. Nutrients 14(24):5296

Gopalakrishnan V, Spencer CN, Nezi L, Reuben A, Andrews MC, Karpinets TV et al (2018) Gut microbiome modulates response to anti–PD-1 immunotherapy in melanoma patients. Science 359(6371):97–103

Heydari Z, Rahaie M, Alizadeh AM, Agah S, Khalighfard S, Bahmani S (2019) Effects of Lactobacillus acidophilus and Bifidobacterium bifidum probiotics on the expression of microRNAs 135b, 26b, 18a and 155, and their involving genes in mice colon cancer. Prob Antimicrob Prot 11:1155–1162

Iqbal MZ, Qadir MI, Hussain T, Janbaz KH, Khan YH, Ahmad B (2014) Probiotics and their beneficial effects against various diseases. Pak J Pharm Sci 27(2):405–415

Jägerstad M, Skog K (2005) Genotoxicity of heat-processed foods. Mutat Res/Fundamental and Molecular Mechanisms of Mutagenesis 574(1–2):156–172

Jan GBAS, Belzacq AS, Haouzi D, Rouault A, Metivier D, Kroemer G, Brenner C (2002) Propionibacteria induce apoptosis of colorectal carcinoma cells via short-chain fatty acids acting on mitochondria. Cell Death Differ 9(2):179–188

Janney A, Powrie F, Mann EH (2020) Host–microbiota maladaptation in colorectal cancer. Nature 585(7826):509–517

Jenkins G, Mason P (2022) The role of prebiotics and probiotics in human health: a systematic review with a focus on gut and immune health. Food Nutr J 6:245

Kaźmierczak-Siedlecka K, Daca A, Fic M, van de Wetering T, Folwarski M, Makarewicz W (2020) Therapeutic methods of gut microbiota modification in colorectal cancer management–fecal microbiota transplantation, prebiotics, probiotics, and synbiotics. Gut Microbes 11(6):1518–1530

Konishi H, Fujiya M, Tanaka H, Ueno N, Moriichi K, Sasajima J et al (2016) Probiotic-derived ferrichrome inhibits colon cancer progression via JNK-mediated apoptosis. Nat Commun 7(1):12365
Kotzampassi K, Stavrou G, Damoraki G, Georgitsi M, Basdanis G, Tsaousi G, Giamarellos-Bourboulis EJ (2015) A four-probiotics regimen reduces postoperative complications after colorectal surgery: a randomized, double-blind, placebo-controlled study. World J Surg 39:2776–2783
Kumari A, Sudhakaran VA, Warrier AS, Singh NK (2024) Unveiling the health benefits of prebiotics: a comprehensive review. Indian J Microbiol 64:1–13
Kvakova M, Bertkova I, Stofilova J, Savidge TC (2021) Co-encapsulated synbiotics and immobilized probiotics in human health and gut Microbiota modulation. Food Secur 10(6):1297
Kvakova M, Kamlarova A, Stofilova J, Benetinova V, Bertkova I (2022) Probiotics and postbiotics in colorectal cancer: prevention and complementary therapy. World J Gastroenterol 28(27):3370
Lazar V, Ditu LM, Pircalabioru GG, Gheorghe I, Curutiu C, Holban AM et al (2018) Aspects of gut microbiota and immune system interactions in infectious diseases, immunopathology, and cancer. Front Immunol 9:1830
Li N, Lu B, Luo C, Cai J, Lu M, Zhang Y, Chen H, Dai M (2021) Incidence, mortality, survival, risk factor and screening of colorectal cancer: A comparison among China, Europe, and northern America. Cancer Lett 522:255–268.
Liong MT (2008) Roles of probiotics and prebiotics in colon cancer prevention: postulated mechanisms and in-vivo evidence. Int J Mol Sci 9(5):854–863
Liu D, Jiang XY, Zhou LS, Song JH, Zhang X (2016) Effects of probiotics on intestinal mucosa barrier in patients with colorectal cancer after operation: meta-analysis of randomized controlled trials. Medicine 95(15):e3342
López-Gómez L, Alcorta A, Abalo R (2023) Probiotics and probiotic-like agents against Chemotherapy-Induced Intestinal mucositis: a narrative review. J Personal Med 13(10):1487
Maftei N-M, Raileanu CR, Balta AA, Ambrose L, Boev M, Marin DB, Lisa EL (2024) The potential impact of probiotics on human health: An update on their health-promoting properties. Microorganisms 12:234
Marchesi JR, Dutilh BE, Hall N, Peters WH, Roelofs R, Boleij A, Tjalsma H (2011) Towards the human colorectal cancer microbiome. PLoS One 6(5):e20447
Markowiak-Kopeć P, Śliżewska K (2020) The effect of probiotics on the production of short-chain fatty acids by human intestinal microbiome. Nutrients 12(4):1107
Martín R, Langella P (2019) Emerging health concepts in the probiotics field: streamlining the definitions. Front Microbiol 10:1047
Miyake M, Oda Y, Owari T, Iida K, Ohnishi S, Fujii T et al (2023) Probiotics enhances anti-tumor immune response induced by gemcitabine plus cisplatin chemotherapy for urothelial cancer. Cancer Sci 114(3):1118–1130
Nagao M, Sugimura T (1993) Carcinogenic factors in food with relevance to colon cancer development. Mutat Res/Fundamental and Molecular Mechanisms of Mutagenesis 290(1):43–51
Ohira H, Fujioka Y, Katagiri C, Mamoto R, Aoyama-Ishikawa M, Amako K et al (2013) Butyrate attenuates inflammation and lipolysis generated by the interaction of adipocytes and macrophages. J Atheroscler Thromb 20(5):425–442
O'keefe SJ (2016) Diet, microorganisms and their metabolites, and colon cancer. Nat Rev Gastroenterol Hepatol 13(12):691–706
O'Morain VL, Ramji DP (2020) The potential of probiotics in the prevention and treatment of atherosclerosis. Mol Nutr Food Res 64(4):1900797
Österlund P, Ruotsalainen T, Korpela R, Saxelin M, Ollus A, Valta P et al (2007) Lactobacillus supplementation for diarrhoea related to chemotherapy of colorectal cancer: a randomised study. Br J Cancer 97(8):1028–1034
Papaliagkas V, Anogianaki A, Anogianakis G, Ilonidis G (2007) The proteins and the mechanisms of apoptosis: a mini-review of the fundamentals. Hippokratia 11(3):108

Pei B, Peng S, Huang C, Zhou F (2024) Bifidobacterium modulation of tumor immunotherapy and its mechanism. Cancer Immunol Immunother 73(5):94
Quigley EM (2011) Gut microbiota and the role of probiotics in therapy. Curr Opin Pharmacol 11(6):593–603
Rabiei M, Zarrini G, Mahdavi M (2020) Lactobacillus casei UT1 isolated from northwest of Iran traditional curd exerts anti-proliferative and apoptosis inducing effects in human colorectal tumor HCT 116 cells. Adv Pharm Bullet 10(1):125
Ragavan ML, Das N (2020) In vitro studies on therapeutic potential of probiotic yeasts isolated from various sources. Curr Microbiol 77:2821–2830
Raman M, Ambalam P, Kondepudi KK, Pithva S, Kothari C, Patel AT et al (2013) Potential of probiotics, prebiotics and synbiotics for management of colorectal cancer. Gut Microbes 4(3):181–192
Roberfroid M (2007) Prebiotics: the concept revisited1. J Nutr 137(3):830S–837S
Roller M, Femia AP, Caderni G, Rechkemmer G, Watzl B (2004) Intestinal immunity of rats with colon cancer is modulated by oligofructose-enriched inulin combined with Lactobacillus rhamnosus and Bifidobacterium lactis. Br J Nutr 92(6):931–938
Round JL, Mazmanian SK (2009) The gut microbiota shapes intestinal immune responses during health and disease. Nat Rev Immunol 9(5):313–323
Saavedra JM (2001) Clinical applications of probiotic agents. Am J Clin Nutr 73(6):1147S–1151S
Sakurai Y, Watanabe T, Miura Y, Uchida T, Suda N, Yoshida M, Nawa T (2022) Clinical and bacteriologic characteristics of six cases of Bifidobacterium breve bacteremia due to probiotic administration in the neonatal intensive care unit. Pediatr Infect Dis J 41(1):62–65
Sánchez B, Delgado S, Blanco-Míguez A, Lourenço A, Gueimonde M, Margolles A (2017) Probiotics, gut microbiota, and their influence on host health and disease. Mol Nutr Food Res 61(1):1600240
Sankpal UT, Pius H, Khan M, Shukoor MI, Maliakal P, Lee CM et al (2012) Environmental factors in causing human cancers: emphasis on tumorigenesis. Tumour Biol 33:1265–1274
Shamekhi S, Abdolalizadeh J, Ostadrahimi A, Mohammadi SA, Barzegari A, Lotfi H et al (2020) Apoptotic effect of Saccharomyces cerevisiae on human colon cancer SW480 cells by regulation of Akt/NF-κB signaling pathway. Prob Antimicrob Prot 12:311–319
Shang F, Jiang X, Wang H, Chen S, Wang X, Liu Y et al (2020) The inhibitory effects of probiotics on colon cancer cells: in vitro and in vivo studies. J Gastrointest Oncol 11(6):1224
Siegel R, Ma J, Zou Z, Jemal A (2014) Cancer statistics, 2014. CA Cancer J Clin 64(1):9
Sivan A, Corrales L, Hubert N, Williams JB, Aquino-Michaels K, Earley ZM et al (2015) Commensal Bifidobacterium promotes antitumor immunity and facilitates anti–PD-L1 efficacy. Science 350(6264):1084–1089
Sobhani I, Tap J, Roudot-Thoraval F, Roperch JP, Letulle S, Langella P et al (2011) Microbial dysbiosis in colorectal cancer (CRC) patients. PLoS One 6(1):e16393
Sun J, Chen S, Zang D, Sun H, Sun Y, Chen J (2024) Butyrate as a promising therapeutic target in cancer: from pathogenesis to clinic. Int J Oncol 64(4):1–16
Sung H, Ferlay J, Siegel RL, Laversanne M, Soerjomataram I, Jemal A, Bray F (2021) Global cancer statistics 2020: GLOBOCAN estimates of incidence and mortality worldwide for 36 cancers in 185 countries. CA Cancer J Clin 71(3):209–249
Tang G, Zhang L (2022) Update on strategies of probiotics for the prevention and treatment of colorectal cancer. Nutr Cancer 74(1):27–38
Terzić J, Grivennikov S, Karin E, Karin M (2010) Inflammation and colon cancer. Gastroenterology 138(6):2101–2114
Thirabunyanon M, Hongwittayakorn P (2013) Potential probiotic lactic acid bacteria of human origin induce antiproliferation of colon cancer cells via synergic actions in adhesion to cancer cells and short-chain fatty acid bioproduction. Appl Biochem Biotechnol 169:511–525
Ting NL, Lau HC, Yu J (2022) Cancer pharmacomicrobiomics: targeting microbiota to optimise cancer therapy outcomes. Gut 71(7):1412–1425

Torres-Maravilla E, Boucard AS, Mohseni AH, Taghinezhad-S S, Cortes-Perez NG, Bermúdez-Humarán LG (2021) Role of gut microbiota and probiotics in colorectal cancer: onset and progression. Microorganisms 9(5):1021

Tripathy A, Dash J, Kancharla S, Kolli P, Mahajan D, Senapati S, Jena MK (2021) Probiotics: a promising candidate for management of colorectal cancer. Cancers 13(13):3178

Tsuda H, Hara K, Miyamoto T (2008) Binding of mutagens to exopolysaccharide produced by Lactobacillus plantarum mutant strain 301102S. J Dairy Sci 91(8):2960–2966

Vandenplas Y, Veereman-Wauters G, Greef ED, Peeters S, Casteels A, Mahler T et al (2011) Probiotics and prebiotics in prevention and treatment of diseases in infants and children. J Pediatr 87:292–300

Wang D, Margalit O, DuBois RN (2013) Metronomic topotecan for colorectal cancer: a promising new option. Gut 62(2):190–191

Wang Y, Ma R, Liu F, Lee SA, Zhang L (2018) Modulation of gut microbiota: a novel paradigm of enhancing the efficacy of programmed death-1 and programmed death ligand-1 blockade therapy. Front Immunol 9:374

WHO. Colorectal cancer. Accessed at 11 July 2023. https://www.who.int/news-room/fact-sheets/detail/colorectal-cancer

Wierzbicka A, Mańkowska-Wierzbicka D, Mardas M, Stelmach-Mardas M (2021) Role of probiotics in modulating human gut microbiota populations and activities in patients with colorectal cancer—a systematic review of clinical trials. Nutrients 13(4):1160

Wilson KM, Giovannucci EL, Mucci LA (2012) Lifestyle and dietary factors in the prevention of lethal prostate cancer. Asian J Androl 14(3):365

Winawer SJ, St John DJ, Bond JH, Rozen P, Burt RW, Waye JD et al (1995) Prevention of colorectal cancer: guidelines based on new data. WHO Collaborating Center for the Prevention of Colorectal Cancer. Bull World Health Organ 73(1):7

Wirbel J, Pyl PT, Kartal E, Zych K, Kashani A, Milanese A et al (2019) Meta-analysis of fecal metagenomes reveals global microbial signatures that are specific for colorectal cancer. Nat Med 25(4):679–689

Wogan GN, Hecht SS, Felton JS, Conney AH, Loeb LA (2004) Environmental and chemical carcinogenesis. In: Seminars in cancer biology, vol 14, No. 6. Academic Press, pp 473–486

Wood RD, Mitchell M, Sgouros J, Lindahl T (2001) Human DNA repair genes. Science 291(5507):1284–1289

Wu H, Chiou J (2021) Potential benefits of probiotics and prebiotics for coronary heart disease and stroke. Nutrients 13(8):2878

Wu K, Giovannucci E, Byrne C, Platz EA, Fuchs C, Willett WC, Sinha R (2006) Meat mutagens and risk of distal colon adenoma in a cohort of US men. Cancer Epidemiol Biomarkers Prev 15(6):1120–1125

Zhao Y, Jiang Q (2021) Roles of the polyphenol–gut microbiota interaction in alleviating colitis and preventing colitis-associated colorectal cancer. Adv Nutr 12(2):546–565

Zhuo Q, Yu B, Zhou J, Zhang J, Zhang R, Xie J et al (2019) Lysates of Lactobacillus acidophilus combined with CTLA-4-blocking antibodies enhance antitumor immunity in a mouse colon cancer model. Sci Rep 9(1):20128

Zmora N, Zilberman-Schapira G, Suez J, Mor U, Dori-Bachash M, Bashiardes S et al (2018) Personalized gut mucosal colonization resistance to empiric probiotics is associated with unique host and microbiome features. Cell 174(6):1388–1405

Role of Dietary Phytochemicals in Gut Dysbiosis: Implications in Colon Cancer Prevention and Treatment

8

Vasanth Kanth Thasma LoganathBabu, R. M. Vijay Murali, P. Shriya, Raksa Arun, Mashira Rahman, Harin N. Ganesh, Satish Ramalingam, and ArulJothi Kandasamy Nagarajan

Abstract

The human gut is home to a complex and diverse microbial community known as the gut microbiota, which plays a crucial role in various aspects of host health and physiology. An imbalance in the microbiome in the gut causes gut dysbiosis. Gut dysbiosis has been associated with various diseases and health issues, ranging from gastrointestinal disorders to metabolic, immunological, and neurological conditions. It has been found that alterations in the microbiome of the human gut contribute to the development of various diseases such as colon cancer, cardiovascular diseases (CVDs), and diabetes mellitus (DM). Dietary phytochemicals (DPCs) are an abundant source of nutrients to the human body that are known to possess a wider range of beneficial effects. The role of DPCs in managing the gut biome is illustrated through a range of molecular mechanisms and pathways involving inhibitory and inductive events that in turn has a reaching effect on the occurrence and progression of colon cancer. The findings affirm that DPCs offer promising advancements in the field of treating colon cancer due to their potent effects comparable to market drugs with extraction technologies such as supercritical fluid extraction, microwave extraction, and ultrasonic extraction.

Keywords

Gut microbiota · Dysbiosis · Pathways · and Colon cancer

V. K. T. LoganathBabu · R. M. Vijay Murali · P. Shriya · R. Arun · M. Rahman · H. N. Ganesh · S. Ramalingam · A. K. Nagarajan (✉)
Department of Genetic Engineering, College of Engineering and Technology, SRM Institute of Science and Technology, Potheri, Kattankulathur, Chengalpattu, Tamilnadu, India
e-mail: vt7848@srmist.edu.in; vq2201@srmist.edu.in; sp9705@srmist.edu.in; ra9029@srmist.edu.in; mm7110@srmist.edu.in; hn7865@srmist.edu.in; aruljotn@srmist.edu.in

S. Pathak et al. (eds.), *Microbiota and Dietary Mediators in Colon Cancer Prevention and Treatment*, https://doi.org/10.1007/978-981-96-0297-1_8

8.1 Introduction

8.1.1 Gut Dysbiosis

The human gut is home to a complex and diverse microbial community known as the gut microbiota, which plays a crucial role in various aspects of host health and physiology (Huarcaya and Ríos 2023; Salem et al. 2018). These microorganisms, primarily composed of anaerobic bacteria, engage in a delicate balance with the human body, contributing to immune function, nutrient absorption, and even brain health (Shim et al. 2023). Dysbiosis, a term used to describe an imbalance in the gut microbiota, can have far-reaching consequences for the host (Selber-Hnativ et al. 2017). Dysbiosis is often characterized by a reduction in the diversity of the gut microbiome, an overgrowth of potentially harmful bacteria, and a disruption in the overall microbial community structure (Pilla and Suchodolski 2020). Gut dysbiosis can be triggered by various factors, including the use of antibiotics, dietary changes, stress, and certain underlying medical conditions (Choi and Cho 2016). Antibiotic use, in particular, has been linked to both short-term and long-term alterations in the gut microbiota, which can contribute to the development of conditions such as obesity and metabolic disorders (Huarcaya and Ríos 2023). Beyond the immediate effects on the gut, dysbiosis has been associated with a wide range of health issues, including inflammatory bowel diseases, irritable bowel syndrome, metabolic syndrome, obesity, type 1 and 2 diabetes, atopic disorders, and even neurological conditions like multiple sclerosis and autism (Choi and Cho 2016; Salem et al. 2018). The link between gut dysbiosis and these conditions is thought to be mediated by the altered production of metabolites, the disruption of immune function, and the breakdown of the intestinal barrier, which can lead to chronic inflammation and systemic effects. Furthermore, these alterations to the functions of the body may lead to cancer occurrence in the gut especially in the colon region (Chattopadhyay et al. 2021).

8.1.2 Dietary Phytochemicals

Plants have been used as a common source of food ever since humans came into being. These plant-based food or medicines consumed by humans are rich sources of phytochemicals. There are several health benefits known to be associated with these compounds. Several studies have investigated the relationship between the consumption of plant-based foods, which are abundant in phytochemicals, and various diseases, including cancer. Given that the majority of illnesses are linked to oxidative stress resulting from the excessive production of free radicals, phytochemicals are known to be beneficial since they naturally regulate the generation of free radicals. The reactive oxygen species present in the body have a tendency to interact with the electrons of other molecules, leading to the disruption of various enzyme systems and causing damage to DNA. This damage can contribute to oxidative damage and the development of inflammatory diseases and conditions such as cancer, ischemia, aging, adult respiratory distress syndromes, and rheumatoid arthritis. Oxidative stress arises from an in equilibrium between free radicals and the antioxidant defense system. A diet that is primarily composed of plant-based foods provides protection against chronic diseases that are caused by oxidative stress. This is performed by the phytochemicals through their antioxidant potential which plays a key role in preventing many diseases including cancer by suppressing oxidative stress-induced DNA damage. The plant kingdom has been the source of numerous antioxidants including compounds like phenolics, carotenoids, anthocyanins, and tocopherols (Jakubowski and Bartosz 1997).

Plants produce a variety of such phytochemicals to cater the needs of humans like protection against free radical damage and its use as a natural flavonoid. Phytochemicals are classified into 12 major categories such as polyphenols, carotenoids, glucosinolates, polysaccharides, lectins, terpenes, alkaloids, polyacetylenes, allium compounds, chlorophyll, capsaicinoids, and betalains (Kumar et al. 2023). Each of these is associated with its own specific function. For instance, polyphenols are known for their unique property of activation at multiple levels, through the modulation of A mitogen-activated protein kinase (MAPK), protein kinase B (PKB), and NF-κB signaling pathways, inhibiting the production of inflammatory cytokines and chemokines, suppressing the activity of cyclooxygenase (COX) and inducible nitric oxide synthase (iNOS) and thus decreasing the production of free radicals.

Numerous diseases like cardiovascular, neurological and diabetes are prevalent world-wide owing to increased mortality rates. Most of these diseases predominantly arise due to the increased free-radical toxicity and the high oxidative stress created by it (Suffredini et al. 2004). Phytochemicals are compounds that are mostly used in the medical sector to resist fungi, bacteria, and plant virus infections, and also consumption by insects and other animals. Their probable health effects of these have not been established yet but they have been categorized into two major categories such as carotenoids and polyphenols (Carotenoids | Linus Pauling institute | Oregon State University 2024; Makhaik et al. 2021).

The continuous consumption of phytochemicals is associated with protection against chronic diseases, such as cardiovascular disease, cancer, and neurological diseases. To establish a connection between the consumption of phytochemicals from food and beneficial health effects, it is crucial to precisely calculate the amount of phytochemicals consumed through diet. The term "phytochemicals" generally refers to these bioactive compounds found in plants, whether they are consumed through diet or extracted and used in supplements or other products. When we specifically refer to "dietary phytochemicals (DPCs)" we are emphasizing the intake of these compounds through natural food sources as part of a regular diet. The bioavailability of DPCs is enhanced by the food matrix and the presence of other nutrients, which can either increase or decrease their absorption and efficiency. Consuming phytochemicals as whole foods yields more powerful combined effects and provides superior health advantages compared to consuming them in their natural, unprocessed, isolated form. The study of DPCs is therefore important to understand the balanced diet in which these phytochemicals are to be consumed (DeFelice

1995). To summarize, DPCs are a type of phytochemical that, when consumed in a normal diet from natural food sources, contribute to health benefits in a manner that is affected by the total dietary context.

Antioxidant qualities are one of the many medical uses for edible fruits and vegetables. They include a diverse array of phytochemicals that have been scientifically demonstrated to mitigate the effects of several ailments. Fruits such as strawberries, citrus fruits, and green vegetables, as well as cereals, are abundant in vitamins and phytochemicals (Makhaik et al. 2021). It has been shown that the consumption of dietary plants lessens the development of lifestyle-related diseases because they are rich in polyphenols and show synergistic effects on the metabolic pathway of disease cells. The medicinal properties of most of these phytochemicals are immense, although they're still being studied. It is well known that certain phytochemicals, such as resveratrol, anthocyanins, and curcumin, have medicinal qualities that enhance immunity, reduce prostaglandin synthesis, regulate nuclear factor-κB activity, and inhibit lipid peroxidation to mediate protection. Phytochemicals typically exhibit potent antioxidant and anti-inflammatory properties, which help explain other health-promoting bioactivities (Makhaik et al. 2021; Qorbani et al. 2022).

There have been ways devised to analyze the phytochemical content in different dietary patterns. The concept of Dietary Phytochemical Index (DPI) plays a role here. DPI is defined as the percentage of dietary calories derived from phytochemical-rich foods such as fruits, vegetables, legumes, nuts, seeds, whole grains, and foods made with them. There is some evidence to support the effects of dietary quality via DPI on the risk of certain noncommunicable diseases (Lingas 2023). Traditional methods of dietary assessment require a recall or documentation of food intakes from a given time period in either a prospective or retrospective manner. To determine the nutrient composition of individual or group intakes, this dietary intake data needs to be converted to an estimated consumption of the "nutrient" of interest. This can be performed by applying a phytochemical Food Composition Database (FCDB) to the dietary intake data.

8.1.3 Colon Cancer

Colon cancer, also known as bowel cancer is a neoplasm that originates in the colon, which is a segment of the large intestine or rectum. Often, it begins as a small outgrowth like a polyp which are just small clumps of cells. Some of these polyps become cancerous overtime but most of these polyps are non-symptomatic. The large intestine is the final segment of the digestive system. The digestive system is responsible for the breakdown of food in order to facilitate its utilization by the body. It is a multifactorial disease and the third most common type of cancer diagnosed (Colorectal cancer statistics I WCRF International 2024). There are various factors that may increase the risk of colon cancer. These factors are listed below as follows:

(a) History of polyps
(b) Inflammatory bowel disease: Any conditions that result in pain and swelling of the intestines are called inflammatory bowel diseases and can increase the risk of colon cancer. These conditions include ulcerative colitis and Crohn's disease.
(c) Inherited syndromes like familial adenomatous polyposis and Lynch syndrome that increase the risk of colon cancer run in families.
(d) Colon cancer and rectal cancer may be associated with a conventional Western diet. This diet is typically deficient in fiber and abundant in fat and calories.
(e) Lifestyle diseases like diabetes could also influence the incidence of this
(f) Smoking, drinking alcohol, no regular exercises and obesity are other factors.

Although the exact cause of colon cancer is still a question mark, the above are the possible factors that can lead to the manifestation of this cancer. Colon cancers typically arise by a complex process that involves a sequence of histological, morphological, and genetic alterations that build gradually.

As per recent data, colon cancer has a high incidence of over 1.9 million new cases diagnosed each year (Simon 2016). Recent data suggests an intricate trend where certain nations have observed a rise or consistent occurrence and death rates, specifically in Eastern Europe, Latin America, and certain regions of Asia. On the other hand, certain areas, especially those with a significantly high Human Development Index (HDI), have experienced a decrease in the occurrence and death rates of a certain condition as a result of improved methods for examining and identifying it at an early stage.

Screening and diagnosis of early-stage precancerous polyps in individuals at average risk for colon cancer has been made possible. This can help prevent the development of cancer and potentially lead to significant reductions in the occurrence of colon cancer. Regarding the prevention of these issues, it is still uncertain and unstable. Certain medications have the ability to decrease the likelihood of developing colon polyps or colon cancer. So, for instance, there is data suggesting that regular use of aspirin or aspirin-like drugs is associated with a decreased risk of polyps and colon cancer. However, the specific dosage and duration required to effectively decrease the likelihood of developing colon cancer remain uncertain. Regular consumption of aspirin carries potential hazards, such as the development of ulcers and gastrointestinal bleeding. In addition to these fundamental lifestyle modifications, one can mitigate the likelihood of developing colon cancer by adhering to a nutritious diet that includes a diverse range of fruits, vegetables, and whole grains. It is also advisable to abstain from alcohol and smoking, engage in consistent and appropriate physical activity, and maintain a favorable body mass index (BMI) (Zmora et al. 2018).

8.2 Road to Colon Cancer Through Gut Microbiota Imbalance

8.2.1 Gut Dysbiosis and Its Implications in the Metabolic Network of the Human Body/Its Biological Implications

Gut dysbiosis is characterized by shift in equilibrium in the gut microbiome population, which is characterized either by loss of beneficial microbes or excess growth of harmful microbes. This shift is proven to possess a wide range of health complications and can prove to be fatal under extreme conditions. The cause of imbalance is influenced by a variety of factors including genetic composition, infections, inflammations, lifestyle, diets that include large amounts of sugar and less fiber, medications, xenobiotics as well as lack of hygiene. These factors contribute to compromised gut barrier integrity, thereby enabling harmful molecules from the diet as well as the microbiota to affect the immune system as well as the metabolism of the host (Hrncir 2022; Zeng et al. 2017). The effect of this underlying dysbiosis is found to alter the metabolism of the host in various ways, where it was found that the effect of gut dysbiosis led to alterations in the bile acid composition, affecting short-chain fatty acid (SCFA) production, and further converting choline to harmful compounds, and influencing glucose and lipid metabolism. This change in the bile acid composition due to dysbiosis disrupts the metabolic processes, since bile acids play a crucial role in lipid metabolism and glucose homeostasis (Dalile et al. 2019). The effect of gut dysbiosis is further proven to increase the proteolytic activity of the gut, thereby causing the intestinal barrier to disrupt and cause visceral hypersensitivity. It was also known that certain host serine proteases are upregulated in individuals with high gut proteolytic activity, contributing to digestive disturbances. Gut dysbiosis has also been studied to cause alterations in enzymes such as alkaline phosphatase, aminopeptidase, and maltase, which are the key enzymes that are involved in digestion process. The alterations caused in these enzymes due to gut dysbiosis is proven to impede the effect as well as the rate of digestion in host organisms (Gromova et al. 2021). Moreover, dysbiosis can trigger systemic inflammation, affecting the expression of tight junction proteins and proinflammatory cytokines, further disrupting intestinal integrity and function, a study conducted by Zhao et al. studied that luminal microbial dysbiosis contributes to mastitis development by inducing systemic inflammation and low-grade endotoxemia, which impairs host anti-inflammatory enzymes like alkaline phosphatase (ALP) through Neu activation. It was further revealed that low-grade endotoxemia caused by gut dysbiosis-derived LPS (lipopolysaccharide) led to mastitis due to the activation of specific inflammatory pathways such as TLR4, cGAS, STING, NF-kB/NLRP3. The findings of this study highlight the significance of gut microbiota in mastitis pathogenesis and suggest regulating gut microbiota and host anti-inflammatory enzymes as a potential strategy for mastitis prevention and other related diseases (Zhao et al. 2022). Gut dysbiosis possesses side effects that can extend beyond the digestive system, thereby affecting the functions of other organs as well. Patients with non-alcoholic fatty liver disease (NAFLD) are found to be more susceptible to inflammation and liver damage. Violi et al. had examined patients with NAFLD and the effect of low-grade endotoxemia by lipopolysaccharide (LPS) in liver inflammation as well as thrombosis and its relation with gut dysbiosis. From the results it was found that the effect of gut dysbiosis led to increased permeability of LPS, making them translocate into the systemic circulation wherein, it can accumulate further in the hepatocytes and endothelial lining. The interaction of LPS with toll-like receptor-4 (TLR-4) that are present in the hepatocytes contribute to the onset and progression of inflammation and thrombosis (Violi et al. 2023). Changes in the gut microbiota also contribute to chronic inflammations in patients especially of older age, leading to chronic kidney disease (CKD), that causes intoxication of the bloodstream with uremic toxins. Even though, there are not many studies that have been conducted to understand the effect of gut dysbiosis in causing CKD, preliminary studies have revealed that changes caused in the immune system due to gut dysbiosis causes abnormal growth and proliferation of lymphocytes which results in the production of self-antibodies which are associated with CKD (Honda and Littman 2016; Feng et al. 2021). The rise in the risk of cardiovascular diseases (CVDs) are found to be in association with gut dysbiosis, in which alterations in the gut microbiome causes diseases such as coronary artery diseases (CAD) and acute viral myocarditis (AVMC). Specific bacterial biomarkers have been identified via whole genome analysis which can differentiate CVD patients from the control group with a greater degree of specificity. The bacteria belonging to the genus Fusicatenibacter has been widely studied to be in close association with the onset of CVDs in patients with gut dysbiosis. *Fusicatenibacter* bacteria are found to be linked with the expression of CVD-related genes such as GBP2, MLKL, and CPR65 that are proven to play a crucial role in the onset of CVDs. The effect of these bacteria along with others belonging to the genus of *Blautia*, *Eubacterium*, and *Monoglobus* possess a major threat towards immune system alterations and CVDs (Chen et al. 2023). The influence of gut dysbiosis in the gut–brain axis plays an important role in the onset of neurodegenerative diseases such as Alzheimer's disease (AD) and Parkinson's disease (PD) (Solanki et al. 2023; Degruttola et al. 2016). The gut–brain axis comprises the dual direction of communication between the gut and the brain involving the metabolic, endocrine, neural, and immune pathways that are essential for maintaining brain homeostasis. The relationship between the gut and the brain highlights the importance of the microbial composition as well as the variations in the gut that can affect brain functioning and health. A variety of physiological factors such as mood, behavior, cognition are controlled by the gut–brain axis and any further alterations in this system causes disruption in the entire neurological system. During gut dysbiosis, the synthesis of bacterial amyloids and lipopolysaccharides causes macrophage dysfunction; increases gut permeability, which causes neuroinflammation, and the deposition of amyloid beta fibrils in the brain (Liu et al. 2020) (Fig. 8.1).

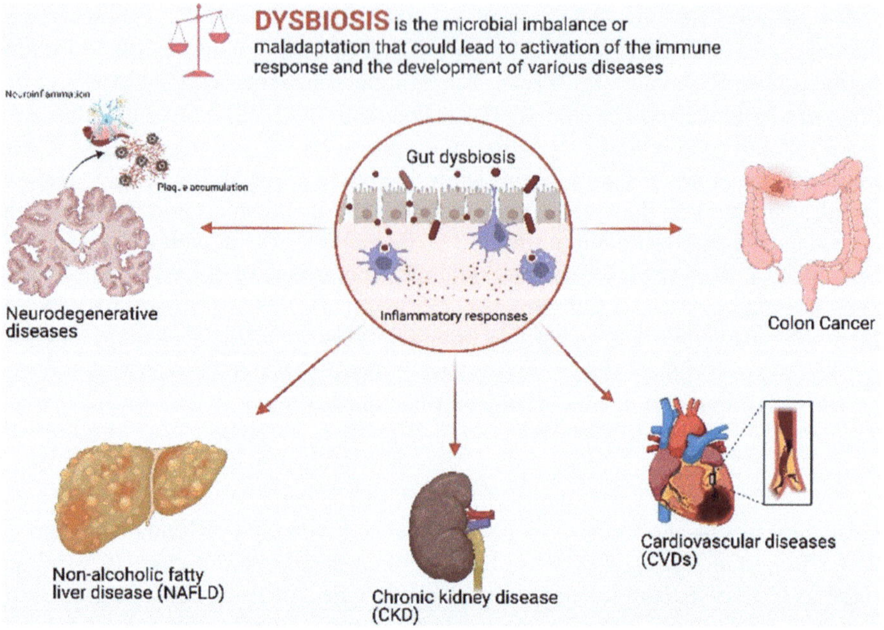

Fig. 8.1 Effect of gut dysbiosis in the host's system and its functioning in the onset of diseases

8.2.2 Gut Dysbiosis and Its Association with Colon Cancer

Gut dysbiosis is a condition which is referred to as an imbalance in the microbial community in the gastrointestinal tract, leading from mild to severe health effects. This is capable of disrupting the normal functions of the gut microbiome, which plays a crucial role in digestion, immune function, and overall health. Gut dysbiosis has been associated with various diseases and health issues, ranging from gastrointestinal disorders to metabolic, immunological, and neurological conditions. Gut dysbiosis as mentioned earlier can be caused by a lot of factors like poor dietary habits, excessive or inappropriate use of antibiotics causing inflammation, diseases like the inflammatory bowel disease and other lifestyle factors (Zmora et al. 2018). A unique aspect of colon cancer is its association with the gut microbiota which has been an important part of the tumor microenvironment (TME). A study by Wong et al. has shown that the development of colorectal tumors is accompanied by an imbalance in the composition of gut bacteria, fungi, viruses, and Archaea, a condition known as dysbiosis. These alterations in the microbial community may potentially contribute to the initiation and progression of colorectal tumors. The various mechanisms for microbiome–human interactions that impact host homeostasis include bacterial metabolites/toxin production, biofilm formation and host immune system modulation. The majority of this exchange of information happens at the epithelial layer of the gut. Therefore, there is growing interest in understanding the impact of these interactions on the development of colorectal cancer, a widespread illness that is currently experiencing considerable changes in its patterns of occurrence. There have been hypotheses that a number of gut microbiota can be potential organisms that contribute to the development of colon cancer, none have been conclusively identified as the direct cause. This indicates that factors such as the individual's lifestyle, interactions within the microbiome community, and the immune response in the mucosal and/or systemic systems may be crucial in the development of the disease (Wong and Yu 2023).

Gut dysbiosis has also been known to be linked to genetic mutations like the TP53 gene that has led to gut dysbiosis by altering the sialic acid metabolism. This subsequently increases inflammation in the intestine and since inflammation and gut dysbiosis are closely interlinked, any minute imbalance in the microbial communities residing in the gastrointestinal tract can also be harmful. The elevated inflammation due to TP53 can lead to a rise in pathobionts, which are microorganisms capable of causing disease in particular cases, illustrates the connection between genetic abnormalities and microbial imbalances in the promotion of gut inflammation (Lee et al. 2022). Specific genetic abnormalities which affect the gut microbiota are also linked to inflammatory bowel illnesses (IBD) such as Crohn's disease and ulcerative colitis. There has also been an increase in specific bacteria such as Enterobacteriaceae and Bacteroides in individuals with inflammatory bowel disease (IBD). These bacteria are known for their role in promoting inflammation in the GI tract. The knowledge of these underlying genetic factors is necessary for developing efficient treatment for this condition (Boccuto et al. 2023).

About 18 quantitative trait loci (QTLs) of the host presented significant linkage with the abundances of specific microbial taxa in the gut. In particular, these host loci control individual microbial species, groups of related taxa, or groups of phylogenetically distant microorganisms. Genes involved in this host genome-driven taxonomic effect on gut microbiome were those responsible for immune signaling like *Irak3*, *Lyz1*, *Lyz2*, *IFN-gamma*, and *IL-22*. These QTLs influence the gut microbial composition. There has also been evidence by Markowitz et al., that there exists a triangular association between human genome variations, gut microbiome composition and human health. A thousand gut microbiome associated variants (MAVs) were screened for their impact on phenotypes on human health. Of these, five MAVs were known to be highly associated with relative abundance of microbes in the human intestine. MAVs occur in all 22 pairs of autosomes, in coding and noncoding regions of the genome, including also the expression quantitative trait loci (eQTLs). These two are important as they determine the genetic and nucleotide variation associated with the differential expression of the target gene (Fig. 8.2).

miRNAs are small endogenous noncoding RNA of about 18–25 nucleotides that regulates gene expression though the degradation of mRNAs or inhibition of translation. Significantly, miRNAs also play a role in regulating the commensal microbiota-dependent intestinal epithelial cells, which are crucial for maintaining a healthy balance (eubiosis) vs an unhealthy imbalance (dysbiosis). Through data analysis, 14 circulating miRNAs' expression and abundance was determined to be particularly higher in case of obese patients. These miRNAs are as follows: miR-107, miR-103a-3p, miR-142-5p, miR-222-3p, miR-221-3p, miR-183-5p, miR-130b-3p,

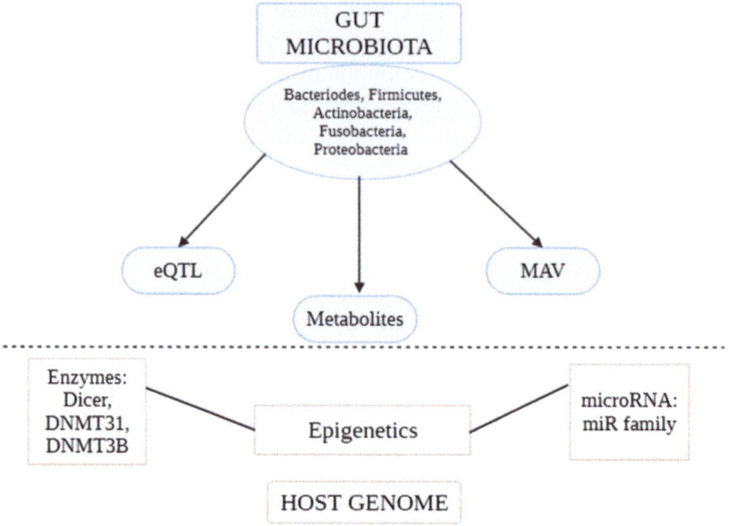

Fig. 8.2 Interaction between environment, gut microbiota and the host with its genome involving the various enzymes and other epigenetic factors. eQTL expression quantitative trait loci, MAV microbiome-associated genetic variants

miR-15a-5p, miR-33a-5p, miR-210-3p, miR-144-3p, miR-185-5p, miR-130a-3p, and miR-21-5p. The final component of the factors influencing gut microbiota is the metabolites which are effective host metabolic regulators and can also be affected by epigenetic mechanisms (Le Chatelier et al. 2013; Markowitz et al. 2022; Assmann et al. 2020). Colon cancer develops through a series of genetic and molecular changes involving multiple genes and proteins which transforms colonic epithelial cells to tumor cells. There are three crucial pathways that aid in this change, the chromosomal instability (CIN) pathway, microsatellite instability (MSI) pathway and CpG island methylator phenotype (CIMP) Pathway (Pino and Chung 2010) (Fig. 8.3).

8.2.2.1 Chromosomal Instability (CIN) Pathway

A majority of cases of colorectal cancer arise through the CIN pathway, which is characterized by widespread imbalances in chromosome number (aneuploidy) and loss of heterozygosity (LOH). Chromosomal instability (CIN) is observed in 70–85% of colon cancers and is often described in terms of MSS-colon cancers (microsatellite stability; MSS) (Worthley and Leggett 2010). It results from defects in chromosomal segregation, telomere stability, and DNA damage. A multistep pathway was proposed by Fearon and Vogelstein in 1990 that serves as a paradigm for solid tumor progression. The tumor suppressor gene APC (adenomatous polyposis coli) is suppressed first followed by activating mutations in the KRAS gene. Additional pathways of cancer like TGF-beta, PIK3CA, and TP53 also have crucial roles in this pathway. The model proposed a minimum of 7 mutations for tumor

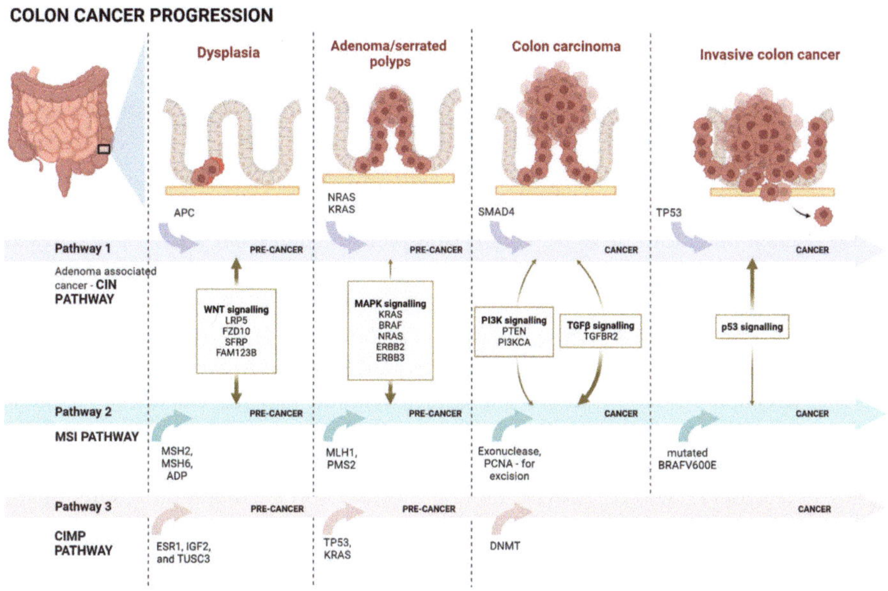

Fig. 8.3 The figure depicts all the three pathways like chromosomal instability (CIN), microsatellite instability (MSI), and CpG island methylator phenotype (cimp) pathway including their various contributing factors like TP53, KRAS, APC, exonuclease, DNMT and mutated BRAFV600E. KRAS Kirsten rat sarcoma viral oncogene homolog, MSH MutS homolog, MLH MutL homolog

progression. The progression of tumor begins with APC playing a role in initiating the adenoma formation. Quite opposite to APC is the KRAS mutant that cannot initiate a tumor by itself unless it is associated with a mutant APC (Fearon and Vogelstein 1990). The tumorigenesis of the colon starts with the formation of aberrant crypt foci (ACF). At this stage, the mutations in the APC gene lead to the activation of the Wnt signaling pathway. This pathway involves a complex interplay of proteins that is important in cell growth, differentiation, and embryonic development. They undergo Wnt ligand binding, inhibition of beta-catenin degradation by a complex of Axin, APC, GSK-3 beta and CK-1, followed by beta-catenin translocation (Koni et al. 2020). Post this stage, the development towards larger adenomas and earl carcinomas requires the activation of proto-oncogene KRAS along with some factors like EFGR and COX2. Mutational activation of PIK3CA occurs later in this sequence of adenoma-carcinoma. Due to the observation of chromosomal and genomic instability in benign adenomas in tandem with tumor progression, this pathway is named so (Wood et al. 2007).

8.2.2.2 Microsatellite Instability (MSI) Pathway

Microsatellites are tiny segments of DNA that repeat themselves and are commonly found scattered across the tumor genome. These sections are particularly susceptible to alterations. The primary faults commonly observed in microsatellites include base–base mismatches, as well as insertions and deletions in DNA coding areas, which result in frameshift mutations that can cause protein truncations. The most common form of MSI colon cancer is hereditary non-polyposis colon cancer (HPNCC or Lynch syndrome) caused by germline mutations in the mismatch repair genes MLH1, PMS2, MSH6, or MSH2, and accounts for about 3–5% of all colon cancer cases (Boland and Goel 2010). In comparison to MSS, MSI (microsatellite instability) tumors are often located in the proximal colon, are poorly differentiated, and possess a mucinous or signet ring histology (Pino and Chung 2011). The MMR system operates by rectifying errors that occur in microsatellites. This process involves a sequence of stages where MMR proteins engage as heterodimers with a complex produced by a MutS and a MutL (Boland and Goel 2010). MSI usually develops as a result of germline mutations in MMR or due to the epigenetic silencing of MLH1 in sporadic tumors due to methylation of CpG islands in the gene promoter region. As mentioned earlier, most of the colon cancers are a result of chromosomal instability. On the other hand, an MMR in DNA due to the MMR genes like MLH1, MSH2, MSH6, or PMS2 some sort of mutation results in the deactivation of MMR with a build-up of errors within microsatellite regions termed as MSI (Thibodeau et al. 1993).

This pathway starts with MSH2 being associated with either MSH6 or MSH3 on detecting a mismatch and it forms MutSα or MutSβ complexes, respectively. Similarly, MLH1 interacts with PMS2, PMS1, or MLH3 to form MutLα, MutLβ, or MutLγ complexes, respectively. The MSH3 gene has also been shown to encode the MutS homologue that dimerizes with MSH2. Upon these interactions the mismatch is excised by recruiting exonuclease 1 followed by resynthesis and religation of the DNA strand. The removal of the mismatch takes place when the DNA MMR protein sliding clamp interacts with exonuclease-1, proliferating cell nuclear antigen (PCNA), and DNA polymerase. This process involves the removal of the daughter strand up to the location of the mismatch. The complex dissociates from the DNA, followed by resynthesis which corrects the error. MSI malignancies exhibit mutations in mononucleotide sequences inside the coding areas of many genes, such as BRAFV600E, TGFβRII, BAX, and IGFIIR. A deficient MMR repair mechanism along with an MSI (frameshift mutations in microsatellites) and BRAFV600E mutation contributes to the progression of colon cancer (Sinicrope and Sargent 2012; Grady and Carethers 2008).

8.2.2.3 CpG Island Methylator Phenotype (CIMP) Pathway

The CpG island methylator phenotype (CIMP) was explained by Toyota et al. in 1999 as another pathway of importance in the development of colon cancer. CIMP is a subset of colon cancers that develop through an epigenetic instability pathway and are characterized by hypermethylation of promoter CpG island sites inactivating a number of tumor suppressor genes (Lao and Grady 2011). CpG (cytosine preceding guanine) islands are regions within the genome that are typically found in promoter sites rich in CpG dinucleotides. Over 50% of human genes have been discovered to be controlled through this mechanism, specifically by promoters that contain CpG islands. In cancer, several CpG dinucleotides that are methylated normally remain unmethylated (Toyota et al. 1999). In some cases, these may also be hypermethylated causing gene expression silencing. This type of aberrant genomic methylation is thought to lead to tumorigenesis by the deregulation of key genes like methyltransferases. DNA methylation is a process that adds methyl (-CH3) groups to the 5′‘position of cytosine using DNA methyltransferases (DNMTs) producing 5′‘-methylcytosine. This uses a substrate—GG dinucleotide and therefore the name CpG pathway. CpG are short repeats of CpG rich regions correlating to the promoter region of the genes. Methylation of these can lead to transcriptional silencing (Lao and Grady 2011). Research has shown that colorectal cancer is linked to two different types of methylation that contribute to the progression of the disease: age-related methylation (type A) and cancer-specific methylation (type C). Initially, age-related type A methylation occurs in normal colon epithelial cells. Methylation can influence the genes that control the proliferation and differentiation of cells, perhaps leading to a susceptibility for colon tumor formation (Toyota and Issa 1999).

8.3 Dietary Phytochemicals-Regulation of Colon Cancer by Gut-Health Mediation

8.3.1 Regulation of Gut Dysbiosis by Dietary Phytochemicals

Dietary phytochemicals play an important role in regulating gut dysbiosis by influencing the composition and metabolic function of the gastrointestinal microbial community. Biologically active phytochemicals include polyphenols like curcumin, gingerol, ferulic acid (FA) and hydroxytyrosol; alkaloids such as piperine, and protopine; terpenoids like artemisinin (terpenes), and astaxanthin (carotenoids); iridoid glycoside such as picroside II, and morroniside and flavonoid such as quercetin. These phytochemicals have been established to possess therapeutic potential in IBD by regulating antioxidant enzymes, regulating gut microbiota, and reducing inflammation markers. Constant consumption of phytochemical-rich diets can selectively stimulate the inhibition or proliferation of specific microbial communities in the intestine, profiting gut microbiota function and diversity (Ballway and Song 2021). Curcumin, derived from *Curcuma longa* (turmeric), showed that its supplementation enhanced the abundance of beneficial bacteria, resulting in improved modulation of gut microbiota, promotion of a healthier gut environment, which contributes to enhancing the intestinal barrier function and reducing intestinal inflammation (Guo et al. 2022). Gingerol is a polyphenol which is found in *Zingiber officinale* (ginger), increases the levels of SCFAs, for overall maintenance of gut health and improves gut microbiota structure by enhancing beneficial bacteria (Wang et al. 2020). Ferulic acid (FA), a polyphenol derived from several substances like cereals (wheat, oats, and rice); fruits (oranges, apples, and tomatoes); vegetables (spinach, parsley, and beets); seeds (flaxseeds and sesame seeds); and coffee (coffee beans). FA mitigates gut dysbiosis by interacting with the gut microbiota and improving intestinal integrity. Studies have shown that nonalcoholic fatty liver disease

(NAFLD) can be reduced by administration of FA, which helps in enhancing the production of SCFAs in the colon, as FA has anti-inflammatory properties and contributes to overall improvement of NAFLD. It helps in restoring the balance of the gut microbiota and increasing the beneficial short-chain fatty acid (SCFA)-producing bacteria, which plays a role in maintaining gut health and reducing inflammation (Fu et al. 2023). Hydroxytyrosol (HT), a polyphenol found in *Olea europaea* (olive oil), also improves the intestinal barrier, and modulates the composition of the gut microbiota. Studies demonstrated that HT intervention leads to a transformation of the gut microbiota, increasing the levels of beneficial bacteria that produce SCFAs and reducing inflammation related microbes (Micheli et al. 2023). Piperine, found in *Piper nigrum* (black pepper), enhances beneficial bacteria and inhibits harmful bacteria which leads to balanced and diverse gut microbiota. This balance strengthens immunity, promotes digestion, helping in maintaining an overall gut health (Dacrema and Ali 2022). Protopine is an isoquinoline alkaloid that is found in *Macleaya cordata* (a herbaceous plant, plume poppy). It plays a vital role in reducing gut dysbiosis by inhibiting the growth of harmful bacteria like *Proteobacteria, Escherichia–Shigella*, and *Enterococcus* in the gut microbiota, which helps in rebalance of gut microbiota. By regulating the composition of intestinal bacteria, protopine contributes to the maintenance of a diverse and beneficial microbial community in the gut, which is essential for overall gut health (Yue et al. 2023). Artemisinin, a terpene belonging to the class of terpenoids, is derived from *Artemisia annua* (a herbaceous plant, sweet wormwood), has shown therapeutic potential in conditions like malaria, and ulcerative colitis (UC) by altering gut microbiota composition, which includes changes in the relative abundance of major phyla like *Bacteroidetes* and *Firmicutes (*Katsi et al. 2018*)*. The combination of artemether-lumefantrine and probiotics like *Arthrospira platensis* has established significant modifications in gut bacteria communities, potentially influencing parasite burden and disease severity in malaria patients (Awah-Nanzdum et al. 2023). Astaxanthin, a carotenoid belonging to the class of terpenoids, is found in *Haematococcus pluvialis* (a freshwater microalga). Astaxanthin alters the composition of gut microbiota and influences the production of SCFAs by enhancing beneficial bacteria. It also modulates gut health by attenuating stress-induced intestinal hypermotility, which is a condition where stress causes the muscles in the intestines to move more quickly than usual. This can lead to digestive issues such as diarrhea, abdominal cramps, and bloating (Yasuda et al. 2022). Picroside II is an iridoid glycoside derived from *Picrorhiza kurroa* which is a herb named "kutki." Picroside II modulates the gut microbiota composition and improves intestinal barrier function. Research specifies that picroside II treatment can protect against pancreatitis-induced intestinal barrier injury by affecting inflammatory and oxidative signaling pathways. Picroside II reduces the UC by repressing the production of NLRP3 inflammasomes via NF-κB signaling pathway (Piao et al. 2020). Morroniside, a bioactive compound found in *Cornus officinalis* (a shrub named, Japanese cornel), can attenuate inflammatory and oxidative stress in an IBD cell model, reducing oxidative damage and inflammation (Zhang et al. 2022). Quercetin, a flavonoid that is found in a variety of plant-based foods like fruits (apples, berries, grapes, and citrus fruits); vegetables (onions (particularly red onions), broccoli, kale, and tomatoes); beverages (green tea and red wine); capers, seeds, and grains, when given as a supplementation can reduce high-fat diet-induced obesity by regulating the abundance of specific gut microbiota, such as *Adlercreutzia*, *Allobaculum*, *Coprococcus*, *Lactococcus*, *Faecalibacterium*, and *Akkermansia*. It can improve abdominal obesity by modulating gut microbiota dysbiosis, downregulating liver retinol metabolism, and reducing hypothalamic damage, highlighting its potential as a prebiotic agent (Su et al. 2022).

8.3.1.1 Molecular Mechanisms Underlying Gut Dysbiosis

Through several molecular processes, gut dysbiosis—which is characterized by changes in the makeup of the gut microbiota—is associated with a number of disorders. Research indicates that the SARS-CoV-2 virus causes enteric infection by binding to the ACE2 receptor; it enters the enterocytes and suppresses the antiviral response and then replicates to form more viral particles; this infection induces the secretion of pro-inflammatory mediators like interferons and cytokines such as interleukins (IL) and tumor necrosis factor (TNF) alpha, which then recruit inflammatory cells like macrophages, neutrophils, and lymphocytes (Clerbaux et al. 2022). Moreover, microbial dysbiosis influences transcriptome-level gene regulatory patterns, which is a critical factor in the emergence of severe depressive illness like depression, anxiety, and stress. Bacteria present in the gut releases endotoxin that incites powerful inflammatory responses in the host organism. Furthermore, toxic metabolites produced by the gut microbiota during protein fermentation including p-cresol and indoxyl sulfate can lead to disruption in gut barrier, this will allow translocation of endotoxin and bacterial metabolites to systemic circulation which contributes to uremic toxicity, inflammation which will result in many diseases like chronic kidney disorder (CKD). Also, chronic neurological and neurodegenerative illnesses are linked to gut microbial dysbiosis; possible explanations for this association include autoimmune dysregulation, misfolded proteins, and persistent inflammation (Mousa et al. 2022). Comprehending these molecular pathways is essential for formulating focused therapies to address ailments linked to dysbiosis of the gut. Gut dysbiosis can affect different pathways and ensuing problems as mentioned above. Furthermore, the use of antibiotics can cause gut dysbiosis that alters metabolic profiles by influencing pathways such the fatty acid biosynthesis, histidine breakdown, and purine and pentose metabolism. Moreover, dysbiosis can result in the overabundance of some bacteria like *Bacteroidetes*, *Firmicutes*, *Actinobacteria*, and *Proteobacteria*, raising levels of lipopolysaccharide (LPS) and trimethylamine-N-oxide (TMAO) in the bloodstream, which can cause inflammatory reactions, and chronic inflammation is a known risk factor for cardiovascular diseases (Manolis et al. 2021). These are some examples of how the gut microbiome interacts with different pathways and how gut dysbiosis causes different diseases gives us insight on how gut microbiome functions at a molecular level.

Dietary phytochemicals also have a significant impact on the composition and metabolic processes of the microbial community in the gastrointestinal tract, which helps to regulate gut dysbiosis. Artemisinin, protopine, ferulic acid (FA), picroside II, quercetin, hydroxytyrosol, morroniside, piperine, curcumin, gingerol, and astaxanthin are phytochemicals that are not biologically active. By altering the gut microbiota, controlling antioxidant enzymes, and lowering inflammatory indicators, these have shown therapeutic promise in the treatment of inflammatory bowel disease (IBD). Regular consumption of diets high in phytochemicals can favorably influence the variety and function of the gut microbiota by either selectively stimulating the growth of certain microbial communities or inhibiting others. Given below are examples of how these phytochemicals play a role in regulating gut microbiota.

Artemisinin

The herb Artemisia species yields artemisinin, prized for its anti-inflammatory and anticancer qualities. It activates the Nrf2 pathway by reducing Nrf2 ubiquitination and improving its stability, essential for cellular defense mechanisms against oxidative damage, and has an impact on pathways linked to inflammation and oxidative stress. Additionally, artemisitene affects the NF-κB signaling system by inhibiting NF-κB activation by suppressing the phosphorylation and degradation of I-κα and p65 nuclear translocation; this is implicated in the development of cancer and inflammatory reactions (Gao et al. 2022; Huang et al. 2019).

Morroniside

It has been demonstrated that the main iridoid glycoside from *Cornus officinalis*, morroniside, influences pathways linked to neuroprotection and anti-inflammatory reactions. It stimulates the PI3K/Akt signaling pathway by increasing p-AKT expression, which improves neuronal cell survival and lowers apoptosis (Deng et al. 2021). Furthermore, morroniside affects the MAPK and NF-κB pathways by inhibiting ERK (extracellular signal-regulated kinase), JNK (c-Jun N-terminal kinase), and p38 in MAPK pathway and TRAF6 (tumor necrosis factor receptor-associated factor 6) in NF-kB pathway, which are essential for inflammation and cell stress reactions (Xiao et al. 2023).

Protopine

Protopine is an isoquinoline alkaloid with anti-inflammatory, antioxidant, and neuroprotective properties that can be found in a variety of plants. It lowers oxidative damage and inflammation by influencing the NF-κB pathway by hindering the degradation of IkB and downregulating JNK, ERK, and p38 proteins in MAPK signaling pathways, which are involved in apoptosis and cell division (Alam et al. 2019).

Ferulic Acid

One phenolic compound with well-known anti-inflammatory and antioxidant qualities is ferulic acid. It increases the expression of antioxidant enzymes such as heme oxygenase-1 and thermal shock protein 70 (HSP-70), which scavenges the reactive oxygen species (ROS)/reactive nitrogen species (RNS) thus enhancing activation of the Nrf2/ARE pathway (Singh et al. 2022). Additionally, by blocking the NF-κB signaling pathway by preventing IkBα phosphorylation, lowers inflammation and guards against oxidative damage. It also influences the MAPK pathway by suppressing JNK/ERK1/2, which adds to its anti-tumor properties (Park and Han 2024).

Quercetin

The flavonoid quercetin is being investigated extensively for its potential anti-inflammatory, antioxidant, and anticancer properties. It alters multiple signaling pathways, such as the NF-κB, MAPK, and PI3K/Akt pathways. It inhibits NF-kB activation through IkB degradation; reduces activation of phosphorylated ERK kinase and p38 MAP kinase in MAPK pathway and overexpressing SIRT5 to inhibit PI3K/Akt (Cho et al. 2003). Quercetin reduces inflammation by inhibiting the NF-κB pathway and increases the Nrf2 pathway by increasing the expression of Nrf2 protein, which promotes antioxidant defense.

Picroside II

Derived from *Picrorhiza kurroa*, picroside II has anti-inflammatory and hepatoprotective properties. It modifies the pathways that are involved in inflammatory reactions, NF-κB and MAPK. It inhibits IKK activation, stabilizes IκBα, prevents NF-κB nuclear translocation in NF-kB pathway exerting an anti-inflammatory effect and reduces the phosphorylation and activation of ERK, JNK, and p38 MAPK, leading to decreased inflammatory responses and cell proliferation (Wang et al. 2015). Additionally, picroside II affects the Nrf2 pathway by activating Nrf2, which strengthens the antioxidant capacity of cells and guards against oxidative stress (Li et al. 2020).

Hydroxytyrosol

Olive oil contains hydroxytyrosol, a phenolic molecule known for its strong antioxidant capabilities. By stimulating the Nrf2 pathway by enhancing Nrf2 protein expression, it lowers oxidative stress and increases the expression of antioxidant enzymes (Han et al. 2023). Moreover, hydroxytyrosol has anti-inflammatory properties by blocking the NF-κB pathway. Hydroxytyrosol inhibits the IKK complex, preventing the degradation of IκBα, which results in decreased nuclear translocation of NF-κB and reduced transcription of pro-inflammatory cytokines (Zhang et al. 2020). Hydroxytyrosol activates the PI3K/Akt pathway, promoting cell survival, proliferation, and anti-apoptotic effects through the phosphorylation of Akt and its downstream targets (Pei et al. 2016).

Piperine

Piperine inhibits the NF-κB signaling pathway by inhibiting p65 and decreases the production of pro-inflammatory cytokines such as TNF-α, IL-1β, and IL-6, thus alleviating inflammation (Duan et al. 2022). It affects the MAPK pathway by decreasing the expression of p-JNK leading to decreased inflammatory response.

Curcumin

Curcumin inhibits the NF-kB pathway by degradation and suppression of phosphorylation of IkB (Olivera et al. 2012). It also inhibits p38 expression in the MAPK pathway which plays a role in inflammation, cell differentiation, and motility (Shamsnia et al. 2023).

Gingerol

Gingerol degrades IκBα degradation and translocation of the p65 NF-κB subunit core, intercepting signal activation through IKK and reducing the synthesis of inflammatory cytokines (Baptista et al. 2022). It inhibits p38 MAP kinase in the MAPK pathway, this inhibition of NF-kB and p38 MAP in turn inhibits COX2.

Astaxanthin

Astaxanthin inhibits NF-kB and Wnt signaling pathways by downregulating IKKβ and GSK-3β. This induces intrinsic apoptosis and decreases inflammation reaction as NF-kB pathway is affected (Kavitha et al. 2013).

These phytochemicals demonstrate significant potential in modulating key signaling pathways involved in oxidative stress, inflammation, and cancer, highlighting their therapeutic promise in various diseases.

8.3.1.2 Effect of Dietary Phytochemicals on Gut Microbiome Regulation

The mechanisms involved in the gut microbiome growth by the dietary phytochemicals which include polyphenols like curcumin, gingerol, ferulic acid (FA) and hydroxytyrosol; alkaloids such as piperine, and protopine; terpenoids like artemisinin (terpenes), and astaxanthin (carotenoids); iridoid glycoside such as picroside II, and morroniside and flavonoid such as quercetin and the bacterial species involved are demonstrated in researches. Feini Zhou et al. showed that curcumin supplementation in animal models enhances *Akkermansia* and *Muribaculum* which are beneficial bacteria. These bacteria inhibit gut dysbiosis by restoring composition of gut microbiota and promoting a healthy balance. The restoration of healthy gut microbiota contributes to stabilizing the gut–liver axis (a bidirectional relationship between gut and liver where gut microbiota and their metabolites such as SCFA, LPS, bile acids etc. influence liver health and function, and vice versa) and improving overall gut health (Zhou et al. 2023). Gingerol enhances beneficial bacteria like *Lactobacillus*, and *Bifidobacterium*, promoting gut health and healthy balance in gut microbiota. It inhibits *Clostridium* and *Escherichia coli*, which causes gut inflammation, leading to gut dysbiosis. Inflammation is caused by inhibition of pro-inflammatory cytokine (TNF-α, IL-1β, and IL-6), which can be inhibited by gingerol to reduce inflammation. Reactive oxygen species (ROS) are highly reactive chemicals formed from oxygen that contribute to inflammation. ROS can activate the NF-κB pathway by promoting phosphorylation. Gingerol reduces ROS by its antioxidant properties, thereby inhibiting the activation of NF-κB pathway. Inhibiting the activation of the pathway, reduces the expression of genes involved in inflammation (TNF-α, IL-1β, IL-6, IL-8) (Shen et al. 2022). FA supplementation increases the levels of SCFA-producing bacteria like *Lactobacillus* and *Bifidobacterium*, which reduces inflammation and enhances gut health. Rodríguez-Morató et al. showed that supplementation of hydroxytyrosol increased the abundance of beneficial bacteria like *Bifidobacterium* and *Faecalibacterium*, which leads to reduced oxidative stress and improved gut health (Rodríguez-Morató et al. 2015). Piperine enhances beneficial bacteria like *Lactobacillus* and *Bifidobacterium*, which are essential for supporting the immune system, promoting digestion and preventing the overgrowth of harmful bacteria like *Escherichia coli* in the gut. It inhibits pro-inflammatory cytokines which lead to decrease in inflammation in gut, reducing the severity of diarrhea (Satitsri et al. 2023). Protopine enhances the growth of beneficial bacteria and lessens pathogenic bacteria, thereby promoting gut health. A study established that protopine treatment in animal models showed higher relative abundance of *Bacteroides* and *Prevotella*, which are involved in maintaining a balanced gut microbiome and reducing inflammation (Aizawa et al. 2019). Artemisinin promotes beneficial bacteria like *Lactobacillus* and *Bifidobacterium* and reduces harmful bacteria, therefore modulating gut microbiota composition (Abdelhamid et al. 2019). The potential of artemisinin results in improved immune response, gut health and anti-inflammatory effects. Picroside II could increase the abundance of beneficial bacteria like *Lactobacillus* and *Ruminococcus*, which are involved in anti-inflammatory responses and gut health (Yang et al. 2020). A study by Meng Wang et al. showed that supplementation of astaxanthin in the gut microbiota of mice enhances beneficial bacteria like *Allobaculum* and *Akkermansia*, which helps maintain the overall metabolic balance and gut health. Astaxanthin modulates gut–liver axis, and production of SCFAs, which have anti-inflammatory properties, is increased by enhancing the beneficial bacteria. It inhibits the growth of harmful bacteria like *Parabacteroides* and *Desulfovibrio*, leading to prevention of release of pro-inflammatory substances like LPS, which can trigger systemic inflammation (Wang et al. 2022). Morroniside impacts gut microbiota by increasing the gut bacteria, which helps in gut barrier integrity and reducing inflammation. A study showed that administration of morroniside in mice led to an increase in *Akkermansia muciniphila* and *Faecalibacterium prausnitzii*. These bacteria are known for their role in gut barrier function and their anti-inflammatory properties. Quercetin can increase the abundance of beneficial bacteria such as *Bifidobacterium* and *Lactobacillus*, which provides better immune function and gut health (Kim et al. 2020).

8.3.2 Regulation of Colon Cancer by Dietary Phytochemicals

The gut microbiota, consisting of bacterial communities, plays a vital role in maintaining immune homeostasis and influencing systemic immune responses. Dysbiosis disrupts this delicate balance, leading to inflammation and tumorigenesis, which is linked to cancer progression in various studies (Kim and Lee 2022). Research specifies that immune crosstalk between type 3 innate lymphoid cells (ILC3s) and gut microbiota maintains intestinal homeostasis influencing immune responses, inflammation in intestines and contributes to immune surveillance against colon cancer. Dysregulation of this cross talk can lead to dysbiosis, chronic inflammation and can impact immune surveillance mechanisms such as activation of T-cells, detection of tumor antigens, leading to development and progression of colon cancer (Drommi et al. 2023). Fecal microbiota transplantation emerged as a potential therapeutic strategy to inhibit cancer development by reducing inflammation, reversing gut microbial disorder, and enhancing anticancer immune responses (Wong and Yu 2019). Antibiotic-induced dysbiosis has been shown to enhance tumor progression by affecting the tumor microenvironment, suppressing cytokine profiles, and reducing anticancer immune surveillance (Gopalakrishnan et al. 2018).The abovementioned dietary phytochemicals also play an important role in regulating colon cancer. Hydroxytyrosol is a potent antioxidant found in olives that enhances the growth of beneficial bacteria like *Bifidobacterium* and *Faecalibacterium* which helps in reducing oxidative stress and inflammation, contributing to a lower risk of colon cancer. FA exerts anticancer effects by scavenging free radicals, modulating gene expression related to cancer progression, and inhibits growth of cancer cells by inducing cell cycle arrest at G1 phase and promoting apoptosis. Quercetin has anti-cancer properties; it modulates pathways related with cancer (Asgharian et al. 2022). Protopine exhibits anti-inflammatory properties and can inhibit cancer cell growth by apoptosis, ferulic acid increases SCFA-producing bacteria also plays a role in reducing colon cancer risk by improving gut health and reducing inflammation. Artemisinin and its derivatives can target tumor cells given their iron-dependent mechanisms, leading to oxidative stress and cell death by apoptosis (Kiani et al. 2020). Picroside II enhances the growth of beneficial bacteria that helps in maintaining gut microbiota that in turn reduces the risk of colon cancer. Morroniside shows anti-inflammatory and antioxidant properties by increasing the growth of beneficial bacteria such as *Akkermansia muciniphila* and *Faecalibacterium prausnitzii*. An imbalance in the gut microbiota leads to gut dysbiosis, which leads to the development of colon cancer.

Recent research highlights several mechanisms through which gut dysbiosis contributes to colon cancer:

- *Inflammation*: A critical factor in colon cancer is the chronic inflammation, which is induced by the gut dysbiosis. *Fusobacterium nucleatum* and *Escherichia coli* are pro-inflammatory bacteria, which can produce genotoxins and activate inflammatory pathways, promoting tumor growth and development (Rubinstein et al. 2013).
- *Immune Response Modulation*: Dysbiosis modifies the immune response, weakening the body's ability to fight tumor cells. Beneficial bacteria such as *Bifidobacterium* and *Lactobacillus* are reduced in dysbiosis, impairing the immune surveillance against cancer cells (Gopalakrishnan et al. 2018).
- *Metabolic Changes:* Dysbiosis can result in the production of secondary bile acids and polyamines, which are harmful metabolites. These are linked to DNA damage and cancer cell proliferation. Beneficial metabolites like short-chain fatty acids (SCFAs) are often decreased in dysbiosis, since they have anti-inflammatory and anti-carcinogenic properties (Kwao-Zigah et al. 2024).
- *Microbial Toxins and Carcinogens:* Certain pathogenic bacteria produce toxins and carcinogenic compounds that directly damage DNA and promote mutations.

It can facilitate the adherence and invasion of tumor cells, enhancing cancer progression (Ivleva and Grivennikov 2022).

- *Gut Barrier Dysfunction:* Dysbiosis can compromise the intestinal barrier, which leads to increased permeability or "leaky gut." This allows microbial products and toxins to enter the bloodstream, causing systemic inflammation and contributing to cancer progression (Genua et al. 2021).

Gut dysbiosis influences colon cancer progression through mechanisms such as inflammation, immune modulation, metabolic changes, production of toxins, and gut barrier dysfunction. These insights underline the importance of maintaining a balanced gut microbiota for cancer prevention and therapy (Saus et al. 2019).

8.3.2.1 Colon Cancer Regulation Through Gut Microbiota Management

Phytochemicals normally found in fruits, vegetables, and other plant-based foods play a major role in the regulation of gut microbiota management and thereby influencing colon cancer outcomes. They do this by maintaining microbial balance of the gut, enhancing immune function by restoring beneficial bacteria, improving gut barrier function, and tackling inflammation. DPCs such as curcumin, piperine and gingerol, are found in household staples that are used as ingredients in all types of cuisine. They play a significant role in the inhibition of major cancer pathways NF-κB, Wnt/β-catenin and MAPK pathway which further contribute to preventing the occurrence of colon cancer. These DPCs, as seen in Table 8.1, not only help fight against colon cancer but also play a crucial role in maintaining overall gut health and improved digestion and absorption. DPCs also scavenge ROS found in the gut which then results in reduced inflammation (Fig. 8.4). As a result of a healthy gut, it also creates a thriving environment for the gut microbiota which in turn creates positive feedback to the gut environment.

8.4 Conclusion

8.4.1 Summary of Key Findings

This study underscores the significant impact of these dietary phytochemicals (DPCs) on maintaining gut microbiota balance, reducing the inflammation caused and colon carcinogenesis. Dietary phytochemicals, such as flavonoids, polyphenols, and terpenoids, have the ability to modulate the gut environment, thereby preventing the onset and metastasis of colon cancer. DPCs are a good alternative to the current market drugs as they are obtained naturally with considerably few negative effects. These also affect the cancer formation, progression, and signaling pathways thereby boosting their therapeutic potential. The extraction of DPCs particularly stands out as a crucial area for development. The extraction cost is relatively low compared to the synthesis of market drugs, making DPCs a cost-effective option for large-scale application. Additionally, the raw ingredients required for these

Table 8.1 Classifications, mechanisms in gut microbiome, and implementation of dietary phytochemicals in colon cancer

S. no.	Classification	Dietary phytochemicals	Mechanism	Implication in colon cancer
1.	Polyphenols	Curcumin	Acts as anti-inflammatory agent, has antioxidant property, induces apoptosis, inhibits cell proliferation	Reduces inflammation and oxidative stress, induces apoptosis, and inhibits proliferation in colon cancer cells
		Gingerol	Reduces inflammation, reduces oxidative stress, increases the expression of pro-apoptotic proteins, and reduces the expression of angiogenic factors	Inhibits pathways such as NF-kB and MAPK, essential for cancer development Promotes beneficial bacteria that produce anti-cancer SCFAs like *Bifidobacterium,* and *Lactobacillus*
		Hydroxytyrosol	Has prebiotic properties, also exerts anti-inflammatory and antioxidant effects	A potent antioxidant inhibits colon cancer cell proliferation and induce apoptosis by increasing beneficial bacteria like *Bifidobacterium* and *Faecalibacterium* which reduces oxidative stress and inflammation
		Ferulic Acid (FA)	Acts as a prebiotic, also reduce the abundance of harmful bacteria, thus maintaining a balanced gut microbiota	Exerts anticancer effects by scavenging free radicals, modulating gene expression related to cancer progression, and inducing apoptosis in colon cancer cells
2.	Flavonoid	Quercetin	Acts as a prebiotic, also has anti-inflammatory properties that help maintain gut health	Modulates signaling pathways like NF-κB and Wnt/β-catenin essential for cancer development, also increases beneficial bacteria contributing to its protective effects against colon cancer
3.	Alkaloids	Piperine	Enhances drug bioavailability, has anti-inflammatory and antioxidant properties, induces apoptosis and inhibits cell proliferation	Reduces inflammation by inhibiting NF-kB pathways which is a risk factor in colon cancer, enhances absorption of nutrients and drugs helping in prevention of colon cancer
		Protopine	Inhibits pathogenic bacterial strains, also enhances gut barrier function and reduces inflammation	Enhances growth of beneficial gut bacteria, which helps in maintaining a healthy gut environment and reducing the risk of colon cancer progression

4.	Terpenoids	Terpene	Artemisinin	Maintains gut homeostasis and improve intestinal health	Targets cancer cells selectively due to their iron-dependent mechanisms, leading to oxidative stress and cell death in cancerous tissues
		Carotenoid	Astaxanthin	Acts as an Antioxidant, anti-inflammatory agent, induces apoptosis, inhibits cell proliferation, and helps in immune modulation	Promoting the growth of beneficial bacteria like *Lactobacillus* and *Bifidobacterium* which produces SCFA preventing colon cancer, reduces inflammation by inhibiting pathways like NF-kB, also can induce apoptosis and inhibit proliferation of colon cancer cells
5.	Iridoid glycoside		Picroside II	Has a modulatory effect on gut microbiota, also enhances intestinal barrier function	Its anti-inflammatory and antioxidant properties show potential in preventing colon cancer
			Morroniside	Promotes beneficial bacteria and inhibits pathogens, also improves gut barrier integrity and reduces inflammation	Supports gut barrier integrity and reduces inflammation, potentially lowering colon cancer risk

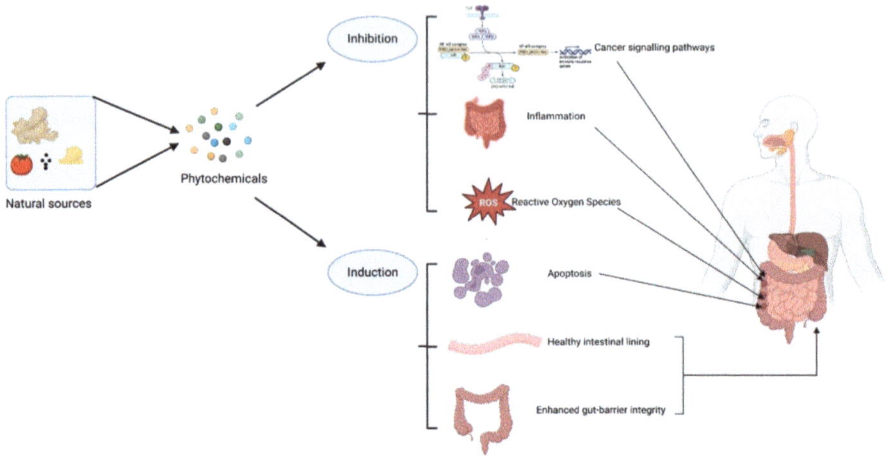

Fig. 8.4 The regulatory impact of DPCs on the development of colon cancer is illustrated through a range of molecular mechanisms and pathways involving inhibitory and inducive events

extractions are often abundant and easily procured, further reducing costs and ensuring a steady supply chain. Therapeutically DPCs offer a multifaceted approach interacting with the gut microbiota and exerting anti-inflammatory and anticancer effects. Overall, the findings affirm that DPCs are an effective strategy for the efficient and cost-effective treatment of colon cancer.

8.4.2 Future Prospect and Limitations

DPC extraction offers promising advancements in the field of treatment of colon cancer with extraction technologies such as supercritical fluid extraction, microwave extraction and ultrasonic extraction. These methods improve the efficiency and yield of DPCs while preserving their bioactivity. Despite the fact that extraction of facilities at the earlier stage involves high costs, the global cost of production is significantly lower when compared to market drugs. This could be considered a con if extraction costs were to be higher due to specific phytochemical requirements, but based on today's perspective, dietary phytochemicals are relatively cheap. Some of the phytochemicals are synthesized from commonly used plants and agricultural residues, thus making the sources readily available and renewable. This makes the costs of procurement relatively lower than in the case of the specialty, and many times rare, materials used in pharmaceutical drug synthesis. Market drugs demand post-Industrial revolution processing due to the improved quality for the same product compared to the industrialization of dietary phytochemicals. This reduces processing and in turn brings down the cost of production lessening the time it takes for the product to come to the market.

8.4.3 Therapeutic Approaches and Treatment

DPCs operate through diverse ways such as exerting antioxidant properties, altering the genes that are expressed, and working with the bacteria present in the gastrointestinal tract. These compounds suppress the growth of cancer cells, trigger the process of apoptosis and decrease inflammation making the compounds useful in handling cancer. In this regard, dietary phytochemicals are adjuncts to conventional therapies for cancer that can increase the efficacy of standard treatments all the while decreasing the toxicity related to treatments. All these compounds have the ability to suppress the growth of cancer cells as well as trigger cancer cell death and reduce inflammatory reactions which are critical in cancer management. Also, dietary phytochemicals can act synergistically with conventional treatment and also minimize the side effects of the drugs making it a sort of boost to the primary treatment. For instance, integration of phytochemicals into chemotherapy has rated positive outcomes in enhancing treatment effectiveness and the general wellbeing of patients. In conclusion, the role of DPCs in managing gut dysbiosis presents a promising avenue for the prevention and treatment of colon cancer.

References

Abdelhamid AG, El-Masry SS, El-Dougdoug NK (2019) Probiotic lactobacillus and bifidobacterium strains possess safety characteristics, antiviral activities and host adherence factors revealed by genome mining. EPMA J 10(4):337–350. https://doi.org/10.1007/S13167-019-00184-Z/FIGURES/1

Aizawa E et al (2019) Bifidobacterium and Lactobacillus counts in the gut microbiota of patients with bipolar disorder and healthy controls. Front Psychiatry 10:418081. https://doi.org/10.3389/FPSYT.2018.00730/BIBTEX

Alam MB, Ju MK, Kwon YG, Lee SH (2019) Protopine attenuates inflammation stimulated by carrageenan and LPS via the MAPK/NF-κB pathway. Food Chem Toxicol 131. https://doi.org/10.1016/J.FCT.2019.110583

Asgharian P et al (2022) Potential mechanisms of quercetin in cancer prevention: focus on cellular and molecular targets. Cancer Cell Int 22(1):1–20. https://doi.org/10.1186/S12935-022-02677-W

Assmann TS, Cuevas-Sierra A, Riezu-Boj JI, Milagro FI, Martínez JA (2020) Comprehensive analysis reveals novel interactions between circulating MicroRNAs and gut microbiota composition in human obesity. Int J Mol Sci 21(24):1–17. https://doi.org/10.3390/IJMS21249509

Awah-Nanzdum AJ et al (2023) Inter-relationship between malaria, probiotic food intake and gut microbiota status among malaria patients. J Biomed Res Environ Sci 4(4):600–609. https://doi.org/10.37871/JBRES1715

Ballway JW, Song B-J (2021) Translational approaches with antioxidant phytochemicals against alcohol-mediated oxidative stress, gut dysbiosis, intestinal barrier dysfunction, and fatty liver disease. Antioxidants *10*(3):*384*. https://doi.org/10.3390/ANTIOX10030384

Baptista BG, Ribeiro M, Cardozo LF, Leal V d O, Regis B, Mafra D (2022) Nutritional benefits of ginger for patients with non-communicable diseases. Clin Nutr ESPEN 49:1–16. https://doi.org/10.1016/J.CLNESP.2022.04.017

Boccuto L, Tack J, Ianiro G, Abenavoli L, Scarpellini E (*2023*) Human genes involved in the interaction between host and gut microbiome: regulation and pathogenic mechanisms. *Genes 14*(4):*857*. https://doi.org/10.3390/GENES14040857

Boland CR, Goel A (2010) Microsatellite instability in colorectal cancer. Gastroenterology 138(6):2073. https://doi.org/10.1053/J.GASTRO.2009.12.064

Carotenoids | Linus Pauling institute | Oregon State University. [Online]. Available: https://lpi.oregonstate.edu/mic/dietary-factors/phytochemicals/carotenoids. Access: 01 Jul 2024

Chattopadhyay I et al (2021) Exploring the role of gut microbiome in colon cancer. Appl Biochem Biotechnol 193(6):1780–1799. https://doi.org/10.1007/S12010-021-03498-9

Chen L, Mou X, Li J, Li M, Ye C, Gao X, Liu X, Ma Y, Xu Y, Zhong Y (2023) Alternations in gut microbiota and host transcriptome of patients with coronary artery disease. medRxiv. https://doi.org/10.1101/2023.07.14.23292642

Cho SY et al (2003) Quercetin suppresses proinflammatory cytokines production through MAP kinases and NF-κB pathway in lipopolysaccharide-stimulated macrophage. Mol Cell Biochem 243(1–2):153–160. https://doi.org/10.1023/A:1021624520740/METRICS

Choi HH, Cho YS (2016) Fecal microbiota transplantation: current applications, effectiveness, and future perspectives. Clin Endosc 49(3):257–265. https://doi.org/10.5946/CE.2015.117

Clerbaux LA et al (2022) Mechanisms leading to gut dysbiosis in COVID-19: current evidence and uncertainties based on adverse outcome pathways. J Clin Med 11(18):5400. https://doi.org/10.3390/JCM11185400

Colorectal cancer statistics | WCRF International. [Online]. Available: https://www.wcrf.org/cancer-trends/colorectal-cancer-statistics/. Accessed 01 Jul 2024

Dacrema M, Ali A et al (2022) Spice-derived bioactive compounds confer colorectal cancer prevention via modulation of gut microbiota. Cancers 14(22):5682. https://doi.org/10.3390/CANCERS14225682

Dalile B, Van Oudenhove L, Vervliet B, Verbeke K (2019) The role of short-chain fatty acids in microbiota–gut–brain communication. Nat Rev Gastroenterol Hepatol 16(8):461–478. https://doi.org/10.1038/s41575-019-0157-3

DeFelice SL (1995) The nutraceutical revolution: its impact on food industry R&D. Trends Food Sci Technol 6(2):59–61. https://doi.org/10.1016/S0924-2244(00)88944-X

Degruttola AK, Low D, Mizoguchi A, Mizoguchi E (2016) Current understanding of dysbiosis in disease in human and animal models. Inflamm Bowel Dis 22(5):1137. https://doi.org/10.1097/MIB.0000000000000750

Deng D, Yan J, Wu Y, Wu K, Li W (2021) Morroniside suppresses hydrogen peroxide-stimulated autophagy and apoptosis in rat ovarian granulosa cells through the PI3K/AKT/mTOR pathway. Hum Exp Toxicol 40(4):577–586. https://doi.org/10.1177/0960327120960768

Drommi F et al (2023) Crosstalk between ILC3s and microbiota: implications for colon cancer development and treatment with immune check point inhibitors. Cancers 15(11):2893. https://doi.org/10.3390/CANCERS15112893

Duan Z, Xie H, Yu S, Wang S, Yang H (2022) Piperine derived from piper nigrum L. inhibits LPS-induced inflammatory through the MAPK and NF-κB signalling pathways in RAW264.7 cells. Foods 11(19):2990. https://doi.org/10.3390/FOODS11192990

Fearon ER, Vogelstein B (1990) A genetic model for colorectal tumorigenesis. Cell 61(5):759–767. https://doi.org/10.1016/0092-8674(90)90186-I

Feng Z et al (2021) Association between gut dysbiosis and chronic kidney disease: a narrative review of the literature. J Int Med Res 49(10). https://doi.org/10.1177/03000605211053276

Fu J et al (2023) Ferulic acid alleviates hepatic lipid accumulation and inflammation by improving proximal and distal intestinal barriers in NAFLD mice. Tohoku J Exp Med 260(2):149–163. https://doi.org/10.1620/TJEM.2023.J023

Gao W et al (2022) Dissecting the crosstalk between Nrf2 and NF-κB response pathways in drug-induced toxicity. Front Cell Dev Biol 9:809952. https://doi.org/10.3389/FCELL.2021.809952/BIBTEX

Genua F, Raghunathan V, Jenab M, Gallagher WM, Hughes DJ (2021) The role of gut barrier dysfunction and microbiome dysbiosis in colorectal cancer development. Front Oncol 11:626349. https://doi.org/10.3389/FONC.2021.626349/BIBTEX

Gopalakrishnan V et al (2018) Gut microbiome modulates response to anti-PD-1 immunotherapy in melanoma patients. Science (1979) 359(6371):97–103. https://doi.org/10.1126/SCIENCE.AAN4236

Grady WM, Carethers JM (2008) Genomic and epigenetic instability in colorectal cancer pathogenesis. Gastroenterology 135(4):1079. https://doi.org/10.1053/J.GASTRO.2008.07.076

Gromova LV et al (2021) Gut digestive function and microbiome after correction of experimental dysbiosis in rats by indigenous bifidobacteria. Microorganisms 9(3):522. https://doi.org/10.3390/MICROORGANISMS9030522

Guo X, Xu Y, Geng R, Qiu J, He X (2022) Curcumin alleviates dextran Sulfate sodium-induced colitis in mice through regulating gut microbiota. Mol Nutr Food Res 66(8):2100943. https://doi.org/10.1002/MNFR.202100943

Han H et al (2023) Hydroxytyrosol attenuates diquat-induced oxidative stress by activating Nrf2 pathway and modulating colonic microbiota in mice. J Nutr Biochem 113:109256. https://doi.org/10.1016/J.JNUTBIO.2022.109256

Honda K, Littman DR (2016) The microbiota in adaptive immune homeostasis and disease. Nature 535(7610):75–84. https://doi.org/10.1038/NATURE18848

Hrncir T (2022) Gut microbiota dysbiosis: triggers, consequences, diagnostic and therapeutic options. Microorganisms 10(3). https://doi.org/10.3390/MICROORGANISMS10030578

Huang X-T, Liu W, Zhou Y, Hao C-X, Zhou Y, Zhang C-Y, Sun C-C, Luo Z-Q, Tang S-Y (2019) Dihydroartemisinin attenuates lipopolysaccharide-induced acute lung injury in mice by suppressing NF-κB signaling in an Nrf2-dependent manner. Int J Mol Med 44(6):2213–2222. [Online]. Available: https://www.spandidos-publications.com/10.3892/ijmm.2019.4387. Accessed 02 Jul 2024

Huarcaya LRD, Ríos MFD (2023) Microbiota and depression an update. Microenviron and Microecol Res 5(2):8. https://doi.org/10.53388/MMR2023008

Ivleva EA, Grivennikov SI (2022) Microbiota-driven mechanisms at different stages of cancer development. Neoplasia 32:100829. https://doi.org/10.1016/J.NEO.2022.100829

Jakubowski W, Bartosz G (1997) Estimation of oxidative stress in saccharomyces cerevisae with fluorescent probes. Int J Biochem Cell Biol 29(11):1297–1301. https://doi.org/10.1016/S1357-2725(97)00056-3

Katsi V et al (2018) GUT microbiome-GUT dysbiosis-arterial hypertension: new horizons. Curr Hypertens Rev 15(1):40–46. https://doi.org/10.2174/1573402114666180613080439

Kavitha K, Kowshik J, Kishore TKK, Baba AB, Nagini S (2013) Astaxanthin inhibits NF-κB and Wnt/β-catenin signaling pathways via inactivation of Erk/MAPK and PI3K/Akt to induce intrinsic apoptosis in a hamster model of oral cancer. Biochim Biophys Acta Gen Subj 1830(10):4433–4444. https://doi.org/10.1016/J.BBAGEN.2013.05.032

Kiani BH, Kayani WK, Khayam AU, Dilshad E, Ismail H, Mirza B (2020) Artemisinin and its derivatives: a promising cancer therapy. Mol Biol Rep 47(8):6321–6336. https://doi.org/10.1007/S11033-020-05669-Z/FIGURES/4

Kim J, Lee HK (2022) Potential role of the gut microbiome in colorectal cancer progression. Front Immunol 12:807648. https://doi.org/10.3389/FIMMU.2021.807648/BIBTEX

Kim JH, Lee S, Cho EJ (2020) The protective effects of Acer okamotoanum and isoquercitrin on obesity and amyloidosis in a mouse model. Nutrients 12(5):1353. https://doi.org/10.3390/NU12051353

Koni M, Pinnarò V, Brizzi MF (2020) The Wnt signalling pathway: a tailored target in cancer. Int J Mol Sci *21*(20):7697. https://doi.org/10.3390/IJMS21207697

Kumar A et al (2023) Major phytochemicals: recent advances in health benefits and extraction method. *Molecules 28*(2):887. https://doi.org/10.3390/MOLECULES28020887

Kwao-Zigah G et al (2024) Microbiome dysbiosis, dietary intake and lifestyle-associated factors involve in epigenetic modulations in colorectal cancer: a narrative review. Cancer Control 31. https://doi.org/10.1177/10732748241263650

Lao VV, Grady WM (2011) Epigenetics and colorectal cancer. Nat Rev Gastroenterol Hepatol 8(12):686–700. https://doi.org/10.1038/NRGASTRO.2011.173

Le Chatelier E et al (2013) Richness of human gut microbiome correlates with metabolic markers. Nature 500(7464):541–546. https://doi.org/10.1038/nature12506

Lee JG et al (2022) Host tp53 mutation induces gut dysbiosis eliciting inflammation through disturbed sialic acid metabolism. Microbiome 10(1):1–25. https://doi.org/10.1186/S40168-021-01191-X/FIGURES/6

Li T et al (2020) Picroside II protects against cholestatic liver injury possibly through activation of farnesoid X receptor. Phytomedicine 68:153153. https://doi.org/10.1016/J.PHYMED.2019.153153

Lingas EC (2023) Early-onset colon cancer: a narrative review of its pathogenesis, clinical presentation, treatment, and prognosis. Cureus 15(9). https://doi.org/10.7759/CUREUS.45404

Liu S, Gao J, Zhu M, Liu K, Zhang HL (2020) Gut microbiota and dysbiosis in Alzheimer's disease: implications for pathogenesis and treatment. Mol Neurobiol 57(12):5026–5043. https://doi.org/10.1007/S12035-020-02073-3

Makhaik MS, Shakya AK, Kale R, Makhaik MS, Shakya AK, Kale R (2021) Dietary phytochemicals: as a natural source of antioxidants. In: Antioxidants – benefits, sources, mechanisms of action. https://doi.org/10.5772/INTECHOPEN.99159

Manolis AA, Manolis TA, Melita H, Manolis AS (2021) Gut microbiota and cardiovascular disease: symbiosis versus dysbiosis. Curr Med Chem 29(23):4050–4077. https://doi.org/10.2174/0929867328666211213112949

Markowitz RHG et al (2022) Microbiome-associated human genetic variants impact phenome-wide disease risk. Proc Natl Acad Sci USA 119(26):e2200551119. https://doi.org/10.1073/PNAS.2200551119/SUPPL_FILE/PNAS.2200551119.SD11.XLSX

Micheli L et al (2023) Role of hydroxytyrosol and oleuropein in the prevention of aging and related disorders: focus on neurodegeneration, skeletal muscle dysfunction and gut microbiota. Nutrients 15(7):1767. https://doi.org/10.3390/NU15071767

Mousa WK, Chehadeh F, Husband S (2022) Microbial dysbiosis in the gut drives systemic autoimmune diseases. Front Immunol 13:906258. https://doi.org/10.3389/FIMMU.2022.906258/BIBTEX

Olivera A et al (2012) Inhibition of the NF-κB signaling pathway by the curcumin analog, 3,5-Bis(2-pyridinylmethylidene)-4-piperidone (EF31): anti-inflammatory and anti-cancer properties. Int Immunopharmacol 12(2):368–377. https://doi.org/10.1016/J.INTIMP.2011.12.009

Park JE, Han JS (2024) Improving the effect of ferulic acid on inflammation and insulin resistance by regulating the JNK/ERK and NF-κB pathways in TNF-α-treated 3T3-L1 adipocytes. Nutrients 16(2):294. https://doi.org/10.3390/NU16020294

Pei YH et al (2016) Hydroxytyrosol protects against myocardial ischemia/reperfusion injury through a PI3K/Akt-dependent mechanism. Mediat Inflamm 2016(1):1232103. https://doi.org/10.1155/2016/1232103

Piao X et al (2020) Picroside II improves severe acute pancreatitis-induced intestinal barrier injury by inactivating oxidative and inflammatory TLR4-dependent PI3K/AKT/NF-κB Signaling and improving gut microbiota. Oxidative Med Cell Longev 2020(1):3589497. https://doi.org/10.1155/2020/3589497

Pilla R, Suchodolski JS (2020) The role of the canine gut microbiome and metabolome in health and gastrointestinal disease. Front Vet Sci 6:502799. https://doi.org/10.3389/FVETS.2019.00498/BIBTEX

Pino MS, Chung DC (2010) The chromosomal instability pathway in colon cancer. Gastroenterology 138(6):2059–2072. https://doi.org/10.1053/J.GASTRO.2009.12.065

Pino MS, Chung DC (2011) Microsatellite instability in the management of colorectal cancer. Expert Rev Gastroenterol Hepatol 5(3):385–399. https://doi.org/10.1586/EGH.11.25

Qorbani M, Angoorani P, Ejtahed H-S (2022) Dietary phytochemical index as a biomarker in nutritional studies: features and applications. In: Biomarkers in nutrition. Springer, Cham, pp 307–328. https://doi.org/10.1007/978-3-031-07389-2_18

Rodríguez-Morató J, Xicota L, Fitó M, Farré M, Dierssen M, De La Torre R (2015) Potential role of olive oil phenolic compounds in the prevention of neurodegenerative diseases. Molecules 20(3):4655–4680. https://doi.org/10.3390/MOLECULES20034655

Rubinstein MR, Wang X, Liu W, Hao Y, Cai G, Han YW (2013) Fusobacterium nucleatum promotes colorectal carcinogenesis by modulating E-cadherin/β-catenin signaling via its FadA adhesin. Cell Host Microbe 14(2):195–206. https://doi.org/10.1016/j.chom.2013.07.012

Salem AE, Singh R, Ayoub YK, Khairy AM, Mullin GE (2018) The gut microbiome and irritable bowel syndrome: state of art review. Arab J Gastroenterol 19(3):136–141. https://doi.org/10.1016/j.ajg.2018.02.008

Satitsri S, Akrimajirachoote N, Nunta K, Ruennarong N, Amnucksoradej O, Muanprasat C (2023) Piperine as potential therapy of post-weaning porcine diarrheas: an in vitro study using a porcine duodenal enteroid model. BMC Vet Res 19(1):1–13. https://doi.org/10.1186/S12917-022-03536-6/FIGURES/7

Saus E, Iraola-Guzmán S, Willis JR, Brunet-Vega A, Gabaldón T (2019) Microbiome and colorectal cancer: roles in carcinogenesis and clinical potential. Mol Asp Med 69:93–106. https://doi.org/10.1016/J.MAM.2019.05.001

Selber-Hnativ S et al (2017) Human gut microbiota: toward an ecology of disease. Front Microbiol 8:243683. https://doi.org/10.3389/FMICB.2017.01265/BIBTEX

Shamsnia HS et al (2023) Impact of curcumin on p38 MAPK: therapeutic implications. Inflammopharmacology 31(5):2201–2212. https://doi.org/10.1007/S10787-023-01265-2/FIGURES/3

Shen CL et al (2022) Dietary supplementation of gingerols- and shogaols-enriched ginger root extract attenuate pain-associated behaviors while modulating gut microbiota and metabolites in rats with spinal nerve ligation. J Nutr Biochem 100:108904. https://doi.org/10.1016/J.JNUTBIO.2021.108904

Shim JA, Ryu JH, Jo Y, Hong C (2023) The role of gut microbiota in T cell immunity and immune mediated disorders. Int J Biol Sci 19(4):1178. https://doi.org/10.7150/IJBS.79430

Simon K (2016) Colorectal cancer development and advances in screening. Clin Interv Aging 11:967–976. https://doi.org/10.2147/CIA.S109285

Singh S, Arthur R, Upadhayay S, Kumar P (2022) Ferulic acid ameliorates neurodegeneration via the Nrf2/ARE signalling pathway: a review. Pharmacol Res Modern Chinese Med 5:100190. https://doi.org/10.1016/J.PRMCM.2022.100190

Sinicrope FA, Sargent DJ (2012) Molecular pathways: microsatellite instability in colorectal cancer: prognostic, predictive, and therapeutic implications. Clin Cancer Res 18(6):1506–1512. https://doi.org/10.1158/1078-0432.CCR-11-1469/84712/AM/MOLECULAR-PATHWAYS-MICROSATELLITE-INSTABILITY-IN

Solanki R, Karande A, Ranganathan P (2023) Emerging role of gut microbiota dysbiosis in neuroinflammation and neurodegeneration. Front Neurol 14. https://doi.org/10.3389/FNEUR.2023.1149618/PDF

Su L, Zeng Y, Li G, Chen J, Chen X (2022) Quercetin improves high-fat diet-induced obesity by modulating gut microbiota and metabolites in C57BL/6J mice. Phytother Res 36(12):4558–4572. https://doi.org/10.1002/PTR.7575

Suffredini IB et al (2004) Screening of antibacterial extracts from plants native to the Brazilian Amazon rain Forest and Atlantic Forest. Braz J Med Biol Res 37(3):379–384. https://doi.org/10.1590/S0100-879X2004000300015

Thibodeau SN, Bren G, Schaid D (1993) Microsatellite instability in cancer of the proximal colon. Science 260(5109):816–819. https://doi.org/10.1126/SCIENCE.8484122

Toyota M, Issa JPJ (1999) CpG Island methylator phenotypes in aging and cancer. Semin Cancer Biol 9(5):349–357. https://doi.org/10.1006/SCBI.1999.0135

Toyota M, Ahuja N, Ohe-Toyota M, Herman JG, Baylin SB, Issa JPJ (1999) CpG Island methylator phenotype in colorectal cancer. Proc Natl Acad Sci USA 96(15):8681. https://doi.org/10.1073/PNAS.96.15.8681

Violi F, Cammisotto V, Bartimoccia S, Pignatelli P, Carnevale R, Nocella C (2023) Gut-derived low-grade endotoxaemia, atherothrombosis and cardiovascular disease. Nat Rev Cardiol 20(1):24–37. https://doi.org/10.1038/S41569-022-00737-2

Wang L, Liu X-H, Chen H, Chen Z-Y, Weng X-D, Qiu T, Liu L (2015) Picroside II protects rat kidney against ischemia/reperfusion-induced oxidative stress and inflammation by the TLR4/NF-κB pathway. Exp Ther Med 9(4):1253–1258. [Online]. Available: https://www.spandidos-publications.com/10.3892/etm.2015.2225. Accessed 02 Jul 2024

Wang J, Chen Y, Hu X, Feng F, Cai L, Chen F (2020) Assessing the effects of ginger extract on polyphenol profiles and the subsequent impact on the fecal microbiota by simulating digestion and fermentation in vitro. Nutrients 12(10):3194. https://doi.org/10.3390/NU12103194

Wang M et al (2022) Astaxanthin from Haematococcus pluvialis prevents high-fat diet-induced hepatic steatosis and oxidative stress in mice by gut-liver axis modulating properties. Front Nutr 9:840648. https://doi.org/10.3389/FNUT.2022.840648/BIBTEX

Wong SH, Yu J (2019) Gut microbiota in colorectal cancer: mechanisms of action and clinical applications. Nat Rev Gastroenterol Hepatol 16(11):690–704. https://doi.org/10.1038/s41575-019-0209-8

Wong CC, Yu J (2023) Gut microbiota in colorectal cancer development and therapy. Nat Rev Clin Oncol 20(7):429–452. https://doi.org/10.1038/s41571-023-00766-x

Wood LD et al (2007) The genomic landscapes of human breast and colorectal cancers. Science 318(5853):1108–1113. https://doi.org/10.1126/SCIENCE.1145720

Worthley DL, Leggett BA (2010) Colorectal cancer: molecular features and clinical opportunities. Clin Biochem Rev 31(2):31. [Online]. Available: /pmc/articles/PMC2874430/. Accessed 01 Jul 2024

Xiao J et al (2023) Morroniside inhibits inflammatory bone loss through the TRAF6-mediated NF-κB/MAPK signalling pathway. Pharmaceuticals (Basel) 16(10). https://doi.org/10.3390/PH16101438

Yang M et al (2020) Absorption difference between hepatotoxic pyrrolizidine alkaloids and their N-oxides – mechanism and its potential toxic impact. J Ethnopharmacol 249:112421. https://doi.org/10.1016/J.JEP.2019.112421

Yasuda R et al (2022) Astaxanthin attenuated the stress-induced intestinal motility disorder via altering the gut microbiota. Int J Vitam Nutr Res 93(5):427–437. https://doi.org/10.1024/0300-9831/A000756

Yue M, Huang J, Ma X, Huang P, Liu Y, Zeng J (2023) Protopine alleviates dextran sodium sulfate-induced ulcerative colitis by improving intestinal barrier function and regulating intestinal microbiota. *Molecules* 28(13):5277. https://doi.org/10.3390/MOLECULES28135277

Zeng MY, Inohara N, Nuñez G (2017) Mechanisms of inflammation-driven bacterial dysbiosis in the gut. Mucosal Immunol 10(1):18. https://doi.org/10.1038/MI.2016.75

Zhang L et al (2020) Hydroxytyrosol inhibits LPS-induced neuroinflammatory responses via suppression of TLR-4-mediated NF-κB P65 activation and ERK signaling pathway. Neuroscience 426:189–200. https://doi.org/10.1016/j.neuroscience.2019.12.005

Zhang S, Lai Q, Liu L, Yang Y, Wang J (2022) Morroniside alleviates lipopolysaccharide-induced inflammatory and oxidative stress in inflammatory bowel disease by inhibiting NLRP3 and NF-κB signaling pathways. Allergol Immunopathol (Madr) 50(6):93–99. https://doi.org/10.15586/AEI.V50I6.674

Zhao C et al (2022) Gut dysbiosis induces the development of mastitis through a reduction in host anti-inflammatory enzyme activity by endotoxemia. Microbiome 10(1). https://doi.org/10.1186/S40168-022-01402-Z

Zhou F et al (2023) The improvement of intestinal dysbiosis and hepatic metabolic dysfunction in dextran sulfate sodium-induced colitis mice: effects of curcumin. J Gastroenterol Hepatol 38(8):1333–1345. https://doi.org/10.1111/JGH.16205

Zmora N, Suez J, Elinav E (2018) You are what you eat: diet, health and the gut microbiota. Nat Rev Gastroenterol Hepatol 16(1):35–56. https://doi.org/10.1038/s41575-018-0061-2

Probiotics: Immune Responses and Gut Microbiome

9

Mrithika Ajay, Abilash Navaneethan, Vasukidevi Ramachandran, Dhanavathy Gnanasampanthapandian, and Kanagaraj Palaniyandi

Abstract

Probiotics are live microorganisms that offer various health benefits when consumed. These beneficial bacteria, commonly found in the gastrointestinal tract of animals and humans, contribute to improved gut health, nutrient absorption, and immune modulation in the host. Human understanding of probiotics is growing rapidly. Understanding the interplay between our bodies and diets is crucial for future advancements. This chapter provides an insight into various topics, the origin and development of probiotics over many decades, different classes of probiotics and their effects on the gut to understand the effect of different strains and to develop disease-specific and effective treatment, the interaction between the host and the microbes and how it directly affects nutritional and immune responses, the interaction between the gut and the brain known as the gut–brain axis, their function in relation to each other and their pathways, the role and function of probiotics and their efficacy in treating various diseases, mechanism and mode of action of probiotics. In addition to exploring probiotics, this chapter will also delve into antibiotics, their impact on the gut, and their interaction with the gut microbiome.

Keywords

Probiotics · Gut microbiota · Gut–brain axis · Diseases

M. Ajay · A. Navaneethan · D. Gnanasampanthapandian (✉) · K. Palaniyandi (✉)
Cancer Science Laboratory, Department of Biotechnology, School of Bioengineering, SRM Institute of Science and Technology, Kattankulathur, India

V. Ramachandran
Department of Microbiology and Biotechnology, Bharath Institute of Higher Education and Research, Selaiyur, Chennai, India

S. Pathak et al. (eds.), *Microbiota and Dietary Mediators in Colon Cancer Prevention and Treatment*, https://doi.org/10.1007/978-981-96-0297-1_9

9.1 Introduction

The term "probiotics" is derived from the Greek word "pro-bios," which translates to "for life." The histories of probiotics and humans are interconnected deeply. The Greeks and Romans were already familiar with fermented milk and cheese and often advocated their consumption, especially by children and people recovering from illness (Soccol 2010). In 1974, Parker put forward an important definition, suggesting that probiotics are "the organisms that promote a healthy balance of gut microbes" (Fuller 1989). The definition created by the Food and Agriculture Organization of the WHO and recognized by the International Scientific Association for Probiotics and Prebiotics, which is regarded widely as the most comprehensive and up-to-date, describes probiotics as "living bacteria that, when given in suitable proportions, provide a health benefit on the host" (FAO 2006). Despite varying theoretical definitions, it is difficult to determine whether a microbe is a probiotic. It has been suggested that probiotics should be selected based on strict criteria, such as safety for the host, resistance to digestive processes, adherence to intestinal cells, antibacterial activity, and food stability (Fijan 2014). Although not all current probiotics meet all these requirements, the most commonly used lactic acid bacteria, such as *Bifidobacterium*, *Lactobacillus*, and *Streptococcus*, especially *Streptococcus thermophilus*, are able to survive the gastrointestinal tract, adhere to the intestinal mucosa, and colonize effectively (Dempsey and Corr 2022). Probiotics are widely found in dairy and nondairy products and are often consumed after antibiotic therapy to replenish the microbial flora in the digestive tract, promoting a good balance of gut microbes. Probiotics, often consumed through natural and fermented foods like fruits, yogurt, and cheese, are now recognized as a vital part of a healthy diet (Raheem et al. 2021). Extensive research has shown that probiotics are effective in the treatment of diseases, such as Crohn's disease and chronic inflammation, and can reduce cancer, obesity, and diabetes. There is also growing evidence that they can improve immune function and overall health (Wang et al. 2021). Probiotics are considered dietary components that can influence the human gut microbiota, by having a regulatory effect on its composition and structure, which in turn significantly influences the immune system through functions, such as maintaining the integrity of the mucosal barrier, providing nutrients, and fighting pathogens (Wang et al. 2021). Researchers worldwide are looking for alternatives to the widespread use of antibiotics that indiscriminately attack the natural bacteria of the gut and the genital tract and do not weaken the body's defense mechanisms (Tegegne and Kebede 2022). Recent research suggests that altering beneficial bacteria may play an important role in preventing and combating various gastrointestinal (GI) infections. These infections are caused by a diverse population of good bacteria that reside in the respiratory tract, GI tract, cutaneous, oral, and urogenital tract of humans (Tegegne and Kebede 2022). The total weight of the human microbiota is estimated to be 1–2 kg. While the respiratory system has the largest surface area colonized by microbes, the GI tract harbors the most abundant and diverse microbial community. Each individual harbors a unique profile of ~100–1000 microbial species in the GI tract, where microbial cells account for about half of the wet

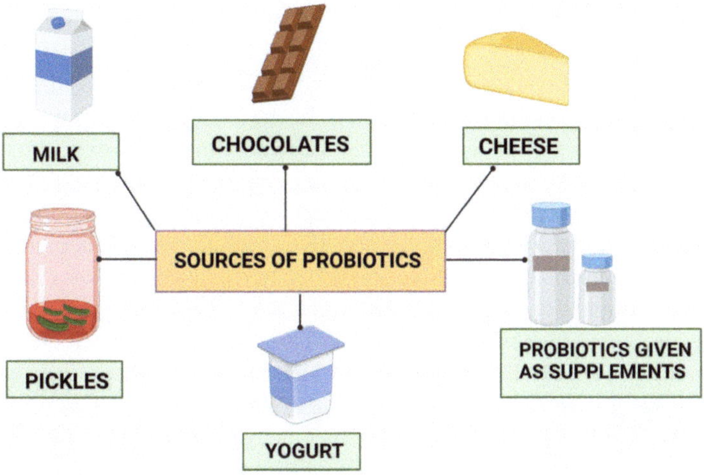

Fig. 9.1 Various sources of probiotics: probiotics play a prominent role in regulating the response of the immune system. They also influence the release of hormones in the gastrointestinal tract and the behaviour of the brain through bidirectional neuronal signaling. In addition, probiotics promote a healthy microbial environment and assist in the treatment of overweight and obesity

weight of colonic material and outnumber human tissue cells tenfold (Kerry et al. 2018). In general, the stomach contains about 10^3 bacterial species, whereas the total microbial population in the colon is ~10^{11}–10^{12} cfu/g (Kerry et al. 2018). Microbial antagonism, one of the main mechanisms of action of probiotics, is that the normal flora inhibits pathogenic colonization and benefits the host by eliminating harmful bacteria, improving nutrient supply and strengthening the immune system (Tegegne and Kebede 2022). Intestinal bacterial colonization begins at birth, such as exposure to a nonsterile environment, and develops throughout life due to complex interactions involving the diet, genome, lifestyle, and antibiotic use (Kerry et al. 2018). In effective therapies for GI disorders, the use of competing probiotic bacteria is essential to ensure that they are harmless and do not disrupt the natural gut biome (Hemarajata and Versalovic 2013). The antibacterial effect of probiotics is associated with their synthesis of organic acids, ethanol, hydrogen peroxide, and protein-based substances such as bacteriocins (Tegegne and Kebede 2022). As antibiotics are becoming less effective due to resistance, probiotics serve as the second most critical immune defense system, as recognized by the WHO, and this treatment is referred to as microbial interference therapy (Tegegne and Kebede 2022) (Fig. 9.1).

Probiotics play a prominent role in response regulation of the immune system and thus affect GI hormone release and influence brain behavior through bidirectional neuronal signaling (Kerry et al. 2018). In addition, a probiotic must exhibit antagonism against pathogens, stimulate the immune system, and demonstrate observable benefits to the host while also maintaining its activity, viability, and growth efficacy after undergoing technological treatment (Plaza-Diaz et al. 2019).

9.2 History of Probiotics

Lilly and Stillwell coined the term "probiotics" in 1965 and provided the first definition of probiotics, by explaining their role as bacterial substances that stimulate the growth of other microorganisms (Lilly and Stillwell 1965). However, the history of probiotics dates back to the early twentieth century, when Ilya Ilyich Metchnikoff, a Nobel Prize winner, introduced the concept of probiotics (Schmalstieg Jr. and Goldman 2008). According to him, putrefactive or proteolytic pathogenic microorganisms cause "intestinal autointoxication," or toxins in the colon (Dutta 2016). Metchnikoff's main hypothesis in his brilliant work "The Prolongation of Life" was that toxins produced by consumed food and feces in the gut contribute to a shortened human lifespan. He also suggested a connection between the consistent consumption of lactic acid fermented milk in the Bulgarian and Caucasian population and their significantly longer average life expectancy (Zommiti et al. 2020). Zommiti et al. suggested that the consumption of yogurt containing *Lactobacillus* species promotes well-being and possibly extends the life expectancy of Bulgarian farmers (Zommiti et al. 2020). Tissier and Metchnikoff were among the pioneers who hypothesized the probiotic potential of good bacteria, suggesting that people with diarrhea could rebuild a healthy gut microbiota by ingesting these bacteria (Zommiti et al. 2020). Grigoroff played a crucial role by cultivating *Lactobacillus bulgaricus* from podkvassa, which was used as a starter culture for the production of Bulgarian kiselomleko, commonly known as sour milk or yogurt (Dutta 2016). Grigoroff's discovery also led to the identification of another organism, *Streptococcus thermophilus*, which did not receive much attention at the same time as it was considered a pathogenic microorganism (Dutta 2016). In the early 1900s, Louis Pasteur discovered the microbes responsible for fermentation. In 1907, Metchnikoff proposed that the bacteria responsible for the fermentation of yogurt, such as *Lactobacillus bulgaricus* and *Streptococcus thermophilus*, help inhibit harmful fermentation in the gut, which also has a positive effect on overall health (Soccol 2010; Tegegne and Kebede 2022). In the early 1920s, *L. acidophilus* milk was recognized to have therapeutic properties, particularly in facilitating digestion. For centuries, live microorganisms, especially lactic acid-producing bacteria, have been used in foods to preserve and improve health (Tegegne and Kebede 2022). Interestingly, older civilizations, such as the Romans and Greeks, often consumed fermented dairy products to live a healthy life. In early 76 BC, a historian from ancient Rome introduced the concept of consuming fermented dairy products to alleviate dysentery (Tegegne and Kebede 2022). The practice of fermenting foods, such as milk, has ancient origins dating back to the Neolithic period (Raheem et al. 2021). Historical records show that the fermentation of milk was practiced as early as 10,000 BC in regions such as the Middle East and India, and between 7000 and 5000 BC in areas such as Rome, Egypt and Greece, as well as other parts of Europe (Raheem et al. 2021). However, research into the microorganisms involved in the effects of fermented foods on human health is relatively new (Tegegne and Kebede 2022). In the early 1930s, Shirota focused his research on identifying strains of intestinal bacteria strains that could withstand the gastrointestinal tract (Wong et al.

2022). Using these strains, he developed fermented milk, which was distributed in his clinic. With his first initial product, *L. acidophilus* Shirota, which was later renamed *L. casei* Shirota, he laid the foundation for the company Yakult Honsha (Wong et al. 2022).

9.3 Classification of Probiotics

Probiotics are live microorganisms that have positive properties on the health of the host by improving or restoring the microbiome. They are classified based on various criteria, such as genus, species, and strain (Amirreza et al. 2016). Probiotics can also be categorized according to their mode of action and health benefits. Understanding the classification of probiotics is essential for identifying the most appropriate strains for specific health conditions and for formulating effective probiotic products. In the following sections, we will delve into the different classification criteria for probiotics and explore their importance in the context of promoting overall health.

9.3.1 Lactobacilli

Lactobacillus, a group of lactic acid-producing gram-positive rod-shaped bacteria that drive under both obligate and facultative anaerobic conditions in the human gastrointestinal and urogenital tracts, plays a crucial role in stabilizing the mucosal barrier and reducing intestinal permeability (Amirreza et al. 2016). Taking lactobacillus probiotics during antibiotic treatment may prevent or minimize the depletion of normal flora and colonization by pathogenic bacteria. Moreover, research suggests that Lactobacillus has the ability to bind dietary carcinogens, thereby reducing the development of tumors in the colon following carcinogen exposure. In addition, preliminary studies suggest that the efficacy of lactobacilli, as with other probiotics, depends on their ability to colonize specific tissue areas, which requires the presence of live and viable organisms in lactobacillus preparations (Amirreza et al. 2016). However, efficacy may be influenced by differences between lactobacilli strains in adherence to epithelial cells, which are regulated by host factors such as hormone levels. Lactobacilli and other probiotic bacteria have the ability to bind bile acids to cholesterol, which could reduce the concentration of fatty acids in the bloodstream through mechanisms, such as inhibition of hepatic cholesterol synthesis or redistribution of cholesterol. In adults, lactobacilli make up only a small proportion of the fecal bacterial population, typically ranging from 0.01% to 0.06% or 10^5 to 10^8 CFU/g. Among the indigenous species in the adult gut, *Lactobacillus gasseri*, *L. reuteri*, *L. crispatus*, *L. salivarius*, and *L. ruminis* are predominant. In contrast to the microbiota of adults, infants have a more unstable and variable fecal microbiota. Newborn feces generally contain about 10^5 CFU/g, with the amount increasing to 10^6–10^8 CFU/g in infants older than 1 month (Zhang et al. 2018a). The ingestion of probiotic lactobacilli is known to be effective in the treatment of certain infectious diseases caused by pathogens, such as adenoviruses, herpes simplex virus type 1, and rotaviruses. Lactobacilli also play a role in immunomodulation by triggering immunoregulatory responses and inducing the development of regulatory dendritic cells and T cells (Zhang et al. 2018a).

Research on certain strains of lactobacilli has shown that they can influence cholesterol levels, an important factor in cardiovascular health. These strains, *L. reuteri* NCIMB 701089, have been identified as particularly capable of assimilating cholesterol in the gut, making them a promising candidate for the prevention and treatment of cardiovascular disease (Zhang et al. 2018a).

9.3.2 Bifidobacteria

Bifidobacterium is a species of gram-positive, anaerobic bacteria with a rod-shaped form characterized by its ability to produce both lactic and acetic acid when it consumes glucose (Amirreza et al. 2016). This genus of bacteria is particularly important for its health benefits. In the context of menopause, for example, natural measures aimed at increasing estrogen levels have been shown to promote recolonization with *lactobacilli* without the need for additional probiotic supplements (Amirreza et al. 2016). Furthermore, the combination of *Bifidobacterium* with *Lactobacillus* strains and the probiotic yeast *Saccharomyces boulardii* has been shown to reduce the side effects associated with the treatment of Helicobacter infections, although it does not affect patient compliance.

Remarkably, *Bifidobacteria* have been associated with a reduction in oral inflammation and have been observed to positively influence inflammatory markers when consumed by patients in the form of fermented yogurt. Psychological well-being could also be influenced by the content levels of these bacteria. For example, people who are more stress-resistant have lower levels of *Bifidobacteria* in their gut (Kazemifard et al. 2022). However, it has been shown that a combination of *B. longum* and *L. helveticus* significantly alleviates depressive symptoms in patients with severe depression disorder and has a positive effect on the psychological health of healthy volunteers (Kazemifard et al. 2022). *Bifidobacteria* strains used in fermented milk are known for their high survival rate in the GI tract and exert their probiotic effect in the large intestine, which fulfills the essential criteria for probiotics (Picard et al. 2005). The taxonomy of these strains has evolved considerably with the introduction of modern molecular techniques, such as polymerase chain reaction and genotyping methods, which are crucial for the accurate identification of species and differentiation between *Bifidobacteria* strains. In particular, the *Bifidobacterium infantis* strain exhibits a broad spectrum of antimicrobial effects and produces acids as well as other compounds that inhibit the growth of pathogens effectively (Picard et al. 2005). Furthermore, in vitro studies show that the antimicrobial substances detected in the supernatants of spent cultures are secreted by the bifidobacterial strains and are not associated with metabolism by-products. In addition, studies indicate that *bifidobacteria* are resistant to low concentrations of nitrite, with their acidic by-products playing a role in nitrite reduction (Picard et al. 2005).

Moreover, experiments conducted on rats have shown that bacteria, such as *Lactobacillus, Bifidobacterium*, and *Streptococcus thermophilus*, are able to minimize DNA damage in the colon in response to the genotoxic agent 1,2-dimethylhydrazine (Picard et al. 2005). Baricault et al. used the human colon cancer cell line HT-29 to study the effects of fermented milk on the proliferation and differentiation of tumor cells, with *Bifidobacterium* being particularly effective in suppressing the growth rate of HT-29 cells (Baricault et al. 1995). These findings emphasize the potential health benefits of *bifidobacteria*, not only in enhancing gut health, but also in preventing the development of colon cancer (Baricault et al. 1995).

Several studies have investigated the potential of *bifidobacteria* to prevent and/or treat colorectal cancer (O'Callaghan and van Sinderen 2016). Most of this research uses animal models, and their results suggest that a combination of prebiotics and *bifidobacteria* could reduce the incidence of cancer cells arising from carcinogens in mice (O'Callaghan and van Sinderen 2016). *Bifidobacterium longum* BB-536, a dietary supplement, has shown significant potential to reduce colon and liver tumors in male rats exposed to 2-amino-3-methylimidazo [4,5-f]quinoline (IQ) and to suppress the development of breast and liver cancer in female rats (Chen et al. 2021). Another strain, BCRC 910051, effectively prevents colon cancer induced by 1,2-dimethylhydrazine. In addition, laboratory studies indicate that *Bifidobacterium bifidum* BGN4 inhibits various types of human colon cancer cells in vitro (Chen et al. 2021). *Bifidobacteria* can prevent inflammation in the mouth by inhibiting inflammation in the gut. Yogurt fermented with *B. lactis* DN-173010 can improve gingival inflammatory parameters by decreasing IL-1β levels and the total amount in gingival cancer fluid. Oral administration of *B. lactis* HN019 can improve scaling and root planing (SRP) in the treatment of experimental periodontitis in rats and patients (Chen et al. 2021). Bifidobacteria are used in the treatment of many gastrointestinal disorders, including diarrhea, inflammatory bowel disease, and ulcerative colitis (O'Callaghan and van Sinderen 2016). The efficacy of treating diarrhea with *B. longum* subspecies infantis CECT 7210 and *B. breve* K-110 has been attributed to their ability to block rotavirus, which is the leading cause of sporadic diarrhea in children (O'Callaghan and van Sinderen 2016). *Bifidobacterium longum* BB536 promotes the production of nutrients, such as pimelate, biotin, and butyrate, through enhanced fermentation processes facilitated by interactions between *B. longum* BB536 and the microbiota in the human gut (O'Callaghan and van Sinderen 2016). Furthermore, *Bifidobacterium pseudocatenulatum* is often present in human fecal samples throughout life, with certain strains exhibiting beneficial properties, such as the synthesis of enterolignan, urolithin, and conjugated linoleic acid (O'Callaghan and van Sinderen 2016). *Bifidobacteria*, which are primarily found in the oral cavity and digestive tract, are used to alleviate and prevent certain inflammations in these regions, such as irritable bowel syndrome, intestinal barrier disorders, and infant colic (Chen et al. 2021). In addition, certain strains of *bifidobacteria* have shown the ability to reduce skin inflammation. For example, a combination of *B. breve* M-16 V and *B. longum* BB536 has been shown to reduce the incidence of eczema and atopic dermatitis in infants (Chen et al. 2021). Genetically modified probiotic strains are considered a promising strategy as they offer various benefits by modifying certain strains to improve their properties.

9.3.3 Saccharomyces Species

Saccharomyces boulardii, a well-researched yeast probiotic, is known for its distinct efficacy demonstrated in double-blind studies, making it a patented and widely trusted remedy for GI disorders (Czerucka et al. 2007). In particular, it is used for both the prevention and the treatment of diarrhea caused by antimicrobial agents. Although *S. boulardii* has genetic and karyotypic similarities with its relative *S. cerevisiae*, it has unique characteristics; in particular, increased growth performance and increased resistance to acid and temperature stress, which differentiates it from *S. cerevisiae* (Czerucka et al. 2007). Yeast thrives optimally between pH 4.5 and 6.5, but can generally tolerate environments as low as pH 1.5, making them robust candidates for probiotic applications due to their inherent resistance to GI stressors, such as enzymes, bile salts, and varying pH and temperature. The natural antibiotic resistance of *S. boulardii* allows for safe co-administration with antibiotics, potentially preventing recurrent *Clostridium difficile* infections (Czerucka et al. 2007). Studies suggest that a higher number of yeasts in the stool due to ingestion of *S. boulardii* correlates with a significantly lower relapse rate in patients. Certain strains, such as *Saccharomyces boulardii* 17 and *Saccharomyces cerevisiae* UFMG 905 and BY4741, exhibit the ability to adhere to some pathogenic bacteria, an attribute not observed in the *S. cerevisiae* W303 strain used as a control, highlighting the potential of specific yeast probiotics for maintaining gut health (Czerucka et al. 2007).

9.3.4 Decoding the Gut Microbiome–Human Synergism

The microbiota can be described as the group of microorganisms present in a particular environment, whereas the microbiome refers to the entire population, its genetic make-up, and the surrounding environmental conditions in that habitat (Ramirez et al. 2020). Although individuals have diverse compositions and types of microbiota, there is a common set of genes encoded by the microbiota, which implies that diverse compositions lead to similar functional outcomes (Ahern and Maloy 2020). The gut microbiome plays an essential role in maintaining our overall health by contributing to physiological processes and their balance, facilitating growth and protecting us from invasive pathogens. Although multiple microbial populations reside in various parts of our body, including the gut, lungs, vagina, and mouth, the gut microbiota has received the most attention in research (Michaudel and Sokol 2020). The influence of the gut microbiome on the human body can be both beneficial and detrimental, which is mainly determined by its interactions with the human body.

The colonization of microbes in our gut is a complex process (Ahern and Maloy 2020). Studies have shown that the meconium microbiota has contributed to our understanding that colonization of the microbiota occurs directly during birth, but other studies also suggest that ingestion of amniotic fluid may be one of the possible routes for colonization (Kosiewicz et al. 2011). The gut microbiome performs a wide range of therapeutic actions in the body by helping to defend against harmful pathogens by colonizing mucosal surfaces and producing antimicrobial substances, thereby strengthening the immune system (Gomaa 2020). The gut microbiota has a major influence on many aspects of human health. It aids in digestion and metabolism, regulates epithelial cell growth, influences insulin sensitivity and secretion, and helps with gut–brain communication, thereby impacting cognitive and neurological function (Gomaa 2020). The microbiota acts as a protective barrier by colonizing the intestinal surface and creating an environment that prevents the invasion of harmful microorganisms (Gomaa 2020). The gut microbiota plays an integral part in metabolism by metabolizing many dietary fibers into biologically active food components. In this environment, these bacteria have the ability to break down complex carbohydrates, such as cellulose, hemicellulose, retrograde starch, pectin, oligosaccharides, and lignin, into short-chain fatty acids (SCFAs), such as acetic, propionic, and butyric acids (Gomaa 2020). These SCFAs are thought to have promising anti-inflammatory and chemopreventive properties, so they are considered as tumor suppressors. In addition, the anticarcinogenic and anti-inflammatory effects of propionate and butyrate have been confirmed (Gomaa 2020). The fatty acids, which are mainly produced by Firmicutes, Bacteroidetes, and other anaerobic gut microbiota, are not broken down in the stomach. Instead, they enter the large intestine where they are metabolized (Gomaa 2020). Similarly, the gut microbiota plays a role in the production of cholesterol, bile acids, and conjugated fatty acids (Gomaa 2020). The gut microbiota can also make a positive contribution to the host organisms. For example, it contributes significantly to the synthesis of vitamins, such as biotin, thiamine, cobalamin, riboflavin, niacin, pantothenic acid, and vitamins B and K (Gomaa 2020). Commensal bacteria perform a crucial role in the defense against autoimmune conditions by controlling inflammation in the gut. They do this either by promoting immune regulation or by controlling bacteria that directly contribute to intestinal inflammation (Kosiewicz et al. 2011). In particular, commensal bacteria from the phyla Bacteroidetes and Firmicutes are characterized by their ability to promote the production of regulatory T cells (Tregs), which in turn can regulate inflammatory Th17 cells in the gut (Kosiewicz et al. 2011) (Fig. 9.2).

The immunological response and role of the gut microbiota can be traced back to the early twentieth century when Elie Metchnikoff, a zoologist, became interested in unravelling the detrimental effects of the gut microbiota on health and how it can be restored through proper care using probiotics (Salvo-Romero et al. 2020). The gut is the largest surface area of the GI tract that is exposed constantly to dietary and bacterial antigens. The possibility of pollutants and infectious pathogens jeopardizes the balance between health and disease (Salvo-Romero et al. 2020). The gut microbiota resides and proliferates on the intestinal surface, creating a stable environment that prevents invasion by pathogenic microbes. Disruption of this

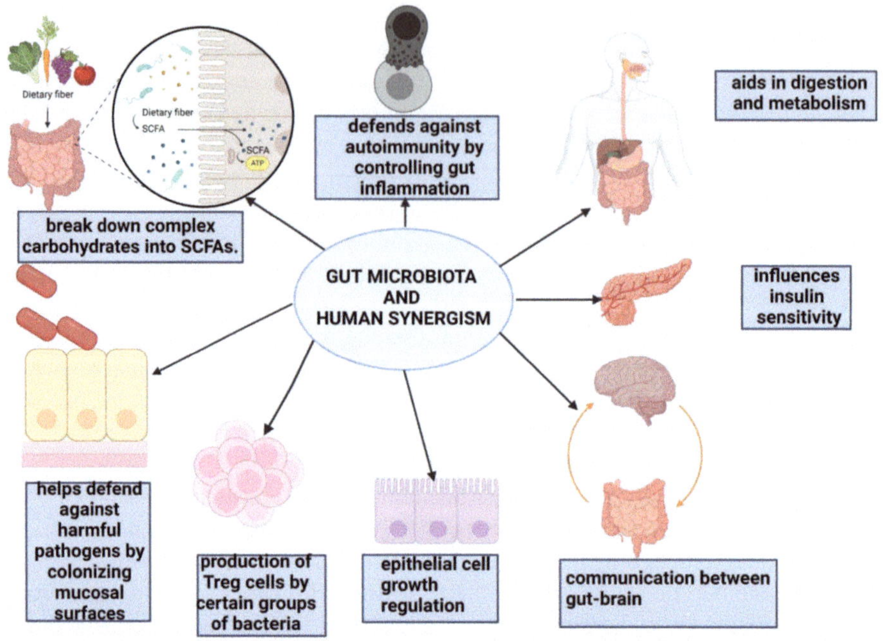

Fig. 9.2 The interactions of the gut microbiota with cells and tissues in the human system

commensal bacteria can lead to a variety of diseases including diabetes, obesity, inflammatory bowel disease (IBD), allergies, autoimmune diseases, and cardiovascular disease (Gomaa 2020).

9.3.5 The Gut–Brain Axis

The gut–brain axis is a complex bidirectional communication system in which the gut and brain interact via neural pathways, including the enteric nervous system, vagus nerve, sympathetic, and spinal nerves, as well as humoral pathways, such as cytokines, hormones, and neuropeptides (Bercik et al. 2012). As the enteric nervous system (ENS) controls many gastrointestinal functions, such as motility and absorption, it is an important mediator between the gastrointestinal tract and the brain (Chen et al. 2013). The gut microbiota has been shown to influence brain function via this axis. Disruption of the gut microbiota caused by factors, such as diet, genetics, stress, and medications, can disrupt gut homeostasis and lead to adverse effects. This disruption of the gut–brain axis can contribute to various metabolic disorders (Chen et al. 2013). The gut microbiota produces glycoside hydrolases that convert complex polysaccharides into monosaccharides and SCFAs. The SCFAs, especially those that bind to the G-protein-coupled receptors Gpr41 and Gpr43 on enteroendocrine cells, trigger the release of hormones such as peptide YY, which, in turn, affect gut transit time and nutrient absorption and link the gut microbiota to obesity via changes in host energy balance and body mass gain (Chen et al. 2013) (Fig. 9.3).

Furthermore, probiotics influence the regulation of neurotransmitters such as γ-aminobutyric acid, which are central to mood and cognitive function. Certain probiotics, known as psychobiotics, can produce neuroactive molecules such as serotonin and GABA, which act on the gut–brain axis and contribute to cognitive health (Liang et al. 2018). This suggests that modulation of the gut microbiota with probiotics may influence mood and behavioral regulation via the GBA.

Probiotics also have the potential to support skin health via the gut–brain–skin axis. They can reduce systemic inflammation in the brain, improve nutrient absorption, and consequently support the synthesis of the skin barrier (Wieers et al. 2019). They have been studied in various skin conditions, including atopic dermatitis and improvement of skin aging. Beyond the skin, there is evidence that probiotics can relieve migraine headaches by improving gut barrier function via the gut–brain axis (Tegegne and Kebede 2022). In animal studies, the interaction between vagal signaling and the immune response has been shown to attenuate inflammation, potentially alleviating conditions similar to human gut inflammation. Oral consumption of *Lactobacillus reuteri* also diminishes perifollicular dermatitis.

Probiotics are thought to play a role in relieving migraine headaches by enhancing the functionality of the intestinal barrier, which is influenced by the complex interactions of the gut–brain axis (Wieers et al. 2019). In studies using a mouse model of murine enteritis, it was observed that efferent vagal signaling triggered by corticotropin-releasing hormone is able to inhibit inflammation triggered by an

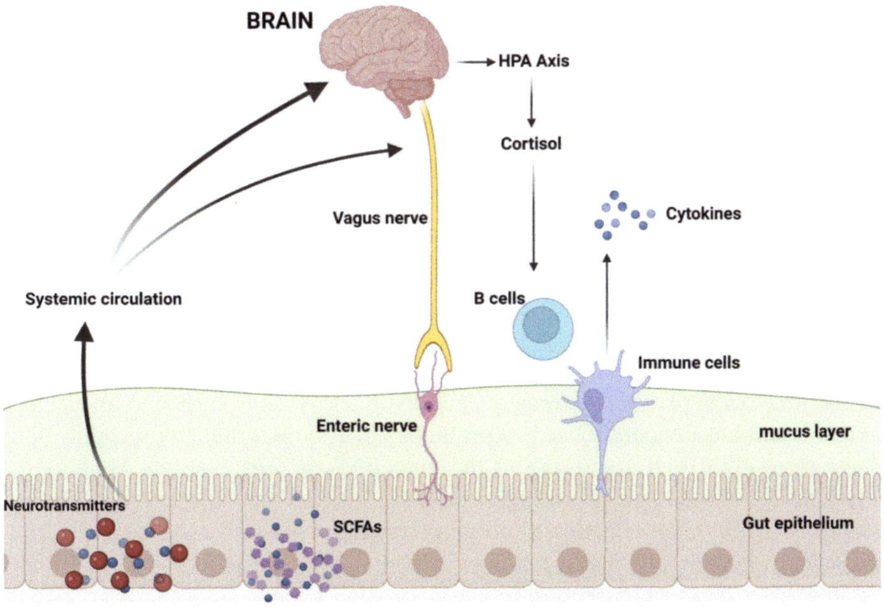

Fig. 9.3 An illustration of the communication between the gut and the brain

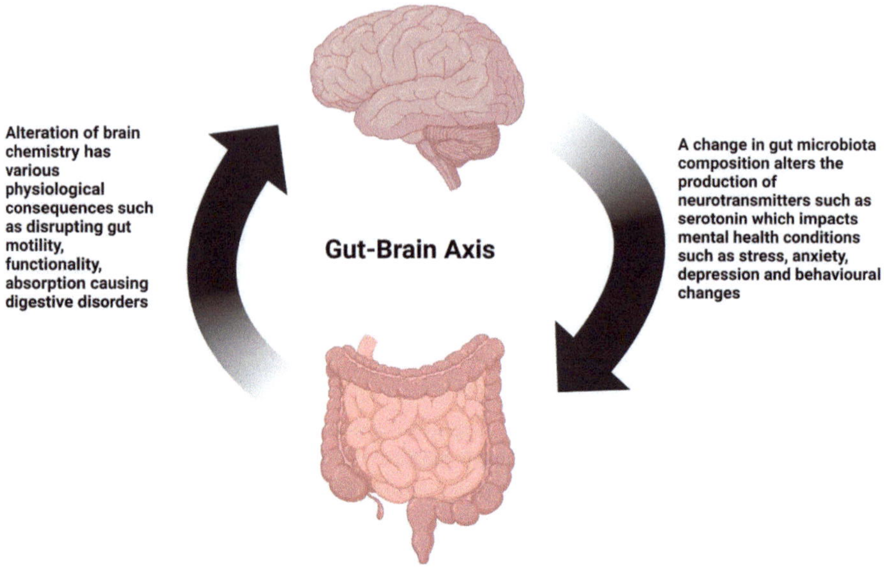

Fig. 9.4 The interaction between the gut and the brain

inflammasome, effectively reducing the incidence of intestinal enteritis. This process demonstrates the ability of gamma-aminobutyric acid (Bamigbade et al. 2022) to regulate the activation of innate immune cells in the gut via the hypothalamic–pituitary axis, resulting in the suppression of inflammation in the intestinal system. In addition, it has been suggested that probiotics may provide protection against brain injury by interfering with the transport of harmful biomolecules through the GBA. These promising research findings suggest that probiotics may be a valuable therapeutic strategy to control inflammation affecting both gut health and neurological disorders, such as migraine (Sherman et al. 2015) (Fig. 9.4).

9.3.6 Selection Criteria for Probiotics

Numerous microbial species have probiotic properties, with lactic acid bacteria (LAB) being the most prominent group in the gut microbiota. They are characterized by their appearance as gram-positive rods or cocci and the absence of spores, are adapted to low pH values, ferment carbohydrates by using carbon as an electron acceptor, and are frequently used as probiotics due to their numerous applications (Raheem et al. 2021). According to the guidelines of organizations, such as the WHO, the Food and Agriculture Organization (FAO), and the European Food Safety Authority (EFSA), the selection of probiotic microbes is based on three main criteria: safety, functionality, and technological benefit (Zommiti et al. 2020). Hill and his coworkers discovered that the probiotic capabilities of a bacterium are specifically linked to certain strains and not to the entire genus or species (Hill et al. 2014).

The bacterial strains are tested *in vitro* to assess their alignment with the requirements (Zommiti et al. 2020). Using these established criteria from laboratory research allows microorganisms to be evaluated to determine if they qualify as probiotic strains (Zommiti et al. 2020). Several critical considerations must be made before selecting a probiotic strain for the purpose of genetic modification. These include evaluation of plasmid longevity and reliability of genotype and phenotype including variables, such as patterns of protein or carbohydrate utilization, tolerance to bile or acid, growth and survival rates, resistance to antibiotics, adherence to intestinal epithelial cells, production of antimicrobials, ability to inhibit pathogens, and immunogenicity (Mazhar et al. 2020). Probiotic encapsulation technology (PET) has received considerable attention in the last decade due to its ability to preserve bacterial cells and improve their vitality during production and storage (Mazhar et al. 2020). This method ensures the regulated and efficient delivery of bacterial strains into the GI tract, and improves their viability and protection in the acidic atmosphere of the stomach. In their live form, probiotics can therefore provide greater health benefits to the host (Mazhar et al. 2020). Although some strains, such as *Bifidobacterium animalis*, are not exactly derived from human sources, they can still be quite beneficial. However, it is generally preferred that probiotics used for humans are derived from human sources (Mazhar et al. 2020). Other properties may also be considered when selecting probiotic strains, including cognitive impacts such as lowering cholesterol levels, antioxidant activity, or cytotoxic effects on cancer cells (Raheem et al. 2021). Confirmation of the criteria can be done through both *in vitro* and *in vivo* experiments, with human trials being fundamental for the use of probiotics in humans. Regarding the consumption of probiotics, the lack of sufficient data on minimum effective levels contributes to the general acceptance that probiotic products should have a minimum concentration of 10^6 CFU per milliliter or gram, and it is recommended that consumers ingest a total daily intake of approximately 10^8 to 10^9 probiotic microorganisms to experience the full benefits of probiotics (Zommiti et al. 2020). Clear selection criteria of probiotics include improving host health, surviving intestinal transit, adhering to the intestinal epithelial cell membrane, releasing antibiotic compounds to fight infection, and regulating the gut microbiota (Tegegne and Kebede 2022).

9.4 Mechanisms of Probiotics in the Immune System

A study has shown that the effectiveness of probiotic microbes depends on various factors. These factors include their ability to resist colonization, stimulate phagocytosis, produce antimicrobial compounds, exert antimutagenic effects, stimulate the production of chemokines, and influence enzyme activity and delivery (Zommiti et al. 2020). Recently, particular attention has been focused on the ability of probiotics to prevent the attachment of pathogens to the intestinal surface, support the integrity of the epithelial barrier, and regulate the immune system (Wang et al. 2021). In addition, probiotics also have a regulating effect on the microecology of the gut and provide benefits to the human body (Wang et al. 2021). The mechanism involves four major steps: (i) microbial antagonism through the release of antimicrobial compounds, (ii) pathogenic bacteria compete for epithelial adhesion and vital nutrients (Cohen et al. 2018), (iii) modulation and maturation of the immune system, and (iv) suppression of bacterial toxin.

9.4.1 Microbial Antagonism Is Accomplished by the Release of Antimicrobial Compounds

For a microbial strain to be considered a probiotic, it must fulfill certain criteria. These include colonization and interaction with the host, for which it must have the ability to adhere to the intestinal mucosa (Plaza-Diaz et al. 2019). This interaction is crucial for modulating antagonism against pathogens and for influencing the immune system (Plaza-Diaz et al. 2019). The combined effect of the probiotic strains creates a protective barrier that prevents the colonization of the epithelium with pathogens (Sharma 2021). Certain Lactobacillus proteins increase mucosal adhesion, support epithelial cell attachment and inhibit pathogen invasion, all of which have a significant impact on host health (Sharma 2021). In addition, probiotics that bind to epithelial cells trigger a signaling cascade that leads to immunomodulation and stimulates immune cells directly or indirectly via epithelial cells (Sharma 2021). *L. rhamnosus* ATCC 7469 and various *L. plantarum* strains were tested on porcine intestinal epithelial J2 cells (Plaza-Diaz et al. 2019). *E. coli* increased the expression of TLR-4 and NOD2, which was counteracted by treatment with *L. rhamnosus* (Plaza-Diaz et al. 2019). *L. plantarum* TIFN101 affected gene pathways by promoting matrix metalloproteinase and tissue inhibitors of metalloproteinase 1 and 3 and downregulating tricarboxylic acid cycle II-related genes (Plaza-Diaz et al. 2019). According to various studies, probiotics may play a function in the elimination of cancer cells (Sharma 2021).

9.4.2 Pathogenic Bacteria Compete for Epithelial Adhesion and Vital Nutrients

Various criteria are used to classify antimicrobial peptides (AMPs) based on their origin (e.g., animals, fungi, plants, and bacteria), their mechanism of action (e.g., targeting cell surface molecules or intracellular components), their structure (such as covalent bonding patterns), and their biosynthetic pathway (e.g., nonribosomal or ribosomal synthesis) (Raheem et al. 2021). Bacteriocins are antimicrobial peptides (AMPs) derived from prokaryotes, especially lactic acid bacteria (LAB), and can be categorized into three different classes: Class I includes bacteriocins that have undergone post-translational modifications, such as antibiotics (Raheem et al. 2021). Class II includes unmodified peptides that are heat-stable and have a size less than 10 kDa. Examples of Class II peptides are pediocin PA1, leucocin A, plantaricin A, and enterocin X. Class III comprises large peptides that are heat-labile and have a size more than 30 kDa. An example of a class III peptide is crustins (Raheem et al. 2021). Bacteriocins, which have a low molecular weight, form openings in the membranes of certain cells, thereby leading to the death of harmful microbes. They also have an anti-cancer effect (Raheem et al. 2021). Bacteriocins are small, cationic molecules consisting of approximately 30–60 amino acids that act on bacterial cytoplasmic membranes, where they hinder proton motive force by targeting energized membrane vesicles (Plaza-Diaz et al. 2019). In addition, bacteriocins exhibit immunomodulatory effects, particularly through their anti-inflammatory action in pathogenic infections (Raheem et al. 2021). These bacteriocins are nontoxic, especially those derived from LAB, and due to this property they are used for food preservation (Raheem et al. 2021). Many studies have shown that the action of bacteriocins inhibits pathogenic bacteria. Specific bacteriocins synthesized by *L. plantarum* and *L. acidophilus* are able to inhibit the proliferation of Helicobacter, *C. difficile*, rotaviruses, and drug-resistant strains of *Shigella* species and *E. coli* in gastrointestinal diseases (Plaza-Diaz et al. 2019). These compounds also show efficacy against various uropathogens (Plaza-Diaz et al. 2019). The bacteriocins synthesized by *Lactobacillus salivarius* have the ability to inhibit the proliferation of food-borne and clinically important bacteria, such as *Listeria monocytogenes*, various strains of staphylococcus, *Neisseria gonorrhoeae*, Bacillus, and Enterococcus, by creating pores in the membranes of these bacteria, causing the leakage of cell contents (Raheem et al. 2021). In addition, certain bacteriocins derived from *L. plantarum* show efficacy against yeasts and molds, such as Fusarium, Candida, Aspergillus, and Mucor (Raheem et al. 2021). *L. rhamnosus* was tested together with *E. coli* intestinal epithelial cells of pigs. The presence of *E. coli* increased the expression of Toll-like receptor (TLR)-4 and nucleotide-binding oligomerization domain-containing protein 2 (NOD2), which was reduced by treatment with *L. rhamnosus* (Sharma 2021). The probiotics contributed to the maintenance of the epithelial barrier and promoted the activation of intestinal epithelial cells, suggesting a possible role in the elimination of cancer cells (Sharma 2021). Bacteriocins from various probiotic species have been shown to be important in inducing apoptosis and suppressing tumor development while inhibiting cancer cell proliferation. They also cause depolarization of cancer cell membranes during treatment (Raheem et al. 2021).

9.4.3 Immune Modulation and Proper Maturation of the Immune System

For a microbial strain to be considered a probiotic, it must fulfil certain criteria. These include colonization and interaction with the host, for which it must have the ability to adhere to the intestinal mucosa (Plaza-Diaz et al. 2019). This interaction is crucial for modulating antagonism against pathogens and for influencing the immune system (Plaza-Diaz et al. 2019). The combined effect of the probiotic strains creates a protective barrier that prevents the colonization of the epithelium with pathogens (Sharma 2021). Certain Lactobacillus proteins increase mucosal adhesion, support epithelial cell attachment and inhibit pathogen invasion, all of which have a significant impact on host health (Sharma 2021). In addition, probiotics that bind to epithelial cells trigger a signaling cascade that leads to immunomodulation and stimulates immune cells directly or indirectly via epithelial cells (Sharma 2021). *L. rhamnosus* ATCC 7469 and various *L. plantarum* strains were tested on porcine intestinal epithelial J2 cells (Plaza-Diaz et al. 2019). *E. coli* increased the expression of TLR-4 and NOD2, which was counteracted by treatment with *L. rhamnosus* (Plaza-Diaz et al. 2019). *L. plantarum* TIFN101 affected gene pathways by increasing matrix metalloproteinase and tissue inhibitors of metalloproteinase 1 and 3 and downregulating genes associated with the tricarboxylic acid cycle II pathway (Plaza-Diaz et al. 2019). According to various studies, probiotics may have a function in the elimination of cancer cells (Sharma 2021).

One of the most important mechanisms of action of probiotics is the regulation of the host immune response (Sharma 2021). The gut microbiota influences the immune system by producing molecules with immunomodulatory and anti-inflammatory functions, which in turn stimulate immune cells through interactions with epithelial cells, dendritic cells (DCs), macrophages, and lymphocytes (Plaza-Diaz et al. 2019). The immune system is mainly divided into two types: the innate and the adaptive system. Adaptive immunity relies on B and T lymphocytes to recognize specific antigens, whereas innate immunity responds to pathogen-associated molecular patterns (PAMPs) (Plaza-Diaz et al. 2019). Probiotics can influence both the innate and the acquired immune system via metabolites and cell wall components (Sharma 2021). Intestinal immune cells and epithelial cells play a crucial role in the immune response (Sharma 2021). The cell wall components of LAB activate macrophages. Probiotic bacteria can modulate the production of cytokines by immune cells in the GI system (Sharma 2021). Pattern recognition receptors (PRRs), such as Toll-like receptors (TLRs), nucleotide-binding oligomerization domain-like receptors (NLRs), and C-type lectin receptors, engage with PAMPs to initiate cell signaling (Raheem et al. 2021). PRRs are transmembrane proteins found in many immune and nonimmune cells, including B cells, natural killer cells, DCs, macrophages, fibroblasts, and epithelial and endothelial cells (Plaza-Diaz et al. 2019). Toll-like receptors (TLRs) are among the well-studied PRRs. They can activate signaling pathways that cause cell proliferation and the release of cytokines, thereby influencing the immune system (Cristofori et al. 2021).

Intestinal epithelial cells are the main point of contact with probiotics, while DCs also engage with them, linking innate and adaptive immune responses (Raheem et al. 2021). Intestinal epithelial cells and DCs use PRRs to establish communication and respond to microorganisms in the gut (Sharma 2021). Activated DCs, influenced by probiotics/commensal microbiota, promote an appropriate immune response (Raheem et al. 2021). This includes the development of naïve CD4 T cells into regulatory T cells (Tregs), which normally reduce Th1-, Th2-, and Th17-mediated inflammation (Raheem et al. 2021). An imbalance between effector T helper (Th) cells and regulatory T cells influences the immunological response (Plaza-Diaz et al. 2019).

Furthermore, probiotics attenuate inflammation in the gut by reducing the expression of TLR through the secretion of metabolites that inhibit TNF-α and by inhibiting NF-κB signaling in enterocytes (Raheem et al. 2021). In addition, probiotics regulate the synthesis of various cytokines (Raheem et al. 2021). Probiotic bacteria regulate immunological responses in the stomach by producing anti-inflammatory cytokines (Cristofori et al. 2021). However, the specific molecular processes that control the interactions between probiotics and the host are not yet fully understood (Cristofori et al. 2021). Various Lactobacillus strains can influence cytokine production by immune cells, while Bifidobacterium promotes the acquisition of tolerance (Cristofori et al. 2021). These regulatory functions are associated with the structural characteristics of the bacteria, the spectrum of mediators involved, and the signaling pathways they activate simultaneously (Cristofori et al. 2021).

9.4.3.1 Cytokine-Mediated Immune Response

Probiotics are being researched extensively for their ability to modulate the immune system in various diseases (Raheem et al. 2021). The main mechanism by which these effects occur is the secretion of cytokines, such as interleukins, transforming growth factors (TGFs), tumor necrosis factors (TNFs), interferons (IFNs), and chemokines, by immune cells with cytokines then playing a role in regulating immunological responses (Raheem et al. 2021). Immunostimulatory probiotics fight infections and cancer cells by triggering the secretion of IL-12, activating NK cells, and promoting the production of Th1 cells (Azad et al. 2018). They have also been shown to reduce allergens by harmonizing Th1 and Th2 cells (Raheem et al. 2021). On the other hand, immunoregulatory probiotics stimulate the development of Treg cells and the secretion of IL-10, which suppresses the excessive inflammation observed in diseases, such as inflammatory bowel disease and autoimmune disorders (Raheem et al. 2021). The effect of probiotics on cytokine production is distinctive and differs from strain to strain. This has been observed in laboratory experiments in vitro (Raheem et al. 2021). For example, a combination of *Lactobacillus paracasei* and *L. reuteri* reduced colitis and the production of proinflammatory cytokine in *Helicobacter hepaticus* IL-10-deficient mice (Raheem et al. 2021).

9.4.3.2 Toll-Like Receptor-Mediated Immune Response

Probiotics interact with the TLRs of the immune system, a type of PRR that can recognize microbial components (Plaza-Diaz et al. 2019). TLR is a family of 11 proteins in mammals that are activated after binding to extracellular leucine-rich repeats (Sharma 2021). TLRs can activate various signaling pathways that lead to immune responses. TLR1, TLR2, TLR4, TLR5, TLR6, and TLR10 respond primarily to bacterial surface-associated PAMPs, whereas TLR3, TLR7, TLR8, and TLR9 respond primarily to virus- and bacterial nucleic acid-based PAMPs on endosomal surfaces (Plaza-Diaz et al. 2019). The TLR pathway, apart from TLR3, activates the MAPK and NF-κB signaling pathways via the myeloid differentiation primary response 88, while TLR3 induces the expression of type 1 IFNs via an adaptor protein and TLR9 signaling is crucial for moderating the anti-inflammatory effect of probiotics (Sharma 2021). Probiotics contribute to the prevention of intestinal inflammation by reducing the regulation of TLR expression, by secreting of metabolites that block the entry of TNFα into blood mononuclear cells, and by inhibiting NF-κB signaling in enterocytes (Sharma 2021). For example, Lactobacillus can reduce inflammation in the gut by reducing TLR expression, TNF-α, and NF-κB signaling (Plaza-Diaz et al. 2019). Lactobacilli components bind to TLR2 and TLR6, inducing cytokine production (Plaza-Diaz et al. 2019). Immune cells respond by upregulating CD-206 and TLR2 receptors (Sharma 2021). Several strains, such as *L. CCFM634*, *L. CCFM734*, *L.CCFMM381*, and *L. fermentum* CCFM137, stimulate TLR2/TLR6, which emphasizes their importance for immune regulation (Sharma 2021). Strains such as *L. casei* 431 communicate with epithelial cells via the TLR2 (Plaza-Diaz et al. 2019). Furthermore, lactobacilli activate TLR2/TLR6 signaling, which has an impact on immunological modulation (Plaza-Diaz et al. 2019). A study on rats with induced dysbiosis found that a probiotic mixture of four strains improved mucosal barrier damage, reduced proinflammatory factors and cytokines, and minimized neutrophil infiltration (Plaza-Diaz et al. 2019). This was associated with the restoration of microbial balance in the gut and changes in TLR2 and TLR4 signaling pathways (Plaza-Diaz et al. 2019). TLR9, another important TLR activated by probiotics, shows anti-inflammatory effects at the epithelial surface (Plaza-Diaz et al. 2019). Different probiotic species have different abilities to stimulate TLR9 and trigger different intracellular signaling pathways (Plaza-Diaz et al. 2019). *L. rhamnosus* GG, a transplant from the human gut microbiota, influences gut bacterial communities and modulates immune signaling pathways in response to an oral attenuated human rotavirus vaccine (Plaza-Diaz et al. 2019). *L. paracasei* CNCM I-4034 and its culture supernatant modulate Salmonella-induced inflammation in a novel co-culture model with human intestinal dendritic cells and Caco-2 cells (Plaza-Diaz et al. 2019). *L. rhamnosus* HN001 has been shown to be beneficial in intestinal diseases such as NEC by activating TLR9 and attenuating NEC without reported toxicity (Plaza-Diaz et al. 2019).

9.4.4 Suppression of Bacterial Toxin Production

Inhibition of bacterial toxins involves deactivating toxins and assisting their removal from the body (Sharma 2021). However, due to their uniqueness, not all probiotics have the ability to detoxify. Further research is needed to identify strains with strong detoxification properties. Suppression of certain metabolic processes can produce various toxins associated with the activation of metabolic pathways, leading to the production of internal antimicrobial compounds, enzymes, and vitamins (Sharma 2021). The intestinal epithelial barrier plays a critical role in preventing the entry of harmful substances such as toxins and bacteria into the body. Many studies have shown that LAB can help strengthen the intestinal epithelial barrier and reduce the damage caused by pathogens (Raheem et al. 2021). The effect of probiotics is exerted through various mechanisms. These include the presence of a heat shock protein 27 (HSP27)-inducible polyphosphate (poly P) fraction in Lactobacillus

(Raheem et al. 2021). These polyphosphates strengthen epithelial barrier function by activating the integrin-p38 MAPK signaling pathway, thereby promoting intestinal homeostasis (Raheem et al. 2021).

9.5 Therapeutics of Immune Disorders and Metabolic Disorders

9.5.1 Diarrhea

Probiotics are often recommended for the treatment and prevention of diarrhea. This condition can be acute or chronic and can have various origins, including infections and the use of antibiotics. Clinical trials and meta-analyses have generally shown that probiotics can be of great benefit in acute diarrhea, especially in children. For example, the administration of probiotics in acute diarrhea has been shown to shorten its duration (Krysiak et al. 2021). In cases of persistent diarrhea, particularly in developing areas, probiotics have resulted in a reduction in duration of approximately 4 days, along with improvements in growth parameters (Krysiak et al. 2021). Antibiotic-associated diarrhea is often due to *Clostridium difficile*, which can overgrow in the gut when the natural microbial balance is disturbed by antibiotics. The standard treatments for this condition are metronidazole or vancomycin, but probiotics have been shown to help rebuild the protective "barrier microbiota" that can prevent *C. difficile* colonization and recurrence of infection (Liu et al. 2018). Several studies have reported that certain strains, such as *Saccharomyces boulardii* and *Lactobacillus rhamnosus* GG, can reduce the rate of nosocomial diarrhea, antibiotic-associated diarrhea, and the recurrence of *C. difficile*-associated diarrhea (Alagawany et al. 2018). The beneficial effects of probiotics on AAD include a 40–60% reduction in the frequency of diarrhea. However, there are fewer studies specifically documenting a reduction in *C. difficile*-associated diarrhea, and some aspects of this use remain controversial. It is important to note that while there is promising evidence, not all probiotic strains are effective for all types of diarrhea, and efficacy may vary depending on the patient group and the strains used (Alagawany et al. 2018). Moreover, the use of synthetic molecules derived from probiotic or protective gut species is being explored. For example, immunoactive products, such as p40 from *L. rhamnosus* GG or polysaccharide A from *Bacteroides fragilis*, could be used due to their defined biological effects (Krysiak et al. 2021). Prebiotics such as inulin or fructo-oligosaccharides also play a role by increasing the number of beneficial Bifidobacterium species and the concentrations of SCFAs, which are crucial for colon health. Conversely, refined sugars and certain food additives can promote the growth of harmful bacterial species (Krysiak et al. 2021). The global decline in diarrhea-related deaths is attributed to improved sanitation, increased breastfeeding, and the introduction of the rotavirus vaccine, rather than the use of probiotics.

9.5.2 Necrotizing Enterocolitis

Necrotizing enterocolitis (NEC) is a significant cause of neonatal morbidity and mortality, but its specific etiology remains unclear (Milner et al. 2021). Contributing factors may include the immature immune system of neonates and disruption of the intestinal microbiome following antibiotic treatment. Symptoms suggestive of NEC include feeding intolerance, lethargy, bloating, and bloody stools. Premature infants are more susceptible to NEC, and this vulnerability may be related to marked differences in their gut microbiota compared to full-term infants (Milner et al. 2021). In particular, infants diagnosed with NEC often show a reduction in Firmicutes and an increase in Gamma Proteobacteria in their gut microbiota. The frequent use of antibiotics in these preterm infants may contribute to further the alteration of their microbiota, which would increase their risk of NEC (Liu et al. 2018). NEC is associated with a high mortality rate of around 30%, despite extensive medical and surgical interventions, and survivors often face serious long-term complications (Sanders and Klaenhammer 2001). Early diagnosis of NEC can be difficult and is often not made until intestinal perforation has occurred.

Research using meta-analyses of probiotics, such as strains of *Bifidobacterium, Lactobacillus, Saccharomyces*, and *Streptococcus thermophilus*, have shown promising results in the prevention of NEC. These studies indicate a decrease in the incidence and overall mortality due to NEC (Sanders and Klaenhammer 2001). Although the American Academy of Pediatrics acknowledges that probiotics have been shown to prevent NEC in very low birth weight infants, it calls for further research to determine the most effective probiotic strains and dosages before making a formal clinical recommendation (Sanders and Klaenhammer 2001). Current treatment for NEC includes stabilizing the patient with fluid replacement, providing nutritional support, administering anti-infectives, and surgical intervention if necessary (Liu et al. 2018).

9.5.3 Irritable Bowel Syndrome

Irritable bowel syndrome (IBS) is a common functional GI disorder characterized by symptoms, such as abdominal pain and irregular bowel habits such as diarrhea and/or constipation (Butel 2014). The incidence of IBS is higher in women and younger people under the age of 50, with prevalence rates ranging from 7% to 21% worldwide. The development of IBS can be influenced by various factors including psychological stress, low-grade inflammation of the intestinal mucosa, abnormal gut motility, and GI infections (Butel 2014). The pathophysiology of IBS is complex and involves a combination of altered brain–gut interactions, changes in the gut microbiome, visceral hypersensitivity, abnormal GI motility, and psychosocial factors (Compare et al. 2022). Despite extensive research, the complete mechanism of IBS is still unclear.

Research on the treatment of IBS with probiotics has shown potential benefits. A systematic review by Brenner et al. reported that *Bifidobacterium infantis* was effective in improving IBS symptoms compared to a placebo (Brenner et al. 2009). In another study, significant improvements in health-related quality of life and IBS symptoms were observed after treatment with *Bifidobacterium bifidum* MIMBb75 (Compare et al. 2022). Nonviable bacteria are also being considered as they offer potential advantages over live prebiotics in terms of standardization, stability, and safety.

The treatment of IBS remains challenging as there is no validated treatment algorithm (Compare et al. 2022). Probiotics have been proposed as an effective approach to modulate the gut microbiota, stimulate the immune system, and improve gut permeability. For example, *Bifidobacterium coagulans* Unique IS2 has been shown to significantly alleviate abdominal discomfort/pain and improve bowel movements in a clinical trial involving 136 patients. In addition, probiotics have shown positive effects on visceral pain reflexes in rodent models. However, further research is needed to understand their effects in humans with or without IBS (Compare et al. 2022). Antibiotics, particularly the nonabsorbable rifaximin, have been shown to be effective in improving IBS symptoms. However, the conclusions of meta-analyses on probiotics in IBS vary due to factors such as small sample sizes, different study designs, and the use of different strains of probiotics (Compare et al. 2022). Overall, probiotics represent a promising but yet to be explored option for the treatment of IBS, with bifidobacteria or mixed strains being among the most potentially effective. However, further clinical studies are needed to confirm their efficacy and establish treatment protocols (Compare et al. 2022).

9.5.4 Crohn's Disease

Crohn's disease (CD) and ulcerative colitis are two forms of chronic inflammatory bowel disease. Ulcerative colitis (UC) is characterized by inflammation of the mucosa and submucosa of the colon, whereas CD can affect any part of the GI tract from the mouth to the anus and involves inflammation that can permeate the entire thickness of the intestinal wall (Iqbal et al. 2014). The pathogenesis of IBD is multifactorial and includes dietary factors, gut barrier dysfunction, alterations in the microbiome, and activation of both innate and adaptive immune responses in genetically predisposed individuals. Probiotic treatments have varying degrees of success in the treatment of IBD. Compared to expectations derived from mechanistic and animal studies, such treatments have produced substandard results, particularly in CD (Compare et al. 2022). However, there are encouraging results with certain probiotic strains for the treatment of UC. For example, *E. coli* Nissle 1917 was shown to be effective in maintaining remission of UC, with a meta-analysis showing that it induced remission in 61.6% of patients. In the current guidelines of the European Crohn's and Colitis Organisation, this strain is recognized as effective for the initiation and maintenance of UC remission (Compare et al. 2022). The strain *Bifidobacterium longum* 536 has shown a significant reduction in the disease activity index of UC in patients compared to a placebo group over an 8-week treatment period. In addition, *Saccharomyces boulardii*, a nonpathogenic yeast with anti-inflammatory properties, has been studied for prevention of relapse in CD patients and maintenance of UC remission. In a study on pouchitis, patients treated with *Saccharomyces boulardii* showed a significantly lower relapse rate.

Murine models of IBD suggest that an upregulated Th1 cell response and a loss of mucosal tolerance to native gut flora may play a role in the development of IBD. Genetic factors such as polymorphisms in IBD-related genes, including those regulating mucus glycosylation (e.g., Fut2), could also contribute by altering bacterial profiles and influencing the response to dietary interventions such as prebiotics (Compare et al. 2022). A potential therapeutic approach for IBD could be to induce rapid clinical remission and mucosal healing with corticosteroids and/or biologics, followed by probiotic and prebiotic interventions to support sustained remission (Sanders and Klaenhammer 2001). Currently, there is no solid evidence for the use of probiotics in the treatment of active IBD and limited data on their efficacy in maintaining remission (Sanders and Klaenhammer 2001).

9.5.5 Cancer

Colorectal cancer (CRC) is one of the most prevalent cancers worldwide, and its pathogenesis involves several molecular and cellular steps. Environmental factors, such as obesity and diet, have been shown to have a significant impact on the risk of developing CRC, with both factors associated with changes in the gut microbiome (Sanders and Klaenhammer 2001). The bacterium *Bacteroides fragilis*, for example, can stimulate E-cadherin catabolism, causing intestinal inflammation, and thus increasing the risk of CRC. Studies on the microbiota of CRC patients suggest lower diversity or alterations in the microbial population, along with higher levels of *Fusobacterium nucleatum* sequences (Sanders and Klaenhammer 2001). Several mechanisms have been proposed to explain the influence of the microbiota on colon carcinogenesis, including microbial species and metabolic changes, changes in colonic pH, binding or inactivation of carcinogens, enhanced immune response, decreased inflammation, decreased proliferation of epithelial cells, and increased apoptosis (Sanders and Klaenhammer 2001). A specific human study indicated a lower recurrence rate of adenomatous polyps with atypical features after 4 years of taking *Lactobacillus casei*. In addition, a long-term observational study of more than 45,000 volunteers showed a lower incidence of CRC in people with high yogurt consumption, although the study lacked a control group for comparison. Probiotics have also been studied for their role in managing the adverse effects of radiotherapy and chemotherapy in the treatment of pelvic cancer (Sanders and Klaenhammer 2001). *Lactobacillus rhamnosus* GG and its conditioned medium showed a protective effect against radiation-induced intestinal damage and apoptosis in mice, which was dependent on the TLR2, COX2, and MyD88 signaling pathways. Moreover, probiotics may have beneficial effects in mitigating radiation and chemotherapy damage to the GI tract in cancer patients (Sanders and Klaenhammer 2001). Certain fecal bacterial enzymes, such as beta-glucuronidase, nitroreductase, azoreductase, and steroid 7-alpha-dehydroxylase, have been associated with the conversion of procarcinogens to carcinogens in the colon (Wise and Cummings 2022). Despite the various beneficial properties attributed to probiotics, there is currently no evidence confirming their protective effect against the development of CRC.

9.5.6 Diabetes Mellitus

In type I diabetes, the immune system attacks the beta cells of the pancreas. Various genetic and environmental factors influence the development. Consumption of probiotics increases anti-inflammatory cytokines, such as TGF-β and IL-10, while reducing pro-inflammatory cytokines, such as TNF-α, IL-6, and IL-1, suggesting a possible role in the prevention of type 1 diabetes (Mishra et al. 2019). Oral administration of probiotics has been found to prevent autoimmune diabetes in nonobese diabetic mice (Kim et al. 2020). These compounds, including Lactobacilli (*L. acidophilus*, *L. casei*, *L. delbrueckii* subspecies *L. bulgaricus*, and *L. plantarum*) and Bifidobacteria (*B. longum*, *B. infantis*, and *B. breve*), showed antidiabetic effects (Tegegne and Kebede 2022). The intake of Lactobacillus genus and *L. plantarum* in healthy people could delay or prevent autoimmune diabetes (Tegegne and Kebede 2022).

Type 2 diabetes (T2DM) is a major health challenge worldwide. Obesity, which is influenced by the gut microbiome, is associated with T2DM, with changes in the gut microbiota in turn affecting energy production, metabolism, and inflammation (Boulange et al. 2016). Metabolites produced by the gut microbiota, such as the SCFAs, play a role in metabolic processes and provide a link between the microbiota and glucose regulation (Liu et al. 2022). Probiotics enhance gut integrity, reduce lipopolysaccharides, increase incretins, alleviate endoplasmic reticulum stress, and improve insulin sensitivity (Salles et al. 2020). They also improve glucose tolerance, regulate lipid metabolism, increase antioxidants, and modulate gut flora (Tegegne and Kebede 2022). Probiotics prevent inflammation, autoimmunity, and oxidative stress with strains such as *Lactiplantibacillus plantarum* and *Bifidobacterium lactis* reducing the unfavorable effects of high-fat meals and modulating immune responses (Tegegne and Kebede 2022). The combination of *Lacticaseibacillus rhamnosus* with *Lactobacillus gasseri* and *Bifidobacterium lactis* inhibits weight gain, in particular reduces adipose tissue mass and increases the potential of probiotics in the treatment of diabetes (Tegegne and Kebede 2022).

Advances in the field of nanotechnology and biotechnology have contributed to the discovery of novel drugs with significant properties, but whose side effects raise major concerns (Kerry et al. 2018). As a result, natural sources such as probiotics have attracted attention due to their potential anticarcinogenic properties (Gorska et al. 2019). In laboratory studies, probiotic strains, such as *Lactobacillus fermentum* NCIMB-5221 and NCIMB -8829, showed significant inhibition of CRC cells while promoting normal epithelial colon cell proliferation (Kerry et al. 2018). Although these results are encouraging, further studies are needed to confirm the anti-cancer efficacy of probiotics in vivo to move on to animal and clinical studies (Kerry et al. 2018). Collaboration among clinical nutritionists, scientists, and industrialists is critical for the development of effective drugs with little or no side effects (Kerry et al. 2018).

Fecal microbiota transplantation (FMT) involves injecting feces into a patient's gastrointestinal tract to treat diseases associated with gut microflora dysbiosis (Biazzo and Deidda 2022). It is a promising approach for the treatment of gastrointestinal disorders, such as IBS, IBD, and constipation, as well as other metabolic syndromes, neurodevelopmental disorders, allergies, and autoimmune diseases (Zommiti et al. 2020). FMT is a highly effective procedure against recurring *Clostridium difficile* infections, alleviating early-onset colitis, and it also helps suppress multidrug-resistant gut bacteria in immunocompromised patients, making it a crucial healthcare procedure (Zommiti et al. 2020).

9.6 Therapeutics of Probiotic Bacteria

It is fascinating to know that our ancestors had an idea of scientific and nutritional concepts long before the internet age and had a comprehensive understanding of various concepts that took several decades to be proven scientifically. For example, in India, it has been customary for generations to consume curd after or with meals, and we know that curd is a rich source of the probiotic lactobacillus. Probiotics have a variety of therapeutic effects for many diseases and disorders, which are also shown to be therapeutic with the FMT technique.

Probiotics are commonly used in the treatment of autoimmune diseases, such as systemic lupus erythematosus (SLE), rheumatoid arthritis (RA), celiac disease, Crohn's disease, ulcerative colitis, multiple sclerosis, Sjogren's syndrome, systemic sclerosis, antiphospholipid syndrome, myasthenia gravis, and type 1 diabetes (Tegegne and Kebede 2022). In randomized controlled trials, probiotics have demonstrated efficacy in reducing GI symptoms and inflammation in various organs, potentially improving well-being by regulating immune responses (Tegegne and Kebede 2022).

Helicobacter pylori infection is often found in the gastric epithelium. Conventional triple therapy may be ineffective and disrupt the gut flora (Tegegne and Kebede 2022). Probiotics can significantly regulate the immune system and fight *H. pylori* infection, reducing treatment-related side effects and GI inflammation (Tegegne and Kebede 2022). Lactobacillus is considered one of the most promising probiotics (Tegegne and Kebede 2022). However, due to the conflicting test results, it is difficult to assess its efficacy (Tegegne and Kebede 2022). Probiotic bacteria can influence the activity of *H. pylori* via immunological and nonimmunological mechanisms by exerting bacteriostatic or bactericidal effects, reducing cytokine production, strengthening the gastric defense system, and interacting with the host's innate immune system (Tegegne and Kebede 2022). Zhang et al. developed a vaccine against *H. pylori* using modified *L. lactis* to produce the *H. pylori* lipoprotein Lpp20 (Zhang et al. 2018b).

Probiotics, such as *L. casei* strain Shirota found in fermented milk, can help treat hypertension by lowering luminal pH, secreting antimicrobial peptides, and promoting the growth of beneficial bacteria (Tegegne and Kebede 2022). Fermented milk produces bioactive peptides, such as Ile-Pro-Pro and Val-Pro-Pro, which inhibit angiotensin-converting enzymes (Tegegne and Kebede 2022). The intake of fermented dairy milk products containing LcS has been associated with a reduction in risk of hypertension by promoting the production of prostaglandin I2 and reducing peripheral vascular resistance (Tegegne and Kebede 2022). IBD, including Crohn's disease and UC, is associated with chronic inflammation of the gut, disruption of barrier function, and imbalance of the gut microbiome. Bacteria contribute to IBD through imbalanced immune responses and alterations in the gut microbiota (Tegegne and Kebede 2022). SCFAs, such as acetate, butyrate, and propionate, are essential for colonic homeostasis and have anti-inflammatory properties, which are often deficient in IBD (Kerry et al. 2018). Supplementation with indigestible carbohydrates and fiber (prebiotics), either alone or in combination with probiotics, is a promising therapeutic method to increase the synthesis of SCFA (Kerry et al. 2018).

Researchers are working on genetically engineered probiotic strains that can produce immunomodulators, such as interleukin-10 and lipoteichoic acid, to restore levels of protective commensal bacteria. *Lactobacillus*, *Bifidobacterium*, *Enterobacter*, and *E. coli* are popular probiotics, but new or genetically engineered strains are also being researched for the treatment of IBD (Kerry et al. 2018). In a colitis model, *Lactiplantibacillus plantarum* Lp91 reduced inflammatory markers while increasing anti-inflammatory IL-10 expression (Tegegne and Kebede 2022). Lactic acid bacteria release antioxidant enzymes that could be helpful in the treatment of IBD. Consuming probiotic yogurt can improve gut function by increasing the levels of probiotic bacteria (Tegegne and Kebede 2022). Probiotics effectively treat acute infectious diarrhea and antibiotic-associated diarrhea and prevent traveler's diarrhea (Tegegne and Kebede 2022). Long-term therapy with probiotics restores normal levels of fecal bacteria and normal cytokine expression, although efficacy in antibiotic-associated diarrhea depends on strain and duration of treatment (Tegegne and Kebede 2022).

9.7 Safety of Probiotics

Probiotics are widely known for their health benefits, particularly for improving gut health and aiding digestion. However, it is important to consider the potential negative effects they can carry (Yan and Polk 2011). People with weakened immune systems or with intestinal permeability, commonly referred to as "leaky gut," may be susceptible to infection or sepsis when taking probiotics (Yan and Polk 2011). Furthermore, the probiotics market suffers from a lack of regulation and quality control, resulting in products that vary in consistency and can potentially cause harm (Yan and Polk 2011). Contamination of probiotic supplements is a concern as it can sometimes lead to sepsis (Sanders et al. 2010). The probiotics themselves can move from the GI tract to other parts of the body, known as translocation, which can lead to infections (Yan and Polk 2011). There is also a theoretical risk of horizontal gene transfer, where antibiotic resistance genes could be passed from probiotic bacteria to the normal gut flora, turning the gut a repository for these genes (Toth et al. 2021). This is of particular concern when this flora is exposed to antibiotics or physical contact that facilitates gene transfer. Despite their risks, probiotics have been studied for their cholesterol-lowering potential. In some studies, certain probiotic strains showed a decrease in serum cholesterol after 6 weeks. However, the benefits must be weighed against the theoretical risks, including infections and the spread of antibiotic resistance (Yan and Polk 2011). In pediatric cases, probiotics have shown some efficacy in reducing the duration of diarrhea and associated fever, although their impact on hospitalization and frequency of bowel movements appears to be marginal. The administration of probiotics has had limited impact on persistent diarrhea but has shown promise in reducing the incidence of nosocomial rotavirus diarrhea (Sanders et al. 2010).

9.8 Antibiotics: Depletion of Gut Microbiota and the Rise of Antibiotic Resistance

Antibiotics not only serve to control pathogenic bacteria, but also disrupt beneficial bacterial populations in the gut, causing a major biological imbalance that has been linked to various diseases, such as obesity, type II diabetes, asthma, and Crohn's disease (Altves et al. 2020). The use of broad-spectrum antibiotics can lead to alterations in ~30% of the bacteria in the gut ecosystem, resulting in a rapid and substantial decline in taxonomic richness, diversity, and evenness, with notable consequences such as the occurrence of diarrhea disease due to opportunistic infection by *Clostridium difficile*, which is primarily attributed to the use of such antibiotics. Moreover, the gut environment serves as an optimal setting for the horizontal transmission of numerous resistance genes, increasing the risk associated with indiscriminate antibiotics use (Altves et al. 2020; Francino 2015). In addition, the use of antibiotics, especially in the early stages of life, contributes to negative outcomes as it plays a crucial role in the formation of a healthy microbiome. Studies focusing on infants indicated a link between antibiotics use, including exposure during the prenatal period, and an increase in the strain *Proteobacteria phylum*, which are known to harbor genes resistant to various antibiotics, while at the same time there is a decrease in *Bifidobacteria*, which are recognized as beneficial microbes (Altves et al. 2020). Antibiotics use leads to a reduction in the overall biodiversity of the gut microbiota, which depletes important beneficial bacteria, triggers metabolic shifts, increases susceptibility to colonization by beneficial bacteria in the gut, and promotes the development of antibiotic resistance (Ramirez et al. 2020). For example, by reducing biodiversity, antibiotics can contribute to the growth and colonization of pathogenic organisms such as the toxigenic *C. difficile*. It is significant to recognize that a reduction in biodiversity is not always accompanied by a decrease in the overall number of bacteria (Ramirez et al. 2020). After cessation of antibiotic treatment, the microbiota may show some degree of resistance, and possibly return to a composition similar to the original state. However, full recovery is challenging and often not achieved, as the changes induced by antibiotics persist for a prolonged period of time, which may take months or even years (Francino 2015). After antibiotic treatment, there was a persistent change in the gut microbiota that lasted over 2 years and was characterized by decreased biodiversity and an increase in clindamycin resistance genes among the gut bacteria. On the contrary, microorganisms have short life cycles and are often subject to genetic mutations, which increases their ability to adapt to parasitic challenges (Altves et al. 2020). As a result, microorganisms respond swiftly to parasites compared to their host by competing with each other through toxin or antibiotic production, hyperparasitism, or competing for resources to prevent parasite colonization. The microbiota may also indirectly induce the host to initiate a stronger immune response against parasites or strengthen the host resistance, ultimately reducing the host susceptibility to parasite damage and maintaining tissue integrity (Altves et al. 2020). *Clostridium difficile* is a gram-positive, spore-producing bacterium that lives in an anaerobic environment. These spores are extremely resistant to desiccation, chemicals, and high temperatures and survive for months. *C. difficile* secretes toxins A and B, which can injure the gut lining (Ramirez et al. 2020).

H. pylori is a spiral-shaped bacterium that colonizes the gastric mucosa where it can thrive in an anaerobic environment. It frequently triggers inflammation, with most cases being asymptomatic but some developing ulcers, intestinal metaplasia, or even gastric cancer, while failure of eradication therapy may be associated with poor compliance due to the negative consequences of antibiotic use (Ramirez et al. 2020).

Studies suggest an association between antibiotic use in childhood, particularly multiple courses of antibiotics in the first 3 months of life, and an increased risk of IBD, especially Crohn's disease, which is higher in boys than in girls and is mainly associated with cephalosporins (Ramirez et al. 2020). This association may be due to the changes in the composition of the gut microbiota caused by antibiotics, which is also increasingly recognized as a contributing factor to obesity (Ramirez et al. 2020).

9.9 Probiotics and Vaccine

The potential of certain probiotics to influence and enhance the immune response to vaccines has been documented in various studies. For example, supplementation with *Bifidobacterium longum* has been associated with an increased T-cell response and higher serum IgG and fecal polio-specific IgA levels, suggesting an improvement in both innate and adaptive immunity, which may lead to lower infection rates (Al Mahmud et al. 2023). Furthermore, the consumption of probiotics containing *Lactobacillus* strains, such as *L. paracasei* and *Lactobacillus casei* 431, did not significantly alter the immune response to the influenza vaccine, but these strains contributed to a shorter duration of respiratory symptoms (Xu et al. 2024). Other studies have reported that the intake of *B. lactis* and *L. paracasei* over a 6-week period can increase influenza-specific IgG, IgG1, and IgG3 levels, and seroconversion rates of these immunoglobulins, while simultaneously increasing influenza-specific IgA levels in saliva following trivalent inactivated influenza vaccination (Jespersen et al. 2015). Genetically engineered probiotics are also promising, such as *Lactobacillus casei*, which has been modified to target dendritic cells and has been shown to increase specific IgG and IgA antibody responses against anti-Porcine epidemic diarrhea virus in mice (Kazemifard et al. 2022). In parallel, human studies in infants have shown that administration of *Lactobacillus casei* GG prior to rotavirus vaccination, followed by continuous treatment, resulted in higher serum antiviral IgM, IgG, and IgA titers. Similarly, the use of *L. rhamnosus* GG or *L. acidophilus* CRL431 increased neutralizing antibodies and anti-polio serum IgG and IgA levels after a 7-week period (de Vrese et al. 2005). These results highlight the ability of specific probiotic strains to potentially act as adjuvants and thus improve vaccine efficacy, and suggest a significant role for probiotics in immunization strategies. It should be noted that while these early results are promising, further research is needed to confirm the benefits and understand the mechanisms of action.

9.10 Conclusion

Probiotics is a concept whose definition is constantly changing as they become more advanced because of the growing awareness on the importance of probiotics to health. Regardless of our contemporary understanding of the pivotal role gut microbiome plays in human health and disease, developing strains to treat diseases also requires deeper understanding on a molecular basis and the potential benefits and harm it can cause to people. Understanding the mechanistic role of gut microbiota and probiotics will help us target many diseases as a future therapeutic or preventive strategy. Since we have an insight into the gut–human symbiosis, this would help the further development of microbial-based targeted therapy, which can help prevent or treat major infections. For an effective synergism, diet is equally significant in the flourishing and the survival of the gut microbiota. Gut microbiota modulation in the future such that to favor the growth of beneficial metabolites-producing bacteria, thereby increasing the production of favorable metabolites as well as the hampering of detrimental metabolites like bacterial toxins. Usage of probiotics as immune promoters alone with vaccines can lead to better health. Clinical studies are not sufficient to confirm the role of probiotics in cancer with certainty. The development of genetically modified probiotics (GM probiotics) is promising, but further research is needed to develop affordable therapies.

References

Ahern PP, Maloy KJ (2020) Understanding immune-microbiota interactions in the intestine. Immunology 159(1):4–14. https://doi.org/10.1111/imm.13150

Al Mahmud A, Ahmed SS, Karim MR, Al-Mamun MR, Akhter S, Sohel M, Hasan M, Bellah SF, Amin MN (2023) Clinically proven natural products, vitamins and mineral in boosting

up immunity: a comprehensive review. Heliyon 9(4):e15292. https://doi.org/10.1016/j.heliyon.2023.e15292
Alagawany M, Abd El-Hack ME, Farag MR, Sachan S, Karthik K, Dhama K (2018) The use of probiotics as eco-friendly alternatives for antibiotics in poultry nutrition. Environ Sci Pollut Res Int 25(11):10611–10618. https://doi.org/10.1007/s11356-018-1687-x
Altves S, Yildiz HK, Vural HC (2020) Interaction of the microbiota with the human body in health and diseases. Biosci Microbiota Food Health 39(2):23–32. https://doi.org/10.12938/bmfh.19-023
Amirreza K, Reza B, Shabnam K (2016) Probiotics: a comprehensive review of their classification, mode of action and role in human nutrition. In: Venketeshwer R, Leticia GR (eds) Probiotics and prebiotics in human nutrition and health. IntechOpen, Rijeka, p Ch 2
Azad MAK, Sarker M, Wan D (2018) Immunomodulatory effects of probiotics on cytokine profiles. Biomed Res Int 2018:8063647. https://doi.org/10.1155/2018/8063647
Bamigbade GB, Subhash AJ, Kamal-Eldin A, Nystrom L, Ayyash M (2022) An updated review on prebiotics: insights on potentials of food seeds waste as source of potential prebiotics. Molecules 27(18). https://doi.org/10.3390/molecules27185947
Baricault L, Denariaz G, Houri JJ, Bouley C, Sapin C, Trugnan G (1995) Use of HT-29, a cultured human colon cancer cell line, to study the effect of fermented milks on colon cancer cell growth and differentiation. Carcinogenesis 16(2):245–252. https://doi.org/10.1093/carcin/16.2.245
Bercik P, Collins SM, Verdu EF (2012) Microbes and the gut-brain axis. Neurogastroenterol Motil 24(5):405–413. https://doi.org/10.1111/j.1365-2982.2012.01906.x
Biazzo M, Deidda G (2022) Fecal microbiota transplantation as new therapeutic avenue for human diseases. J Clin Med 11(14). https://doi.org/10.3390/jcm11144119
Boulange CL, Neves AL, Chilloux J, Nicholson JK, Dumas ME (2016) Impact of the gut microbiota on inflammation, obesity, and metabolic disease. Genome Med 8(1):42. https://doi.org/10.1186/s13073-016-0303-2
Brenner DM, Moeller MJ, Chey WD, Schoenfeld PS (2009) The utility of probiotics in the treatment of irritable bowel syndrome: a systematic review. Off J Am Coll Gastroenterol | ACG 104(4):1033–1049
Butel MJ (2014) Probiotics, gut microbiota and health. Med Mal Infect 44(1):1–8. https://doi.org/10.1016/j.medmal.2013.10.002
Chen X, D'Souza R, Hong ST (2013) The role of gut microbiota in the gut-brain axis: current challenges and perspectives. Protein Cell 4(6):403–414. https://doi.org/10.1007/s13238-013-3017-x
Chen J, Chen X, Ho CL (2021) Recent development of probiotic bifidobacteria for treating human diseases. Front Bioeng Biotechnol 9:770248. https://doi.org/10.3389/fbioe.2021.770248
Cohen CW, Fontaine KR, Arend RC, Alvarez RD, Leath CA III, Huh WK, Bevis KS, Kim KH, Straughn JM Jr, Gower BA (2018) A ketogenic diet reduces central obesity and serum insulin in women with ovarian or endometrial cancer. J Nutr 148(8):1253–1260. https://doi.org/10.1093/jn/nxy119
Compare D, Sgamato C, Nardone OM, Rocco A, Coccoli P, Laurenza C, Nardone G (2022) Probiotics in gastrointestinal diseases: all that glitters is not gold. Dig Dis 40(1):123–132. https://doi.org/10.1159/000516023
Cristofori F, Dargenio VN, Dargenio C, Miniello VL, Barone M, Francavilla R (2021) Anti-inflammatory and immunomodulatory effects of probiotics in gut inflammation: a door to the body. Front Immunol 12:578386. https://doi.org/10.3389/fimmu.2021.578386
Czerucka D, Piche T, Rampal P (2007) Review article: yeast as probiotics — saccharomyces boulardii. Aliment Pharmacol Ther 26(6):767–778. https://doi.org/10.1111/j.1365-2036.2007.03442.x
Dempsey E, Corr SC (2022) Lactobacillus spp. for gastrointestinal health: current and future perspectives. Front Immunol 13:840245. https://doi.org/10.3389/fimmu.2022.840245
Dutta JA (2016) Review on probiotics. Scholars Acad J Biosci 4(6):549–554. https://doi.org/10.36347/sajb.2016.v04i06.018
FAO (2006) Probiotic in food. WHO, Rome.

Fijan S (2014) Microorganisms with claimed probiotic properties: an overview of recent literature. Int J Environ Res Public Health 11(5):4745–4767. https://doi.org/10.3390/ijerph110504745

Francino MP (2015) Antibiotics and the human gut microbiome: dysbioses and accumulation of resistances. Front Microbiol 6:1543. https://doi.org/10.3389/fmicb.2015.01543

Fuller R (1989) Probiotics in man and animals. J Appl Bacteriol 66(5):365–378

Gomaa EZ (2020) Human gut microbiota/microbiome in health and diseases: a review. Antonie Van Leeuwenhoek 113(12):2019–2040. https://doi.org/10.1007/s10482-020-01474-7

Gorska A, Przystupski D, Niemczura MJ, Kulbacka J (2019) Probiotic bacteria: a promising tool in cancer prevention and therapy. Curr Microbiol 76(8):939–949. https://doi.org/10.1007/s00284-019-01679-8

Hemarajata P, Versalovic J (2013) Effects of probiotics on gut microbiota: mechanisms of intestinal immunomodulation and neuromodulation. Ther Adv Gastroenterol 6(1):39–51. https://doi.org/10.1177/1756283X12459294

Hill C, Guarner F, Reid G, Gibson GR, Merenstein DJ, Pot B, Morelli L, Canani RB, Flint HJ, Salminen S et al (2014) Expert consensus document. The International Scientific Association for Probiotics and Prebiotics consensus statement on the scope and appropriate use of the term probiotic. Nat Rev Gastroenterol Hepatol 11(8):506–514. https://doi.org/10.1038/nrgastro.2014.66

Iqbal MZ, Qadir MI, Hussain T, Janbaz KH, Khan YH, Ahmad B (2014) Review: probiotics and their beneficial effects against various diseases. Pak J Pharm Sci 27(2):405–415

Jespersen L, Tarnow I, Eskesen D, Morberg CM, Michelsen B, Bugel S, Dragsted LO, Rijkers GT, Calder PC (2015) Effect of lactobacillus paracasei subsp. paracasei, L. casei 431 on immune response to influenza vaccination and upper respiratory tract infections in healthy adult volunteers: a randomized, double-blind, placebo-controlled, parallel-group study. Am J Clin Nutr 101(6):1188–1196. https://doi.org/10.3945/ajcn.114.103531

Kazemifard N, Dehkohneh A, Baradaran Ghavami S (2022) Probiotics and probiotic-based vaccines: a novel approach for improving vaccine efficacy. Front Med (Lausanne) 9:940454. https://doi.org/10.3389/fmed.2022.940454

Kerry RG, Patra JK, Gouda S, Park Y, Shin HS, Das G (2018) Benefaction of probiotics for human health: a review. J Food Drug Anal 26(3):927–939. https://doi.org/10.1016/j.jfda.2018.01.002

Kim TK, Lee JC, Im SH, Lee MS (2020) Amelioration of autoimmune diabetes of NOD mice by immunomodulating probiotics. Front Immunol 11:1832. https://doi.org/10.3389/fimmu.2020.01832

Kosiewicz MM, Zirnheld AL, Alard P (2011) Gut microbiota, immunity, and disease: a complex relationship. Front Microbiol 2:180. https://doi.org/10.3389/fmicb.2011.00180

Krysiak K, Konkol D, Korczynski M (2021) Overview of the use of probiotics in poultry production. Animals (Basel) 11(6). https://doi.org/10.3390/ani11061620

Liang S, Wu X, Jin F (2018) Gut-brain psychology: rethinking psychology from the microbiota-gut-brain axis. Front Integr Neurosci 12:33. https://doi.org/10.3389/fnint.2018.00033

Lilly DM, Stillwell RH (1965) Probiotics: growth-promoting factors produced by microorganisms. Science 147(3659):747–748. https://doi.org/10.1126/science.147.3659.747

Liu Y, Tran DQ, Rhoads JM (2018) Probiotics in disease prevention and treatment. J Clin Pharmacol 58 Suppl 10(Suppl 10):S164–S179. https://doi.org/10.1002/jcph.1121

Liu J, Tan Y, Cheng H, Zhang D, Feng W, Peng C (2022) Functions of gut microbiota metabolites, current status and future perspectives. Aging Dis 13(4):1106–1126. https://doi.org/10.14336/AD.2022.0104

Mazhar SF, Afzal M, Almatroudi A, Munir S, Ashfaq UA, Rasool M, Raza H, Munir HMW, Rajoka MSR, Khurshid M (2020) The prospects for the therapeutic implications of genetically engineered probiotics. J Food Qual 2020(1):9676452. https://doi.org/10.1155/2020/9676452

Michaudel C, Sokol H (2020) The gut microbiota at the service of immunometabolism. Cell Metab 32(4):514–523. https://doi.org/10.1016/j.cmet.2020.09.004

Milner E, Stevens B, An M, Lam V, Ainsworth M, Dihle P, Stearns J, Dombrowski A, Rego D, Segars K (2021) Utilizing probiotics for the prevention and treatment of gastrointestinal diseases. Front Microbiol 12:689958. https://doi.org/10.3389/fmicb.2021.689958

Mishra SP, Wang S, Nagpal R, Miller B, Singh R, Taraphder S, Yadav H (2019) Probiotics and prebiotics for the amelioration of type 1 diabetes: present and future perspectives. Microorganisms 7(3). https://doi.org/10.3390/microorganisms7030067
O'Callaghan A, van Sinderen D (2016) Bifidobacteria and their role as members of the human gut microbiota. Front Microbiol 7:925. https://doi.org/10.3389/fmicb.2016.00925
Picard C, Fioramonti J, Francois A, Robinson T, Neant F, Matuchansky C (2005) Review article: bifidobacteria as probiotic agents — physiological effects and clinical benefits. Aliment Pharmacol Ther 22(6):495–512. https://doi.org/10.1111/j.1365-2036.2005.02615.x
Plaza-Diaz J, Ruiz-Ojeda FJ, Gil-Campos M, Gil A (2019) Mechanisms of action of probiotics. Adv Nutr 10(suppl_1):S49–S66. https://doi.org/10.1093/advances/nmy063
Raheem A, Liang L, Zhang G, Cui S (2021) Modulatory effects of probiotics during pathogenic infections with emphasis on immune regulation. Front Immunol 12:616713. https://doi.org/10.3389/fimmu.2021.616713
Ramirez J, Guarner F, Bustos Fernandez L, Maruy A, Sdepanian VL, Cohen H (2020) Antibiotics as major disruptors of gut microbiota. Front Cell Infect Microbiol 10:572912. https://doi.org/10.3389/fcimb.2020.572912
Salles BIM, Cioffi D, Ferreira SRG (2020) Probiotics supplementation and insulin resistance: a systematic review. Diabetol Metab Syndr 12(1):98. https://doi.org/10.1186/s13098-020-00603-6
Salvo-Romero E, Stokes P, Gareau MG (2020) Microbiota-immune interactions: from gut to brain. LymphoSign J 7(1):1–23. https://doi.org/10.14785/lymphosign-2019-0018
Sanders ME, Klaenhammer TR (2001) Invited review: the scientific basis of lactobacillus acidophilus NCFM functionality as a probiotic. J Dairy Sci 84(2):319–331. https://doi.org/10.3168/jds.S0022-0302(01)74481-5
Sanders ME, Akkermans LM, Haller D, Hammerman C, Heimbach J, Hormannsperger G, Huys G, Levy DD, Lutgendorff F, Mack D et al (2010) Safety assessment of probiotics for human use. Gut Microbes 1(3):164–185. https://doi.org/10.4161/gmic.1.3.12127
Schmalstieg FC Jr, Goldman AS (2008) Ilya Ilich Metchnikoff (1845–1915) and Paul Ehrlich (1854–1915): the centennial of the 1908 Nobel Prize in Physiology or Medicine. J Med Biogr 16(2):96–103. https://doi.org/10.1258/jmb.2008.008006
Sharma H (2021) Potential role and mechanism of probiotics. Ann Roman Soc Cell Biol 25(1):3616–3624
Sherman MP, Zaghouani H, Niklas V (2015) Gut microbiota, the immune system, and diet influence the neonatal gut-brain axis. Pediatr Res 77(1–2):127–135. https://doi.org/10.1038/pr.2014.161
Soccol CR (2010) The potential of probiotics: a review. Food Technol Biotechnol 48(4):413–434
Tegegne BA, Kebede B (2022) Probiotics, their prophylactic and therapeutic applications in human health development: a review of the literature. Heliyon 8(6):e09725. https://doi.org/10.1016/j.heliyon.2022.e09725
Toth AG, Csabai I, Judge MF, Maroti G, Becsei A, Spisak S, Solymosi N (2021) Mobile antimicrobial resistance genes in probiotics. Antibiotics (Basel) 10(11). https://doi.org/10.3390/antibiotics10111287
de Vrese M, Rautenberg P, Laue C, Koopmans M, Herremans T, Schrezenmeir J (2005) Probiotic bacteria stimulate virus-specific neutralizing antibodies following a booster polio vaccination. Eur J Nutr 44(7):406–413. https://doi.org/10.1007/s00394-004-0541-8
Wang X, Zhang P, Zhang X (2021) Probiotics regulate gut microbiota: an effective method to improve immunity. Molecules 26(19). https://doi.org/10.3390/molecules26196076
Wieers G, Belkhir L, Enaud R, Leclercq S, Philippart de Foy JM, Dequenne I, de Timary P, Cani PD (2019) How probiotics affect the microbiota. Front Cell Infect Microbiol 9:454. https://doi.org/10.3389/fcimb.2019.00454
Wise JL, Cummings BP (2022) The 7-alpha-dehydroxylation pathway: an integral component of gut bacterial bile acid metabolism and potential therapeutic target. Front Microbiol 13:1093420. https://doi.org/10.3389/fmicb.2022.1093420
Wong WY, Chan BD, Sham TT, Lee MM, Chan CO, Chau CT, Mok DK, Kwan YW, Tai WC (2022) Lactobacillus casei strain Shirota ameliorates dextran sulfate sodium-induced colitis in mice by increasing taurine-conjugated bile acids and inhibiting NF-kappaB signaling via stabilization of IkappaBalpha. Front Nutr 9:816836. https://doi.org/10.3389/fnut.2022.816836

Xu C, Aqib AI, Fatima M, Muneer S, Zaheer T, Peng S, Ibrahim EH, Li K (2024) Deciphering the potential of probiotics in vaccines. Vaccines 12(7):711

Yan F, Polk DB (2011) Probiotics and immune health. Curr Opin Gastroenterol 27(6):496–501. https://doi.org/10.1097/MOG.0b013e32834baa4d

Zhang Z, Lv J, Pan L, Zhang Y (2018a) Roles and applications of probiotic lactobacillus strains. Appl Microbiol Biotechnol 102(19):8135–8143. https://doi.org/10.1007/s00253-018-9217-9

Zhang R, Wang C, Cheng W, Duan G, Shi Q, Chen S, Fan Q (2018b) Delivery of helicobacter pylori HpaA to gastrointestinal mucosal immune sites using Lactococcus lactis and its immune efficacy in mice. Biotechnol Lett 40(3):585–590. https://doi.org/10.1007/s10529-017-2502-3

Zommiti M, Feuilloley MGJ, Connil N (2020) Update of probiotics in human world: a nonstop source of benefactions till the end of time. Microorganisms 8(12). https://doi.org/10.3390/microorganisms8121907

Probiotics Formulation and Cancer Nanovaccines in Gastrointestinal (GI) Tract Cancer

10

Shuvomoy Banerjee, Sakshi Kumari, Debarati Paul, Juni Banerjee, and Sudeep Bose

Abstract

Probiotics are well known to execute crucial functions in the regulation of the immune system. The combination of probiotics with cancer vaccines has the potential to yield a synergistic outcome. To optimize the effectiveness of probiotics, the utilization of a suitable probiotic formulation becomes imperative. Within this context, the encapsulation of *Lacticaseibacillus rhamnosus* and *Bifidobacterium longum* with a lipid membrane is implemented to pursue the objective of minimizing loss of activity and enhancing colonization in the colon. In a murine model of subcutaneously transplanted colon cancer, the probiotics formulation exhibited substantial preventive and therapeutic efficacy, a trait that could be further enhanced through the confluence with cancer nanovaccines. The probiotics formulation may act as adjuvants to amplify the innate immune response or function as in-situ cancer vaccines. In several studies focusing on the prevention of chemically induced orthotopic colon cancer, the probiotics formulation in isolation effectively decreased the number of tumors in the colon, with an augmented efficacy observed when combined with cancer nanovaccines. Collectively, these investigations underscore that probiotics formulation can optimize the efficacy of cancer nanovaccines. In this book chapter, we comprehensively discuss the recent development of cancer nanovaccine research in

S. Banerjee · J. Banerjee
Department of Biotechnology and Bioengineering, Institute of Advanced Research (IAR), Gandhinagar, Gujarat, India

S. Kumari · D. Paul · S. Bose (✉)
Amity Institute of Biotechnology, Amity University Uttar Pradesh, Noida, Uttar Pradesh, India
e-mail: sbose1@amity.edu

S. Pathak et al. (eds.), *Microbiota and Dietary Mediators in Colon Cancer Prevention and Treatment*, https://doi.org/10.1007/978-981-96-0297-1_10

gastrointestinal cancers and the prospect of introducing probiotic formulations in combination with cancer nanovaccines as potential preventive and therapeutic interventions.

Keywords

Cancer nanovaccine · Probiotics · Cancer prevention · Cancer therapeutics · Cancer nano-medicine

10.1 Introduction

Cancer, a lethal malignancy of high clinical significance, remains a notable contributor to morbidity as well as death on a global scale. It ranks as the second leading cause of death worldwide. The incidence and fatality rates have shown a consistent increase over the last century, according to Sung et al., the global cancer burden is anticipated to be 28.4 million new cases in 2040 which is almost a 47% rise from the 2020 global cancer scenario (Sung et al. 2021). Notable types of cancer responsible for substantial mortality include lung, prostate, stomach, colorectal, liver, and breast cancers (Mattiuzzi and Lippi 2019). The onset of cancer can be traced back to defects in DNA repair mechanisms or mutations that arise during DNA replication (Vassilev and DePamphilis 2017). Characterized by unregulated cell proliferation, cancer typically manifests at an advanced stage of progression. It originates from a single aberrant cell and disseminates throughout the organism. The transformation of healthy cells into a neoplastic state is characterized by a cascade of genetic modifications, culminating in the emergence of tumorigenic and ultimately malignant cells. These malignant cells manifest diverse pathophysiological changes, such as self-sustained growth signaling, insensitivity to growth-inhibitory cues, evasion of cell death, unregulated proliferation, continuous angiogenesis, and metastatic potential. The primary approach to managing cancer clinically revolves around the utilization of conventional pharmaceuticals. The administration of several drugs and hormonal therapies not only targets cancer cells but also inflicts harm on healthy tissues, fostering the development of drug resistance. Moreover, the cytotoxic impacts of these medications are linked to severe, life-threatening adverse reactions that frequently surpass the severity of the cancer itself (Vinogradova et al. 2020).

Gastrointestinal (GI) cancers, which encompass malignancies of the esophagus, stomach, liver, pancreas, colon, rectum, and anus, epitomize a substantial burden in global health with increasing prevalence and mortality rates. According to the Global Cancer Observatory (GLOBOCAN) 2020 data, GI cancers account for over 4.8 million new cancer cases annually, comprising more than 26% of the global cancer incidence, and resulting in approximately 3.4 million deaths, which is nearly 35% of all cancer-related mortalities worldwide (Bray et al. 2024). Colorectal cancer is particularly prevalent, being the third most common cancer globally and the second leading cause of cancer deaths. The impact of GI cancers is profound, with significant disparities observed across different regions and socioeconomic groups, influenced by factors such as diet, lifestyle, genetic predisposition, and access to healthcare. The high morbidity and mortality associated with GI cancers are compounded by late-stage diagnoses, limited availability of early screening programs, and variations in treatment efficacy (Liang et al. 2022).

Current treatment modalities for GI cancers include chemotherapy, radiation therapy, surgery, and immunotherapy, each with specific applications and limitations. Surgical resection remains the cornerstone of treatment for localized GI cancers, but its effectiveness is limited in progressive stages where metastasis has arisen. Chemotherapy, often used in conjunction with surgery, aims to eradicate residual cancer cells but is associated with significant side effects such as nausea, fatigue, and increased susceptibility to infections. Radiation therapy can be effective, particularly for rectal and esophageal cancers, but its use is constrained by the potential for damage to surrounding healthy tissues and resultant complications (Anand et al. 2022). Targeted therapies, which inhibit specific molecules involved in tumor growth and progression, have shown promise but are only applicable to cancers with certain genetic mutations and can lead to resistance over time. Immunotherapy, which harnesses the body's immune system to fight cancer, has developed advanced-mode treatments for some GI cancers, particularly progressive melanoma and certain types of colon cancer with high rates of microsatellite instability. However, its efficacy is limited to a subset of patients, and it can cause severe immune-related adverse effects. Furthermore, the economic burden is substantial, as these cancers require extensive medical interventions, and long-term care, and result in loss of productivity (Smith et al. 2019). Research advancements are focused on improving early detection, understanding molecular mechanisms, and developing targeted therapies, yet the global challenge remains formidable. Concerted efforts in public health policies, awareness campaigns, and equitable healthcare access are essential to mitigate the impact of GI cancers (Kale et al. 2023). Early screening programs, personalized treatment approaches, and addressing healthcare disparities are critical steps toward reducing the global burden of gastrointestinal cancers.

10.2 Probiotics in Gastrointestinal Tract Cancer

10.2.1 Probiotics and Their Mechanism of Action

Probiotics are well defined as live microorganisms that confer important health benefits on the host. This category includes some beneficial bacteria, mostly from the genera *Lactobacillus and Bifidobacterium*, along with certain fungi like yeasts (*Saccharomyces boulardii*), which are increasingly becoming familiar for their role in maintaining gut health and immunity (Shi et al. 2016). The mechanism of action of these beneficial microorganisms is multifaceted and involves modulation of the gut microbiota, improvement of intestinal immune functions, and production of different bioactive compounds (Mazziotta et al. 2023). One main mechanism of action is the modulation of the gut microbiota in terms of their numbers and functions. The human gastrointestinal tract harbors a wide range of complex and dynamic microbial communities that play a crucial role in maintaining the good health of gut epithelium and homeostasis. Importantly, probiotics can increase the variety and stability of the gut microbiota, thereby outcompeting several disease-causing bacteria through the mechanism of "competitive exclusion" (Petrariu et al. 2024). Probiotics were known to inhibit the growth and colonization of pathogenic microorganisms by adhering to the mucosal lining of gut epithelium and generating antimicrobial substances including bacteriocins, organic acids, hydrogen peroxide, etc. Therefore, probiotics play a major role in preventing infections and maintaining a balanced gut ecosystem. Probiotics also regulate the intestinal barrier function, which is very crucial for averting the translocation of potential pathogens and toxins from the gut lumen into the body's circulatory systems. They attain this by inducing the expression of tight junction proteins including occludin and zonula occludens-1 (ZO-1), which close the intercellular spaces therefore reducing the intestinal permeability (Chelakkot et al. 2018). Such a mechanism helps to sustain gut integrity and prevent specific pathological conditions such as leaky gut syndrome, which has been a concern in different inflammatory and autoimmune disorders (Liu et al. 2018). Also, probiotics show significant immunomodulatory effects by influencing both the innate and adaptive immune systems. Research indicated that they interact with gut epithelial cells as well as immune cells in the gut-associated lymphoid tissue (GALT) and play an important role in modulating the production of cytokines and promoting the differentiation of regulatory T cells (Tregs) (Guo and Lv 2023). This interaction helps in maintaining immune tolerance and reducing inflammation. For instance, probiotics can stimulate the production of anti-inflammatory cytokines such as interleukin-10 (IL-10) and transforming growth factor-beta (TGF-β), while inhibiting pro-inflammatory cytokines like tumor necrosis factor-alpha (TNF-α) and interleukin-6 (IL-6) (Kwok et al. 2022). These immune-modulating properties are particularly beneficial in managing conditions such as inflammatory bowel disease (IBD), irritable bowel syndrome (IBS), and allergies. In addition to these mechanisms, probiotics produce various bioactive compounds, including short-chain fatty acids (SCFAs) such as acetate, propionate, and butyrate, through the fermentation of dietary fibers (Markowiak-Kopeć and Śliżewska 2020). SCFAs serve as an energy source for colonocytes, promote epithelial cell proliferation, and exhibit anti-inflammatory properties. Butyrate, in particular, has been shown to enhance mucin production, which reinforces the mucus layer protecting the intestinal epithelium. Moreover, SCFAs can regulate host metabolism by influencing glucose and lipid homeostasis, thus playing a role in preventing metabolic disorders like obesity and type 2 diabetes (Anachad et al. 2023). The beneficial effects of probiotics extend beyond the gut, impacting systemic health. Recent research showed the connections between the gut-brain axis (a bidirectional communication network between the gut and the brain). Of note, probiotics can impact this regulatory axis by modulating the composition of gut microbiota. They produce neurotransmitters including serotonin and gamma-aminobutyric acid (GABA), and impede systemic inflammatory reactions (Sabit et al. 2023). These molecular mechanisms have been allied with improvements in different mental conditions like mood, anxiety, and cognitive functions, signifying a potential role for probiotics in the management of mental health conditions. Apart from the promising benefits, the effectiveness of probiotics was found strain-specific and mainly dependent on several factors such as dosage, time frame of administration, and individual host characteristics (Tegegne and Kebede 2022). Interestingly, not all probiotics deliberate the same health benefits, and therefore, their effects may show individual-centric results. Consequently, it is essential to identify and select "clinically proven" probiotic strains to address particular health conditions and related pathological conditions. Moreover, the safety and quality of probiotic products are crucial considerations. Probiotics must be safe for consumption, free from contaminants, and capable of surviving the acidic environment of the stomach to reach the intestines in sufficient numbers. Standardization and rigorous quality control measures are necessary to ensure the viability and efficacy of probiotic supplements.

10.2.2 Role of Probiotics in Cancer Prevention

The potential role of probiotics in cancer prevention has garnered significant interest, driven by evidence from both preclinical and clinical studies, as well as mechanistic insights into their anti-carcinogenic properties. Probiotics, primarily from the genera *Lactobacillus* and *Bifidobacterium*, are known to modulate gut microbiota, enhance immune responses, and produce bioactive metabolites that can impede cancer development and progression (Górska et al. 2019). This section delves into the evidence from preclinical and clinical studies and explores the mechanisms through which probiotics may contribute to cancer prevention, highlighting recent scientific discoveries.

10.2.2.1 Evidence from Preclinical and Clinical Studies

Preclinical studies using animal models have provided robust evidence supporting the anti-cancer potential of probiotics. Several studies validated that probiotics can prevent the development of various human malignancies. For instance, *Lactobacillus rhamnosus GG* and *Bifidobacterium lactis* have been shown to lessen the incidence of chemically induced colon tumors in rodent species (Lactobacillus 2016). The identified mechanisms include variation of the gut microbiota, decrease of inflammation, enhancement of mucosal barrier functionalities, and production of anti-carcinogenic metabolites (Wang et al. 2021). In another study, researchers found that *Lactobacillus casei* significantly decreased the number of azoxymethane-induced aberrant crypt foci (ACF), which are precursor molecules to colorectal cancer induction in rats. Such observations suggest that probiotics might restrict the early processes of cancer development (Orlando et al. 2008). Moreover, probiotics have also established their protective role against hepatocellular carcinoma in different animal models, perhaps by dropping the level of liver inflammation and oxidative stress (Wan and El-Nezami 2018). Importantly, few clinical trials are going on to investigate the role of probiotics in cancer prevention but have shown promising results. A randomized controlled trial (RCT) including patients with colorectal adenomas showed that supplementation with a mixture of *Lactobacillus* and *Bifidobacterium* strains can reduce the relapse of these malignancies. Another report demonstrated that a daily intake of probiotics could decline the levels of fecal carcinogens as well as genotoxicity markers in our bodies, representing a reduced risk of cancer. A meta-analysis involving several clinical trial results showed that probiotics could potentially reduce the risk of colorectal cancer by modifying the gut microbiome and enhancing immune responses (Fong et al. 2020). However, the authors also mentioned the necessity for more large-scale sample sizes, and long-term RCTs to confirm these observations and launch clear guidelines for the use of probiotics in the prevention of cancer.

10.2.2.2 Mechanistic Insights into Cancer Prevention with Recent Discoveries

Modulation of Gut Microbiota

One of the primary mechanisms by which probiotics may prevent cancer is through the modulation of the gut microbiota. Dysbiosis, an imbalance in the gut microbial community, has been implicated in the development of various cancers, particularly colorectal cancer. Probiotics help restore a healthy microbiota balance by promoting beneficial bacteria and suppressing pathogenic species. This balance reduces the production of carcinogenic compounds such as secondary bile acids and nitrosamines while enhancing the production of short-chain fatty acids (SCFAs) like butyrate, which has anti-inflammatory and anti-carcinogenic properties.

10.2.2.3 Immune Modulation

Probiotics play a crucial role in modulating the immune system, which is vital for cancer prevention. They enhance the activity of natural killer (NK) cells, macrophages, and cytotoxic T lymphocytes, all of which are critical for identifying and eliminating cancer cells (Aziz and Bonavida 2016). Probiotics also influence the production of cytokines, promoting anti-inflammatory cytokines like IL-10 and reducing pro-inflammatory cytokines such as IL-6 and TNF-α. This immunomodulatory effect helps maintain a state of immune surveillance against tumor cells, preventing their proliferation and metastasis.

10.2.2.4 Enhancement of Barrier Function

Probiotics strengthen the intestinal barrier function, which prevents the translocation of carcinogenic substances and pathogenic bacteria from the gut lumen into the systemic circulation. They enhance the expression of tight junction proteins and mucins, thereby maintaining intestinal integrity. This barrier function is crucial in preventing chronic inflammation, a known risk factor for various cancers, including colorectal cancer (Genua et al. 2021).

10.2.2.5 Production of Anti-carcinogenic Metabolites

Probiotics produce several metabolites that have direct anti-carcinogenic effects. For instance, SCFAs like butyrate induce apoptosis in cancer cells, inhibit their proliferation, and promote differentiation. Recent studies have also highlighted the role of specific probiotic strains in producing bioactive compounds such as conjugated linoleic acid (CLA) and bacteriocins, which possess anti-tumor properties (Song et al. 2021). Additionally, probiotics can metabolize dietary components into compounds that inhibit carcinogenesis, such as the conversion of glucosinolates from cruciferous vegetables into cancer-protective isothiocyanates (Bouranis et al. 2021).

While the evidence supporting the role of probiotics in cancer prevention is compelling, several challenges and limitations need to be addressed. One major limitation is the strain-specific nature of probiotics. Not all probiotics confer the same benefits, and their effects can vary significantly depending on the strain, dosage, and duration of administration. This variability necessitates the identification of specific probiotic strains that are most effective in cancer prevention and the establishment of standardized dosing regimens. Another critical issue is the translation of preclinical findings into clinical practice. While animal studies provide valuable insights, human trials are essential to confirm the efficacy and safety of probiotics in cancer prevention (Śliżewska et al. 2020). Many clinical trials to date have been small-scale and short-term, highlighting the need for larger, long-term studies to validate the benefits observed in preclinical models. Furthermore, the mechanisms through which probiotics exert their anti-carcinogenic effects are complex and not fully understood. Recent discoveries have shed light on some pathways, but more research is needed to elucidate the precise molecular mechanisms involved. Understanding these mechanisms will enable the development of targeted probiotic therapies and personalized approaches to cancer prevention. Of note, safety and quality control are paramount. Probiotics must be free from contaminants and capable of surviving the gastrointestinal tract to reach the intestines in sufficient numbers. Rigorous quality control measures and standardization are necessary to ensure the efficacy and safety of probiotic supplements.

10.3 Cancer Nanovaccines in GI Tract Cancer

10.3.1 Concept of Nanovaccines

The existence of any signals originating from cells and tissues associated with the preservation of gut-microbial synergy is yet to be determined. According to recent research, cancer and the gut microbiota are intimately related. Different tumor forms exhibit distinct immunological and metabolic enrichment states from their respective local tumor microenvironments (Shi et al. 2020). It has been shown that in patients with colon adenocarcinomas, certain intratumor microbiota compositional patterns are closely related to different stages of colorectal carcinogenesis (Tortora et al. 2022). Additionally, it has been found that the intra-tumor tissues have a particular, advantageous microbial makeup that varies depending on the degree of colorectal neoplasia. These discoveries could deepen our comprehension of the roles played by these microbes in local cancer signaling, and possibly pave the way for the development of probiotics intended to quicken the Most cancer vaccines maintain significant potential for the treatment of stable cancers. Nanomedicine offers various techniques to increase the efficacy of vaccines. Especially, or (sub)cell vaccines may be brought to the target lymphoid tissues and cells by using nanocarriers and nanoplatforms to grow the potency and durability of antitumor immunity and minimize poor aspect consequences. Nanoparticles (NPs) are used in conjunction with or as adjuvants in nanovaccines (Bezbaruah et al. 2022). NPs offer the benefits of optimal nanoscale size, high stability, sufficient antigen loading, high immunogenicity, and adjustable antigen presentation. Increased lymph node retention as well as immunity marketing. Cancer vaccines rely on tumor antigens to elicit antitumor immunity; these antigens can be given as nucleic acids, extracellular vesicles (EVs), or mobile membrane-encapsulated nanoparticles (NPs) (Zhang et al. 2022). Perfect most cancer vaccines overcome tumor-induced immune suppression while boosting both humoral and cellular immunity. In this book chapter, we examine the key components of nanovaccines for immunotherapy against the majority of malignancies and emphasize the current developments in their enhancement are mostly dependent on the structure and makeup of various nanocarriers, including semi-biogenic and synthetic ones. Furthermore, we talk about tumor cellular–derived vaccines (together with complete-tumor-cellular components, EVs, cellular membrane–encapsulated NPs, and hybrid membrane–lined NPs), nanovaccine action mechanisms, and the challenges of immune cancer remedy and their translation to clinical programs.

10.3.2 Nanovaccine Composition

Due to their limited duration of protective immunity and inadequate immune activation, subunit vaccines cannot provide immunity against deadly illnesses on their own. The drawbacks of traditional vaccination adjuvants may be addressed by employing NP-based delivery systems, such as virosomes, liposomes, micelles, microemulsions, dendrimers, and nanogels. The transport of adjuvants and antigens, the presentation of antigens by APCs, the stimulation of innate immune responses, and the development of strong effector T-cell responses can all be enhanced by nanovaccines (Est-Witte et al. 2021). Thus, the development of potent immunotherapeutic formulations for human malignancies can greatly benefit from the use of nanovaccines, which typically contain antigens, adjuvants, and nanocarriers. Dendritic cells (DCs) absorb the antigens to cause their maturation and activation once the adjuvant and antigen are delivered to lymphoid tissues together with the nanovaccine. Major Histocompatibility Complexes (MHCs) are used by the DCs to transmit antigens to CD8 + T lymphocytes and stimulate their proliferation.

10.3.3 Antigens

Through MHC-I, tumor antigens can be expressed on the surface of tumor cells, activating T cells. These antigens are classified into two categories: tumor-associated antigens (TAAs; generated at lesser levels in both cancerous and noncancerous cells) and tumor-associated antigens (TSAs; present only in malignant cells). Because of their immune evasion and immunosuppressive mechanisms, endogenous cancer antigens are unable to elicit immunologically meaningful responses. Exogenous tumor-associated antigens are helpful in the creation of cancer vaccines because they trigger antigen-specific immune responses against tumor cells during cancer immunotherapy. TAAs are expressed in a variety of cancer types by both normal and malignant cells. The most significant TAAs utilized in cancer vaccines include HER-2/neu, melanoma-associated antigen (MAGE)-A, TTK protein kinase (TTK), LAGE-1, and the gene encoding NY-ESO (New York esophageal squamous cell carcinoma) 1 (Fang et al. 2022).

10.3.4 Adjuvants

Adjuvants activate the immune system and augment direct and immunological reactions; they are utilized to elicit robust responses in vaccine recipients. Since subunit antigens are usually weakly immunogenic, these characteristics are essential. Adjuvants are generally classified into two groups according to their main modes of operation: (a) adjuvants for the administration of vaccines, including virosomes, mineral salts, emulsions, and liposomes which control the release and storage of antigens and deliver more effectively; and (b) adjuvants for immunostimulatory molecular vaccines, such as cytokines, immunological cells like APCs are activated by toll-like receptor (TLR) agonists, sting agonists, costimulatory ligands, and cytokines, which improve antigen-specific immune responses (Chakraborty et al. 2023). When loaded into NPs, adjuvants, and antigens can be protected from being broken down by phosphatases, nucleases, and proteases (Chauhan et al. 2021).

10.3.5 Nanocarriers

The most crucial part of nanovaccines is their nanocarriers, which have a high surface area, a surface-to-volume ratio, effective loading, and the ability to carry a range of payloads, such as nucleic acids, big proteins, and small compounds, among other benefits (Din et al. 2017). Because of the relatively large nano-size, which restricts efflux transit from cells, effective interactions with the cell membrane are made possible, which improves and encourages in-cell accumulation. Furthermore, NP synthesis can improve the payload drugs' solubility, in vivo stability, systemic circulation, and bio-distribution inside the body. To effectively regulate transcription, nucleic acids can be protected from enzymatic degradation by nanocarriers, which can also prolong in vivo circulation and improve targeted distribution and cellular absorption (Chauhan et al. 2021).

10.4 Nanocarrier Types

10.4.1 OMVs

OMVs or outer membrane virus-produced nano-sized proteoliposomes, are of great interest in cancer immunotherapy because they are excellent carriers for the intracellular delivery of antigens to adenocarcinomas (APCs) and lymph nodes. OMVs have an average size of 50–300 nm. OMVs are safe and devoid of live bacterial cells, making them non-replicating from a safety standpoint. Typical instances of biochemical signaling involving the transport of bacterial OMVs are OMVs containing RNA, DNA, endotoxins, proteins, and virulence chemicals. Bacterial OMVs can be employed as carriers and adjuvants for vaccine production because they include immune-stimulating danger signals such as lipopolysaccharides, lipoproteins, and flagellin, which trigger TLR4 and TLR5, respectively (Zheng et al. 2024). To increase antitumor immune responses and eradicate tumor cells with the fewest possible adverse results, OMVs can be added with TLR agonists and antitumor cytokines in cancer vaccines (Ren et al. 2023). OMVs have been utilized to generate vaccinations against bacteria and cancer. It is feasible to boost these APCs' maturity and the survival of antigen cross-presentation to CD8 + T cells by creating OMVs made by programming DCs. Recently, cancer antigens on the surface of OMVs were fused with cytolysin A. To expedite the display method targeting tag/catcher protein pairs was used. When OMVs are adorned with multiple protein catchers, they can express multiple tumor antigens at the same time, which will work in concert to stimulate anticancer immune responses. Multiple tumor antigen-containing engineered OMVs were utilized to stop melanoma-induced lung metastasis and subcutaneous colorectal cancer growth (Guo et al. 2022) (Fig. 10.1).

10.4.2 Exosomes

Exosomes are 30–150 nm-sized nanovesicles. Exosomes' phospholipid bilayer architectures facilitate steady drug transport, increase the half-life of medications during administration, and shield pharmaceuticals from enzyme breakdown. Furthermore, exosome membranes have an easy time binding to target cells, which increases the medications' bioavailability. In contrast to traditional drug delivery systems like liposomes, exosomes exhibit lower levels of immunogenicity and toxicity. Exosomes can efficiently cure tumors because of their small size, which enables them to travel throughout the tumor tissue and extravasate into tumor arteries. Consequently, exosomes have the potential to enhance medication efficacy and surmount multiple physiological barriers, making them highly promising for the delivery of vaccines and immunotherapy (Elliott and He 2021) (Fig. 10.1).

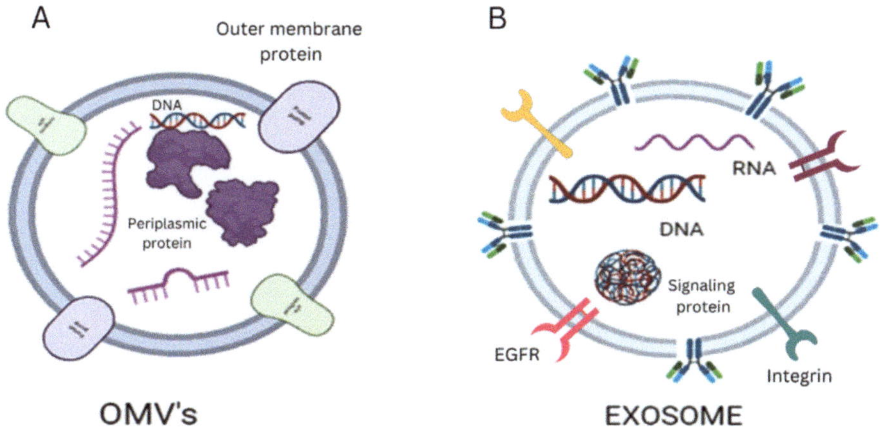

Fig. 10.1 Structures of outer membrane vesicles (OMVs) and exosomes. (**a**) Characteristics and benefits of exosomes and outer membrane vesicles (OMVs) as nanocarriers. Panel on the left OMVs are nanoparticles (NPs) that contain bioactive substances such as peptidoglycans, metabolites, DNA, RNA, periplasmic proteins, cytoplasmic proteins, lipopolysaccharides, porins, flagellin, outer membrane proteins, lipoproteins, toxins, and enzymes. As a result, OMVs can be quickly identified and taken up by DCs, and they can trigger a variety of innate immune signaling pathways. This functions as an organic adjuvant, delaying the breakdown of loaded antigens. (**b**) Exosomes are NPs with a lipid bilayer membrane structure that include a variety of functional components, including proteins, lipids, microRNA (miRNA), long non-coding RNA (lnc RNA), and other elements like transporters, nucleic acids, and signaling molecules. Exosomal components function as potential biomarkers and are essential to the course of cancer

10.5 Kinds of Cancer Immunization

Vaccines can be distinguished from one another by the technologies they employ to trigger an "immune response." We refer to this technological innovation as a "vaccine platform." There are four types of cancer vaccines: cell, peptide, viral, and nucleic acid-based (Fig. 10.2).

10.5.1 Cancer Vaccines Based on Cells

Cell-based cancer vaccines are typically made from whole or divided cells, those frequently include tumor antigens and elicit stronger immune responses. Among the cell-based vaccinations are DC vaccines. Clinical investigations have demonstrated encouraging anticancer benefits from DCs (Abakushina et al. 2021). The development of DC vaccines is impeded by expensive and time-consuming processes. Allogeneic or autologous cancer cells obtained from patients can be used to create cellular vaccines. Adding genes to tumor cell lines can increase the immune response against entire tumor cells. The method has its drawbacks, such as the difficulty of gathering a sufficient number of cells to prompt an efficient immune response.

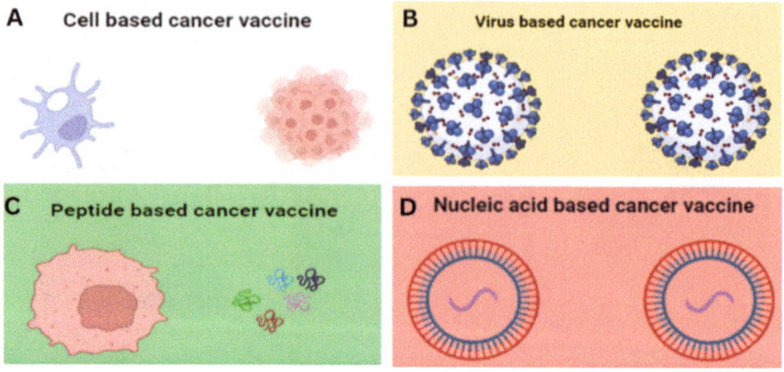

Fig. 10.2 Types of cancer vaccines. (**a**) Tumor-associated antigens are engaged by tumor vaccines. Dendritic cells are infused with various tumor antigens. (**b**) Tumor antigens are transported into the immune system by engineered viruses. (**c**) Specific cancer epitopes contained by the lengthy & precise are allocated to the immune system. (**d**) DNA & RNA concealing tumor antigens to the immune system

Cancer vaccines based on cells are produced using DCs. DCs are given different tumor antigens in the form of DNA, RNA, tumor lysates, or peptides to produce vaccines based on DCs (Cintolo et al. 2012). Numerous DC vaccinations have been created. DCs derived from monocytes and leukemia cells are used in DC vaccines. A sufficient number of DCs can be cultured to investigate DC cancer vaccines in phase I, II, and III trials (Santos and Butterfield 2018).

10.5.2 Cancer Vaccines Based on Viruses

Cancer vaccines use tumor-specific peptide antigens that are highly immunogenic to elicit an optimal immune response. Strong immune responses can be elicited by synthetic and chemical variants of known or suspected specific tumor antigens. An effective way to prevent and cure viral infections is to administer a vaccine with adjuvants. Studies have shown that vaccines based on Virus-like particles (VLPs) elicit cellular immune responses (Kheirvari et al. 2023). Synthetic peptides and peptide vaccination techniques are being used in the development of personalized cancer vaccines. Antigenic peptides are picked up by the APCs and exposed on the cell surface. Surface antigens that trigger an immunological response are detectable by T-cells. Peptide-based vaccinations offer some advantages over other vaccine types, most notably ease of manufacture and safety. Some of the vaccines that are examples of peptide-based are the HBV and HPV vaccines (Petrosky et al. 2017).

10.5.3 Cancer Vaccines Based on Nucleic Acids

Due to their ease of usage and capacity to produce strong MHC I-mediated CD8 + T-cell responses, nucleic acid vaccines have great potential as cancer vaccine platforms. Multiple vaccines can be delivered at the same time to increase humoral and cellular protection. Nucleic acid vaccines can be used to give full-length tumor antigens, allowing them to cross-present different epitopes. Nucleic acid vaccines are easy to create and hold promise for the development of tailored neo-antigen vaccines. Safety, specificity in generating immune responses to target antigens, activation of both humoral and cellular immunological responses, low production costs, and ease of manufacture are some of the benefits of conjugate acid vaccines (Wen et al. 2023).

10.6 Nanovaccine Types

10.6.1 Neoantigens

Tumor cells develop self-antigens called neoantigens as a result of genetic anomalies, such as somatic mutations. Another source of neoantigens is dysregulated RNA-derived novel proteins or peptides. Neoantigen vaccines made of synthetic peptides, mRNA, or DNA show potential as a cancer treatment. A unique approach to tumor immunotherapy is provided by tumor neoantigens, which are based on mutant proteins that are only expressed in tumor cells and antigens produced by tumor viruses that induce genome-integrated cancers. These novel antigens can be produced by virally encoded open reading, are immunogenic, and vary in tumor composition frames in virally infected cancers, like EBV-associated nasopharyngeal carcinoma and HPV-positive cervical cancer. Neoantigens can be recognized by sequencing the genomes of healthy and malignant cells, or by using MS-based proteomics (Shemesh et al. 2021).

10.6.2 Nanovaccines Based on STING Agonist

Immunotherapies have revolutionized the treatment of cancer, but a significant percentage of patients do not respond to them. Recent research has focused on targets that initiate or enhance anticancer immune responses. STING, an endoplasmic protein known as the stimulator of interferon (IFN) genes, is known to promote the synthesis of pro-inflammatory cytokines such as type I IFNs. It is one example of a new target. STING is a signaling, protein that is associated with the endoplasmic reticulum and plays a major role in the innate immune system's response to bacteria and viruses by controlling the transcription of several genes related to the immune system (Barber 2015). Cytosolic cyclic GMP-AMP synthase (cGAS) detects cytolytic DNA, and STING triggers the production of pro-inflammatory cytokines and type I IFNs to start inflammatory processes and get rid of pathogens. Furthermore, after injecting a DNA vaccine, adaptive immunity can be induced by STING-dependent signaling. Through the STING pathway, CD8 + DCs can produce type I IFN to activate the presentation of antigens and prime CD8 + T-cells. Furthermore, tumor-derived STING-activating components can be recognized by B cells and other CD11b+ host APCs that infiltrate the tumor. This recognition triggers the release of STING-mediated type I IFNs produced by leukocytes and primes cytotoxic NK cells for the elimination of tumor cells (Vitale et al. 2019).

10.6.3 Artificial APCs (aAPCs)

Between innate and adaptive immune responses, APCs act as a link. Antigen-specific T cells, which are crucial for the adaptive immune response, are activated when antigen-specific APCs display antigen class I and II MHCs on the membrane along with costimulatory signals after an antigen has been ingested. APCs, like DCs, can transfer information from vaccines to macrophages, DCs, and T or B cells, which are in charge of capturing, processing, and presenting antigens via MHC-II. These cells are then activated by MHC-II, which triggers the production of adaptive immune responses necessary for cancer immunotherapy. Research interest in autologous DC-based cancer vaccines has increased significantly due to the significance of DCs in cancer immunotherapy. To make these vaccines, separate patient-derived DCs and stimulate them with cancer antigens (Vitale et al. 2019).

10.7 Cancer Cell-Based Vaccines

10.7.1 Vaccines Generated from Tumor Cells

To effectively prevent tumor emergence, progression, and recurrence, tumor immunotherapy entails activating the immune system to produce an anticancer immune response. Vaccines produced from tumor cells can stimulate T lymphocytes specific to tumors and efficiently transport TSAs to APCs. Generally speaking, tumor vaccines are categorized as either therapeutic or preventative. With the help of antigens unique to tumor cells, therapeutic tumor vaccines can destroy tumor cells by inducing the immune system to produce antibodies or immune cells that match the target antigens. Tumor vaccination Provenge is used to treat metastatic hormone-resistant prostate cancer that shows little to no symptoms.

10.7.2 Vaccines Utilizing Whole Tumor Cells

Whole tumor cell vaccines have generated a lot of interest because of the diverse array of antigens present in tumor cells. Compared to single-epitope vaccines, whole-cell immunization can reduce tumor escape at a higher rate (Chiang et al. 2010). These vaccinations can be classified as either autologous or allogeneic based on where the tumor cells are located. Autologous vaccines, in which the patient's tumor cells are used as the antigen source, provide personalized therapy. Allogeneic immunizations employ tumor cell lines produced from individuals of the same race, offering the advantages of consistent quality control and large production. On the other hand, unmodified tumor cells do not elicit a robust immune response. The genetically Modified vaccine GVAX Pancreas uses tumor cells that have been engineered to express GM-CSF, an immune adjuvant that increases DC survival and antigen-presenting capacity (Yu et al. 2016). Apart from inhibiting the action of regulatory T-cells (Tregs), GVAX Pancreas also produces T-cell immunity against tumor antigens and mesothelin. According to clinical research, patients with metastatic pancreatic cancer have longer lifetimes while taking GVAX Pancreas in addition to CRS-207. Using vaccination-targeted tumor cells and CAFs, autologous tumor whole-cell vaccination with high FAP expression has been created, leading to a significant decrease in CAFs, the elimination of the immunosuppressive TME, and a rise in the recruitment of effector T cells to improve anticancer efficacy. Enhancing the range of antigens not only boosts the synthesis of immune stimulants but also the immunogenicity of tumor vaccines (Liu et al. 2022).

10.7.3 Vaccines Based on Tumor Cell Lysate

The generation of a diverse array of TAAs that may augment antitumor immunity is made possible by the creation of homologous tumor lysates. Tumor lysate antigen-based vaccinations seem to be more clinically helpful than peptide-based ones in the context of cancer immunotherapy. Moreover, stronger anticancer responses and longer-lasting T-cell memory are anticipated from the concurrent MHC class I and class II-restricted antigens extracted from lysates of tumor cells. Because these lysates have low immunogenicity, have few TSAs, and TCLs break down quickly in the body, APCs are unable to absorb them efficiently. As a result, technical efforts have mostly concentrated on modifying TCLs' immunogenicity. Tumor cells have undergone genetic modification to regulate the essential elements of the tumor immune response, hence enhancing immunogenicity (Ilinskaya and Dobrovolskaia 2016).

10.7.4 RNA Vaccines for Whole Tumors

To fully activate certain tumor immune responses, tumor mRNA vaccines can transcribe the entirety of the antigen seen on tumor cells, and they are generally safer than DNA and live attenuated vaccines. The ability of the tumor mRNA to simultaneously encode multiple tumor antigens is one of the advantages of whole tumor RNA vaccines over conventional tumor vaccines (Chen et al. 2024). Moreover, mRNA can avoid the restrictions imposed by MHC classification, obtain immunogenicity without the need for adjuvant or insertional mutations, and not integrate into the host genome. On the other hand, RNA vaccines are unstable by nature. RNases external to the cell quickly degrade naked RNA, preventing it from localizing to APCs and making it inactive. Furthermore, bare RNA's negative charge may prevent it from entering APCs.

10.7.5 Microvesicle Vaccines Produced from Tumors

Tumor-derived microvesicles (TMVs) are EVs that, when released by tumor (or other) cells in the tumor microenvironment (TME), have a diameter of 100–1000 nm. They are composed of a variety of bioactive chemicals, such as proteins, nucleic acids, and other materials that are liberated from the plasma membrane and connected to malignancies (Zhu et al. 2021). These biological entities have an impact on the angiogenesis, invasion, immune response, metastasis, and angiogenesis of malignancies. After being exposed to ionizing radiation, C6 glioma cells create TMVs, which are therapeutic tumor vaccines. This vaccination significantly reduces the tumor volume by releasing tumor antigens upon radiation exposure, which aids T lymphocytes in infiltrating immunized rat tumors and promoting tumor cell death. It has been demonstrated that TMVs carry the tumor antigen profile and innate DNA signals, which encourage DC maturation, T-cell activation, and tumor rejection by activating type I IFNγ via the cGAS/STING pathway (Bose 2017).

10.8 Activation Action of Nanovaccines

When it comes to getting antigens to tumor locations, NPs are crucial. NP-delivered antigens are absorbed through a variety of endocytic routes. However, several variables, including NP type, size, surface charge, and surface functionalization, affect how antigens are presented, internalized, and delivered to APCs. For instance, cationic NPs often facilitate intracellular trafficking by endosomal escape and are absorbed by APCs more quickly than other types. Catalytic dendrimer NPs containing adsorbed antigens activate DCs and release cytokines, such as IL-1β and IL-12, in addition to enhancing antigen delivery to DCs. The innate and adaptive immune systems must work together, and DCs are crucial for this. DCs are necessary for vaccine-activated cellular immune responses against viral and neoplastic diseases because most vaccines used in clinical practice today are exogenous to cells. Therefore, utilizing nanovaccines for DC targeting has been the focus of many approaches. The development of mannose receptors on DCs' cell surfaces, which encourage the mannosylation-mediated internalization of antigens, stimulates CD4 + and CD8 + T-cell responses (Al-Barwani et al. 2014). The application of the same methodology to a dextran-based nanovaccine containing LPS was successful. When soluble antigen and LPS were combined, nanoformulations generated stronger, more antigen-specific CD8 + T-cell responses. Targeting Langerhans cells in human skin with liposomes combined with Langerin ligands was successful because the focus was on Langerins (CD207), which are only expressed in these cells. DCs can process antigens and present them via the MHC I pathway, activating CD8 + T cells in a process known as cross-presentation that takes place via the cytosolic route as an alternative to the standard MHC II presentation and CD4 + T-helper-cell activation pathway (Dhatchinamoorthy et al. 2021) (Fig. 10.3).

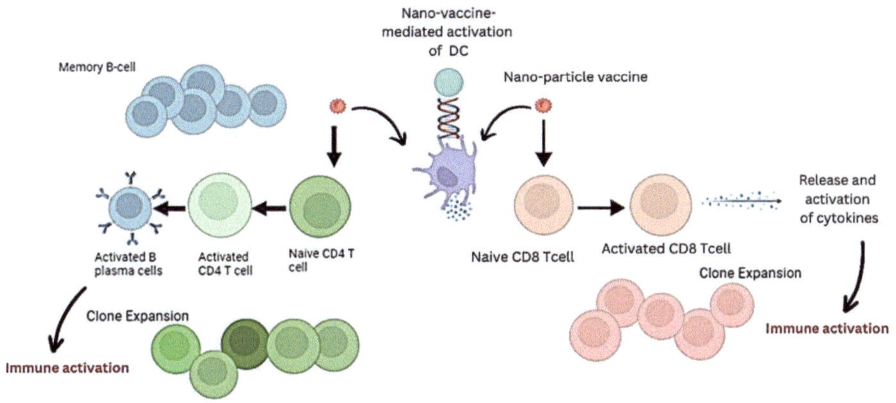

Fig. 10.3 Mechanism of action of nano vaccines. Different kinds of NP-attached antigens drive antigen-presenting cells (APCs) to digest and distribute these antigens in different ways. Following their destruction by APCs, these antigens are subsequently presented by MHC I or MHCII. Through the secretion of cytokines, the APCs influence the cytokine milieu and determine whether an inflammatory response is pro- or anti-inflammatory during this process. The immune response is strengthened by the activated T and B cells' clonal expansion. Activated plasma B cells react to the specific NP-coupled antigen by releasing antibodies

10.9 Safety, Innovation, and Mitigation Methods for Cancer Nanovaccines

When it comes to cancer treatments, in particular, side effects or toxicity might have disastrous results. Therefore, careful preclinical research using animal models is necessary to assess the safety profile before moving forward with human clinical trials. Concerns about specificity, linkers, unexpected disruptions in cell signaling, overburdening the machinery of cellular breakdown, immunological reactions, biodistribution and pharmacokinetics, long-term effects, and the emergence of resistance are among the safety issues that need to be addressed. The unique benefits of nanovaccines, including enhanced specificity, efficacy, and multi-functionality as well as the potential for customized treatment, have enabled precision medicine (Svensson et al. 2023). Using biocompatible and biodegradable nanoparticles is one possible delivery strategy that could improve the particles' stability and biodistribution, effects, and enable controlled and prolonged release at the tumor site. The goal of developing nanovaccines should be to tailor them to the unique characteristics of each patient's tumor while also optimizing selectivity and efficacy. The development of personalized nanovaccines will improve the prognosis and quality of life for cancer patients by providing them with an effective and long-lasting treatment. Personalized cancer nanovaccines can be developed using targeted protein degradation and nanotechnology, providing highly effective, patient-specific, and selective treatments. This approach has the potential to overcome the limitations of current treatments by allowing the degradation of previously unreachable targets, reducing systemic toxicity, and enhancing the immune response against cancer cells.

10.10 Synergistic Effects of Probiotics and Nanovaccines

Probiotics and nanovaccines work synergistically to enhance immune responses and prevent disease (Xu et al. 2023). This has great promise. Probiotics are good bacteria that help maintain intestinal health. They can also affect the immune system, making it more effective and prepared. The total immunogenic effect can be increased when paired with nanovaccines, which are vaccinations designed at the nanoscale to enhance delivery, stability, and efficacy. Because nanovaccines provide precise antigen delivery and probiotics increase mucosal immunity, this combination may result in a more robust and tailored immune response. As a result, this synergy may lead to decreased side effects, increased vaccination efficacy, and the possibility of developing novel treatments for a variety of illnesses.

10.10.1 The Immune System and Probiotics Interactions

Live bacteria known as probiotics have been shown to provide health advantages. The major interactive site for the probiotics with the immune system was found through the gut-associated lymphoid tissue (GALT). By enhancing the functionalities of natural killer cells, dendritic cells, and macrophage activity, probiotic bacteria expand the innate immune response to maintain gut health. Moreover, probiotics also affect T-cell differentiation and proliferation, which ultimately enhances the synthesis of anti-inflammatory cytokines while decreasing the synthesis of pro-inflammatory cytokine molecules. Such equilibrium helps in the maintenance of homeostasis of normal gut flora and the defense of pathogenic microorganisms (Kim et al. 2018). They achieve this beneficial function by increasing the expression of tight junction proteins, like mucin, and activating the antimicrobial peptide synthesis pathways. In addition, pathogenic bacteria and probiotics both compete for the nutrition and sites of their adhesion, thereby reducing the frequency of gut infections. Together, these interactions support immune system function, minimize infection risk, and may even reduce the prevalence of immunological-related disorders including inflammatory bowel disease and allergies.

10.10.2 Immune Response Augmentation

Probiotics and nanovaccines are examples of cutting-edge methods for boosting immunity. Live beneficial bacteria or probiotics, can modify the gut microbiota and boost immunity by boosting the synthesis of antimicrobial compounds, improving barrier function, and inducing both adaptive and innate immune responses. They affect the function of immune cells, including lymphocytes, dendritic cells, and macrophages, which increase the synthesis of cytokines and antibodies. Stronger and more efficient immune defense is produced as a result of this modulation, which aids in improved pathogen identification and removal.

Nanotechnology is used in nanovaccines to enhance vaccine effectiveness and delivery (Chowdhury and Kundu 2023). These vaccines contain antigens in nanoparticles, which prevents them from breaking down and guarantees a regulated release. Through improved antigen uptake by antigen-presenting cells, T and B cells are strongly activated as a result of this technique. Additionally, nanovaccines can be designed to specifically target cells or tissues, enhancing the immune response's potency and accuracy. Researchers hope to develop more effective and durable vaccinations that offer better defense against infectious diseases and other medical disorders by merging these technologies.

10.11 Challenges and Future Directions for Probiotic Nanovaccines

To date, a few of the many immunotherapeutic techniques for cancer have been investigated including immune checkpoint inhibition, adoptive cell transfer therapy, oncolytic virotherapy, etc. (Muthukutty and Yoo 2023). Among them, cancer therapeutic vaccines can enhance the immune responses specific to any cancer and significantly modulate the pathways of immunosuppression. These unique qualities may help in addressing the drawbacks of the immunotherapies that are now in use, like poor response rates and hypersensitivity-related adverse effects. The potential of nanovaccines was found to boost the efficacy of cancer-treating vaccinations. Nanovaccines successfully co-deliver adjuvant molecules and multi-valent molecular antigens which enter inside the immune cells and lymphoid organs to improve antigen-specific adaptive immune responses during the cancer treatment. Such an approach, thereby significantly increases the immunogenicity of molecular vaccines. Through the combination of some nano-engineering techniques, a wide range of nanovaccines have been created for cancer immunotherapy with proper evaluation rarely in humans and mostly in preclinical experimental models.

Tumor antigens are delivered to APCs in a way that promotes APC maturation and activation, boosts the infiltration of functional CD8 + T cells that fight cancer, helps DCs activate and build antigen depots, keeps things stable, and aids in optimizing the potential of therapeutic vaccinations for cancer, cancer nanovaccines can effectively treat tumors. The most current developments in nanocarrier types, nanoplatforms, and nanovaccines—such as mRNA nanovaccines, artificial antigen-presenting cells, STING-activating nanovaccines, and cancer neoantigen nanovaccines—were covered in this study. Moreover, we focused on vaccination procedures mainly targeting tumor cells, such as EVs, cell membrane-encapsulated NPs, etc. Vaccines generated from tumor cells have a diversity of tumor antigens that can cause strong immune responses in the patient's body. In recent times, the development of nanovaccine techniques could advance the field of cancer immunotherapy in a revolutionary way. Even nanovaccines have demonstrated a lot of promise for cancer immunotherapy but more research needs to be carried out to improve antigen immunogenicity, prevent cancer cells from evading the immune system, and develop potential tumor vaccine delivery strategies. Importantly, cancer cell-derived vaccines have drawn significant attention previously for their unique application in enhancing the long-term "immunological memory" that affects cancer progression (Fan et al. 2023). Several research teams also demonstrated positive preclinical outcomes related to cancer nanovaccine development. Therefore, the advantages of NPs with tunable features for nanovaccine design should be taken into account during cancer vaccine production. The generation of strong, safe, and customized nanovaccines that deliver potent and long-lasting anti-tumor immune responses is quite anticipated to be a prerequisite for future cancer vaccine research. Vaccines generated from tumor cells have a wide range of antigens; nonetheless, they still need to be significantly improved to induce tumor immunogenicity. The foundation for the creation of novel exosome-based biomarkers and effective cancer immunotherapy medications will be laid by a better knowledge of the interactions between cell-specific molecules on exosomes and immune and tumor cells. This is particularly true for exosomes and OMVs, two types of EV-derived tumor vaccines. TEVs need to overcome some challenges to proceed with characterization, purification, fabrication, and scale-up manufacturing. These challenges are mostly related to EV differentiation and isolation. TMVs and TEXs are similar in size, and a common biomarker to differentiate between TEV isoforms is not available. It is therefore challenging to evaluate the clinical efficacy of vaccinations due to their isolation. Since EVs can only be successfully extracted once, we can then study the characteristics and labeling of each subpopulation and how each subgroup functions in physiological processes and cellular communication (Diao and Liu 2023). Before the official use of EV vaccinations in clinical settings, biopharmaceuticals' quality classification and standards need to be addressed. To ensure the safety of exosome treatments, the FDA should implement specific GMP regulations as soon as possible. There have already been attempts at EV quality management, and some studies have shown how to increase the production of exosomes. TEVs could be produced in the future as engineering tools advance.

10.12 Conclusion

Cancer immunotherapy, including techniques like adoptive cell transfer, immune checkpoint inhibition, and cancer vaccines, shows promise for treating gastrointestinal (GI) tract cancers. Nanovaccines, particularly, can enhance immune responses by co-delivering adjuvants and antigens, boosting vaccine immunogenicity. Recent advancements include mRNA nanovaccines and tumor cell-derived vaccines, which have shown potential in preclinical studies. Probiotic formulations can also modulate the gut microbiota, potentially enhancing the effectiveness of these nanovaccines. Despite their promise, challenges remain in improving immunogenicity, preventing immune evasion, and optimizing delivery. Developing robust, safe, and personalized nanovaccines for GI tract cancers, along with addressing regulatory and manufacturing challenges for exosome-based vaccines, is crucial for future clinical application.

References

Abakushina EV, Popova LI, Zamyatnin AA Jr, Werner J, Mikhailovsky NV, Bazhin AV (2021) The advantages and challenges of anticancer dendritic cell vaccines and NK cells in adoptive cell immunotherapy. Vaccines (Basel) 9(11):1363. Published 2021 Nov 19. https://doi.org/10.3390/vaccines9111363

Al-Barwani F, Young SL, Baird MA, Larsen DS, Ward VK (2014) Mannosylation of virus-like particles enhances internalization by antigen presenting cells. PLoS One 9(8):e104523. Published 2014 Aug 14. https://doi.org/10.1371/journal.pone.0104523

Anachad O, Taouil A, Taha W, Bennis F, Chegdani F (2023) The implication of short-chain fatty acids in obesity and diabetes. Microbiol Insights 16:11786361231162720. Published 2023 Mar 25. https://doi.org/10.1177/11786361231162720

Anand U, Dey A, Chandel AKS et al (2022) Cancer chemotherapy and beyond: current status, drug candidates, associated risks and progress in targeted therapeutics [published correction appears in Genes Dis. 2024 Jan 20;11(4):101211. doi: 10.1016/j.gendis.2024.101211]. Genes Dis 10(4):1367–1401. Published 2022 Mar 18. https://doi.org/10.1016/j.gendis.2022.02.007

Aziz N, Bonavida B (2016) Activation of natural killer cells by probiotics. For Immunopathol Dis Therap 7(1–2):41–55. https://doi.org/10.1615/ForumImmunDisTher.2016017095

Barber GN (2015) STING: infection, inflammation and cancer. Nat Rev Immunol 15(12):760–770. https://doi.org/10.1038/nri3921

Bezbaruah R, Chavda VP, Nongrang L et al (2022) Nanoparticle-based delivery systems for vaccines. Vaccines (Basel) 10(11):1946. Published 2022 Nov 17. https://doi.org/10.3390/vaccines10111946

Bose D (2017) cGAS/STING pathway in cancer: Jekyll and Hyde story of cancer immune response. Int J Mol Sci 18(11):2456. Published 2017 Nov 18. https://doi.org/10.3390/ijms18112456

Bouranis JA, Beaver LM, Ho E (2021) Metabolic fate of dietary Glucosinolates and their metabolites: a role for the microbiome. Front Nutr 8:748433. Published 2021 Sep 22. https://doi.org/10.3389/fnut.2021.748433

Bray F, Laversanne M, Sung H et al (2024) Global cancer statistics 2022: GLOBOCAN estimates of incidence and mortality worldwide for 36 cancers in 185 countries. CA Cancer J Clin 74(3):229–263. https://doi.org/10.3322/caac.21834

Chakraborty S, Ye J, Wang H et al (2023) Application of toll-like receptors (TLRs) and their agonists in cancer vaccines and immunotherapy. Front Immunol 14:1227833. Published 2023 Oct 23. https://doi.org/10.3389/fimmu.2023.1227833

Chauhan DS, Dhasmana A, Laskar P et al (2021) Nanotechnology synergized immunoengineering for cancer. Eur J Pharm Biopharm 163:72–101. https://doi.org/10.1016/j.ejpb.2021.03.010

Chelakkot C, Ghim J, Ryu SH (2018) Mechanisms regulating intestinal barrier integrity and its pathological implications. Exp Mol Med 50(8):1–9. Published 2018 Aug 16. https://doi.org/10.1038/s12276-018-0126-x

Chen B, Yang Y, Wang X et al (2024) mRNA vaccine development and applications: a special focus on tumors (review). Int J Oncol 65(2):81. https://doi.org/10.3892/ijo.2024.5669

Chiang CL, Benencia F, Coukos G (2010) Whole tumor antigen vaccines. Semin Immunol 22(3):132–143. https://doi.org/10.1016/j.smim.2010.02.004

Chowdhury N, Kundu A (2023) Nanotechnology platform for advancing vaccine development against the COVID-19 virus. Diseases 11(4):177. Published 2023 Dec 10. https://doi.org/10.3390/diseases11040177

Cintolo JA, Datta J, Mathew SJ, Czerniecki BJ (2012) Dendritic cell-based vaccines: barriers and opportunities. Future Oncol 8(10):1273–1299. https://doi.org/10.2217/fon.12.125

Dhatchinamoorthy K, Colbert JD, Rock KL (2021) Cancer immune evasion through loss of MHC class I antigen presentation. Front Immunol 12:636568. Published 2021 Mar 9. https://doi.org/10.3389/fimmu.2021.636568

Diao L, Liu M (2023) Rethinking antigen source: cancer vaccines based on whole tumor cell/tissue lysate or whole tumor cell. Adv Sci (Weinh) 10(22):e2300121. https://doi.org/10.1002/advs.202300121

Din FU, Aman W, Ullah I et al (2017) Effective use of nanocarriers as drug delivery systems for the treatment of selected tumors. Int J Nanomedicine 12:7291–7309. Published 2017 Oct 5. https://doi.org/10.2147/IJN.S146315

Elliott RO, He M (2021) Unlocking the power of exosomes for crossing biological barriers in drug delivery. Pharmaceutics 13(1):122. Published 2021 Jan 19. https://doi.org/10.3390/pharmaceutics13010122

Est-Witte SE, Livingston NK, Omotoso MO, Green JJ, Schneck JP (2021) Nanoparticles for generating antigen-specific T cells for immunotherapy. Semin Immunol 56:101541. https://doi.org/10.1016/j.smim.2021.101541

Fan T, Zhang M, Yang J, Zhu Z, Cao W, Dong C (2023) Therapeutic cancer vaccines: advancements, challenges, and prospects. Signal Transduct Target Ther 8(1):450. Published 2023 Dec 13. https://doi.org/10.1038/s41392-023-01674-3

Fang X, Lan H, Jin K, Gong D, Qian J (2022) Nanovaccines for cancer prevention and immunotherapy: an update review. Cancers (Basel) 14(16):3842. Published 2022 Aug 9. https://doi.org/10.3390/cancers14163842

Fong W, Li Q, Yu J (2020) Gut microbiota modulation: a novel strategy for prevention and treatment of colorectal cancer. Oncogene 39(26):4925–4943. https://doi.org/10.1038/s41388-020-1341-1

Genua F, Raghunathan V, Jenab M, Gallagher WM, Hughes DJ (2021) The role of gut barrier dysfunction and microbiome dysbiosis in colorectal cancer development. Front Oncol 11:626349. Published 2021 Apr 15. https://doi.org/10.3389/fonc.2021.626349

Górska A, Przystupski D, Niemczura MJ, Kulbacka J (2019) Probiotic bacteria: a promising tool in cancer prevention and therapy. Curr Microbiol 76(8):939–949. https://doi.org/10.1007/s00284-019-01679-8

Guo N, Lv LL (2023) Mechanistic insights into the role of probiotics in modulating immune cells in ulcerative colitis. Immun Inflamm Dis 11(10):e1045. https://doi.org/10.1002/iid3.1045

Guo Y, Wang Z, Shi X, Shen M (2022) Engineered cancer cell membranes: an emerging agent for efficient cancer theranostics. Exploration (Beijing) 2(1):20210171. Published 2022 Jan 25. https://doi.org/10.1002/EXP.20210171

Ilinskaya AN, Dobrovolskaia MA (2016) Understanding the immunogenicity and antigenicity of nanomaterials: past, present and future. Toxicol Appl Pharmacol 299:70–77. https://doi.org/10.1016/j.taap.2016.01.005

Kale S, Hirani S, Vardhan S et al (2023) Addressing cancer disparities through community engagement: lessons and best practices. Cureus 15(8):e43445. Published 2023 Aug 14. https://doi.org/10.7759/cureus.43445

Kheirvari M, Liu H, Tumban E (2023) Virus-like particle vaccines and platforms for vaccine development. Viruses 15(5):1109. Published 2023 May 2. https://doi.org/10.3390/v15051109

Kim HW, Hong R, Choi EY et al (2018) A probiotic mixture regulates T cell balance and reduces atopic dermatitis symptoms in mice. Front Microbiol 9:2414. Published 2018 Oct 15. https://doi.org/10.3389/fmicb.2018.02414

Kwok KO, Fries LR, Silva-Zolezzi I, Thakkar SK, Iroz A, Blanchard C (2022) Effects of probiotic intervention on markers of inflammation and health outcomes in women of reproductive age and their children. Front Nutr 9:889040. Published 2022 Jun 6. https://doi.org/10.3389/fnut.2022.889040

Gamallat Y, Meyiah A, Kuugbee ED et al (2016) Lactobacillus rhamnosus induced epithelial cell apoptosis, ameliorates inflammation and prevents colon cancer development in an animal model. Biomed Pharmacother 83:536–541. https://doi.org/10.1016/j.biopha.2016.07.001

Liang J, Jiang Y, Abboud Y, Gaddam S (2022) Role of endoscopy in management of upper gastrointestinal cancers. Diseases 11(1):3. Published 2022 Dec 27. https://doi.org/10.3390/diseases11010003

Liu Y, Alookaran JJ, Rhoads JM (2018) Probiotics in autoimmune and inflammatory disorders. Nutrients 10(10):1537. Published 2018 Oct 18. https://doi.org/10.3390/nu10101537

Liu J, Fu M, Wang M, Wan D, Wei Y, Wei X (2022) Cancer vaccines as promising immunotherapeutics: platforms and current progress. J Hematol Oncol 15(1):28. Published 2022 Mar 18. https://doi.org/10.1186/s13045-022-01247-x

Markowiak-Kopeć P, Śliżewska K (2020) The effect of probiotics on the production of short-chain fatty acids by human intestinal microbiome. Nutrients 12(4):1107. Published 2020 Apr 16. https://doi.org/10.3390/nu12041107

Mattiuzzi C, Lippi G (2019) Current cancer epidemiology. J Epidemiol Glob Health 9(4):217–222. https://doi.org/10.2991/jegh.k.191008.001

Mazziotta C, Tognon M, Martini F, Torreggiani E, Rotondo JC (2023) Probiotics mechanism of action on immune cells and beneficial effects on human health. Cells 12(1):184. Published 2023 Jan 2. https://doi.org/10.3390/cells12010184

Muthukutty P, Yoo SY (2023) Oncolytic virus engineering and utilizations: cancer immunotherapy perspective. Viruses 15(8):1645. Published 2023 Jul 28. https://doi.org/10.3390/v15081645

Orlando FA, Tan D, Baltodano JD et al (2008) Aberrant crypt foci as precursors in colorectal cancer progression. J Surg Oncol 98(3):207–213. https://doi.org/10.1002/jso.21106

Petrariu O-A et al (2024) Role of probiotics in managing various human diseases, from oral pathology to cancer and gastrointestinal diseases. Front Microbiol 14:1296447

Petrosky EY, Liu G, Hariri S, Markowitz LE (2017) Human Papillomavirus vaccination and age at first sexual activity, National Health and Nutrition Examination Survey. Clin Pediatr (Phila) 56(4):363–370. https://doi.org/10.1177/0009922816660541

Ren C, Li Y, Cong Z, Li Z, Xie L, Wu S (2023) Bioengineered bacterial outer membrane vesicles encapsulated Polybia-mastoparan I fusion peptide as a promising nanoplatform for bladder cancer immune-modulatory chemotherapy. Front Immunol 14:1129771. Published 2023 Mar 14. https://doi.org/10.3389/fimmu.2023.1129771

Sabit H, Kassab A, Alaa D et al (2023) The effect of probiotic supplementation on the Gut-Brain Axis in psychiatric patients. Curr Issues Mol Biol 45(5):4080–4099. Published 2023 May 6. https://doi.org/10.3390/cimb45050260

Santos PM, Butterfield LH (2018) Dendritic cell-based cancer vaccines. J Immunol 200(2):443–449. https://doi.org/10.4049/jimmunol.1701024

Shemesh CS, Hsu JC, Hosseini I et al (2021) Personalized cancer vaccines: clinical landscape, challenges, and opportunities. Mol Ther 29(2):555–570. https://doi.org/10.1016/j.ymthe.2020.09.038

Shi LH, Balakrishnan K, Thiagarajah K, Mohd Ismail NI, Yin OS (2016) Beneficial properties of probiotics. Trop Life Sci Res 27(2):73–90. https://doi.org/10.21315/tlsr2016.27.2.6

Shi R, Tang YQ, Miao H (2020) Metabolism in tumor microenvironment: implications for cancer immunotherapy. MedComm 1(1):47–68. Published 2020 Jun 3. https://doi.org/10.1002/mco2.6

Śliżewska K, Markowiak-Kopeć P, Śliżewska W (2020) The role of probiotics in cancer prevention. Cancers (Basel) 13(1):20. Published 2020 Dec 23. https://doi.org/10.3390/cancers13010020

Smith GL, Lopez-Olivo MA, Advani PG et al (2019) Financial burdens of cancer treatment: a systematic review of risk factors and outcomes. J Natl Compr Cancer Netw 17(10):1184–1192. https://doi.org/10.6004/jnccn.2019.7305

Song NE, Kim NJ, Kim YH, Baik SH (2021) Probiotic properties of lactic acid bacteria with high conjugated linoleic acid converting activity isolated from Jeot-Gal, high-salt fermented seafood. Microorganisms 9(11):2247. Published 2021 Oct 28. https://doi.org/10.3390/microorganisms9112247

Sung H, Ferlay J, Siegel RL et al (2021) Global cancer statistics 2020: GLOBOCAN estimates of incidence and mortality worldwide for 36 cancers in 185 countries. CA Cancer J Clin 71(3):209–249. https://doi.org/10.3322/caac.21660

Svensson E, von Mentzer U, Stubelius A (2023) Achieving precision healthcare through nanomedicine and enhanced model systems. ACS Mater Au 4(2):162–173. Published 2023 Dec 18. https://doi.org/10.1021/acsmaterialsau.3c00073

Tegegne BA, Kebede B (2022) Probiotics, their prophylactic and therapeutic applications in human health development: a review of the literature. Heliyon 8(6):e09725. Published 2022 Jun 22. https://doi.org/10.1016/j.heliyon.2022.e09725

Tortora SC, Bodiwala VM, Quinn A, Martello LA, Vignesh S (2022) Microbiome and colorectal carcinogenesis: linked mechanisms and racial differences. World J Gastrointest Oncol 14(2):375–395. https://doi.org/10.4251/wjgo.v14.i2.375

Vassilev A, DePamphilis ML (2017) Links between DNA replication, stem cells and cancer. Genes (Basel) 8(2):45. Published 2017 Jan 25. https://doi.org/10.3390/genes8020045

Vinogradova Y, Coupland C, Hippisley-Cox J (2020) Use of hormone replacement therapy and risk of breast cancer: nested case-control studies using the QResearch and CPRD databases. BMJ 371:m3873. Published 2020 Oct 28. https://doi.org/10.1136/bmj.m3873

Vitale M, Cantoni C, Della Chiesa M et al (2019) An historical overview: the discovery of how NK cells can kill enemies, recruit defense troops, and more. Front Immunol 10:1415. Published 2019 Jun 19. https://doi.org/10.3389/fimmu.2019.01415

Wan MLY, El-Nezami H (2018) Targeting gut microbiota in hepatocellular carcinoma: probiotics as a novel therapy. Hepatobil Surg Nutr 7(1):11–20. https://doi.org/10.21037/hbsn.2017.12.07

Wang X, Zhang P, Zhang X (2021) Probiotics regulate gut microbiota: an effective method to improve immunity. Molecules 26(19):6076. Published 2021 Oct 8. https://doi.org/10.3390/molecules26196076

Wen Y, Zhang RY, Wang J et al (2023) Novel sialoglycan linkage for constructing adjuvant-protein conjugate as potent vaccine for COVID-19. J Control Release 355:238–247. https://doi.org/10.1016/j.jconrel.2023.01.062

Xu X, Zhang M, Liu X et al (2023) Probiotics formulation and cancer nanovaccines show synergistic effect in immunotherapy and prevention of colon cancer. iScience 26(7):107167. Published 2023 Jun 19. https://doi.org/10.1016/j.isci.2023.107167

Yu TW, Chueh HY, Tsai CC, Lin CT, Qiu JT (2016) Novel GM-CSF-based vaccines: one small step in GM-CSF gene optimization, one giant leap for human vaccines. Hum Vaccin Immunother 12(12):3020–3028. https://doi.org/10.1080/21645515.2016.1221551

Zhang X, Cui H, Zhang W, Li Z, Gao J (2022) Engineered tumor cell-derived vaccines against cancer: the art of combating poison with poison. Bioact Mater 22:491–517. Published 2022 Oct 26. https://doi.org/10.1016/j.bioactmat.2022.10.016

Zheng K, Feng Y, Li L, Kong F, Gao J, Kong X (2024) Engineered bacterial outer membrane vesicles: a versatile bacteria-based weapon against gastrointestinal tumors. Theranostics 14(2):761–787. Published 2024 Jan 1. https://doi.org/10.7150/thno.85917

Zhu S, Li S, Yi M, Li N, Wu K (2021) Roles of microvesicles in tumor progression and clinical applications. Int J Nanomedicine 16:7071–7090. Published 2021 Oct 18. https://doi.org/10.2147/IJN.S325448

Gut Microbiome: Associations with Liver and Colon Cancer

11

Satya Priya Sharma and Ki Tae Suk

Abstract

The human body is home to approximately 100 trillion microorganisms, collectively referred to as the human microbiota, with their genomes forming the human microbiome. Key sites of colonization include the skin, oral cavity, respiratory tract, gastrointestinal tract, and urogenital system, with each site exhibiting a distinct microbial composition due to niche-specific adaptations. The most abundant and diverse microbial community resides in the gastrointestinal tract, particularly the colon. The gut microbiome has emerged as a critical player in human health, influencing metabolic, immunological, and neurological pathways. It is involved in nutrient metabolism, energy harvesting, and the production of short-chain fatty acids (SCFAs) and essential vitamins. The gut microbiome also plays a pivotal role in immune system development, maintaining intestinal homeostasis, and influencing neurodevelopment through the gut–brain axis. Dysbiosis, or alterations in the microbiome composition, is linked to various diseases such as inflammatory bowel disease, liver diseases, metabolic disorders, and cancers. Given its central role in human health, the gut microbiome presents promising therapeutic opportunities, including probiotics, prebiotics, synbiotics, postbiotics, and fecal microbiota transplantation, for disease prevention and treatment.

Authors' current affiliations and institutions participating in the study

S. P. Sharma (✉)
Institute for Liver and Digestive Diseases, Hallym University, Chuncheon, Republic of Korea

K. T. Suk
Institute for Liver and Digestive Diseases, Hallym University, Chuncheon, Republic of Korea

Department of Internal Medicine, Hallym University College of Medicine, Chuncheon, Republic of Korea

S. Pathak et al. (eds.), *Microbiota and Dietary Mediators in Colon Cancer Prevention and Treatment*, https://doi.org/10.1007/978-981-96-0297-1_11

Keywords

Gut microbiota · Dysbiosis · Liver cancer · Colon cancer

11.1 Introduction

Humans are a natural habitat for 100 trillion microbial organisms including bacteria, archaea, fungi, protozoans, and viruses formally known as human microbiota and collective genomes of these microorganisms referred to as the human microbiome. The prime sites of human microbiome colonization are the skin, oral cavity, respiratory tract, gastrointestinal tract, and urogenital system. The microbes that exist at each site are distinct in composition due to the adaptability of specific environmental conditions and interactions with the host related to that specific niche. The most copious and diverse microbial community in human body is found in the gastrointestinal tract, predominantly in the colon, which contain 10^{13}–10^{14} colony-forming units (CFU)/g of feces. The second largest microbial community is found in the oral cavity with an estimate of 10^{8}–10^{11} CFU/mL of saliva, followed by skin microbiome which estimated 10^{2}–10^{7} CFU/cm^{2}, but can significantly fluctuate depending on the anatomical collection site. A gradient alteration in microbial colonization exhibits in the respiratory tract, with higher microbial colonization in the upper respiratory tract (10^{3}–10^{5} CFU/mL) compared to the lower respiratory tract in healthy individuals. Lastly, the urogenital tract microbiome, especially the female vaginal microbiome, approximately contains 10^{8}–10^{9} CFU/mL. These diverse microbiomes coevolved with their human hosts since birth and have established a labyrinthine mutual relationship that can influence several physiological pathways which includes metabolic, immunological, and neurological development. The compositional diversity of these microbial ecology at each anatomical site are influenced through several factors, including host age, eating habits, genetics, environmental and medical exposures, which could change it into the personalized microbial fingerprint of each individual (Acharya and Thirunavukkarasu 2024).

Recently, the gut microbiome has emerged as a very critical factor for humans due to its direct and profound impact on multiple physiological pathways. The importance of the gut microbiome is highlighted by its immunomodulatory effects, vast metabolic capacity, and bidirectional communication with the host. Primarily, the gut microbiome is involved in energy harvest through nutrient metabolism and breaking down indigestible complex dietary components like plant-based polysaccharides and resistant starches that which are undigested by human enzymes. This metabolic breakdown produces short-chain fatty acids (SCFAs), which are the main energy sources for colonocytes and exert extensive health-promoting effects on the host. The gut microbiome also synthesizes essential vitamins, such as vitamin K and multiple B vitamins, which significantly improve the host's nutritional status. Furthermore, the microbiome is pivotal in shaping the host immune system, predominantly in the early stage of life. The gut microbiome promotes regulatory T cells development which is essential for intestinal homeostasis and intestinal barrier function. The gut microbiome also influences the distal part of the human body such as brain through the gut–brain axis, which possesses a bidirectional communication network connecting the enteric and central nervous systems. Thus, the gut microbiome has an effect on neurodevelopment, behavior, and mental health, and compositional alteration in the gut microbiome associated with neurological and psychiatric disorders. Moreover, the gut microbiome's metabolic activities influence the pharmacokinetics and pharmacodynamics of several medicines and has a significant impact on drug efficacy. In the context of liver health, gut microbiota dysbiosis, have been linked to the progression of various liver diseases, including hepatocellular carcinoma (HCC). Dysbiotic changes are characterized by an increase in potentially pathogenic bacteria and a decrease in beneficial microbes, disrupting the delicate balance necessary for maintaining health. For instance, the phyla Proteobacteria and Fusobacteria are often elevated in HCC patients, while the beneficial Firmicutes and Bacteroidetes are reduced. These shifts in microbial composition can instigate pro-inflammatory pathways, exacerbating liver disease severity and contributing to tumorigenesis. Understanding the complex interplay between the gut microbiome and liver cancer is crucial for developing microbiome-based therapeutic strategies aimed at disease prevention and management (Rajapakse et al. 2023). The gut microbiome's central role in promoting human health offers promising avenues for disease prevention and treatment including microbiome-based therapeutic approaches such as probiotics, prebiotics, postbiotics, synbiotics, and fecal microbiota transplantation (Bendriss et al. 2021).

11.2 Healthy Gut Microbiome Composition

The human gut harbors a dynamically complex microbial ecosystem containing a diverse range of microbes, dominated by bacteria, and followed by fungi, viruses, and other types of microorganism including archaea, and protozoa. In a healthy adult, bacterial cells make up to 90–95% of the gut microbiome and the gut microbiome typically contains roughly 10^{13}–10^{14} cells/g microbial cells, which makes the human to bacterial cell ratio probable at 1:1.3. Fungal cells, formally known as mycobiome, make up to 0.1% and cell count estimated up to 10^{7}–10^{8}, and viruses, commonly known as virome, are estimated to be around 10^{8}–10^{9} cells/g. Mostly, viruses are bacteriophages: viruses that infect bacteria. The colon has the highest microbial count in the whole GI tract with a bacterial cell count up to 10^{11}–10^{12} cells/g of intestinal content. The microbiome composition in the gut is influenced by a multitude of factors like age, gender, ethnicity, diet in healthy condition, thus resulting in varied gut microbiomes among individuals. Normally, healthy adults harbor between 300 and 1000 different bacterial species which contain 150-fold microbial genes compared to human genome. Compositional variations based on gender are very evident with higher abundance of Bacteroides and Prevotella in males and higher Firmicutes levels in females. Age-dependent changes are also well observed, as infants and the elderly tend to have a less diverse gut microbiome, with the infant microbiome dominated by *Bifidobacterium* and Bacteroidetes in higher proportions in the elderly microbiome. Whereas, the composition of healthy adult's gut microbiome is dominated by Bacteroidetes and Firmicutes, with minor proportions of Actinobacteria, Proteobacteria, and Verrucomicrobia. Based on race and ethnicity microbial composition also differs even in healthy individuals, in African descent showed higher Prevotella and lower Bacteroides when compared to Europeans. Similarly, African Americans also have higher Prevotella levels compared to the Caucasians.

In healthy adults irrespective of gender, race, and ethnicity, the normal gut microbiome is dominated by two bacterial phyla: Bacteroidetes and Firmicutes, which together constitute more than 90% bacterial population. Firmicutes usually range from 60% to 80% which includes very important genera like *Clostridium*, *Lactobacillus*, *Enterococcus*, and *Ruminococcus*. Bacteroidetes are primarily dominated by *Bacteroides* and *Prevotella* genera and the total Bacteroidetes phylum ratio in bacterial composition ranges from 20% to 40%. Other low-abundance phyla include Actinobacteria with a high ratio of *Bifidobacterium*, Proteobacteria, Verrucomicrobia, and Fusobacteria. At the order level, a bacterial composition dominated by the orders *Lactobacillales*, *Enterobacteriales*, *Clostridiales*, *Bacteroidales*, and *Bifidobacteriales* is considered as a healthy gut. Unlike phylum and order, higher interindividual variation is observed in genus-level composition; still, the healthy gut includes a higher abundance of *Bacteroides*, *Faecalibacterium*, *Blautia*, *Prevotella*, *Roseburia*, *Ruminococcus*, *Eubacterium*, *Akkermansia*, and *Bifidobacterium* genus. The bacterial species belong to these genera are important for various health-promoting physiological pathways including fermentation, bile acid metabolism, and synthesis of vitamins. However, the relative abundance of these taxa could be variable and can be influenced by multiple internal and external factors, but the "core microbiome" has no significant alteration in the healthy gut. The core microbiome signifies that certain functional genes and/or metabolic pathways are conserved among healthy gut rather than particular taxa. The core microbiome includes functional genes indulging in amino acid and carbohydrate metabolism, vitamin biosynthesis, and short-chain fatty acid production, which underline the significance of microbiome importance in human health.

While gut microbiome composition is dominated by bacteria, other microbes like fungi and viruses are also crucial for maintaining a healthy gut ecosystem. The mycobiome (gut fungus) is less explored and not well known in comparison to bacteria, and the ratio is less than 0.1% in overall gut microbiome. In mycobiome, Ascomycota and Basidiomycota are most prevalent fungal phyla and dominated by Ascomycota. The most abundant genera in gut includes *Candida*, *Saccharomyces*, *Aspergillus*, and *Penicillium* which all belong to Ascomycota, whereas, lowered-ratio genera are *Malassezia* and *Cryptococcus* which belong to Basidiomycota phylum. Out of the abovementioned genera, *Candida* is the most abundant genus due to the predominance of the *C. albicans* species. Along with *Candida*, *Saccharomyces* (particularly *S. cerevisiae*), *Cladosporium*, and *Malassezia* are other abundant genera in the gut mycobiome. Furthermore, the gut virome (viral community in the gut) is dominated by bacteriophages which use bacteria as the host for replication. In the healthy gut, major bacteriophages with double-stranded DNA primarily come from

Caudovirales order including viral families *Podoviridae, Siphoviridae*, and *Myoviridae*. The virome community is very critical for maintaining and shaping the bacterial composition and enabling horizontal gene transfer. Additionally, the healthy gut also has eukaryotic viruses but in lesser abundance which include families like Picobirnaviridae, and Anelloviridae. As mentioned previously, healthy gut composition is variable and individual specific which could change under the influence of multiple factors such as host genetics, immune system, and geographical environment (Cheng et al. 2020; Dougherty and Jobin 2023).

11.3 Gut Microbiome Composition in Hepatocellular Carcinoma

In a healthy state, the gut microbiome is considered to be in eubiosis, characterized by high microbial diversity and a balanced equilibrium among microbes at each taxonomic level. However, when microbial diversity is significantly reduced due to the overgrowth of potentially harmful bacteria, the microbiome shifts to a dysbiosis state. In liver cancer, especially in hepatocellular carcinoma (HCC) condition, gut microbiome composition is altered significantly compared to healthy state and exhibits a dysbiotic condition and is the prime contributor in escalating the severity of the diseases. HCC showed an extensive increment in Proteobacteria and Fusobacteria, with a significant decline in Bacteroidetes and Firmicutes. The ratio of Firmicutes to Bacteroidetes, which signifies the healthy state, is also altered by a decline in Firmicutes and an increase in Bacteroidetes. In addition, potentially pathogenic bacteria including *Escherichia* and *Klebsiella* from *Enterobacteriaceae* family exhibited an increase in HCC. These dysbiotical changes are responsible for instigating the constant pro-inflammatory pathways that exacerbate liver disease severity and are possible contributors in tumorigenesis. Furthermore, Fusobacteria, a phylum with high abundance in healthy individuals, increased significantly in HCC patients and contribute to dysbiosis (Greten et al. 2023) (Fig. 11.1).

HCC patients' gut microbial composition has distinctive composition compared to healthy individuals at the lower the taxonomical level such as order and genus levels. The order *Clostridiales* showed overall decline in abundance in HCC condition, however genera like *Veillonella* and *Streptococcus* form the same order displayed prominent upsurge in the HCC condition. The few species which belong to *Veillonella* genus showed prognostic biomarker ability for liver cirrhosis conditions. Similarly, genus *Prevotella* from phylum Bacteroidetes also increased mostly in alcohol induced HCC. The significantly increased phylum Proteobacteria displayed genus-level increment, mainly in genus like *Escherichia*, *Klebsiella*, and *Enterobacter*. Nonetheless, phylum Firmicutes showed a decline in overall abundance; still, genus *Enterococcus* showed an increased overall ratio among all the genera from Firmicutes. On the other hand, bacterial genera recognized as health promoting through their anabolizing capability of short-chain fatty acids such as those belonging to *Lachnospiraceae* and *Ruminococcaceae* showed a decline in the abundance ratio in HCC. Additionally, probiotics bacterial genus like *Bifidobacterium*

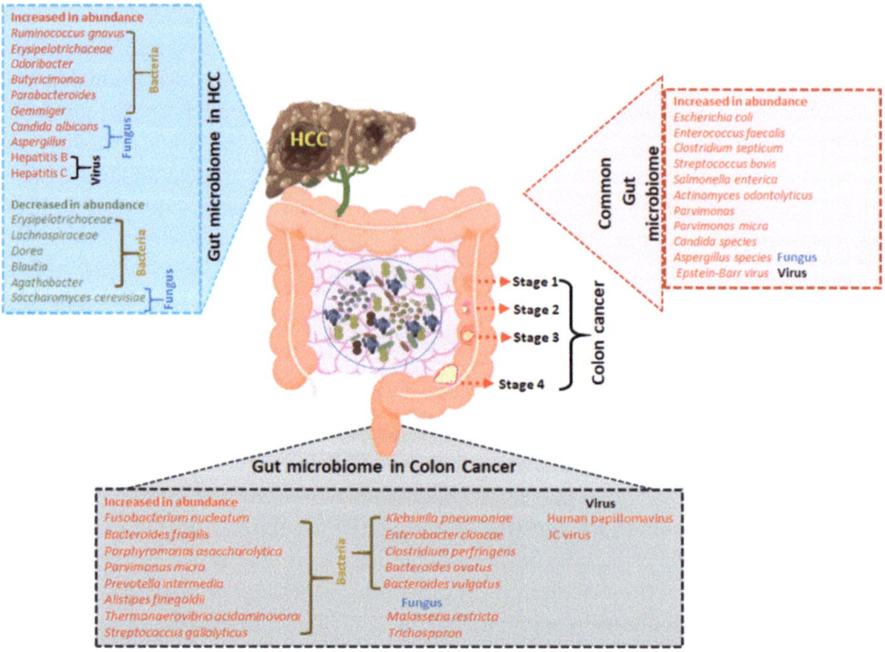

Fig. 11.1 Dysbiosis in the gut microbiome in colonic and hepatocellular carcinoma

often show lower abundance in HCC. These compositional alterations in gut bacteria are primarily responsible for changing the metabolic profile which includes major metabolic pathways that have central role in pathogenesis and progression of HCC. There are multitudes of metabolic and physiological pathways which are influenced through these metabolic alterations affected by change in microbial ecology and which may contribute to the pathogenesis and progression of HCC (Golonka and Vijay-Kumar 2021).

The gut fungal component or mycobiome, also experiences substantial changes in HCC. The gut mycobiome is dominated by the Ascomycota and Basidiomycota in normal healthy humans, with typically Ascomycota being more abundant. Nevertheless, there is often an overall increase in fungal abundance and diversity, in HCC patients. in HCC patients *Candida*, particularly *Candida albicans*, shows a marked increase when compared to healthy controls. Genera like *Aspergillus*, *Saccharomyces*, and *Malassezia* also increased in abundance in HCC. These changes in fungal composition are also connected with increased intestinal permeability and modulation of the immune response and are thought to contribute to the progression of liver cancer. The Basidiomycota-to-Ascomycota ratio is proposed as a potential indicator of disease progression in HCC, with an increased ratio associated with a higher risk of hospitalization. Moreover, some *Candida* species such as *Candida parapsilosis*, *Candida glabrata*, and *Candida tropicalis* have been found in higher abundance in HCC patients with viral hepatitis-related HCC.

The virome or gut viruses are less extensively studied compared to bacteria and fungi in HCC, but emerging studies suggest significant modifications in virome composition as well. In healthy individuals, bacteriophages constitute the predominant part of the gut virome, and are crucial in shaping the bacterial community structure. Frequently, an overall increase in viral diversity and abundance has been seen in HCC patients which include an increase in Pneumoviridae, Anelloviridae, and Circoviridae families. The virome composition in HCC can be influenced by multiple factors like gender, age, ethnicity, and race; however, these influencing factors are not as well characterized as for bacteria and fungi (Greten et al. 2023).

11.4 Gut Microbiome in Colon Cancer

Significant alterations are observed in the gut microbiome composition in colon cancer patients compared to healthy individuals, reflecting a state of dysbiosis that is both a consequence and a potential contributor to colorectal carcinogenesis. At the phylum level, colon cancer patients characteristically display a marked upsurge in the abundance of Proteobacteria and Fusobacteria, while the ratio of Firmicutes and Bacteroidetes usually shrinks. However, in colon cancer, the Firmicutes to Bacteroidetes ratio is often significantly changed, with a prominent reduction in Firmicutes and rise in Bacteroidetes. Precisely, the phylum Firmicutes drops up to 40–60% in colon cancer patients from 60% to 80% in healthy individuals, while Bacteroidetes increases to 30–50% from 20–40%. The Proteobacteria phylum, typically found less than 5% in the gut microbiota in healthy individuals, can grow up to 10–20% in colon cancer patients. Likewise, Fusobacteria, present in very low abundance (<1%) in healthy humans, can increase to 5–15% in colon cancer patients. This shift in gut microbiome composition influenced the pro-inflammatory and potentially oncogenic environment in the colon. Furthermore, Actinobacteria, which generally includes beneficial bacteria like *Bifidobacterium*, display a shrinking from 2–5% in healthy to less than 1% in colon cancer. These phylum-level alterations in composition responsible for changes in microbial diversity at different levels, and patients with colon cancer generally present reduced alpha diversity (within-sample diversity) and altered beta diversity (between-sample diversity) compared to healthy controls (Guo et al. 2022).

The gut microbiome in colon cancer showed further distinguishing changes at the lower taxonomical level including order and genus levels, Within the Firmicutes phylum, the order *Clostridiales* typically dominates healthy gut microbiomes, showing a reduction in colon cancer. Precisely, butyrate-producing bacteria which are considered as beneficial microbes and belong to the *Lachnospiraceae* and *Ruminococcaceae* families (from *Clostridiales* order) are often depleted. For instance, the genus *Faecalibacterium*, which accounts up to 5–10% of the total bacterial abundance in healthy individuals, decreases to below 1% in colon cancer patients. Contrariwise, bacteria considered as potentially pathogenic from Firmicutes, like species from the genera *Streptococcus* and *Enterococcus*, frequently increase in abundance. While the overall abundance of Bacteroidetes increases, there are specific shifts in the composition of explicit genera. The genus

Bacteroides, typically predominant in the healthy gut (comprising 10–30% of total bacteria), displayed modifications in particular species abundances rather than overall genus-level changes in colon cancer. *Bacteroides fragilis*, particularly known for producing enterotoxins, has been found to be increased in some colon cancer patients. Similarly, genera such as *Escherichia*, *Klebsiella*, and *Enterobacter* from the Proteobacteria phylum showed significant increases, particularly *Escherichia coli* (known as adherent-invasive strains) showed significant increase. *Escherichia coli* increases from approximately 1% abundance ratio in healthy individuals to 5–10% in colon cancer patients. Genus *Fusobacterium*, which belongs to the order Fusobacteriales, also showed the most remarkable increase in colon cancer. Particularly, *Fusobacterium nucleatum*, which typically exists in very low abundance in healthy individuals, shows an incredible increase in ratio and is one of the most dominant species in colon cancer, occasionally reaching up to 10–15% in the abundance of the total bacteria. This increase in *F. nucleatum* has been closely connected with tumorigenesis and severity progression with poor prognosis. Importantly, considered to be beneficial bacteria like *Bifidobacterium* from Actinobacteria phylum and *Lactobacillus* from Firmicutes phylum, which are crucial for maintaining the healthy gut and modulating the immune response, showed significant reduction in colon cancer patients. Collectively, these changes at genus-level contribute to a microenvironment that may promote colorectal carcinogenesis through various mechanisms, including increased inflammation, altered metabolite production, and disruption of the intestinal barrier integrity (Hanus et al. 2021).

Similarly, the mycobiome composition was also altered significantly in the colon cancer; however, few studies extensively studied the gut fungal ecology compared to the bacterial microbiome. The mycobiome is dominated by Ascomycota and Basidiomycota phyla in healthy individuals, with Ascomycota more dominant than the other. As in HCC, colon cancer patients also showed overall fungal diversity and abundance increments with a proportional shift at each taxonomical level, especially increments in Basidiomycota-to-Ascomycota ratio.

In terms of genus, *Candida* is the most abundant, due to remarkable increase in *Candida albicans*, in colon cancer patients compared to healthy individuals. In colon cancer, *Candida* species raised up to 1–5% in abundance of the total gut microbiota. Alongside, *Malassezia* and *Saccharomyces* also changed in their abundance in colon cancer. Species like *Malassezia restricta* exhibit increased abundance, and *Saccharomyces cerevisiae* often shows a decrease in abundance in colon cancer patients. Likewise, virome composition is also altered in colon cancer with the composition change in both bacteriophages and eukaryotic viruses. Bacteriophage families such as Microviridae and Siphoviridae increased and influenced the bacterial composition in colon cancer. Moreover, an increase in the abundance of human papillomavirus (HPV) and human polyomaviruses in the gut of colon cancer patients has been reported, indicating the plausible role of eukaryotic viruses in colorectal carcinogenesis.

Interestingly, several different types of patterns emerge when the gut microbiome composition in colon cancer is compared based on demographic factors such as gender, age, ethnicity, and race.

Gender-dependent variances have been observed in the colon cancer gut microbiome, with higher abundance of *Fusobacterium* and certain *Bacteroides* species in male compared to females' patients. Similarly, age-related changes aggravated, with studies showing more pronounced dysbiosis in older colon cancer patients. The changes in gut microbiome of colon cancer patients based on ethnicity and racial have also been reported which reflects the multilayer interplay between genetic, dietary, and environmental factors. Studies showed higher abundances of *Fusobacterium nucleatum* in African Americans colorectal cancer patients compared to Caucasians. Likewise, ethnicity-based difference in the species from certain *Bacteroides* and *Enterobacteriaceae* is observed in colon cancer patients. Therefore, these demographic associated gut microbiome variations in colon cancer highlight the individualistic approach while using the microbiome as a tool for diagnosis and/or for intervention. Importantly, the basis of these demographic differences in the gut microbiome composition is confounded by multiple factors like diet, lifestyle, environmental exposures, and genetic arrangement, which makes it more difficult to draw a distinguished microbiome profile based on gender, age, ethnicity, and race in colon cancer (Kong et al. 2023).

11.5 Gut-Linked Hepatocellular Carcinoma and Colon Cancer

The cross talk between gut and liver happens through the gut–liver axis which presents a critical anatomical and physiological connection among these two organs. About 70% of liver's blood supply comes through the portal vein caring nutrient-rich blood from the intestines. Similarly, bile acids secreted by the liver into the biliary system which are later released into the duodenum and help maintain gut microbiome composition and intestinal barrier function. This bidirectional communication establishes a complex interplay between the gut and liver. Even a small change in microbial composition or their metabolic products profile can have a huge impact on hepatic function and contribute to the pathogenesis of several liver diseases, as these microbiome-dependent metabolic changes impact on gut and could contribute to the colonic diseases pathogenesis (Lee et al. 2021).

In both conditions, HCC and colon cancer exhibit dysbiosis in gut the microbiome which has similarities and distinctions at multi-taxonomical levels. Starting with phylum, both cancer conditions display an amplified Proteobacteria-to-Fusobacteria abundance ratio, though the proportions of Firmicutes and Bacteroidetes are usually reduced. The species that increased in both the conditions include *Fusobacterium nucleatum*, *Escherichia coli*, *Bacteroides fragilis*, *Porphyromonas asaccharolytica*, and *Parvimonas micra*. These bacterial species are well known to activate the immune system by stimulating pro-inflammatory cells which can lead to oncogenic pathways. The *Fusobacterium nucleatum* is linked with immunosuppressive and cellular proliferative responses, whereas toxins producing Escherichia coli (colibactin, a genotoxin) apparently damage the DNA and attribute in the oncogenic. The species *Bacteroides fragilis* also release epithelial barrier-disrupting toxins which promote inflammation. Beside these species, bacteria that also increased in both HCC and colon cancer are *Peptostreptococcus anaerobius*, *Clostridium septicum*, *Streptococcus bovis*, *Salmonella enterica*, *Enterococcus faecalis*, *Atopobium parvulum*, *Actinomyces odontolyticus*, *Prevotella intermedia*, *Alistipes finegoldii*, and *Thermanaerovibrio acidaminovorans*. All these species contribute to the tumorigenic microenvironment through various mechanisms, like oxidative stress, genomic instability, and activation of oncogenic pathways.

The gut microbiome showed more similarities some increment in some specific genera like *Butyricimonas*, *Odoribacter*, *Parvimonas*, *Peptostreptococcus*, *Prevotella*, Alistipes, *Porphyromonas*, and these are involved in the both HCC and colon cancer generation. The bacterial species from these genera are known for their vital role in immune modulation, inflammation triggering and metabolic dysregulations. The species from *Prevotella* are connected with chronic inflammation and implicated to progression of both liver and colorectal cancers. Likewise, bacteria from *Alistipes* genus normally considered as health-promoting bacteria but possibly become pathogens in severe dysbiosis conditions and can contribute to tumorigenesis through the production of harmful metabolites. Additionally, bacteria from *Porphyromonas*, predominantly *Porphyromonas asaccharolytica*, induce cellular proliferation and stimulate tumorigenic pathways through the secretion of butyrate, a short-chain fatty acid with pro-oncogenic properties. There are many other bacterial species which also showed increased in abundance in both condition includes *Streptococcus gallolyticus*, *Klebsiella pneumoniae*, *Enterobacter cloacae*, *Clostridium perfringens*, *Mediterraneibacter gnavus*, *Bacteroides ovatus*, *Phocaeicola vulgatus*, *Escherichia coli*, *Lancefieldella parvula*, *Schaalia odontolytica*, *Fusobacterium nucleatum*, *Bacteroides fragilis*, *Porphyromonas asaccharolytica*, *Parvimonas micra*, *Prevotella intermedia*, *Alistipes finegoldii*, *Thermanaerovibrio acidaminovorans*. All these species contribute to the tumorigenic microenvironment through various pathophysiological mechanisms.

Other than bacterial species, the dysbiosis in mycobiome and virome also exhibits significant similarities in both HCC and colon cancer. Starting with gut fungi, in both liver and colon cancer conditions an overall increase in fungal abundance and diversity has been observed compared to healthy conditions. As reported in clinical studies Basidiomycota-to-Ascomycota ratio increased in both HCC and colon cancer patients compared to healthy controls. Predominantly, genus *Candida* and species *Candida albicans* constantly show a significant proliferation in both HCC and colon cancer. Other genera like *Malassezia* and *Saccharomyces* also showed the same alteration in both HCC and colon cancer. The gut viral component known as virome also altered in both HCC and colon cancer and showed some commonality like increased bacteriophage families like Microviridae and Siphoviridae. Also, an increase in the abundance of human papillomavirus (HPV) and human polyomaviruses in the gut of both HCC and colon cancer patients has been reported by some studies, proposing a probable role for these eukaryotic viruses in carcinogenesis.

Unlike the commonalities, many different patterns are also observed in HCC and colon cancer in terms of gut microbiome composition. *Akkermansia muciniphila*, recognized as a mucin-utilizing bacterium, is often found in reduced abundance in HCC but increased in colon cancer. Also, *Bifidobacterium longum*, considered beneficial, shows a decline in HCC but a rise in colon cancer. Other bacterial species including *Roseburia hominis*, *Blautia obeum*, *Coprococcus comes*, *Ruminococcus bromii*, *Faecalibacterium prausnitzii*, *Clostridium leptum*, *Lactobacillus rhamnosus*, *Lactobacillus plantarum*, *Lacticaseibacillus casei*, *Lactobacillus acidophilus*, *Streptococcus thermophilus*, *Streptococcus salivarius*, *Enterococcus faecium*, *Enterococcus durans*, *Enterococcus hirae*, *Finegoldia magna*, *Parvimonas micra*, *Blautia producta*, *Peptostreptococcus anaerobius*, and *Anaerococcus tetradius* showed different patterns of abundance in both HCC and colon cancer conditions.

11.6 Common Pathophysiological Pathways in HCC and Colon Cancer Regulated by Common Gut Bacteria

Dysbiosis in the gut microbial community plays a vital role in the pathogenesis cancers specifically for both HCC and colon cancer. Gut dysbiosis influences numerous physiological pathways and converts these normally regulated pathways to pathophysiological pathways and is thus attributed to the carcinogenic stage. The altered gut microbiome composition also changes its metabolic profile which damages intestinal barrier function, stimulates inflammation, and tempers immune responses, all of which can generate a pro-tumorigenic environment. These changes in the gut environment equally affect both local and distal organs through the gut–liver axis and systemic circulation. Dysbiosis-induced changes also interfere with several metabolic pathways like bile acid metabolism, short-chain fatty acid production, and bacterial by-product translocation and contribute to the initiation and progression of both HCC and colon cancer. By tempering key signaling pathways and cellular processes, gut dysbiosis creates a permissive environment for tumor development and growth in both the liver and colon. Here we discuss a few key pathways that are commonly operated in HCC and colon cancer and influenced by the microbial species that occur in a similar abundance in both conditions (Fig. 11.2) (Mignini et al. 2024; Muscolino et al. 2023; Qu et al. 2023; Rao et al. 2020).

1. *Chronic Inflammation and NF-κB Activation*

The gut dysbiosis in both conditions, the increment in pro-inflammatory stimulating microbes like *Enterobacteriaceae* and reduction in health-promoting bacteria like *Faecalibacterium prausnitzii*, leads to the production of inflammatory mediators. Lipopolysaccharide (LPS), a toxic component of gram-negative bacterial cell walls, activates the NF-κB pathway. Disruption of gut barrier function permits LPS to enter the circulation and reach the liver and colon, where it binds and activates the toll-like receptor 4 (TLR4), which is present on immune and epithelial cells. This TLR4 triggering activates NF-κB, which is a key transcription factor that regulates the expression of pro-inflammatory cytokines, chemokines, and anti-apoptotic genes. The constant NF-κB activation promotes a chronic inflammatory state, which

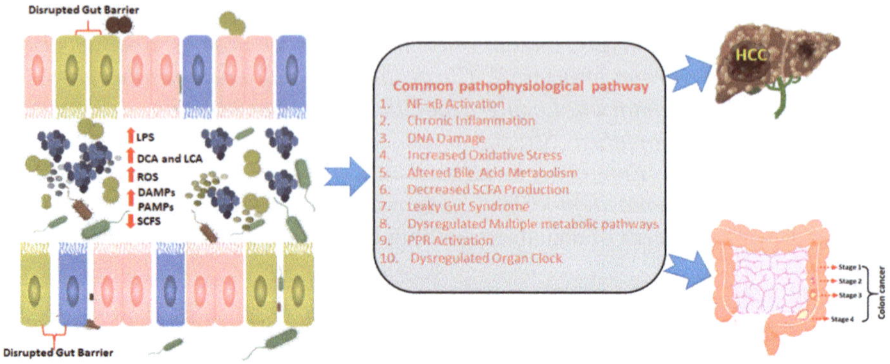

Fig. 11.2 Gut microbiome regulated common pathophysiological pathways involved in colonic and hepatocellular carcinoma

initiates tumorigenesis and progression into both HCC and colon cancer (Rebersek 2021; Sharma et al. 2024).

2. *Oxidative Stress and DNA Damage*

The compositional change in gut microbiome structure leads to an imbalance in the reactive oxygen species (ROS). Certain gut bacterial species, like *Enterococcus faecalis*, produce superoxide, which damages cellular components, also DNA. The increased number of sulfate-reducing microbes like *Desulfovibrio* species leads to higher hydrogen sulfide production, which is a genotoxic biochemical and induces DNA damage. The translocation of these bacteria and their metabolites from gut to liver and intestine to colon exacerbates oxidative stress. Also, the chronic inflammation induced by dysbiosis promotes immune cell-dependent ROS production, and creates a vicious cycle of oxidative stress and inflammation. The accumulation of oxidative DNA damage can lead to mutations in key oncogenes and tumor suppressor genes, promoting the initiation and progression of both HCC and colon cancer (Tataru and David 2020).

3. *Altered Bile Acid Metabolism*

The bile acid metabolism dysregulation is a crucial pathophysiological pathway commonly contributing to both HCC and colon cancer, which is significantly influenced by gut dysbiosis. The conversion from primary to secondary bile acids is mainly performed by the gut microbiome via deconjugation, dehydroxylation, and epimerization. In dysbiosis, bacterial species like *Clostridium scindens* and *Clostridium hylemonae* increased significantly which augmented the production of secondary bile acids like deoxycholic acid (DCA) and lithocholic acid (LCA). These hydrophobic secondary bile acids are known to have pro-carcinogenic effects in both the liver and colon. The higher DCA and LCA concentration induces DNA damage, activates inflammatory pathways, promotes cellular senescence, induces oxidative stress, and promotes cellular proliferation, thus contributing to cancer. Furthermore, altered bile acid profiles can activate the farnesoid X receptor (FXR) and G protein-coupled bile acid receptor 1 (TGR5), leading to changes in lipid and glucose metabolism, inflammation, and cell proliferation, which are also relevant to both HCC and colon cancer pathogenesis (Ternes et al. 2022).

4. *Short-Chain Fatty Acid Production Dysregulation*

The short-chain fatty acid (SCFA) production is a critical physiological pathway influenced by the gut that contributes to marinating good health. The SCFAs, primarily acetate, propionate, and butyrate, are produced by complex dietary fiber via fermentation by gut bacteria like *Faecalibacterium prausnitzii*, *Eubacterium rectale*, and *Roseburia* species. In a state of dysbiosis, SCFA-producing bacterial abundance reduces significantly, thus leading to decreased SCFA levels and having vital implications on both liver and colon health. The butyrate is crucial for maintaining intestinal barrier integrity, promoting apoptosis and regulating inflammation. The decline in butyrate production impairs the normal differentiation and apoptosis of colonocytes, promoting cellular proliferation and potentially leading to colon cancer. This decline made some anatomical alterations like decreasing the gut integrity and increasing the leaky gut which help bacterial metabolites and toxins to pass through from the gut and reach the liver. The dysregulation of SCFA production also affects the balance of T regulatory cells and effector T cells, potentially compromising anti-tumor immune responses in both the liver and colon (Mignini et al. 2024; Wen et al. 2023).

5. *Intestinal Barrier Integrity Disruption*

Breaking the intestinal barrier integrity commonly known as "leaky gut syndrome," considered as a crucial pathophysiological pathway common to both HCC and colon cancer, which is significantly influenced by gut dysbiosis. The primary role of the intestinal barrier is to restrain the translocation of harmful substances from the gut lumen to the circulation and it is achieved by the tight junctions that are formed between single layers of epithelial intestinal cells. With the overgrowth of pathobionts like *Escherichia coli* and the decline in beneficial bacteria like *Akkermansia muciniphila*, degradation of the mucus layer occurs and direct contact with epithelial damaging toxins produced by pathobionts is initiated. This synchronistic process increases the severity of leaky gut conditions and allows for free translocation of bacterial products, toxins, and pro-inflammatory molecules into the portal circulation. A certain strain of *Bacteroides fragilis* produces a metalloprotease, which is a toxin that cleaves E-cadherin, a key component of adherens junctions in the intestinal barrier. Through these leaky gut conditions, the translocated substances reach to the liver and trigger similar tumorigenic pathways as in colon cancer (Wong and Yu 2023).

6. *Tryptophan Metabolism Modulation*

Tryptophan is an essential amino acid that uses both host- and gut microbiome-produced enzymes for its metabolism and any distortion in tryptophan metabolism is associated with both HCC and colon cancer pathophysiology. The dysbiosis that occurs in HCC and colon cancer alters the compositional diversity of some bacterial species like *Clostridium* and *Lactobacillus* species, which are responsible for tryptophan catabolism. This compositional alteration leads to production of tryptophan metabolites, especially kynurenine pathway metabolites and indole derivatives. The indole and its derivatives are crucial for regulating the intestinal barrier function and reducing inflammation. A decline in bacteria-produced indole by-products can increase inflammation and intestinal permeability, which are the major and common pathologies in HCC and colon cancer generation and progression. Contrary to this, a rise in kynurenine production through altered microbial metabolism or by host indoleamine 2,3-dioxygenase (IDO) activation possibly suppresses the immune response toward HCC and colon cancer. Higher kynurenine promotes regulatory T cell differentiation and slows down the effector T cell responses, and attenuates the immune response for tumor progression in HCC and colon cancer. Also, certain tryptophan metabolites are aryl hydrocarbon receptor (AhR) ligands, which regulate immune responses, cell proliferation, and differentiation. Therefore, ill regulation of tryptophan metabolism dysregulates these cellular responses and promotes tumorigenesis, leading to HCC and colon cancer (Wen et al. 2023).

7. *Peculiar Activation of Pattern Recognition Receptors*

Pattern recognition receptors (PRRs) like NOD-like receptors (NLRs), Toll-like receptors (TLRs), are important for activation of innate immune response by early and quick sensing of gut microbiome associated molecular markers such as pathogen-associated molecular patterns (PAMPs) and damage-associated molecular patterns (DAMPs). The gut dysbiosis appeared in HCC and colon cancer has a higher number of certain bacteria that produce PAMPs, such as peptidoglycan and bacterial DNA, which activate TLR2 and TLR9, respectively. In addition, the dysbiosis state also increases the abundance of gram-negative bacteria those are responsible for elevated levels of lipopolysaccharide (LPS), which is a potent TLR4 agonist and activates the Kupffer and hepatic stellate cells promotes inflammation, fibrosis, and eventually hepatocarcinogenesis. Likewise, high LPS concentration in the colon also activates colonic TLRs and induces inflammation and distorted cell proliferation and leads to colonic carcinoma. This gut microbial dysbiosis occurs in HCC and colonic cancer induces a low-grade but persistent activation of PRRs that leads to constant release of pro-inflammatory cytokines and chemokines and creates a tumorigenic microenvironment (Zhou et al. 2020).

8. *Estrogen Metabolism Dysregulation*

The gut microbiome involvement in estrogens metabolism and circulation, is very crucial and irregularity in estrogen metabolism created by gut dysbiosis one of possible contributors of both HCC and colon cancer development. The gut microbial species like Clostridium, *Escherichia coli*, and Bacteroides responsible for producing β-glucuronidase enzymes, which is responsible for deconjugating estrogens in the gut and any alteration in these bacterial species abundance can directly influence estrogen metabolism and biological activity. Interestingly, in postmenopausal women a high risk of HCC and colon cancer is related to elevated circulatory estrogen due to an increase in β-glucuronidase activity produced by gut dysbiosis. Beside this, higher estrogen levels contribute to hepatocytic and colonic epithelial cell proliferation by being positive for the estrogen receptor, thus leading to tumor generation and possibly contributing to HCC and colon cancer. This gut microbiome-dependent estrogen dysregulation can affect multiple other tumorigenic pathways like imbalance in sex hormones and related metabolites which negatively influence the inflammatory and immune pathways helps in HCC and colon cancer pathogenesis. Also, estrogen is indispensable for expression of estrogen-responsive genes, which are involved in various pathways like cell cycle regulation, apoptosis, and angiogenesis, and any alteration in estrogen metabolism can influence these genes' expression and promote a pro-tumorigenic environment (Wong and Yu 2023).

9. *Polyamine Metabolism Alteration*

Polyamine metabolism is an important physiological pathway that is influenced by the gut microbiome and dysbiosis in the gut can negatively influence this pathway and contribute to HCC and colon cancer development. Polyamines such as spermidine, putrescine, and spermine are vital for normal cell growth, proliferation, and differentiation, whereas dysbiosis in the gut often creates the imbalance in polyamine-producing and -metabolizing bacterial species. The compositional change in certain bacterial species abundance like few *Enterococcus* species and *Escherichia coli* creates the imbalance in polyamine levels. This alteration in polyamines interfere with several physiological pathways like change in the rate of cell proliferation and apoptosis, oxidative stress gradient, impact on immunosuppressive activity, influence the anti-tumor immunity in both HCC and colon cancer. Also, polyamines can have an impact on metastasis genes and alteration in polyamines can instigate a pro-tumorigenic environment in both HCC and colon cancer (Rao et al. 2020).

10. *Disruption of Organ Circadian Rhythms*

The gut microbiome is related to body clock and organs circadian rhythms which is linked with microbiome dysbiosis intertwined with diurnal oscillations which can contribute to HCC and colon cancer development. The dysbiosis state leads to altered microbial metabolite production which is responsible for disrupting the host gene's expression related to circadian rhythms. The decline in short-chain fatty acid (SCFA) production induced by dysbiosis has a direct impact on the gene expression related to organs' circadian rhythms in both the liver and colon. Particularly, butyrate is a prime regulator of organs core clock genes like BMAL1 and PER2. The altered expression of these genes possibly has a profound impact on many pathways at the cellular level, such as metabolic pathways, DNA repair, cell cycle regulation, and immune function, all deeply associated with carcinogenesis. The microbiome-directed organ clock disruption altered the lipid and glucose metabolism, potentially connected with the development of fatty liver disease and subsequent HCC. Likewise, in the colon, disturbed circadian rhythms have an impact on colonic epithelial cell's proliferation and differentiation that potentially stimulates colorectal carcinogenesis. Along with this, the circadian rhythms dysregulation upset the DNA repair processes timing that potentially leading to increased accumulation of DNA damage and genomic instability in both the liver and colon. The altered circadian rhythms can also affect the production of hormones such as melatonin, which has anti-inflammatory and antioxidant properties, further contributing to the pro-tumorigenic environment in both HCC and colon cancer (Acharya and Thirunavukkarasu 2024; Zhou et al. 2020).

References

Acharya S, Thirunavukkarasu C (2024) Gut microbiome and cancer: from cancer development to therapeutics with a special focus on hepatocellular carcinoma. Front Biosci (Landmark Ed) 29(2):67. https://doi.org/10.31083/j.fbl2902067. PMID: 38420824

Bendriss G, Al-Ali D, Shafiq A, Laswi I, Mhaimeed N, Salameh M, Burney Z, Pillai K, Chaari A, Zakaria D, Yousri NA (2021) Targeting the gut microbiome: a brief report on the awareness, practice, and readiness to engage in clinical interventions in Qatar. Qatar Med J 2020(3):47. https://doi.org/10.5339/qmj.2020.47. PMID: 33598417; PMCID: PMC7863707

Cheng Y, Ling Z, Li L (2020) The intestinal microbiota and colorectal cancer. Front Immunol 11:615056. https://doi.org/10.3389/fimmu.2020.615056. PMID: 33329610; PMCID: PMC7734048

Dougherty MW, Jobin C (2023) Intestinal bacteria and colorectal cancer: etiology and treatment. Gut Microbes 15(1):2185028. https://doi.org/10.1080/19490976.2023.2185028. PMID: 36927206; PMCID: PMC10026918

Golonka RM, Vijay-Kumar M (2021) Atypical immunometabolism and metabolic reprogramming in liver cancer: deciphering the role of gut microbiome. Adv Cancer Res 149:171–255. https://doi.org/10.1016/bs.acr.2020.10.004. Epub 2020 Nov 19. PMID: 33579424

Greten TF, Villanueva A, Korangy F, Ruf B, Yarchoan M, Ma L, Ruppin E, Wang XW (2023) Biomarkers for immunotherapy of hepatocellular carcinoma. Nat Rev Clin Oncol 20(11):780–798. https://doi.org/10.1038/s41571-023-00816-4. Epub 2023 Sep 19. PMID: 37726418

Guo X, Okpara ES, Hu W, Yan C, Wang Y, Liang Q, Chiang JYL, Han S (2022) Interactive relationships between intestinal Flora and Bile acids. Int J Mol Sci 23(15):8343. https://doi.org/10.3390/ijms23158343. PMID: 35955473; PMCID: PMC9368770

Hanus M, Parada-Venegas D, Landskron G, Wielandt AM, Hurtado C, Alvarez K, Hermoso MA, López-Köstner F, De la Fuente M (2021) Immune system, microbiota, and microbial metabolites: the unresolved triad in colorectal cancer microenvironment. Front Immunol 12:612826. https://doi.org/10.3389/fimmu.2021.612826. PMID: 33841394; PMCID: PMC8033001

Kong C, Liang L, Liu G, Du L, Yang Y, Liu J, Shi D, Li X, Ma Y (2023) Integrated metagenomic and metabolomic analysis reveals distinct gut-microbiome-derived phenotypes in early-onset colorectal cancer. Gut 72(6):1129–1142. https://doi.org/10.1136/gutjnl-2022-327156. Epub 2022 Aug 11. PMID: 35953094

Lee KA, Luong MK, Shaw H, Nathan P, Bataille V, Spector TD (2021) The gut microbiome: what the oncologist ought to know. Br J Cancer 125(9):1197–1209. https://doi.org/10.1038/s41416-021-01467-x. Epub 2021 Jul 14. PMID: 34262150; PMCID: PMC8548300

Mignini I, Piccirilli G, Galasso L, Termite F, Esposto G, Ainora ME, Gasbarrini A, Zocco MA (2024) From the colon to the liver: how gut microbiota may influence colorectal cancer metastatic potential. J Clin Med 13(2):420. https://doi.org/10.3390/jcm13020420. PMID: 38256554; PMCID: PMC10815973

Muscolino P, Granata B, Omero F, De Pasquale C, Campana S, Calabrò A, D'Anna F, Drommi F, Pezzino G, Cavaliere R, Ferlazzo G, Silvestris N, Speranza D (2023) Potential predictive role of gut microbiota to immunotherapy in HCC patients: a brief review. Front Oncol 13:1247614. https://doi.org/10.3389/fonc.2023.1247614. PMID: 37692859; PMCID: PMC10486017

Qu R, Zhang Y, Ma Y, Zhou X, Sun L, Jiang C, Zhang Z, Fu W (2023) Role of the gut microbiota and its metabolites in tumorigenesis or development of colorectal cancer. Adv Sci (Weinh) 10(23):e2205563. https://doi.org/10.1002/advs.202205563. Epub 2023 Jun 1. PMID: 37263983; PMCID: PMC10427379

Rajapakse J, Khatiwada S, Akon AC, Yu KL, Shen S, Zekry A (2023) Unveiling the complex relationship between gut microbiota and liver cancer: opportunities for novel therapeutic interventions. Gut Microbes. 15(2):224003. https://doi.org/10.1080/19490976.2023.2240031. PMID: 37615334; PMCID: PMC10454000

Rao BC, Lou JM, Wang WJ, Li A, Cui GY, Yu ZJ, Ren ZG (2020) Human microbiome is a diagnostic biomarker in hepatocellular carcinoma. Hepatobiliary Pancreat Dis Int 19(2):109–115. https://doi.org/10.1016/j.hbpd.2020.01.003. Epub 2020 Jan 30. PMID: 32037278

Rebersek M (2021) Gut microbiome and its role in colorectal cancer. BMC Cancer 21(1):1325. https://doi.org/10.1186/s12885-021-09054-2. PMID: 34895176; PMCID: PMC8666072

Sharma SP, Gupta H, Kwon GH, Lee SY, Song SH, Kim JS, Park JH, Kim MJ, Yang DH, Park H, Won SM, Jeong JJ, Oh KK, Eom JA, Lee KJ, Yoon SJ, Ham YL, Baik GH, Kim DJ, Suk KT (2024) Gut microbiome and metabolome signatures in liver cirrhosis-related complications. Clin Mol Hepatol. https://doi.org/10.3350/cmh.2024.0349. Epub ahead of print. PMID: 39048520

Tataru CA, David MM (2020) Decoding the language of microbiomes using word-embedding techniques, and applications in inflammatory bowel disease. PLoS Comput Biol 16(5):e1007859. https://doi.org/10.1371/journal.pcbi.1007859. Erratum in: PLoS Comput Biol. 2020 Nov 2;16(11):e1008423. doi: 10.1371/journal.pcbi.1008423. PMID: 32365061; PMCID: PMC7244183

Ternes D, Tsenkova M, Pozdeev VI, Meyers M, Koncina E, Atatri S, Schmitz M, Karta J, Schmoetten M, Heinken A, Rodriguez F, Delbrouck C, Gaigneaux A, Ginolhac A, Nguyen TTD, Grandmougin L, Frachet-Bour A, Martin-Gallausiaux C, Pacheco M, Neuberger-Castillo L, Miranda P, Zuegel N, Ferrand JY, Gantenbein M, Sauter T, Slade DJ, Thiele I, Meiser J, Haan S, Wilmes P, Letellier E (2022) The gut microbial metabolite formate exacerbates colorectal cancer progression. Nat Metab 4(4):458–475. https://doi.org/10.1038/s42255-022-00558-0. Epub 2022 Apr 18. Erratum in: Nat Metab. 2023 Sep;5(9):1638. doi: 10.1038/s42255-023-00898-5. PMID: 35437333; PMCID: PMC9046088

Wen X, Ye X, Yang X, Jiang R, Qian C, Wang X (2023) The crosstalk between intestinal bacterial microbiota and immune cells in colorectal cancer progression. Clin Transl Oncol 25(3):620–632. https://doi.org/10.1007/s12094-022-02995-5. Epub 2022 Nov 14. PMID: 36376701

Wong CC, Yu J (2023) Gut microbiota in colorectal cancer development and therapy. Nat Rev Clin Oncol 20(7):429–452. https://doi.org/10.1038/s41571-023-00766-x. Epub 2023 May 11. PMID: 37169888

Zhou A, Tang L, Zeng S, Lei Y, Yang S, Tang B (2020) Gut microbiota: a new piece in understanding hepatocarcinogenesis. Cancer Lett 474:15–22. https://doi.org/10.1016/j.canlet.2020.01.002. Epub 2020 Jan 7. PMID: 31917160

Prebiotic Fortification: Millets in Building Resilience Against Colon Cancer

12

Maitri Chakraborty, Shrishti Madhan, and Arikketh Devi

Abstract

Millets are a significant cereal consumed globally, ranking sixth in importance. Additionally, components derived from millets exhibit crucial roles in disease prevention. Innumerable studies have proven that specific nutrients found in millet bran, particularly prebiotics, displayed remarkable inhibitory effects on the growth of colon cancer cells in humans. However, these effects were not observed in normal colon epithelial cells. The mechanistic investigations revealed that the selective anti-cancer properties of the different nutritional components of millets and other millet products were primarily achieved by increasing the accumulation of reactive oxygen species (ROS) in colon cancer cells compared to normal cells. Developing snack products with increased prebiotic content in millets and making them commercially available could be a valuable initiative in the fight against and prevention of colon cancer by regulating certain factors and pathways involved. Therefore, from the research studies conducted we can conclude that millet bran-derived products have therapeutic potential for managing colon cancer.

Keywords

Prebiotics · Millets · Colon cancer · Reactive oxygen species

M. Chakraborty · S. Madhan · A. Devi (✉)
Cancer and Stem Cell Biology Laboratory, Department of Genetic Engineering, School of Bioengineering, SRM Institute of Science and Technology, Kattankulathur, Chengalpattu District, Tamil Nadu, India
e-mail: devia@srmist.edu.in

S. Pathak et al. (eds.), *Microbiota and Dietary Mediators in Colon Cancer Prevention and Treatment*, https://doi.org/10.1007/978-981-96-0297-1_12

12.1 Introduction to Prebiotics and Colon Cancer

Prebiotics, a specialized group of biological nutrients, play a crucial role in maintaining a healthy gastrointestinal tract (Davani-Davari et al. 2019). These nutrients are primarily degraded by beneficial microflora such as Lactobacilli and Bifidobacteria. Upon ingestion, either as a food additive or as a supplement, prebiotics are metabolized by colonic microflora. This metabolic process results in the production of short-chain fatty acids, which are then released into the colon and absorbed into the bloodstream, contributing to various health benefits (Bamigbade et al. 2022). With increasing awareness of the health benefits linked to prebiotic consumption, their inclusion in daily diets has become more common. Colon cancer (CC) is the third leading cause of cancer-related deaths among adults worldwide. In 2022, an estimated 106,180 new cases of CC and 52,580 deaths were anticipated round the globe. Despite advancements in treatments, the five-year relative survival rate remains low at 64.7%. For patients with metastatic disease, the five-year survival rate drops to less than 20% (Fabregas et al. 2022). Various factors contribute to the development of colon cancer (Sawicki et al. 2021). A family history of colorectal cancer significantly raises the risk due to inherited genetic predispositions and lifestyle factors. Key determinants for future colon cancer risk include the generational distance of affected relatives, the age at which first-degree relatives were diagnosed, the number of family members with colorectal cancer, family history of other cancers (e.g., endometrial, ovarian, urinary tract, pancreatic), and personal cancer history. It is estimated that 2-8% of colorectal cancer cases are due to inherited syndromes (Sawicki et al. 2021). The most common hereditary syndromes predisposing individuals to colon cancer are hereditary nonpolyposis colorectal cancer (HNPCC), also known as Lynch syndrome, and familial adenomatous polyposis coli (FAP). HNPCC is caused by a mismatch repair error leading to an autosomal dominant mutation, with most cases linked to mutations in the MLH1(mutL homolog 1) and MSH2 (mutS homolog 2) genes (Steinke et al. 2013). Other gene mutations involved include MSH6 (mutS homolog 6), MLH3 (MutL homolog 3), TGFBR2 (transforming growth factor beta receptor 2), PMS1 (PMS homolog 1), and PMS2 (PMS homolog 2). Additionally, inflammatory bowel disease (IBD) ranks as the third highest risk condition for colon cancer development. IBD encompasses chronic, incurable diseases characterized by uncontrolled inflammation of the gastrointestinal tract due to immune system effects. Patients with type 2 diabetes are also at a higher risk of developing colorectal cancer, with their risk being two to three times greater than that of the nondiabetic population (Yu et al. 2022). The development of colorectal cancer in these patients is believed to be associated with increased insulin levels and accompanying inflammatory conditions. Recent studies have shown that changes in the composition and functionality of the normal gut microbiota can initiate, promote, and progress colon cancer (Siddiqui et al. 2022). Toxic metabolites produced by bacteria can cause DNA damage, disrupt cell cycles, stimulate immune responses, and disturb the function of the intestinal barrier (Ma et al. 2022). These disruptions create a microenvironment that favors the development of colon cancer. Understanding the relationship between gut microbiota and colon cancer is essential for developing new prevention and treatment strategies. This has led to researchers focusing on prebiotics as an important therapeutic angle.

12.2 How Can Our Diet Help Cancer Prevention?

The daily diet of an individual has proved to be a major determinant in the occurrence of colon cancer (Vernia et al. 2021). In 2015, the International Agency for Research on Cancer (IARC) classified processed meat as carcinogenic (Group 1) and red meat as probably carcinogenic (Group 2A) (Turesky 2018). It is recommended to limit red meat intake to less than three portions weekly, and processed meat, particularly those with nitrites, should be avoided entirely due to their carcinogenic nature. Mechanisms contributing to this association include the presence of growth-promoting components like heme and arginine, along with the formation of genotoxic substances during meat preservation, curing, and cooking processes. Additionally, protein fermentation products and lipid peroxidation play roles in colon cancer development. A high-fat diet appears to influence gut microbiota and bile acid metabolism, potentially promoting colon cancer. This dietary change may increase bile acids and alter their composition, as seen in studies comparing high-fat, low-fiber diets to low-fat, high-fiber diets. Secondary bile acids may then activate various cellular signaling pathways, promoting colon cancer cell proliferation. Additionally, a gut receptor called FXR (farnesoid X receptor) seems to play a role in regulating cell growth in the colon, and high-fat diets may disrupt its function (Ding et al. 2015). Studies show a strong link between high dietary fiber intake, particularly whole grains, and reduced risk of colon cancer. This protective effect may be due to various mechanisms. Fiber can dilute carcinogens, reduce its exposure to colonocytes by speeding up digestion, and alter gut bacteria to produce beneficial short-chain fatty acids like butyrate. Butyrate specifically has anti-inflammatory properties and may even induce apoptosis in colon cancer cells. While research suggests these benefits, some human trial results are still inconclusive. Gut bacteria play a complex and evolving role in colon cancer (Siddiqui et al. 2022). Certain microbes produce harmful toxins and metabolites that can directly damage colon cells or indirectly fuel cancer growth. However, the gut is not a one-way street. Colon cancer treatment can alter gut bacteria, and some bacterial products, like short-chain fatty acids, may protect against colon cancer. The picture is further muddied by the fact that many bacterial metabolites are produced by multiple species working together, making it difficult to isolate the effects of specific bacteria. Diet also plays a role, influencing not only the types of bacteria present but also their metabolite production.

Prebiotics are special, indigestible short-chain carbohydrates that act as a food source for the good bacteria residing in our gut. Unlike regular carbohydrates that break down in the small intestine, prebiotics travel all the way to the large intestine, or colon, where they encounter these friendly microbes. These prebiotics are often referred to as "resistant starches" or "nondigestible oligosaccharides" (Megur et al. 2022), and commonly include lactulose, inulin, and resistant starches found in chicory root, onions, and bananas. The benefit of prebiotics lies in their ability to selectively promote the growth and activity of certain beneficial gut bacteria. This targeted approach differs from dietary fiber, which can be fermented by a wider range of microbes in the gut, both beneficial and potentially harmful. By providing a targeted food source, prebiotics nurture the growth of these good bacteria, allowing them to thrive and compete with less desirable ones. Once they reach the colon, prebiotics are fermented by these friendly bacteria, producing short-chain fatty acids (SCFAs) as a by-product. These SCFAs, including acetic acid, propionic acid, and butyric acid, serve a dual purpose. They act as an energy source for the gut bacteria themselves and benefit the host organism by promoting gut health and potentially influencing metabolism. Prebiotics can be obtained naturally from various plant sources like chicory root, onions, and bananas. They can also be manufactured and are often incorporated into food products for their health benefits. In some cases, prebiotics are combined with probiotics, which are beneficial bacteria. This combination, known as a synbiotic (Pandey et al. 2015), offers a potentially synergistic effect, providing both the food source (prebiotic) and the friendly bacteria (probiotic) to further enhance gut health.

Beta-glucan, found in oats, barley, mushrooms, and algae, is one such prebiotic fiber (Carlson et al. 2018). Its impact on gut health can vary depending on its structure. Fructan fibers come in two forms: inulin, a natural fiber found in various plants, and oligofructose, a manufactured version. Both promote the growth of beneficial gut bacteria. Galacto-oligosaccharides (GOSs), another prebiotic, are derived from lactose and consists of chains of galactose molecules linked to a single glucose molecule. The length and purity of these chains influence their prebiotic effects. Isomalto-oligosaccharides (IOSs), produced by breaking down cornstarch with enzymes, are well-tolerated fibers that may promote gut health at doses of 5-10 grams daily. Guar gum, a gel-forming fiber derived from guar beans and commonly used as a food additive, is not broken down in the small intestine but can be fermented as an effective product by the gut bacteria. Lactulose, a man-made sugar not found naturally in food, acts as a prebiotic, promoting positive changes in the gut microbiome through bacterial fermentation. Resistant starches (RSs) are starches that resist digestion in the small intestine, and reach the colon for bacterial fermentation. Maltodextrin, another fermentable fiber, can also be broken down by gut bacteria. Some RSs occur naturally, while others are manufactured. Additionally, xylo-oligosaccharides (XOSs) and arabino-oligosaccharides (AOSs) are prebiotics found in certain dairy products, cereals, and beverages. Well-tolerated at doses up to 12 grams daily, XOSs are particularly popular in Japan, where they are recognized as a food ingredient with specific health benefits.

12.3 All About Millets

Millets, tiny powerhouses of nutrition, boast a rich history as a staple food. These gluten-free grains are classified into two main categories: major millets (sorghum, pearl millet, finger millet) and minor millets (foxtail, proso, kodo millet) (Nigam et al. 2024). Millets surpass common cereals like wheat and rice in their content of protein, fiber, essential amino acids, and antioxidants. Recent research suggests exciting health benefits associated with millets, including better blood sugar control due to the presence of high resistant starch (RS) and slowly digestible starch (SDS), which lead to lower blood sugar spikes after meals. Additionally, millets may aid in weight management and reduce the risk of heart diseases and many more health benefits which is schematically shown in Fig. 12.1. The unique protein composition of millets, with higher levels of essential amino acids compared to other cereals and containing prolamin for improved digestibility, further adds to their nutritional value.

Cultivated since the dawn of civilization, millets have a rich history (Veerabadhran et al. 2023). Evidence suggests their consumption in the Indus Valley civilization and by early inhabitants of the Ganges river plains. These versatile grains were not just a staple food but also held significance in traditional medicine. In ancient India, millets like finger millet, sorghum, and pearl millet were prominent crops, and the Vedas and Puranas mention their use in rituals, food, and animal fodder. The millet grains available now are unhulled, polished, unpolished, depending on the requirement in the markets and can be identified easily by their appearance. Ayurveda, the traditional Indian medical system, recognizes the diverse qualities of millets, describing them as easy to digest and beneficial for specific conditions. Millets were also valued in China, with the ruler of the Shang Dynasty even titled "The ruler of Millet." Traditional Chinese medicine considers millets as cooling and diuretic agents, perfect for summer consumption. Archaeological evidence shows extensive cultivation of foxtail and proso millets in China. Millets were a staple food in the Middle East and Northern Africa as well, finding mention in the Hebrew Bible and possibly being grown in the Hanging Gardens of Babylon. With their rich history

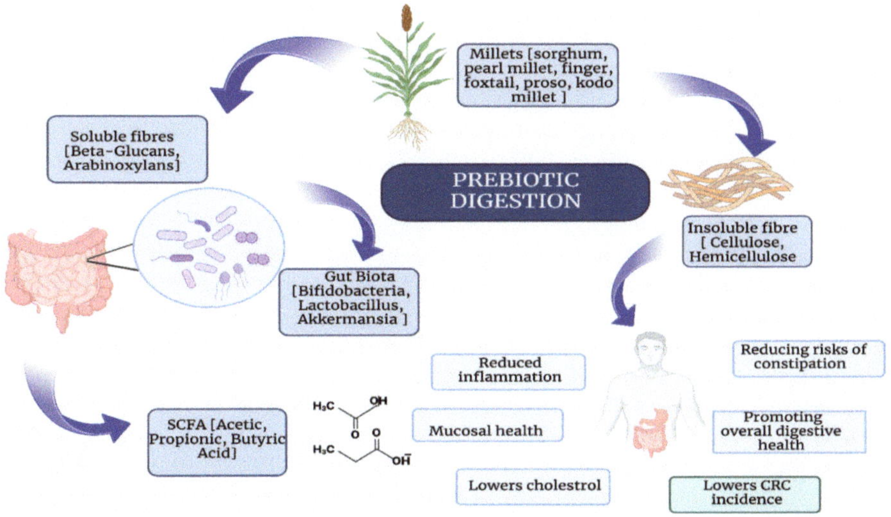

Fig. 12.1 Illustration of fiber digestion and fermentation by gut microbiota, producing SCFAs that improve gut health and reduce disease risks

and promising health benefits linked to their high nutrient content, millets continue to be a valuable food source and offers a unique and nutritious addition to modern diets (Jacob et al. 2024) (Fig. 12.1).

Nutritionally, millets have proved to be a gold mine for dietary fiber, antioxidants, proteins, and carbohydrates (Hassan et al. 2021). However, the nutritional composition for each millet varies considerably. A detailed view of the few commonly used millets is given in Table 12.1.

Millets have shown considerable promise in managing type 2 diabetes mellitus (T2DM) (Anitha et al. 2021). This disease is characterized by elevated blood sugar levels due to insufficient insulin action or production. Millets are ideal for diabetes management because they contain low digestible starch fractions, which slow down carbohydrate digestion, and potent phenolic compounds, which limit the breakdown of carbohydrates. These grains also help reduce reactive oxygen species (ROS), enhance the abundance of probiotic bacteria, and regulate various signaling pathways. For instance, phenolic compounds in millets help protect pancreatic β cells by mitigating oxidative stress. Studies have shown that β-glucan from *Eleusine coracana* (finger millet) seeds can inhibit α-amylase and α-glucosidase, enzymes involved in starch digestion, thereby demonstrating antidiabetic activity. Similarly, phenolic compounds obtained from finger millet and sorghum inhibit these enzymes, effectively reducing postprandial blood glucose levels. Animal studies have supported these findings. Diabetic rats fed with kodo and finger millet-based diets exhibited reduced blood glucose levels and lowered lipid peroxidation, indicating reduced oxidative stress. The hypoglycemic effect of sorghum and foxtail millet extracts was comparable to that of antidiabetic medications. In these studies,

Table 12.1 Nutritional composition of some common millets used around the world

Nutrients	Finger millet	Foxtail millet	Pearl millet	Barnyard millet	Kodo millet	Little millet	Proso millet
Carbohydrates (g)	72.6	63.2	67.0	55.0	66.6	60.9	80.1
Proteins (g)	7.7	11.2	11.8	11.0	9.8	9.7	11.8
Fats (g)	1.5	4.0	4.8	5.2	3.6	5.2	4.9
Fibers (g)	3.6	6.7	2.3	13.6	5.2	7.6	0.7
Ash (g)	2.6	3.3	2.2	4.5	3.3	5.4	NA
Phosphorous (mg)	283	258	285	293	189	207	277
Calcium (mg)	344	31	42	20	27	17	14
Iron (mg)	3.9	1.1	2.8	15	0.6	9.3	3.8
Zinc (mg)	2.8	1.1	1.7	0.4	1.2	1.2	1.7
Vitamin B1 (mg)	0.42	0.59	0.27	0.33	0.15	0.3	0.41
Vitamin B2 (mg)	0.19	0.11	0.15	0.11	0.09	0.09	0.28
Vitamin B3 (mg)	1.1	3.2	0.89	4.2	2.1	3.2	4.6

All values except protein are expressed on a dry weight basis millet-based diets lowered glycemic responses due to their high dietary fiber, peptides, and specific starch content. Moreover, clinical trials have indicated that increased millet consumption can reduce fasting blood glucose levels and improve glycemic control in T2DM (type 2 diabetes mellitus) patients, leading to better HbA1c levels and reduced hyperinsulinemia. Millets also provide significant cardiovascular benefits. They are rich in sterols and policosanols, which prevent cholesterol synthesis. For example, studies on hamsters have shown that sorghum reduces non-HDL cholesterol levels. Millet bran oil (MBO) has been found to improve lipid metabolism in obese mice by reducing hepatic lipid accumulation, oxidative stress, and increasing beneficial gut bacteria like *Akkermansia*. Phenolic extracts from kodo millet effectively inhibit LDL cholesterol oxidation, which is crucial for preventing atherosclerosis and other cardiovascular diseases. Additionally, the high fiber content in sorghum binds bile acids in the small intestine, preventing their reabsorption and thereby lowering cholesterol levels. The antihypertensive properties of millets further highlight their health benefits. Millets are rich in dietary fiber, proteins, minerals, and phytochemicals that help manage hypertension. Studies have shown that foxtail millet protein hydrolysates can reduce blood pressure in spontaneously hypertensive rats by lowering serum ACE (angiotensin I converting enzyme) and angiotensin II levels (Chen et al. 2017). Ethanol extracts from finger millet have also demonstrated antihypertensive effects by modulating the renin-angiotensin system. The details of various millets, the phytochemicals found in them and their health benefits as well has been described in Table 12.2 below. In human studies, diets rich in whole foxtail millets have significantly reduced both systolic and diastolic blood pressure in mildly hypertensive patients. Millets also support thyroid function and help combat obesity (Gaitan et al. 1989). Sorghum is a rich source of manganese, essential for thyroid hormone homeostasis and regulation of fat metabolism. Sorghum's high amylose content forms resistant starch, beneficial for obese and diabetic individuals as it reduces caloric intake by forming complexes with starch. These resistant starches are not digested in the small intestine; they get absorbed there and are fermented in the large intestine, thus delivering the health benefits of dietary fiber. Overall, millets offer multifaceted health benefits, making them a vital dietary staple for promoting global health.

12.4 Millets as Prebiotics

Processing methods can modify fiber structure, impacting their digestibility and bioavailability. Fermentation of soluble fibers converts them into smaller compounds that colon cells absorb and use for energy. The composition of the gut microbiota significantly influences digestion and, consequently, the digestibility and fermentation of these fibers. Insoluble fibers facilitate proper gut transit and overall digestive health. When metabolized in the colon by gut microbiota, dietary fibers act as prebiotics (Portincasa et al. 2022), producing short-chain fatty acids (SCFAs) like acetic, propionic, and butyric acids. Gut microbiotas break down non-digestible millet fibers into monosaccharides, which is converted by the bacteria

Table 12.2 Major millets and their health benefits

S. no.	Millet	Botanical name	Important bioactive compounds	Health benefits
1	Sorghum	*Sorghum bicolor*	Lutein, Zeaxanthin, β-Carotene, α-Tocopherol, α-Tocotrienol, Spermidine, Spermine, Putrescine, Cadaverine, β-Sitosterol, Campesterol, Stigmasterol	Antioxidant activity, reduces risk of cancer, reduces caloric availability, reduces obesity, Antidiabetic
2	Pearl millet	*Pennisetum glaucum*	p-Hydroxybenzoic, protocatechuic, vanillic, caffeic, p-coumaric, cinnamic, sinapic, trans-ferulic, apigenin, myricetin	Antidiabetic, prevents gallstones, high satiety value, reduces obesity, anti-aging, anti-allergic
3	Barnyard millet	*Echinochloa esculenta*	2,3,4-trihydroxybutyric acid, d-ribose, ribonic acid, pentanedioic acid, pentitol, D-mannitol, alpha. glycerophosphoric acid, eicosanoic acid, stigmasterol and beta-sitosterol, malic acid, glutamine	Antidiabetic, helps combat cardiovascular diseases, aids in skin problems, anticancer, supports kidney health, controls blood sugar
4	Finger millet	*Eleusine coracana*	Gallic, p-hydroxybenzoic, protocatechuic, syringic, gentisic, vanillic, caffeic, p-coumaric, cinnamic, sinapic, trans-ferulic, catechin, gallocatechin	Beneficial for anxiety and insomnia, lowers blood pressure, antioxidant, helps combat liver disorders, helps bone health, anticancer
5	Foxtail	*Setaria italica*	Gallic, p-hydroxybenzoic, protocatechuic, syringic, gentisic, vanillic, caffeic, p-coumaric, sinapic, trans-ferulic	Maintains nervous system health, hair growth, antidiabetic, supports bone growth, anticancer, enhances skin
6	Proso millet	*Panicum miliaceum*	Ferulic acid, sinapic acid, chlorogenic acid, p-coumaric acid, syringic acid, vanillic acid, 4-hydroxybenzoic acid, isoferulic acid, p-hydroxycinnamic acid, cryptochlorogenic acid, neochlorogenic acid	Antioxidant, supports respiratory health, anticancer, prevents cardiovascular disease
7	Kodo millet	*Paspalum scrobiculatum*	Gallic, p-hydroxybenzoic, protocatechuic, syringic, vanillic chlorogenic, caffeic, p-coumaric, sinapic, trans-ferulic, cinnamic, kaempferol, apigenin	Antioxidant, Antidiabetic, improves kidney health
8	Little millet	*Panicum sumatrense*	Gallic, protocatechuic, syringic, gentisic, vanillic, caffeic, p-coumaric, sinapic, trans-ferulic, apigenin	Reduces cholesterol, boosts immunity, aids respiratory conditions

into SCFAs. These SCFAs reduce gut permeability, act as signaling molecules to induce satiety, stimulate insulin secretion, and lower glucose levels. Butyric acid, in particular, promotes mucosal healing by increasing epithelial cell migration and reducing colonic inflammation. SCFAs also assist to maintain an anaerobic gut environment, limiting harmful microorganisms like *E. coli* and *Salmonella.*

Animal studies have demonstrated that millet-based supplements can restore gut microbial diversity in acutely malnourished pigs and reduce harmful bacteria (Singh et al. 2023). Consumption of finger millet has been linked to an increased abundance of beneficial bacteria like *Pediococcus* and a decrease in pathogenic bacteria. SCFA production from millet fiber fermentation is associated with a reduced risk of metabolic disorders, such as obesity and type 2 diabetes. Regular consumption of millet fibers enhances gastrointestinal function, improves nutrient absorption, and reduces the risk of digestive disorders. High-fiber millet diets are linked to a decreased risk of colorectal cancer. In experimental studies, foxtail millet consumption has shown a reduction in colonic inflammation and the risk of colitis-associated colorectal cancer in rats. This is achieved through microbial metabolites that activate aryl hydrocarbon and G-protein-coupled receptors while inhibiting STAT3 (signal transducer and activator of transcription 3) phosphorylation. Millets are rich in dietary fibers, with varying fiber content across different types. Soluble fibers in millets, such as beta-glucans and arabinoxylans, have gel-forming properties that increase food viscosity, inhibit macronutrient absorption, lower dyslipidemia, and reduce postprandial glucose response. These fibers are metabolized by gut microflora to produce SCFAs, which decrease bile acid absorption in the small intestine and increase bile acid excretion, thereby lowering blood cholesterol levels. Insoluble fibers, such as cellulose and hemicellulose, add bulk to the stool, enhance fecal volume, and improve intestinal transit by increasing water-holding capacity (Shivam 2024). They are more susceptible to colonic fermentation, which benefits insulin sensitivity. Prebiotic-resistant starch in millets increases beneficial microorganisms like *Bifidobacterium*, *Lactobacillus*, and *Akkermansia*, contributing to antidiabetic and anti-obesity properties. The bran fraction of millet contains a significant amount of dietary fiber, which is often lost during decortication or dehulling. Removing the bran significantly reduces the fiber content, with dehulling over 30% leading to considerable nutritional fiber loss.

Peptic ulcer disease, or gastric ulcers, is a common gastrointestinal disorder resulting from an imbalance between defensive and aggressive factors affecting the gastric mucosa. Millets, particularly foxtail millet, have been reported to protect the gastric mucosa against ulceration. Traditional Chinese medicine has long used foxtail millet to treat conditions such as spleen and stomach heat, stomach reflux, vomiting, and reduced food intake with abdominal distention. Bound polyphenols from foxtail millet bran exhibit anti-inflammatory effects in LPS-induced HT-29 cells and nude mice. Additionally, foxtail millet demonstrates anti-ulcer activity by reducing plasma and mucosal TBARS (thiobarbituric acid reactive substances) levels and increasing gastric non-protein sulfhydryl (NPSH) levels and digestive enzyme activities. Studies show that pretreatment with millet diets significantly prevented the development of gastric mucosal lesions in ulcerated rats by depleting

NPSH levels. Millet diets offer ulcer protection by decreasing ulcer index and TBARS values and increasing NPSH concentrations. Foxtail millet diet has shown preventive and gastroprotective effects in experimental gastric mucosal lesions in rats, protecting against acute gastric mucosal injury and displaying extensive antioxidant effects (Lin et al. 2020). Moreover, heat-treated foxtail millet can alleviate the effects of non-alcoholic fatty liver disease (NAFLD) by reducing hepatic total cholesterol and triglyceride contents and improving NAFLD-related gut microbiota dysbiosis.

Besides their function as probiotics, millets also possess antioxidant properties that help prevent carcinogenesis by halting degenerative processes in the body. Kodo millet phenolic extracts, rich in ferulic and p-coumaric acids, have been effective in inhibiting hydroxyl and peroxyl radicals and reducing the proliferation of HT-29 human colorectal adenocarcinoma cells by 75-100%. Foxtail millet extracts from two varieties, Jingu28 and Jingu34, have demonstrated potential in inhibiting the growth of human breast (MDA) and liver cancer (HepG2) cells in culture. Additionally, foxtail millet bran-derived bound polyphenol (BPIS) has induced apoptosis in human colorectal cancer HCT-116 cells and in HCT-116-bearing nude mice through enhanced ROS production. BPIS triggers ROS generation in HCT-116 cells, leading to ROS-induced apoptosis. A peroxidase enzyme from foxtail millet bran has been found to inhibit cell migration in human colon cancer cells by antagonizing epithelial-mesenchymal transition (EMT) *via* the STAT3 signaling pathway. BPIS also exhibits antiproliferative activity in the colon cancer cell line HT-29 by upregulating miR-149 expression and inhibiting tumor aerobic glycolysis (Shi et al. 2017). Bioactive compounds like vanillin extracted from proso and barnyard millets inhibit cell proliferation and induce apoptosis in colon cancer cell lines (Ramadoss and Sivalingam 2021). Bran from certain sorghum lines, especially black sorghum, contains up to 10 mg/g of 3-deoxyanthocyanidins, which show strong antiproliferative activities against various human cancer cell lines, with potential comparable to quercetin, one of the most potent antioxidant molecules.

12.5 Mechanism of Action

Certain bacteria found in the gut microbiota are more prevalent in the microenvironment of colon cancer, particularly those belonging to the Clostridiales and Bacillales families and more than 100 different types of tumor tissues have been reported containing these bacteria. The majority of bacteria in these two families have the ability to form stress-resistant spores, making them stable and easily storable. In addition, certain probiotic products based on *C. butyricum* and *Bacillus coagulans* have been approved for human use in China and Japan, facilitating the transition from pro-tumor to anti-tumor types. Studies have shown that there are more bacteria from the genus *Clostridium* in cancer tissues than in precancerous tissues (Zheng et al. 2020). A comprehensive list of gut bacteria as well as their specific actions in colorectal cancer cells is available. *Fusobacterium nucleatum*, for instance, promotes myeloid infiltration of intestinal tumors in ApcMin/+ mice and increases the expression of pro-inflammatory genes such as Ptgs2 (prostaglandin-endoperoxide synthase 2) (COX-2) (cyclooxygenase-2), Scyb1 (SCY1 like pseudokinase 1) (IL8) (interleukin-8), Il6 (interleukin-6), Tnf (TNFα) (tumor necrosis factor alpha), and Mmp3 (matrix metallopeptidase 3). This bacterium has also been found to be more prevalent in the colonic mucosa, tissues, and feces of patients with adenomas or adenocarcinomas compared to healthy individuals. Furthermore, its presence has been consistent across different stages of CRCs (Colorectal Cancers) and diverse populations. *Fusobacterium nucleatum* has a complex mechanism for promoting a pro-tumorigenic microenvironment, which has been examined through mouse model experiments by analyzing microRNA (miRNA) expression patterns. It upregulates inflammatory factors and microRNA 21 through toll-like receptors, affecting the downstream targets of miR21, including oncoprotein RAS P21 protein activator 1 (RASA1) and tumor suppressor programmed cell death protein 4 (Pdcd4). Experimental evidence also suggests that aberrant expression of RASA1 in CRC leads to the activation of the RAS-mitogen-activated protein kinase (MAPK) cascade. Therefore, the relationship between miR21 upregulation and MAPK cascade activation is strongly connected, which could play a role in CRC development. *F. nucleatum* can act as a biomarker for this type of cancer (Yang et al. 2017). Sgg (*Streptococcus gallolyticus* subspecies *gallolyticus*) is believed to be linked to CRC and may possess pathogenic characteristics that contribute to the development and spread of colon cancer. Initially, Sgg attaches to the colon's epithelial surfaces and then binds to heparan sulfate proteoglycans on malignant cells or components of the extracellular matrix, such as collagen, using its histone-like protein A. Its pili possesses phase-variable expression, which facilitates the formation of colonies around the tissue and evasion of host immune responses. Once bound to malignant cells, Sgg protein stimulates the production of inflammatory cytokines, which in turn mediates the production of free radicals and expression of NOS2 (nitric oxide synthase 2), leading to the production of NO (nitric oxide) and ultimately inducing cancer in the colon.

Multifactorial diseases are often caused by a combination of various factors, including genetic predispositions, environmental influences, and dietary habits. Currently, a person's lifestyle is considered the most significant factor in either initiating or managing any disease. The gut microbiota can contribute to the development of colon cancer by promoting unhealthy lifestyle choices. By focusing on dietary patterns, it is possible to prevent the gut microbiota from causing cancer or tumors in the colon. A Western-style diet, which is rich in fat, sucrose, red and processed meat, and low in fiber, can alter gut microbiota. On the other hand, consuming whole grains, cereals, and dairy products can decrease the risk of developing colorectal cancer (CRC), as stated by experts from the World Cancer Research Fund and the American Institute for Cancer Research in 2017 (https://www.wcrf.org/dietandcancer/colorectal-cancer). Probiotics, which are rich in fibers from cereals like millet, can help combat CRC when consumed. Studies have shown that including nondigestible carbohydrates in the diet, known as prebiotics, can improve the gut microbiota. Prebiotics are defined as "substrates that are selectively utilized by host microorganisms, conferring health benefits to the host" (Gibson et al. 2017).

The concept of using prebiotics was first introduced by Glenn Gibson and Marcel Roberfroid in 1995 (Gibson and Roberfroid 1995). Prebiotics are fermented by probiotic microorganisms, resulting in the production of SCFAs (acetate, propionate, and butyrate) (Fong et al. 2020). In fact, prebiotic supplementation has been shown to decrease pathogen colonization in human studies, potentially through the action of SCFA. Galacto-oligosaccharides (GOS) and fructo-oligosaccharides (FOS) are commonly used prebiotics that occurs naturally in chicory roots, onions, garlic, asparagus, bananas, and certain cereals, such as millet (Pothuraju et al. 2021). Research on mice have shown that mice fed a diet rich in low fiber, high fat, and high glucose developed colorectal cancer (CRC) over time, whereas mice given a prebiotic-rich diet experienced altered gene expression for CRC and remained free from cancer in the gut, particularly the colon. This could be due to the fact that prebiotic supplementation reduced pathogen colonization in human studies, possibly through the action of short-chain fatty acids (SCFAs). GOS is not metabolized in the small intestine because of its β (1–4) glycosidic linkage and reaches the colon, where it is degraded by the gut microbiota to produce SCFAs, lactate, and several gases (Ohtsuka et al. 1991). On the other hand, healthy human subjects consuming FOS (12.5 g/day) experienced an increase in fecal bifidobacteria with no effect on fecal total anaerobes, pH, nitroreductase, azoreductase, and β-glucuronidase activities, as well as unchanged concentrations of bile acids and neutral sterols, which are involved in CRC progression (Bouhnik et al. 1996). Therefore, adding GOS/FOS to food could be a valuable strategy for treating intestinal disorders. Probiotics are live microorganisms that, when given in adequate amounts, provide health benefits to the host (Hill et al. 2014). Studies have shown that probiotic supplementation in the diet of patients with inflammatory bowel syndrome can reduce its effects by strengthening the mucosal barrier in the affected area (Caballero-Franco et al. 2007).

12.6 Evidence-Based Research on Millets and Colon Cancer Prevention

There are many ongoing research studies considering millet as a potential source in inhibiting colon cancer and preventing it further in the future for those who are prone to developing this deadly condition later in life, as well as for those who have already suffered and recovered.

Various studies to investigate millet's potential in preventing colon cancer have been carried out by many researchers and scientists round the globe, particularly for those at high risk of developing this life-threatening condition later in life or for those who have already been diagnosed and received treatment. In one such study, researchers utilized animal models to show that consuming barley or whole grain oats could reduce the risk of colorectal cancer. However, more conclusive evidence is needed to confirm this finding. On the other hand, foxtail millet (*Setaria italica*), a widely cultivated cereal grain in Asia and Africa and the most commonly consumed coarse cereal in China, surpasses rice in terms of nutritional value (Sharma and Niranjan 2017). Millet contains more protein, fat, and dietary fiber than rice and less starch. It is also an abundant source of tryptophan, an essential amino acid that must be obtained through dietary intake (Islam et al. 2017). Through experiments on mouse models, it was discovered that a diet containing millet does not result in the development of tumors in the colon or any other part of the gut system, whereas a diet containing rice does lead to the development of colorectal cancer (CRC) (Zhang et al. 2020). A comparison of the nutrient content of millet and rice revealed that millet contains similar nutrients but in varying quantities. For instance, millet has 1.2% more fiber and 1.35 times more tryptophan than rice. Additionally, millet contains the highest amount of phenolics and the lowest amount of starch compared to rice, which may contribute to millet's ability to combat or prevent colon cancer. Colon cancer, also known as CRC, is typically caused by aggravated colitis, which involves inflammation.

To determine the effectiveness of a millet-based diet, histopathological tests, including the measurement of markers for inflammation such as MPO, MCP-1, C-peptide, IFN-γ, and IL-1β, were conducted on mice that were sacrificed after consuming either a millet or rice diet. The results showed that mice consuming millet diet had lower levels and counts of these inflammation markers compared to those consuming rice diet. Furthermore, the serum LPS (lipopolysaccharide) levels, which were significantly higher in the other group of mice and can induce inflammation, were reduced in mice consuming a millet diet. A histological examination was also performed to assess the internal damage and the impact of dietary interventions on the colon's morphology. The colons of mice in the normal group displayed normal pathological morphology without any inflammatory damage. Unlike the mice in the negative control group and the group fed with rice diet, those fed with millet diet did not exhibit any signs of inflammatory infiltration, depletion, or distortion of the colonic crypts. Additionally, the administration of millet to these groups resulted in a decrease in the colonic mRNA levels of pro-inflammatory cytokines TNF-α, IL-6, IL-17 (interleukin-17), and COX-2. On the other hand, the mice fed a millet diet had normal levels of these cytokines. Furthermore, the millet diet promoted the expression of IL-22, which is essential for maintaining intestinal integrity. The millet-treated mice also showed a significant decrease in the phosphorylation of STAT3 (Zhang et al. 2020), which regulates cell growth, proliferation, and survival (Fig. 12.2). Additionally, the expression of Bcl-2 (B cell lymphoma-2), an anti-apoptosis factor, and PCNA (Proliferating cell nuclear antigen), a cell-proliferation factor (Fig. 12.2), in the colon tissues were significantly reduced in mice fed a millet diet. Finally, the millet dietary fiber consumption resulted in a decrease in the mRNA expression of VEGF (Vascular endothelial growth factor), a sign of angiogenesis.

12.7 Formulation and Fortification Strategies

Millets are a type of whole grain that are considered essential for consumption due to their high probiotic and prebiotic content, which can enhance the nutritional value of our meals. While the health benefits of millets are well known, their taste is

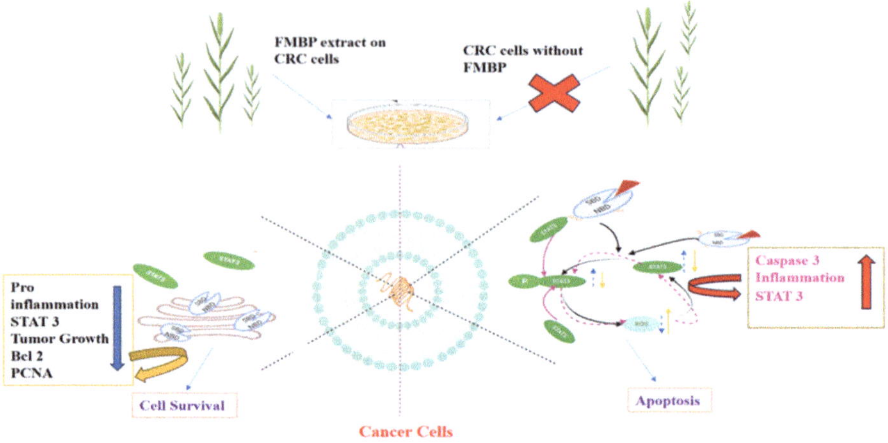

Fig. 12.2 Induction of peroxidase derived from foxtail millet bran (FMBP) can inhibit cancer by regulating STAT3 pathway in CRC cells

not always appealing to consumers. To address this, a trend of millet-based cooking and food has emerged. One potential solution to enhancing the prebiotic content of millets is to formulate them into a snack bar. Currently, there is a high demand for functional foods that are enriched with bioactive plant-based compounds (Martău et al. 2023; Mocanu and Amariei 2022; Badea et al. 2019), such as polyphenols, antioxidants, and probiotics (Simon et al. 2021), which can address conditions like cardiovascular disease, diabetes, obesity, and gastrointestinal disorders (Ahmad and Gani 2021; Behl et al. 2020). However, many snack bars on the market are heavily processed and contain sweeteners, food additives, and saturated fat. Therefore, it is necessary to develop snack bars that are minimally processed, natural, and friendly to the gut microbiota to maximize the health benefits of bioactive compounds (Klerks et al. 2022). One example of such a snack bar is the elderberry snack bar, which has been formulated using three different bases and 0.5% or 1% FDEBP (freeze-dried elderberry powder).

The first, second, and third snack bars were made with a combination of oat flakes, walnuts, shelled hemp seeds, and plum jam. The second base also had millet flakes, peanut butter, shelled hemp seeds, and dates, while the third base used millet instead of oats as its grain source (Haş et al. 2023). Each batch was weighed according to the required percentages and crushed using a food processor, except for the millet flakes, which were added in their intact form and manually incorporated into the mixture. The resulting snack bars were 400 g each and underwent a nutritional assessment that showed they were high in protein, ranging from 13.1 g to 13.8 g per 100 g, and carbohydrates, ranging from 35.7 g to 48.6 g per 100 g. The fat content ranged from 17.6 g to 29.8 g per 100 g, and the fiber content ranged from 5.4 g to 6.9 g per 100 g (Vodnar et al. 2022). The prebiotic potential of FDEBP was tested on five probiotic strains, *L. plantarum* (*Lactobacillus plantarum*), *L. casei* (*Lactobacillus casei*), *L. rhamnosus* (*Lacticaseibacillus rhamnosus*), *L. fermentum*

(*Limosilactobacillus fermentum*), and *S. boulardii* (*Saccharomyces boulardii*), and it was found to have a positive effect on all five strains. These snack bars can be easily made available to the public, providing a rich source of prebiotic-rich millet for people to enjoy and benefit from (Gámbaro and McSweeney 2020).

12.8 Challenges and Future Directions

Research on prebiotic supplementation focusing on health outcomes within cancer patients remains in its infancy. This manifests as numerous gaps and opportunities to enhance both the quality of research and the care of cancer patients. One of the primary challenges lies in the considerable inconsistencies in the strain, dose, and treatment regimen for probiotic supplementation, making it difficult to reach a consensus on effective protocols. Additionally, patients often lack knowledge regarding the use of prebiotics, including potential adverse side effects. This knowledge gap highlights the need for future research to develop educational tools and strategies to ensure patients can safely and effectively use prebiotic supplements, thus advocating for their health. The potential of millets as prebiotics in treating colon cancer is hampered by several factors, including the decline in millet cultivation due to low yields, laborious processing methods, and minimal marketing efforts. Currently, millets are grown on only 15 million hectares compared to 35 million hectares previously, with the low productivity of certain varieties, such as sorghum and finger millet, exacerbating the issue. The lack of awareness about the health benefits of millets among the general population further limits their demand, compounded by their higher cost and limited availability in retail markets. Moreover, the perceived taste and accessibility issues pose additional challenges, discouraging consumer adoption. Agricultural hurdles such as low yields and limited profitability, along with competition from staple foods like rice and wheat, further undermine millet cultivation. Insufficient government support and decentralized processing challenges, including variations in grain size and shelf-life issues due to intrinsic enzyme activity and lipid oxidation, add to the complexity. Marketing and market linkages also suffer due to inconsistent supply and demand, low crop productivity from a lack of access to high-yielding seeds, and suboptimal distribution and price realization. Despite these challenges, millets present emerging opportunities as climate-resilient staple food crops suitable for drought-prone areas due to their drought resistance, low water requirements, and ability to grow in poor soil conditions. Rich in nutrients, gluten-free, adaptable to diverse soils and climates, and often grown using sustainable farming methods, millets hold promise as a versatile crop choice. Future research directions should include comprehensive clinical trials to assess the efficacy of millets as prebiotics in colon cancer treatment, leveraging biotechnological advancements to enhance their beneficial properties, and developing personalized nutrition plans based on individual gut microbiota compositions. Additionally, combining millets with other prebiotic and probiotic foods could create synergistic health benefits, while public awareness campaigns and government support are essential to promote millet cultivation and consumption. Addressing these challenges and harnessing the potential of millets, could play a crucial role in future cancer prevention and treatment strategies, contributing to improved health outcomes and sustainability.

12.9 Conclusion

This chapter focused on the vital role of prebiotics in maintaining a healthy gastrointestinal tract. Beneficial microflora metabolize prebiotics to produce short-chain fatty acids, which offer various health benefits. The relationship between gut microbiota and colon cancer is crucial, with factors such as family history, hereditary syndromes, inflammatory bowel disease, type 2 diabetes, and alterations in gut microbiota composition all playing a role in the development of colon cancer. Biofortified to enhance its prebiotic content through agricultural means, including conventional breeding, genetic engineering, mutagenesis, and agronomic approaches, can be a good dietary source to combat or rather prevent colon cancer. The enhanced prebiotic content promotes the growth of beneficial gut bacteria, which produce short-chain fatty acids that have been shown to protect against colon cancer. Developing this millet into easily available food products can thus be a healthy option for the population.

References

Ahmad M, Gani A (2021) Development of novel functional snacks containing nano-encapsulated resveratrol with anti-diabetic, anti-obesity and antioxidant properties. Food Chem 352:129323. https://doi.org/10.1016/j.foodchem.2021.129323

Anitha S, Kane-Potaka J, Tsusaka TW et al (2021) A systematic review and meta-analysis of the potential of millets for managing and reducing the risk of developing diabetes mellitus. Front Nutr 8:687428. https://doi.org/10.3389/FNUT.2021.687428/PDF

Badea M, di Modugno F, Floroian L, Tit DM, Restani P, Bungau S, Iovan C, Badea GE, Aleya L (2019) Electrochemical strategies for gallic acid detection: potential for application in clinical, food or environmental analyses. Sci Total Environ 672:129–140. https://doi.org/10.1016/j.scitotenv.2019.03.404

Bamigbade GB, Subhash AJ, Kamal-Eldin A et al (2022) An updated review on prebiotics: insights on potentials of food seeds waste as source of potential prebiotics. Molecules 27. https://doi.org/10.3390/MOLECULES27185947

Behl T, Sharma A, Sharma L, Sehgal A, Zengin G, Brata R, Fratila O, Bungau S (2020) Exploring the multifaceted therapeutic potential of Withaferin A and its derivatives. Biomedicines 8(12):571. https://doi.org/10.3390/biomedicines8120571

Bouhnik Y, Flourie B, Riottot M, Bisetti N, Gailing MF, Guibert A, Bornet F, Rambaud JC (1996) Effects of fructo-oligosaccharides ingestion on fecal bifidobacteria and selected metabolic indexes of colon carcinogenesis in healthy humans. Nutr Cancer 26:21–29. https://doi.org/10.1080/01635589609514459

Caballero-Franco C, Keller K, De Simone C, Chadee K (2007) The VSL#3 probiotic formula induces mucin gene expression and secretion in colonic epithelial cells. Am J Physiol Gastrointest Liver Physiol 292(1):G315–G322. https://doi.org/10.1152/ajpgi.00265.2006

Carlson JL, Erickson JM, Lloyd BB, Slavin JL (2018) Health effects and sources of prebiotic dietary fiber. Curr Dev Nutr 2. https://doi.org/10.1093/CDN/NZY005

Chen J, Duan W, Ren X et al (2017) Effect of foxtail millet protein hydrolysates on lowering blood pressure in spontaneously hypertensive rats. Eur J Nutr 56:2129–2138. https://doi.org/10.1007/S00394-016-1252-7

Davani-Davari D, Negahdaripour M, Karimzadeh I et al (2019) Prebiotics: definition, types, sources, mechanisms, and clinical applications. Food Secur 8. https://doi.org/10.3390/FOODS8030092

Ding L, Yang L, Wang Z, Huang W (2015) Bile acid nuclear receptor FXR and digestive system diseases. Acta Pharm Sin B 5:135. https://doi.org/10.1016/J.APSB.2015.01.004

Fabregas JC, Ramnaraign B, George TJ (2022) Clinical updates for colon cancer care in 2022. Clin Colorectal Cancer 21:198–203. https://doi.org/10.1016/J.CLCC.2022.05.006

Fong W, Li Q, Yu J (2020) Gut microbiota modulation: a novel strategy for prevention and treatment of colorectal cancer. Oncogene 39:4925–4943. https://doi.org/10.1038/s41388-020-1341-1

Gaitan E, Lindsay RH, Reichert RD et al (1989) Antithyroid and Goitrogenic effects of millet: role of C-Glycosylflavones. J Clin Endocrinol Metab 68:707–714. https://doi.org/10.1210/JCEM-68-4-707

Gámbaro A, McSweeney MB (2020) Sensory methods applied to the development of probiotic and prebiotic foods. Adv Food Nutr Res 94:295–337. https://doi.org/10.1016/bs.afnr.2020.06.006

Gibson GR, Roberfroid MB (1995) Dietary modulation of the human colonic microbiota: introducing the concept of prebiotics. J Nutr 125:1401–1412. https://doi.org/10.1093/jn/125.6.1401

Gibson GR, Hutkins R, Sanders ME, Prescott SL, Reimer RA, Salminen SJ, Scott K, Stanton C, Swanson KS, Cani PD et al (2017) Expert consensus document: the International Scientific Association for Probiotics and Prebiotics (ISAPP) consensus statement on the definition and scope of prebiotics. Nat Rev Gastroenterol Hepatol 14(8):491–502. https://doi.org/10.1038/nrgastro.2017.75

Haş IM, Vodnar DC, Bungau AF, Tarce AG, Tit DM, Teleky BE (2023) Enhanced elderberry snack bars: a sensory, nutritional, and rheological evaluation. Foods (Basel, Switzerland) 12(19):3544. https://doi.org/10.3390/foods12193544

Hassan ZM, Sebola NA, Mabelebele M (2021) The nutritional use of millet grain for food and feed: a review. Agric Food Secur 10:16. https://doi.org/10.1186/S40066-020-00282-6

Hill C, Guarner F, Reid G, Gibson GR, Merenstein DJ, Pot B, Morelli L, Canani RB, Flint HJ, Salminen S et al (2014) Expert consensus document. The International Scientific Association for Probiotics and Prebiotics consensus statement on the scope and appropriate use of the term probiotic. Nat Rev Gastroenterol Hepatol 11(8):506–514. https://doi.org/10.1038/nrgastro.2014.66

Islam J, Sato S, Watanabe K, Watanabe T, Ardiansyah HK, Aoyama Y, Tomita S, Aso H, Komai M, Shirakawa H (2017) Dietary tryptophan alleviates dextran sodium sulfate-induced colitis through aryl hydrocarbon receptor in mice. J Nutr Biochem 42:43–50. https://doi.org/10.1016/j.jnutbio.2016.12.019

Jacob J, Krishnan V, Antony C et al (2024) The nutrition and therapeutic potential of millets: an updated narrative review. Front Nutr 11:1346869. https://doi.org/10.3389/FNUT.2024.1346869

Klerks M, Román S, Verkerk R, Sanchez-Siles L (2022) Are cereal bars significantly healthier and more natural than chocolate bars? A preliminary assessment in the German market. J Funct Foods 89:104940. https://doi.org/10.1016/j.jff.2022.104940

Lin HC, Sheu SY, Sheen LY et al (2020) The gastroprotective effect of the foxtail millet and adlay processing product against stress-induced gastric mucosal lesions in rats. J Tradit Complement Med 10:336–344. https://doi.org/10.1016/J.JTCME.2020.01.003

Ma J, Piao X, Mahfuz S et al (2022) The interaction among gut microbes, the intestinal barrier and short chain fatty acids. Anim Nutr 9:159. https://doi.org/10.1016/J.ANINU.2021.09.012

Martău GA, Bernadette-Emőke T, Odocheanu R, Soporan DA, Bochiş M, Simon E, Vodnar DC (2023) *Vaccinium* species (Ericaceae): Phytochemistry and biological properties of medicinal plants. Molecules (Basel, Switzerland) 28(4):1533. https://doi.org/10.3390/molecules28041533

Megur A, Daliri EBM, Baltriukienė D, Burokas A (2022) Prebiotics as a tool for the prevention and treatment of obesity and diabetes: classification and ability to modulate the gut microbiota. Int J Mol Sci 23. https://doi.org/10.3390/IJMS23116097

Mocanu ML, Amariei S (2022) Elderberries-A source of bioactive compounds with antiviral action. Plants (Basel, Switzerland) 11(6):740. https://doi.org/10.3390/plants11060740

Nigam P, Sisodiya D, Dashora K (2024) A review on millets and their types. Int J Pharm Sci 02:1–1. https://doi.org/10.5281/ZENODO.10616810

Ohtsuka K, Tsuji K, Nakagawa Y, Ueda H, Ozawa O, Uchida T, Ichikawa T (1991) Utilization and metabolism of [U-14C]4′galactosyllactose(O-betaD-galactopyranosyl-(1—-4)-O-beta-Dgalactopyranosyl-(1—-4)- D-glucopyranose) in rats. J Nutr Sci Vitaminol (Tokyo) 37:173–184. https://doi.org/10.3177/jnsv.37.173

Pandey KR, Naik SR, Vakil BV (2015) Probiotics, prebiotics and synbiotics- a review. J Food Sci Technol 52:7577–7587. https://doi.org/10.1007/S13197-015-1921-1

Portincasa P, Bonfrate L, Vacca M et al (2022) Gut microbiota and short chain fatty acids: implications in glucose homeostasis. Int J Mol Sci 23. https://doi.org/10.3390/IJMS23031105

Pothuraju R, Chaudhary S, Rachagani S, Kaur S, Roy HK, Bouvet M, Batra SK (2021) Mucins, gut microbiota, and postbiotics role in colorectal cancer. Gut Microbes 13(1):1974795. https://doi.org/10.1080/19490976.2021.1974795

Ramadoss D, Sivalingam N (2021) Vanillin extracted from proso and barnyard millets induces cell cycle inhibition and apoptotic cell death in MCF-7 cell line. J Cancer Res Ther 17:1425–1433. https://doi.org/10.4103/JCRT.JCRT_1128_19

Sawicki T, Ruszkowska M, Danielewicz A et al (2021) A review of colorectal cancer in terms of epidemiology, risk factors, development, symptoms and diagnosis. Cancers (Basel) 13. https://doi.org/10.3390/CANCERS13092025

Sharma N, Niranjan K (2017) Foxtail millet: properties, processing, health benefits, and uses. Food Rev Intl 34(4):329–363. https://doi.org/10.1080/87559129.2017.1290103

Shi J, Shan S, Li H et al (2017) Anti-inflammatory effects of millet bran derived-bound polyphenols in LPS-induced HT-29 cell via ROS/miR-149/Akt/NF-κB signaling pathway. Oncotarget 8:74582. https://doi.org/10.18632/ONCOTARGET.20216

Shivam D (2024) A review on usefulness of millets in current prospects. IP J Nutr Metabol Health Sci 6:157–160. https://doi.org/10.18231/J.IJNMHS.2023.026

Siddiqui R, Boghossian A, Alharbi AM et al (2022) The pivotal role of the gut microbiome in colorectal cancer. Biology (Basel) 11. https://doi.org/10.3390/BIOLOGY11111642

Simon E, Călinoiu LF, Mitrea L, Vodnar DC (2021) Probiotics, prebiotics, and synbiotics: implications and beneficial effects against irritable bowel syndrome. Nutrients 13(6):2112. https://doi.org/10.3390/nu13062112

Singh SB, Meena AK, Sisodia BS et al (2023) From farm to gut: Unraveling the role of millets in promoting metabolic well-being via gut microbiota. J Drug Res Ayurved Sci 8:S50–S54. https://doi.org/10.4103/JDRAS.JDRAS_192_23

Steinke V, Engel C, Büttner R et al (2013) Hereditary nonpolyposis colorectal cancer (HNPCC)/ Lynch syndrome. Dtsch Arztebl Int 110:32. https://doi.org/10.3238/ARZTEBL.2013.0032

Turesky RJ (2018) Mechanistic evidence for red meat and processed meat intake and cancer risk: a follow-up on the International Agency for Research on Cancer evaluation of 2015. Chimia (Aarau) 72:718. https://doi.org/10.2533/CHIMIA.2018.718

Veerabadhran V, Palakuru HR, Pratap GP et al (2023) A critical review of the history and practices of millets in India. J Drug Res Ayurved Sci 8:S18–S25. https://doi.org/10.4103/JDRAS.JDRAS_113_23

Vernia F, Longo S, Stefanelli G et al (2021) Dietary factors modulating colorectal carcinogenesis. Nutrients 13:1–13. https://doi.org/10.3390/NU13010143

Vodnar DC, Calinoiu LF, Mitrea L (2022) Editorial: exploiting the effect of dietary fibre on the gut microbiota in patients with pelvic radiotherapy. Br J Cancer 127(9):1575–1576. https://doi.org/10.1038/s41416-022-01993-2

Yang Y, Weng W, Peng J, Hong L, Yang L, Toiyama Y, Gao R, Liu M, Yin M, Pan C, Li H, Guo B, Zhu Q, Wei Q, Moyer MP, Wang P, Cai S, Goel A, Qin H, Ma Y (2017) Fusobacterium nucleatum increases proliferation of colorectal cancer cells and tumor development in mice by activating Toll-Like Receptor 4 signaling to nuclear factor-κB, and up-regulating expression of MicroRNA-21. Gastroenterology 152(4):851–866.e24. https://doi.org/10.1053/j.gastro.2016.11.018

Yu GH, Li SF, Wei R, Jiang Z (2022) Diabetes and colorectal cancer risk: clinical and therapeutic implications. J Diabetes Res 2022. https://doi.org/10.1155/2022/1747326

Zhang B, Xu Y, Liu S, Lv H, Hu Y, Wang Y, Li Z, Wang J, Ji X, Ma H, Wang X, Wang S (2020) Dietary supplementation of Foxtail Millet Ameliorates Colitis-associated colorectal cancer in mice via activation of gut receptors and suppression of the STAT3 pathway. Nutrients 12(8):2367. https://doi.org/10.3390/nu12082367

Zheng DW, Li RQ, An JX, Xie TQ, Han ZY, Xu R, Fang Y, Zhang XZ (2020) Prebiotics-encapsulated probiotic spores regulate gut microbiota and suppress colon cancer. Adv Mater (Deerfield Beach, Fla) 32(45):e2004529. https://doi.org/10.1002/adma.202004529

Probiotics with Preventive and Therapeutic Potential for Colorectal Cancer: An Emphasis on Immunomodulatory and Anticancer Properties

13

Taparia Saloni, Jalahalli M. Siddesha, Basalingappa M. Kanthesh, and Shivaswamy Umamaheshwari

Abstract

Cancer is a growing global health problem, contributing to higher rates of morbidity and mortality with each passing year. The colorectal cancer ranks the third most among the cancer types, caused due to various genetic and epigenetic factors. Currently used cancer therapies are effective; however, their toxicity and side effects diminish the quality of life. Of note, probiotics are known for their beneficial effects on gut health, and immunomodulation, with little or no side effects. This chapter focuses on the immunomodulatory and anti-oncogenic traits of probiotics and their metabolites. The use of probiotics as therapeutic and preventive options might pose challenges and limitations. The selection of appropriate probiotic strains based on their anti-oncogenic effects, proper dosage, and treatment duration is critical. Therefore, future research should focus on clinical trials to evaluate safety and efficacy of probiotics in treatment of colorectal cancer.

T. Saloni · S. Umamaheshwari (✉)
Department of Microbiology, School of Life Sciences, JSS Academy of Higher Education & Research, Mysuru, Karnataka, India
e-mail: umamaheshwari@jssuni.edu.in

J. M. Siddesha
Division of Biochemistry, School of Life Sciences, JSS Academy of Higher Education & Research, Mysuru, Karnataka, India

B. M. Kanthesh
Division of Molecular Biology, School of Life Sciences, JSS Academy of Higher Education & Research, Mysuru, Karnataka, India

S. Pathak et al. (eds.), *Microbiota and Dietary Mediators in Colon Cancer Prevention and Treatment*, https://doi.org/10.1007/978-981-96-0297-1_13

Keywords

Probiotics · Immunomodulation · Anti-oncogenic effects · Gut health

13.1 Cancer–Microbiota–Probiotics

Cancer is a complex life-threatening disease that occurs due to uncontrolled cell proliferation. Genetic alterations are observed due to aging, chemical mutagens, UV rays, and other factors, while epigenetic variations are due to lifestyle and the environmental factors, as shown in Fig. 13.1 (Şener et al. 2021; Chattaraj et al. 2023; Reddy 1998; Legesse Bedada et al. 2020; Pope et al. 2017; Aindelis and Chlichlia 2020; Kumar et al. 2010). Globally, cancer resulting in 7.6 million deaths was the second prevalent cause of mortality in 2018 (Legesse Bedada et al. 2020; Singh et al. 2021; Tripathy et al. 2021). As per WHO 2022 survey, cancer cases have risen to 20 million, leading to 9.7 million deaths. Moreover, from past 10 years, the incidence of cancer has roughly increased by 24% in men, and by 21% in women. It has been estimated that by the end of year 2040, the number of cases will rise to 29.5 million, which calls for early detection and efficient treatment strategies (Singh et al. 2021; Tripathy et al. 2021). Lifestyle characteristics including diet, physical activity, working habits, stress, and consumption of tobacco and alcohol play a prominent role in cancer development. Lifestyle, exposure to radiation, aging, infectious agents, hormonal imbalance, disrupted DNA repair mechanisms, and environmental factors resulting in DNA methylation, histone acetylation, and

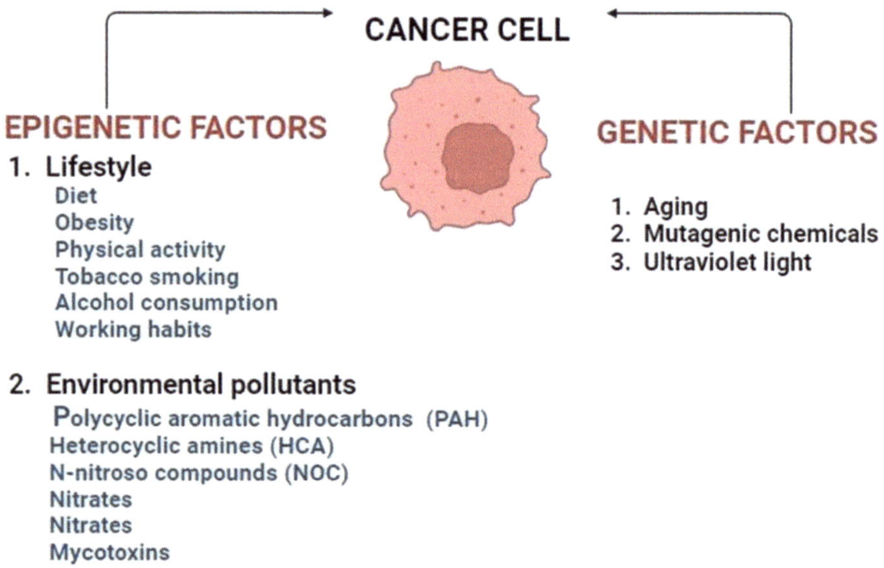

Fig. 13.1 Risk factors associated with cancer

altered microRNA expression collectively contribute to cancer (Chattaraj et al. 2023; Legesse Bedada et al. 2020; Raman et al. 2013).

Notably, colorectal cancer (CRC) being the third most common type of cancer, accounts for ~1.9 million new cases and 930,000 deaths (Tripathy et al. 2021; Uccello et al. 2012; Eslami et al. 2019; Torres-Maravilla et al. 2021; Song and Chan 2019). CRC is a complex neoplastic illness, wherein genetic and environmental factors influence cell proliferation, differentiation, survival, and metabolism. The higher regenerative pressure and persistent contact with nutrients and microbiota make the intestinal epithelia susceptible for malignant transformations. Of note, CRC affect the entire large intestine, and the risk increases with age, especially above 50 years (dos Reis et al. 2017). Other risk factors of CRC include germline mutations in mismatch repair (MMR) genes and adenomatous polyposis coli (APC) genes, genetic and environmental factors, diet and modern lifestyle (Kahouli et al. 2013; Tilg et al. 2018). Besides, food carcinogens, mutagens, and chemicals released during air pollution such as heterocyclic amines, polycyclic aromatic hydrocarbons, N-nitroso compounds, nitrates, mycotoxins, and acrylamide also cause CRC (Pino et al. 2020). Notably, obesity is the leading cause of CRC, as it disturbs metabolic and inflammatory mechanisms via sex hormones, adipokines, and insulin/insulin-like growth factor 1 signaling (Uccello et al. 2012; Song and Chan 2019). Other risk factors include regulation of epithelial cell division, biosynthesis of nutrients, inhibition of harmful microorganisms, and stimulation of intestinal immunity. Therefore, the risk factors contributing to CRC can be abridged through healthy lifestyle including physical activity, and adequate intake of dietary fiber, whole grains, dairy products, and probiotics (Song and Chan 2019). Currently, various therapeutic approaches including surgery, chemotherapy, and radiation therapy are employed to manage CRC (Geier et al. 2006). However, the chemotherapy causes adverse effects, decreases quality of life, and sometimes even the death (Dubey et al. 2016).

Importantly, studies have shown the relationship between the dysbiosis of gut microbiota and CRC progression (Pino et al. 2020). Human body is home to various microorganisms that coordinately play a role in maintaining and preserving health (Singh et al. 2021). It is estimated that ~200 trillion microbial organisms are found in the gut (Torres-Maravilla et al. 2021; Davis and Milner 2009; Zhang et al. 2019). The gut microbiota develops at birth in an organized fashion through interactions between bacteria, immunological processes, and selective blood absorption mediated by the gastrointestinal barrier in humans (Kasarello and Sajdel-Sulkowska 2019; Zhu et al. 2013). The microbiota performs several tasks, including the synthesis of vital components, inhibition of pathogens and bacterial overgrowth that alter intestinal microenvironment, regulation of immune metabolism, and maintenance of epithelial barrier (Sehrawat et al. 2021; Aziz and Bonavida 2016; Sánchez-Alcoholado et al. 2020).

Bacteroides, *Eubacterium*, *Bifidobacterium*, *Fusobacterium*, *Peptostreptococcus*, and *Atopobium* are commonly found in human microbiota. These microbes are associated with digestion and absorption of nutrients, production of essential metabolites such as vitamins and short-chain fatty acids. Of note, these anaerobes compete with pathogens and boost immunity by modulating immune responses (Kumar

Mishra et al. 2008). However, a complex ecosystem develops in the gut microbiota due to the interaction between host physiology and the bacteria. Of note, disturbances in the ecosystem of gut microbiota can impact cancer (Aindelis and Chlichlia 2020; Shida and Nomoto 2013). Dietary imbalance, aging, mental stress, infection, and antibiotic use can disrupt gut microbiota balance. Further, the gut microbiota imbalance leads to immune-related disorders including inflammatory bowel diseases, allergy, and obesity.

The alteration in gut microbiota can be one of the leading factors causing colon cancer (Górska et al. 2019; Aleta and Hrvat n.d.). The gut plays a role in inflammation, activating carcinogens and tumorigenic pathways, and damage to host DNA (Song and Chan 2019; Kaźmierczak-Siedlecka et al. 2021). The imbalance in microbiota–immune system homeostasis results in chronic inflammatory and autoimmune diseases (Singh et al. 2021). Notably, the load of good bacteria has been shown to be lower in CRC patients (Faghfoori et al. 2015). The higher bacterial diversity in large intestine has been shown to increase the risk of colon cancer (Zhu et al. 2013). The gut mucosa functions as a barrier to prevent pathogen invasion (Singh et al. 2021). Importantly, formation and maintenance of healthy microbiota is crucial for the maturation of intestinal mucosa and the gut-associated lymphoid tissue. Moreover, *Fusobacterium nucleatum* and *Bacteroides fragilis* are predominantly found in tumor rather than in normal tissues of CRC patients (dos Reis et al. 2017). Besides, researchers opined that consumption of fermented milk having *Lactobacilli* might help in replacing putrefactive and harmful microbes in the intestine. Importantly, the gut provides a favorable environment for probiotic bacteria and thus protection against carcinogenic factors (Sánchez-Alcoholado et al. 2020; Kaźmierczak-Siedlecka et al. 2021).

Optimal consumption of probiotics might benefit gastrointestinal tract and modulate gut microbiota, by enhancing the immune system (Chattaraj et al. 2023; Khan et al. 2013). Of note, the largely used probiotics include *Lactobacillus* sp. and *Bifidobacterium* sp. (Kaźmierczak-Siedlecka et al. 2021). Other probiotics that can be used are *Lactococcus*, *Leuconostoc, Pediococcus*, *Propionibacterium*, *Bacillus*, and the yeast *Saccharomyces*. Importantly, probiotics might benefit health via lowering cholesterol levels, improving immune system, balancing gut microbiota, preventing constipation, and having anticarcinogenic activity (Kumar Mishra et al. 2008). Probiotics prevent and ameliorate cancer by altering proliferation and apoptosis of cancer cells (Maldonado Galdeano et al. 2019). Moreover, certain gram-positive bacteria activate natural killer cells, suppress tumor growth, and attenuate cancer stem cell differentiation and cytokine release, which collectively fight against cancer. Thus, increased probiotic bacteria and decreased inflammatory cytokine might prevent carcinogenesis and activate phagocytes to eliminate early-stage cancer cells. It is anticipated that probiotics and food containing probiotics support the host's immune homeostasis and control immunological reactions to avert CRC (Shida and Nomoto 2013). Thus, probiotics could be developed as an alternative for the management of CRC, to overcome the adverse effects of chemotherapy, surgery, and radiation.

13.2 Probiotics

The concept of probiotics given by Elie Metchnikoff is gaining importance in recent years, as they play a prominent role in immune function and health. The ideal probiotic possesses the properties including (i) resistance to bile acid and gastric juice, (ii) adherence to mucus and epithelial cells, (iii) antimicrobial activity, (iv) reduce pathogen adhesion, (v) safe, (vi) noninvasive, (vii) nonpathogenic, and (viii) noncarcinogenic (Khan et al. 2013; Aleta and Hrvat n.d.). To use probiotics in food, they must be recognized as safe, with organisms alive and present in adequate amounts. The amount and selection of probiotic variant is decided based on several factors such as age, diet, enzymes, intestinal conditions, stress, drugs, and lifestyle (Eslami et al. 2019). The immune system and colonic epithelium work together to maintain a positive relationship between microbiota and the host. Nevertheless, certain disorders such as obesity, asthma, inflammatory bowel disease, mental illnesses, and colon cancer could disturb the normal microflora. Of note, probiotics play a prominent role in balancing the microbial diversity with little or no side effects and restore the gut health as shown in Fig. 13.2 (Eslami et al. 2019; Sehrawat et al. 2021; Yousefi et al. 2019; De Almeida et al. 2019; Bhatia et al. 2014; Yasmin et al. 2020). Another important feature of probiotics is their anti-oncogenic ability, which is possible via modulation of gut microbiota, production of bacteriocins, inhibition of inflammation, and induction of apoptosis (Chattaraj et al. 2023; Nazir et al. 2018).

Probiotics are considered as functional food due to multiple health benefits, including antioxidant activity, production of health promoting metabolites, and relieving constipation (Chattaraj et al. 2023; Dubey et al. 2016; Aleta and Hrvat n.d.; Maldonado Galdeano et al. 2019). They also prevent urinary tract infections,

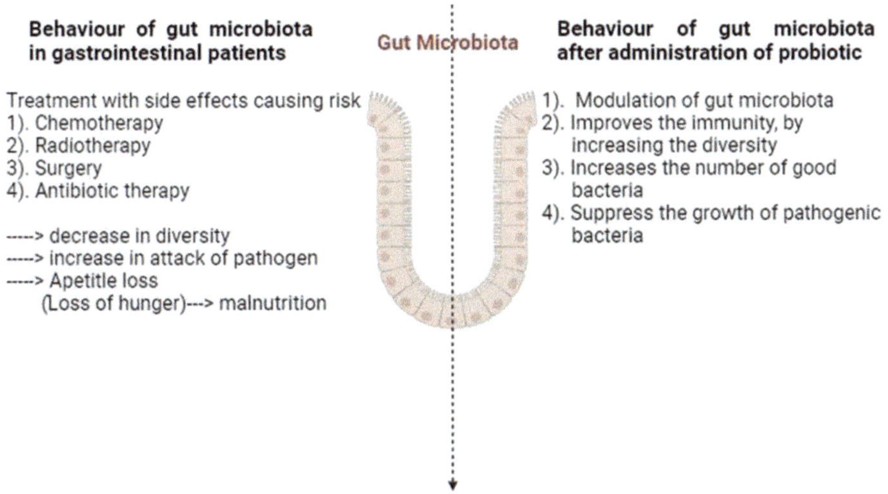

Fig. 13.2 Gut microbiota behavior prior and after administration of probiotics

inhibit pathogen growth, boost up immune response, increase anti-allergic potential, reduce gut inflammation, and exhibit anticarcinogenic effects (Geier et al. 2006; Bhatia et al. 2014; Rafter 2004). It is evident that the consumption of appropriate amounts of probiotics has greatly influenced gut microbiota. Some of the probiotics used as therapeutics against cancer and their mode of action are given in Table 13.1 (Kumar et al. 2010; Davis and Milner 2009; Aziz and Bonavida 2016; Aleta and Hrvat n.d.; Yahfoufi et al. 2018).

Probiotics instantly benefit by restoring homeostasis of intestinal microbiota, reducing antibiotic-associated diarrhea, and enhancing bowel regularity (Yu and Li 2016). They also help in xenobiotics detoxification, vitamin K biosynthesis, dietary fiber fermentation, combating pathogenic microorganisms and immune modulation (Hardy et al. 2013). Besides, they prevent the initiation of cancer, by attenuating the conversion of pro-carcinogens into carcinogens. Moreover, probiotics boost the production of cytokines such as tumor necrosis factor-alpha and gamma-interferons, thereby targeting cancerous cells. Importantly, the statistical human epidemiology reports suggested that the people consuming fermented dairy products had a lower risk for CRC.

13.3 Immunomodulation by Probiotics

Notably, the human intestine is exposed to various pathogenic agents and antigens. Besides, intestinal microflora differs from one individual to another and is basically determined by genotype, primary colonization at the time of birth, and dietary habits (Zhang et al. 2019). In addition, fecal stream is a rich source of cancer inducing agents, as certain carcinogenic compounds get released into the lumen during the bacterial transformation (Commane et al. 2005). Numerous intestinal immune cells fight against infections, and control body's tolerance to ingested antigens and commensal microorganisms. However, the gut microbiome maintains a proper epithelium barrier, by inhibiting adherence of harmful organisms to the intestinal surroundings, altering immune system, and producing various metabolites (Sánchez et al. 2017). The immune system protects against bacteria, viruses, fungi, and parasites during invasion, and activates antigen presentation cells (APC), T cells, B cells, dendritic cells and natural killer cells (Kaźmierczak-Siedlecka et al. 2021; Mazziotta et al. 2023). Of note, it is crucial to identify specific bacteria causing cancer, since their interactions with host system are complex. Remarkably, enterotoxigenic *Bacteroides fragilis*, *Escherichia coli*, and *Fusobacterium nucleatum* have been found to be associated with CRC, through the production of toxins, genotoxins, and specific genes.

Lactobacillus and *Bifidobacterium* are the most common probiotics that reside in the intestinal microbiota (Yahfoufi et al. 2018; Mazziotta et al. 2023). Notably, the intake of probiotics, especially *L. acidophilus* and *B. lactis*, by CRC patients, would increase butyrate-producing bacteria and decrease CRC- associated microbial load. Moreover, *Lactobacilli* have been shown to inhibit colonization of pathogens, stimulate immune system, and exhibit anti-mutagenic and cytotoxic effects on

Table 13.1 List of probiotics and mode of action as an anticancer agent

Probiotic Organism	Cancer type	Mechanism	References
Lactobacillus acidophilus	Breast cancer	Immune regulation by inhibiting tumor cell proliferation; modification of gut microbiota by synthesizing short-chain fatty acids	Plaza-Díaz et al. (2017)
Lactobacillus acidophilus 606	Colon cancer HT-29 cells	Induce Beclin-1 and GRP78 and affect Bcl-2 and Bak regulation	Kuugbee et al. (2016)
Lactobacillus acidophilus	Breast cancer	Inhibit cancer cell growth	Shyu et al. (2014)
Lactobacillus casei Shirota	Bladder cancer, colorectal cancer	Enhance the cytotoxicity of NK cells	Wang et al. (2021)
Lactobacillus casei Shirota	Colorectal cancer, bladder cancer	Immunomodulatory roles to strengthen the immune defense of the host by induction of IL–12 (interleukin 12) production through phagocytes	Molska and Reguła (2019)
Lactobacillus casei, Lactobacillus rhamnosus	Bladder cancer, prostate cancer	Enhance the activity of dendritic cells; Activation of neutrophils; Increase NK activity by the production of TRAIL	Naito et al. (2008)
Lactobacillus fermentum	Colon cancer	Modulating hyperinsulinemia, insulin resistance, hypercholesterolemia, and hypertriglyceridemia	Commane et al. (2005)
Lactobacillus reuteri	Breast cancer	Alternate the gut microbiome and enhance the immune system's ability	Chattaraj et al. (2023)
Lactobacillus reuteri ATCC-PTA 6475	Breast cancer	Block NFk B-p65 nuclear translation and expression in mammary tumor cells	Wang et al. (2021)
Lactobacillus rhamnosus GG	Colon cancer	Increase the pro-apoptotic Bcl-2 associated X-protein (BAX)	Sharaf et al. (2018)
Bifidobacterium bifidum, Bifidobacterium infantum, Lactobacillus acidophilus	Colon cancer	Regulate the apoptotic mediators by increasing toll like receptor-2 (TLR-2) and decreasing Toll like receptor- 4 (TLR-4), caspase 3, cyclooxygenase-2(COX-2), and β-catenin	Kuugbee et al. (2016)
Bifidobacterium bifidum	Colon cancer	As preventive and therapeutic agent	Faghfoori et al. (2021)
Bifidobacterium adolescentis, Lactobacillus acidophilus	Cervix cancer	Suppress the cancer precursors; Improve stool consistency in irradiated patients	Verhoeven et al. (2013)

cancerous cells. They exhibit anticancer activity and enhance immune function by either stimulating B cells or modulating immune effects through macrophages, natural killer cells and T cells. Importantly, probiotics produce antibacterial substances that stimulate immune system through various mechanisms (Bodera and Chcialowski

2009). The probiotics *L. casei*, *L. acidophilus*, *L. rhamnosus*, *L. delbrueckii* subsp. *bulgaricus*, *plantarum*, and *Streptococcus thermophilus* have shown to increase the intestinal IgA-producing cells (Dasari et al. 2017; Mazziotta et al. 2023). They are known to stimulate both specific and nonspecific immunological pathways. Further, probiotics have shown to increase mucus production, activate macrophages through *lactobacilli* signaling, stimulate secretory IgA production, reduce pro-inflammatory cytokine production, and increase peripheral immunoglobulin production (Bodera and Chcialowski 2009). Moreover, probiotics alter immune responses in the gut-associated lymphoid tissue, and interact with intestinal epithelial cells, mucosal dendritic cells, and macrophages in various ways. Of note, the probiotic strains either stimulate immune activation signaling, either by producing IL-12, IL-1β, and TNF-α or activating anti-inflammatory cytokines such as IL-10 and TGF-β. Importantly, efficacy of immunotherapy is greatly influenced by the composition of gut microbiota, especially during the treatment with the immune checkpoint inhibitors (ICIs). Since intestine contains 70–80% of IgA-producing B cells, it is supposed to be the largest immunological organ. IgAs are proteolytic-resistant antibodies, which gives protection by binding and neutralizing antigens. Moreover, probiotics modulate immune system by balancing Th1, Th2, Th17, and Treg cells. Also, probiotics regulate the intestinal gene expression and immune responses via microbial metabolites such as short-chain fatty acids, amino acid metabolites, and bile acids (Zhang et al. 2019). Furthermore, probiotics stop the penetration and adherence of pathogenic microorganisms to the intestinal epithelial cells. Collectively, both innate and adaptive immunity mutually work to protect against the harmful pathogens.

13.4 Probiotics Modulate Immune Checkpoint Inhibitors

Probiotics play a significant role in modulating immune checkpoint inhibitors. The probiotic monoclonal antibodies target immunomodulatory components on T cell surfaces, and trigger an immune response. Besides, the tumor cells induce tolerance by inducing IL10 producing Treg cells and immunosuppressive cytokines. In addition, tumor cells shatter the effector T cells, and reduce immunological response to cytotoxic T lymphocyte associated antigen (CTLA-4; CD152), programmed cell death protein (PD-1) on the surface of activated cytotoxic T cells and regulatory T cells (Aindelis and Chlichlia 2020; Singh et al. 2021). Recently, research has focused on using fully humanized monoclonal antibodies against CTLA-4 and PD-1 to treat cancer patients, which are found to be effective (Kaźmierczak-Siedlecka et al. 2021). Importantly, probiotics block the production of cytokines and promote the production of anti-inflammatory mediators including TGF-β. Of note, TGF-β stimulates the development of immature dendritic cells and induce regulatory T cells (Treg). These cells exert anti-inflammatory effect by regulating Th1, Th2, and maybe Th17 cells. Further, macrophages produce high levels of IL6 in the inflamed mucosa, which could be reduced by the probiotics, while increasing IL10 production. Of note, probiotics could also boost the intestinal production of IL12, which is typically low (Bhatia et al. 2014).

13.5 Immunomodulation by Probiotic Metabolites

Probiotics secrete short-chain fatty acids (SCFAs), conjugated linoleic acid, bacteriocins, and exopolysaccharides that exhibit anticancer properties (Chattaraj et al. 2023; Delgado et al. 2020). SCFAs are known to promote cell proliferation and differentiation in the intestine. Moreover, SCFAs are anti-inflammatory and useful against insulin resistance (Eslami et al. 2019; Tilg et al. 2018). SCFAs such as acetate, butyrate, and propionate are known to suppress cancer cell proliferation and trigger apoptosis in cancer cells. The mechanisms of actions of SCFAs may include (i) activation of G-protein coupled receptors (GPCR) and (ii) suppression of histone deacetylase (HDAC) activity (Torres-Maravilla et al. 2021; Sánchez-Alcoholado et al. 2020; Delgado et al. 2020). The acetate released into the gut, by the intestinal organisms such as *Bifidobacterium* sp., is used as substrate by butyrate- and propionate-producing microbes. These butyrate- and propionate-producing microbes are known to have anti-inflammatory effects and regulate colonic T cells. Moreover, butyrate suppresses HDAC activity of macrophages in intestinal lamina propria and inhibits the secretion of inflammatory mediators such as nitric oxide, IL-6, and IL-1 (Zhang et al. 2019). Besides, propionate produced by Bacteroidetes and Firmicutes is associated with phagocytic activity of dendritic cells and Th2 cell responses. Additionally, SCFAs activate several GPCRs: GPR43, GPR41, and GPR109A. Of note, expression of GPR43 and GPR109A by intestinal epithelial cells suggest their importance in gut homeostasis. They alter immune system by inducing pro-inflammatory cytokine production, affecting gene expression, reducing cell viability, and promoting cancerous cell death (Chattaraj et al. 2023; Eslami et al. 2019; Zhang et al. 2019).

Conjugated linoleic acid (CLA) is another fatty acid produced by probiotics, which have health benefits including antiatherogenic, antidiabetic, anti-inflammatory, and anticarcinogenic effects (Kumar et al. 2010). CLA activates caspases, which initiate apoptosis in cancer cells. Studies have shown that CLA activates caspase-3 and caspase-9, resulting in apoptosis in colon cancer cells (Chattaraj et al. 2023; Śliżewska et al. 2021). CLAs might inhibit eicosanoid production in colonocytes, either via replacement of arachidonic acid in the cell membranes or by interfering with the actions of cyclooxygenase and lipoxygenase. Moreover, it has been suggested that some of the probiotic species such as *Bifidobacterium* (*breve*, *infantis*, *longum*), *Lactobacillus* (*acidophilus*, *casei*, *delbrueckii*, *plantarum*), *Propionibacterium freudenreichii*, and *Streptococcus salivarius* subsp. *thermophilus* have the ability to produce CLAs from linoleic acid and pose beneficial effects.

Furthermore, the bacteria's outer layer consisting repeated monosaccharides create a polymeric matrix of exopolysaccharides (EPs). These EPs loosely coupled to form a slime structure or covalently bound to form capsules. *Bacteroides fragilis* produces zwitterionic polysaccharides, which serve an important role in gut immunological homeostasis. EPs produced by probiotic bacteria have anti-inflammatory and anticancer properties. They exert anticancer effects by inducing apoptosis, inhibiting proliferation, and suppressing angiogenesis. They also increase antitumor effect of natural killer (NK) cells against colon cancer (Chattaraj et al. 2023). Moreover, EPs produced from *Lactobacillus plantarum* KF5 inhibit the growth of colon cancer cells, by inducing apoptosis while inhibiting cell cycle progression. Besides, EPs produced by *Bifidobacterium animalis* subsp. *lactis* CP9 have shown to boost the antitumor effect of NK cells against the colon cancer. Of note, the anticancer activity of EPs is mainly due to its ability to activate NK cells via the regulation of IL-12 and IFN-γ (Chattaraj et al. 2023).

Bacteriocins, the antimicrobial peptides produced by probiotic bacteria, exhibit anticancer properties. They trigger apoptosis in cancer cells by activating caspases, blocking DNA synthesis, and help immune system to recognize and kill cancer cells by inducing apoptosis. Further, bacteriocins can activate caspases in different ways, depending on the type of cancer cells. One of the ways is through the manipulation of mitochondrial membrane potential, which results in the release of cytochrome C and activation of caspase 9. The activated caspase 9 cleaves the downstream effector caspases such as caspase 3, leading to the initiation of apoptosis. Another way is by activating the death receptors or by inhibiting antiapoptotic proteins (Chattaraj et al. 2023).

Other metabolites of probiotics that could lead to immunomodulation include indolic compounds, derived from the breakdown of proteins, specifically from aromatic amino acids such as tryptophan. Thus, the usage of the microbial bioactive molecules instead of whole live microorganisms might be crucial for the treatment of immunocompromised and immunodeficient populations. These bioactive molecules, cellular or metabolic fractions might possess immunomodulatory, antihypertensive, antiproliferative, and antioxidant properties, contributing to the health benefits of gut commensals.

13.6 Probiotics with Preventive and Therapeutic Potential in Colorectal Cancer

Adopting a healthy lifestyle, such as quitting smoking and having a balanced diet, could minimize the risk of cancers. Currently, attention has been given to the dietary habits such as probiotics consumption, which has a direct impact on immune system (Sehrawat et al. 2021; Dasari et al. 2017). The orally taken probiotics have been shown to improve gastrointestinal barrier, limit infections by competing with pathogenic bacteria, build intestinal mucosal barrier by increasing productions of mucin, defensins, and immunoglobulin A. Collectively, these effects modulate and enhance host's immune response through anti-inflammatory pathways (Yu and Li 2016). The probiotics activate cytokines (IL-10), decrease the expression of pro-inflammatory cytokines (IL-6, IL-1β, TNF-α), increase CD4+ CD25+ regulatory responses, and inhibit prostaglandin production (Pino et al. 2020). Further, probiotics are known to increase the production of cytokines (IL-1 and IL-2), antioxidants, antiangiogenic factors and regulate apoptosis in CRC patients. The intake of probiotics increases the population of good bacteria and reduces the side effects caused due to the treatment of cancer, by rebuilding the functions of the gut microbiota. Therefore, probiotic as therapeutics is more patient-friendly option in cancer when compared to chemotherapy and radiation therapy (Sánchez-Alcoholado et al. 2020). The probiotics confer anticancer activities by inducing apoptosis amidst the infection, and trigger both natural and acquired immune responses. Moreover, probiotics reduce the level of carcinogens by neutralizing them at molecular and cellular processes in the intestine (Khan et al. 2013). Therefore, probiotic research to use them as therapeutics is crucial, as it has impact on gut homeostasis, microbial infections, carcinogens, epithelial integrity, and other aspects affecting gut milieu. Notably, probiotics have the potency to alter the intestinal microbiota, reducing chronic inflammation and the effect of carcinogens (Sehrawat et al. 2021).

The probiotic intake (10^9–10^{12} bacterial cells) could exhibit health benefits such as boosting macrophages, lymphocytes, interferon gamma, and immunoglobulin activity in the blood, even after few weeks. Hence, the consumption of probiotic bacteria such as *Lactobacillus* and *Bifidobacterium* species at a dose of 10^{10}–10^{11} CFU/day for at least 4–6 weeks may reduce the risk of CRC (Molska and Reguła 2019). Notably, the probiotics have been demonstrated to restore the quantity and functionality of commensal gut bacteria, which would have been decreased after chemotherapy. *Lactobacillus* sp. and *Bifidobacteria* have profoundly shown to minimize diarrhea and gastrointestinal dysfunction in cancer patients following various treatments. Therefore, based on the health benefits of probiotics, they could be accepted as one of the major approaches for the treatment and/or prevention of CRC, with reduced adverse effects (Sehrawat et al. 2021).

13.7 Conclusion

Probiotics have gained significant popularity in medicine for their potential to prevent and treat various diseases with minimal adverse effects. The probiotic research has highlighted the potential of probiotics in cancer prevention and its potential as an adjuvant treatment during chemotherapy. The currently approved anticancer treatments pose serious adverse effects, which reduce the quality of life. Hence, consumption of probiotics could be an effective therapeutic strategy, with little or no side effects. Despite the above, precise mechanisms by which probiotics prevent CRC remains unclear. Moreover, there are challenges and limitations that need to be addressed, to develop an effective probiotics treatment. It requires the standardization in terms of selection of probiotic strains, which could survive, resist the cancer environment, and still be stable. Henceforth, evidence-based clinical trials proving the efficacy and safety of probiotics, as well as further research to explore their mechanisms of actions of probiotics, are needed.

References

Aindelis G, Chlichlia K (2020) Modulation of anti-tumour immune responses by probiotic bacteria. Vaccine 8:1–17. MDPI AG

Aleta A, Hrvat F. Probiotics review and future aspects [Internet]. Available from: www.ijisrt.com

Aziz N, Bonavida B (2016) Activation of natural killer cells by probiotics. For Immunopathol Dis Therap 7(1–2):41–55

Bhatia A, Lab I, Randhawa MK, Kaur M (2014) Immunomodulatory and therapeutic potential of CPG and probiotic DNA-a review [Internet]. Int J Chem Pharm Anal 1. Available from: http://www.ijcpa.in

Bodera P, Chcialowski A (2009) Immunomodulatory effect of probiotic bacteria. Recent Patents Inflamm Allergy Drug Discov 3(1):58–64

Chattaraj S, Mitra D, Chattaraj A, Chattaraj M, Kundu M, Ganguly A et al (2023) Antioncogenic potential of probiotics: challenges and future prospective. Indian J Microbiol Res 10(1):1–10

Commane D, Hughes R, Shortt C, Rowland I (2005) The potential mechanisms involved in the anti-carcinogenic action of probiotics. Mutat Res – Fund Mol Mech Mutagen 591:276–289

Dasari S, Kathera C, Janardhan A, Praveen Kumar A, Viswanath B (2017) Surfacing role of probiotics in cancer prophylaxis and therapy: a systematic review. Clin Nutr 36:1465–1472. Churchill Livingstone

Davis CD, Milner JA (2009) Gastrointestinal microflora, food components and colon cancer prevention. J Nutr Biochem 20:743–752

De Almeida CV, De Camargo MR, Russo E, Amedei A (2019) Role of diet and gut microbiota on Colorectal cancer immunomodulation. World J Gastroenterol 25:151–162. Baishideng Publishing Group Co

Delgado S, Sánchez B, Margolles A, Ruas-Madiedo P, Ruiz L (2020) Molecules produced by probiotics and intestinal microorganisms with immunomodulatory activity. Nutrients 12:391. MDPI AG

dos Reis SA, da Conceição LL, Siqueira NP, Rosa DD, da Silva LL, Peluzio M d CG (2017) Review of the mechanisms of probiotic actions in the prevention of colorectal cancer. Nutr Res 37:1–19. Elsevier Inc.

Dubey V, Ghosh AR, Bishayee K, Khuda-Bukhsh AR (2016) Appraisal of the anti-cancer potential of probiotic Pediococcus pentosaceus GS4 against colon cancer: in vitro and in vivo approaches. J Funct Foods 23:66–79

Eslami M, Yousefi B, Kokhaei P, Hemati M, Nejad ZR, Arabkari V et al (2019) Importance of probiotics in the prevention and treatment of colorectal cancer. J Cell Physiol 234:17127–17143. Wiley-Liss Inc.

Faghfoori Z, Pourghassem Gargari B, Saber Gharamaleki A, Bagherpour H, Yari Khosroushahi A (2015) Cellular and molecular mechanisms of probiotics effects on colorectal cancer. J Funct Foods 18:463–472. Elsevier Ltd

Faghfoori Z, Faghfoori MH, Saber A, Izadi A, Yari Khosroushahi A (2021) Anticancer effects of bifidobacteria on colon cancer cell lines. Cancer Cell Int 21(1):258

Geier MS, Butler RN, Howarth GS (2006) Probiotics, prebiotics and synbiotics: a role in chemoprevention for colorectal cancer? Cancer Biol Ther 5:1265–1269. Landes Bioscience

Górska A, Przystupski D, Niemczura MJ, Kulbacka J (2019) Probiotic bacteria: a promising tool in cancer prevention and therapy. Curr Microbiol 76:939–949. Springer New York LLC

Hardy H, Harris J, Lyon E, Beal J, Foey AD (2013) Probiotics, prebiotics and immunomodulation of gut mucosal defences: homeostasis and immunopathology. Nutrients 5:1869–1912. MDPI AG

Kahouli I, Tomaro-Duchesneau C, Prakash S (2013) Probiotics in colorectal cancer (CRC) with emphasis on mechanisms of action and current perspectives. J Med Microbiol 62:1107–1123

Kasarello K, Sajdel-Sulkowska E (2019) Developmental significance of early gut-associated lymphoid tissue (GALT)-microbiota interactions in health and disease: creating balance between tolerance and inflammation in children. Open J Pediatr Child Health 4(1):040–046

Kaźmierczak-Siedlecka K, Roviello G, Catalano M, Połom K (2021) Gut microbiota modulation in the context of immune-related aspects of Lactobacillus spp. and Bifidobacterium spp. in gastrointestinal cancers. Nutrients 13:2674. MDPI AG

Khan AA, Khurshid M, Khan S, Alshamsan A (2013) Gut microbiota and probiotics: current status and their role in cancer therapeutics. Drug Dev Res 74:365–375

Kumar Mishra V, Mohammad G, Jha A, Mishra VK (2008) Immunomodulation and anticancer potentials of yogurt probiotic. EXCLI J 7. https://doi.org/10.17877/DE290R-15147

Kumar M, Kumar A, Nagpal R, Mohania D, Behare P, Verma V et al (2010) Cancer-preventing attributes of probiotics: an update. Int J Food Sci Nutr 61(5):473–496

Kuugbee ED, Shang X, Gamallat Y, Bamba D, Awadasseid A, Suliman MA et al (2016) Structural change in microbiota by a probiotic cocktail enhances the gut barrier and reduces cancer via TLR2 signaling in a rat model of colon cancer. Dig Dis Sci 61(10):2908–2920

Legesse Bedada T, Feto TK, Awoke KS, Garedew AD, Yifat FT, Birri DJ (2020) Probiotics for cancer alternative prevention and treatment. Biomed Pharmacother 129:110409. Elsevier Masson SAS

Maldonado Galdeano C, Cazorla SI, Lemme Dumit JM, Vélez E, Perdigón G (2019) Beneficial effects of probiotic consumption on the immune system. Ann Nutr Metab 74(2):115–124

Mazziotta C, Tognon M, Martini F, Torreggiani E, Rotondo JC (2023) Probiotics mechanism of action on immune cells and beneficial effects on human health. Cells 12:184. MDPI

Molska M, Reguła J (2019) Potential mechanisms of probiotics action in the prevention and treatment of colorectal cancer. Nutrients 11:2453. MDPI AG

Naito S, Koga H, Yamaguchi A, Fujimoto N, Hasui Y, Kuramoto H et al (2008) Prevention of recurrence with epirubicin and Lactobacillus casei after transurethral resection of bladder cancer. J Urol 179(2):485–490

Nazir Y, Hussain SA, Abdul Hamid A, Song Y (2018) Probiotics and their potential preventive and therapeutic role for cancer, high serum cholesterol, and allergic and HIV diseases. Biomed Res Int 2018:3428437. Hindawi Limited

Pino A, De Angelis M, Chieppa M, Caggia C, Randazzo CL (2020) Gut microbiota, probiotics and colorectal cancer: a tight relation. World Cancer Res J 7:e1456

Plaza-Díaz J, Ruiz-Ojeda FJ, Vilchez-Padial LM, Gil A (2017) Evidence of the anti-inflammatory effects of probiotics and synbiotics in intestinal chronic diseases. Nutrients 9:555. MDPI AG

Pope JL, Tomkovich S, Yang Y, Jobin C (2017) Microbiota as a mediator of cancer progression and therapy. Transl Res 179:139–154. Mosby Inc.

Rafter J (2004) The effects of probiotics on colon cancer development. Nutr Res Rev 17(2):277–284

Raman M, Ambalam P, Kondepudi KK, Pithva S, Kothari C, Patel AT et al (2013) Potential of probiotics, prebiotics and synbiotics for management of colorectal cancer. Gut Microbes 4(3):181–192

Reddy BS (1998) Prevention of colon cancer by pre- and probiotics: evidence from laboratory studies. Br J Nutr 80(4):S219–S223. CAB International

Sánchez B, Delgado S, Blanco-Míguez A, Lourenço A, Gueimonde M, Margolles A (2017. Wiley-VCH Verlag) Probiotics, gut microbiota, and their influence on host health and disease. Mol Nutr Food Res 61. https://doi.org/10.1002/mnfr.201600240

Sánchez-Alcoholado L, Ramos-Molina B, Otero A, Laborda-Illanes A, Ordóñez R, Medina JA et al (2020) The role of the gut microbiome in colorectal cancer development and therapy response. Cancers 12:1406. MDPI AG

Sehrawat N, Yadav M, Singh M, Kumar V, Sharma VR, Sharma AK (2021) Probiotics in microbiome ecological balance providing a therapeutic window against cancer. Semin Cancer Biol 70:24–36

Şener D, Bulut HN, Güneş Bayir A (2021) Probiotics and relationship between probiotics and cancer types. Bezmialem Sci 9(4):490–497

Sharaf LK, Sharma M, Chandel D, Shukla G (2018) Prophylactic intervention of probiotics (L.acidophilus, L.rhamnosus GG) and celecoxib modulate Bax-mediated apoptosis in 1,2-dimethylhydrazine-induced experimental colon carcinogenesis. BMC Cancer 18(1):1111

Shida K, Nomoto K (2013. Available from: http://journals.lww.com/ijmr) Probiotics as efficient immunopotentiators: translational role in cancer prevention [Internet]. Indian J Med Res 138:808

Shyu PT, Oyong GG, Cabrera EC (2014) Cytotoxicity of probiotics from Philippine commercial dairy products on cancer cells and the effect on expression of cfos and cjun early apoptotic-promoting genes and Interleukin-1 β and Tumor Necrosis Factor-α proinflammatory cytokine genes. Biomed Res Int 2014:491740

Singh RP, Bashir H, Kumar R (2021) Emerging role of microbiota in immunomodulation and cancer immunotherapy. Semin Cancer Biol 70:37–52

Śliżewska K, Markowiak-Kopeć P, Śliżewska W (2021) The role of probiotics in cancer prevention. Cancers 13:1–22. MDPI AG

Song M, Chan AT (2019) Environmental factors, gut microbiota, and colorectal cancer prevention. Clin Gastroenterol Hepatol 17:275–289. W.B. Saunders

Tilg H, Adolph TE, Gerner RR, Moschen AR (2018) The intestinal microbiota in colorectal cancer. Cancer Cell 33:954–964. Cell Press

Torres-Maravilla E, Boucard AS, Mohseni AH, Taghinezhad-S S, Cortes-Perez NG, Bermúdez-Humarán LG (2021) Role of gut microbiota and probiotics in colorectal cancer: onset and progression. Microorganisms 9:1021. MDPI AG

Tripathy A, Dash J, Kancharla S, Kolli P, Mahajan D, Senapati S et al (2021) Probiotics: a promising candidate for management of colorectal cancer. Cancers (Basel) 13(13):3178

Uccello M, Malaguarnera G, Basile F, D'agata V, Malaguarnera M, Bertino G et al (2012) Potential role of probiotics on colorectal cancer prevention [Internet]. Available from: http://www.biomedcentral.com/1471-2482/12/S1/S35

Verhoeven V, Renard N, Makar A, Van Royen P, Bogers JP, Lardon F et al (2013) Probiotics enhance the clearance of human papillomavirus-related cervical lesions: a prospective controlled pilot study. Eur J Cancer Prev 22(1):46–51

Wang FH, Zhang XT, Li YF, Tang L, Qu XJ, Ying JE et al (2021) The Chinese Society of Clinical Oncology (CSCO): clinical guidelines for the diagnosis and treatment of gastric cancer, 2021. Cancer Commun 41(8):747–795

Yahfoufi N, Mallet JF, Graham E, Matar C (2018) Role of probiotics and prebiotics in immunomodulation. Curr Opin Food Sci 20:82–91. Elsevier Ltd

Yasmin I, Saeed M, Khan WA, Khaliq A, Chughtai MFJ, Iqbal R et al (2020) In vitro probiotic potential and safety evaluation (Hemolytic, cytotoxic activity) of bifidobacterium strains isolated from raw camel milk. Microorganisms 8(3):354

Yousefi B, Eslami M, Ghasemian A, Kokhaei P, Salek Farrokhi A, Darabi N (2019) Probiotics importance and their immunomodulatory properties. J Cell Physiol 234:8008–8018. Wiley-Liss Inc.

Yu AQ, Li L (2016) The potential role of probiotics in cancer prevention and treatment. Nutr Cancer 68:535–544. Routledge

Zhang C-X, Wang H-Y, Chen T-X (2019, 2019) Interactions between intestinal microflora/probiotics and the immune system. Biomed Res Int:6764919. Hindawi Limited

Zhu Q, Gao R, Wu W, Qin H (2013) The role of gut microbiota in the pathogenesis of colorectal cancer. Tumour Biol 34:1285–1300. IOS Press BV

Cancer Risk in Alterations of Prebiotics 14

Manikkam Rajalakshmi, Periyasamy Vijayalakshmi, and Sabapathy Indu

Abstract

Prebiotics are indigestible food constituents that selectively induce the growth and activity of beneficiary gut microbiota and have evolved as potent adjuncts in the prevention and therapy of cancer. This chapter gives an in-depth exploration of how cancer is influenced by prebiotics through various biological mechanisms. It signifies the key role of the gut microbiome in immune system modulation, inflammation reduction, and metabolic processes alteration, which are crucial in cancer progression and development. Specific prebiotic substances, such as fructooligosaccharides, galactooligosaccharides, and inulin, have been studied for their effects on various cancers. These were evident from preclinical and clinical studies. This chapter also delves into the pathways associated with prebiotics and its efficiency in enhancing the efficacy of conventional cancer treatment thereby potentially reducing the associated side effects. We included

M. Rajalakshmi (✉)
DBT-BIF Centre, Holy Cross College (Autonomous), Tiruchirappalli, Tamil Nadu, India

PG & Research Department of Zoology, Holy Cross College (Autonomous), Tiruchirappalli, Tamil Nadu, India

PG & Research Department of Biotechnology & Bioinformatics, Holy Cross College (Autonomous), Affiliated to Bharathidasan University, Tiruchirappalli, Tamil Nadu, India
e-mail: rajalakshmi@hcctrichy.ac.in

P. Vijayalakshmi
DBT-BIF Centre, Holy Cross College (Autonomous), Tiruchirappalli, Tamil Nadu, India

PG & Research Department of Biotechnology & Bioinformatics, Holy Cross College (Autonomous), Affiliated to Bharathidasan University, Tiruchirappalli, Tamil Nadu, India

S. Indu
DBT-BIF Centre, Holy Cross College (Autonomous), Tiruchirappalli, Tamil Nadu, India

S. Pathak et al. (eds.), *Microbiota and Dietary Mediators in Colon Cancer Prevention and Treatment*, https://doi.org/10.1007/978-981-96-0297-1_14

the overview of current scientific evidence, enduring the oncological therapeutic potential of prebiotics. However, this chapter addresses challenges and limitations in this field, including the need for rigorous clinical trials and the complexity of host biological interactions with gut microbiome. Moreover, we explore personalized prebiotic therapy concepts, like screening profiles of individual microbiomes and genetic information, to optimize the outcome of treatment. The integration of prebiotics in standard cancer care protocols, along with identifying needed biomarkers, can predict and monitor the response to prebiotic interventions. By provoking the existing gaps in this research, this chapter aims to provide a comprehensive understanding of the role of prebiotics in cancer care along with secondary intervention in various diseases and to pave the way for future studies that can translate these findings into clinical applications.

Keywords

Prebiotics · Cancer prevention and care · Pathway mechanism · Prebiotic therapy · Gut microbiota

14.1 Introduction

A study conducted in 2020 found that cancer is a significant global health issue, impacting more than 18.1 million individuals worldwide. Cancers can be comprehensively characterized by examining the compilation of distinctive features that are developed as a result of many variables, including lifestyle, environment, genetic predisposition, and others. One strategy used by physicians and scientists to develop treatment strategies for patients with cancer is to closely examine the tumor microenvironment. This has facilitated comprehension of the pathophysiology of malignancies, as they do not all advance at the same pace. Cancer treatment methods encompass a diverse range of interdisciplinary fields that provide adverse effects such as peripheral neuropathy, anemia, edema formation, alopecia, chronic tiredness, thrombocytopenia, and psychosis, among others. The impact of these factors on the quality of life is significant, leading to a rise in the popularity of prebiotic treatment (Sung et al. 2021).

Glenn Gibson and Marcel Roberfroid introduced the prebiotics in 1995; they were the pioneers in introducing the prebiotics concepts as a stimulant to induce fermentation in the human microbiome. Prebiotic substances also facilitate beneficial bacterial growth in the gut and enable specific modifications in composition, formation, and gastrointestinal (GI) microflora activity, hence providing various benefits to the hosts. The gut microbiome plays an important role in maintaining human health through its contributions to digestion, synthesis of vital nutrients, and modulation of the immune system (Kim and Park 2023). Prebiotics are substances that undergo fermentation and consist of specific indigestible compounds such as galactooligosaccharides (GOS), starch, inulin, and pectin. These substances promote the good bacterial growth that already exists in the gut, resulting in improved anticancer treatment effects. Prebiotics are thought to facilitate the formation of short-chain fatty acids (SCFAs), which help to reduce byproducts from protein breakdown. In addition, they emit signals that suppress the development of gastrointestinal cancer and possess an anti-adhesive property against harmful microorganisms. The microbiota plays a crucial role in maintaining overall balance and stability within the body. Over 97% of the human microbiota is found in the gastrointestinal tract, primarily in the colon. Gut dysbiosis refers to an imbalance between the microbiota and the human host that has been associated with many physiological problems, including cancer.

14.2 Types of Prebiotics

There are a multitude of prebiotic varieties. Most of them belong to a subset of carbohydrate categories, specifically oligosaccharide carbohydrates (OSCs).

14.2.1 Fructans

This group includes fructooligosaccharide (FOS), oligofructose, and inulin. They are made up of a continuous chain of fructose molecules joined by $\beta(2 \rightarrow 1)$ bonds. They usually have glucose units with a $\beta(2 \rightarrow 1)$ link at the end. The degree of polymerization (DP) of inulin can reach 60, while that of FOSs is less than 10. Previous studies have indicated that fructans can specifically support the growth of lactic acid bacteria. According to recent studies, one of the most important factors in deciding which bacteria can ferment fructans is their chain length. Fructans can, therefore, either directly or indirectly promote the growth of many bacterial species (Kherde et al. 2021; Roy and Dhaneshwar 2023).

14.2.2 Galactooligosaccharides

Two subgroups of GOS are distinguished: (1) GOS containing an excess of galactose at locations C3, C4, or C6, and (2) GOS generated by enzymatic transglycosylation of lactose. As a result of this chemical reaction, a mixture of tri- to pentasaccharides with galactose connected in $\beta(1 \rightarrow 6)$, $\beta(1 \rightarrow 3)$, and $\beta(1 \rightarrow 4)$ configurations is the main product. This specific type of GOS is referred to as trans-GOS or TOS. GOSs have the capacity to greatly accelerate *Lactobacilli* and *Bifidobacteria* growth. *Bifidobacteria* has been shown to significantly assimilate GOS in infants. Similar to *Bifidobacteria, Enterobacteria, Bacteroidetes*, and *Firmicutes* are also susceptible to GOS. There exist certain GOS that are generated from lactulose, which is an isomer of lactose. These prebiotics are generated from lactulose and are also considered as prebiotics. In addition to these types of GOS, there are other varieties known as raffinose family oligosaccharides (RFO) that are based on sucrose extension. The impact of RFO on gut microbiota remains unclear (Davani-Davari et al. 2019).

14.2.3 Starch and Oligosaccharides Derived from Glucose

One kind of starch that is difficult for the upper gastrointestinal tract to digest is called resistant starch (RS). RS can enhance health by generating a substantial amount of butyrate, leading to its potential classification as a prebiotic. Firmicutes have the greatest assimilation of RS, as seen in multiple groups. According to laboratory research, RS can be broken down by *Bifidobacterium adolescentis* and *Ruminococcus bromii*, and to a lesser extent by *Eubacterium rectale* and *Bacteroides thetaiotaomicron*. Nevertheless, in the incubations containing both bacteria and fecal matter, the breakdown of RS cannot occur if *R. bromii* is not present. Polydextrose is an oligosaccharide produced from glucose. The structure is composed of glucan molecules that have several branches and glycosidic connections. There is limited data suggesting that it can promote the growth of Bifidobacteria, although this has not been definitively shown (Davani-Davari et al. 2019).

14.2.4 Additional Oligosaccharides

Pectin is the source of oligosaccharides, which are carbohydrates. Pectic oligosaccharide (POS) is the precise term for this kind of oligosaccharide. They originate from the elongation of either rhamnose (rhamnogalacturonan I) or galacturonic acid (homogalacturonan). Methyl esterification of the carboxyl groups and acetylation of the structure at C2 or C3 are both feasible. Numerous sugars, including arabinose, ferulic acid, xylose, and galactose, are joined to the side chains. Their designs differ significantly depending upon the origination of POSs (de Oliveira et al. 2024).

14.2.5 Noncarbohydrate Oligosaccharides

While the concept of prebiotics often includes carbs more frequently, some substances like flavanols derived from cocoa are not included in this category but are nevertheless advised to be considered prebiotics. It has been shown that flavanols can promote the growth of lactic acid bacteria in both lab and living organisms (Davani-Davari et al. 2019).

14.3 Prebiotics Mechanism of Action

Prebiotics function through a complex interplay between these indigestible chemicals and the intricate gut bacteria ecosystem. Prebiotics are important for the general health of the host since they primarily serve as a food source for the beneficial bacteria that reside in the gut. Incorporating prebiotics into our diet is the most straightforward method to develop or restore a healthy microbiome in cases of dysbiosis. Numerous prebiotics have been proposed to offer a range of health advantages, such as preserving gut health, lowering blood cholesterol, and having anti-inflammatory, antidiabetic, and antihypertensive properties. Prebiotics can improve health in two ways: directly by influencing the immune system and its pro- and anti-inflammatory reactions, or indirectly by acting as a substrate for probiotics, encouraging the synthesis of SCFAs, and preventing infections. Prebiotics primarily act on the epithelium associated with the stomach (Li and Zhou 2023).

Immune cells may be impacted by prebiotics directly or indirectly. Prebiotics are logically used since the big intestine breaks them down instead of the small intestine because humans lack the enzymes needed to break the complex polymer bonds. Prebiotics are transported by the body intact to the large intestine, where they are broken down and fermented specifically to produce particular secondary metabolites by the intestinal flora. These metabolites have a beneficial effect on the physiological functions of the host, including immune regulation, pathogen resistance, intestinal barrier integrity, and mineral absorption. The primary mechanism of prebiotics is to directly promote the proliferation of probiotic microorganisms. The presence of probiotics in the large intestine facilitates an interaction with the intestinal bacteria, hence enhancing the chemical, mechanical, biological, and immunological defenses of the bacterial population. Probiotics interact with intestinal cells once they reach the intestine to stimulate the regrowth of the mucosal lining, enhance the production of mucus, repair the integrity of the intestinal walls, and maintain the strength of the protective barriers (Pujari and Banerjee 2021).

Prebiotics typically stimulate probiotics to generate secondary metabolites, such as SCFAs including butyric, propionic, and acetic acids, as well as various antimicrobial peptides. These metabolites have the ability to decrease the pH of the colon, creating a favorable environment for probiotics to grow and establish themselves. Maintaining an appropriate pH level in the human gut is essential for ensuring that probiotics can effectively cling to surfaces. The SCFA attaches to the G-protein-coupled receptor (GPCR) located in the intestine. This interaction triggers many signaling pathways, such as the Mitogen-activated protein kinase (MAPK) pathway, which leads to the synthesis of antimicrobial peptides. SCFAs stimulate the production of anti-inflammatory cytokines via regulating immune cells such as T cells, dendritic cells, and macrophages, which in turn affects the functioning of epithelial cells (Garcia and Lopez 2023)

It has been observed that butyric acid helps eliminate reactive oxygen species (ROS) in the gut. This creates a protective environment for bacteria that are sensitive to oxygen. Reports indicate that the prebiotics GOS, resistant starch (RS), and inulin-type fructans have the ability to break down bile salts and acids by binding to bile acids. SCFAs have the ability to affect the functioning of intestinal epithelial cells and maintain the integrity and barrier function of the intestines. This leads to a decrease in their reabsorption and an acceleration of their turnover rate in the intestines. Comprehending the methods by which prebiotics work is crucial for understanding their function in supporting a well-balanced and robust gut flora. As research in this field progresses, it creates opportunities to develop specific interventions that can improve gut health and treat numerous health issues related to an imbalance in gut bacteria.

Significant Modes of Action of Prebiotics

Carbohydrates that are indigestible are those that the upper gastrointestinal system is unable to break down. Rather, they travel through the ileum and colon, where the gut flora ferments them.

- Prebiotics do promote intestinal homeostasis.
- Prebiotics coat the surface receptors of the host.
- They synthesize bacteriocins.
- Beneficial bacteria synthesize SCFAs through the use of nondigestible carbohydrates.
- Epithelial cells derive their energy from SCFAs.
- They control metabolic activity and adjust the immune system.

14.4 Targets for Prebiotics: Gut Microbiota and Intestinal Health

SCFAs are widely recognized for their role in promoting gut health. The bacteria in the colon ferment resistant starches and dietary fiber, resulting in the production of SCFAs. However, there is an ongoing dispute regarding their contribution to providing additional energy to the host in cases of obesity. In contrast, SCFAs (SCFA) contribute 10% of the total energy in humans. As per the energy yield theory, dysbiotic microbiomes have an increased ability to extract energy from food. This indicates that the bacteria have the ability to transform dietary fiber and nondigestible carbohydrates into SCFAs (SCFAs), which over time may cause weight growth and supply the host with more energy (Fig. 14.1).

In addition, GPCR has the ability to detect SCFAs (SCFAs) and bind them to the process of lipid and glucose breakdown. Adipocytes have two main proteins, GPR41 and GPR43, that are activated by SCFAs and enter endocrine L cells. Peptide YY is secreted when the intestinal GPR41 is stimulated, resulting in an increase in the time it takes for food to pass through the gut and an increase in the feeling of fullness. Inflammation is reduced by the activation of intestinal GPR43, which also stimulates glucagon-like peptide (GLP), a hormone that plays a role in regulating insulin release. Enteroendocrine L cells both produce and release the gut-trophic hormone GLP-2 in addition to expressing GLP-1. SCFAs aid in the maintenance of glucagon-like peptide-2 (GLP-2). Ingestion of nutrients is the main factor that initiates the release of GLP-2. It functions as a critical protector of the gut barrier, preventing the leakage of lipopolysaccharides (Huang and Tang 2023).

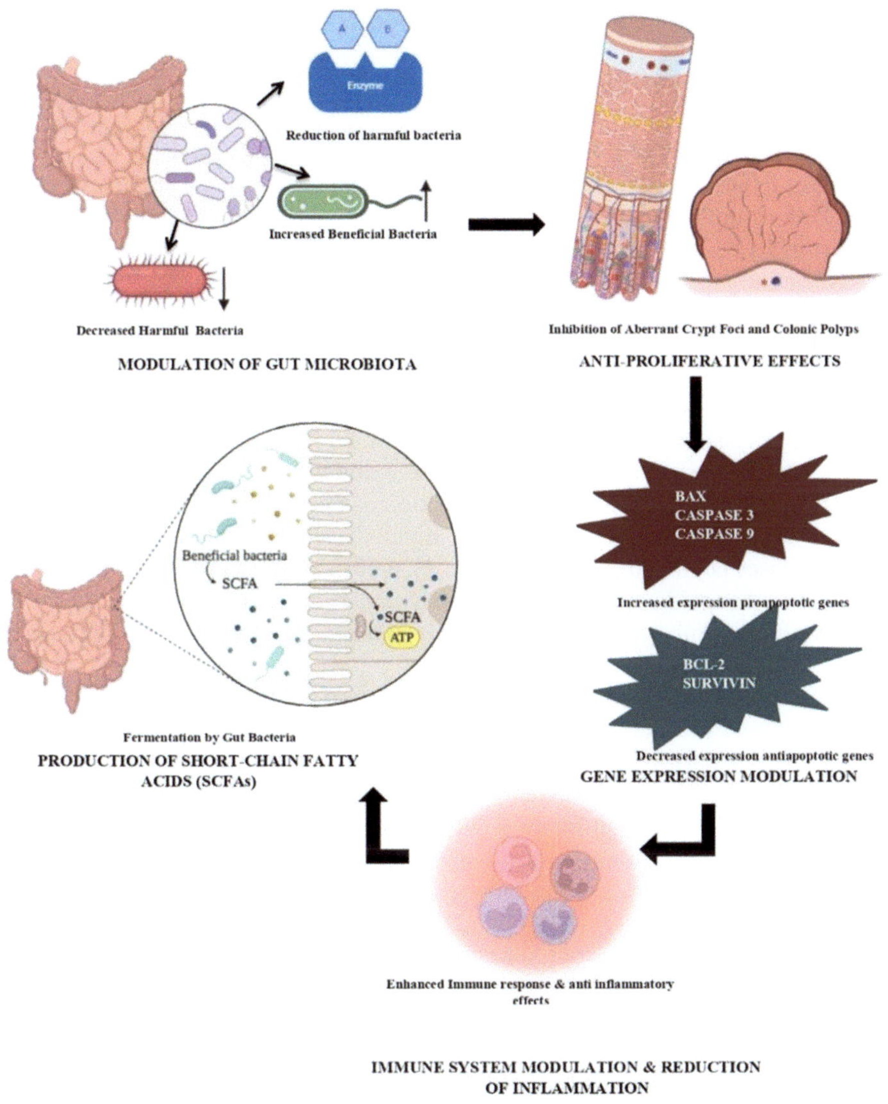

Fig. 14.1 Prebiotics modulate gut microbiota by increasing beneficial bacteria such as *Lactobacillus, Lacticaseibacillus, Lactiplantibacillus, Levilactobacillus, Limosilactobacillus, Bifidobacterium,* and *Bacteroides,* decreasing harmful ones such as *clostridium* spp, reduce harmful enzyme activity such as nitroreductase, ß-glucuronidase, and azoreductase, and inhibit the formation of aberrant crypt foci and colonic polyps. They alter gene expression to promote cancer cell death by increasing proapoptotic genes Bax, Caspase 3, and Caspase 9 and decreasing antiapoptotic genes Bcl-2 and Survivin. In addition, prebiotics enhance the immune response, reduce chronic inflammation in the gut, and produce SCFAs through bacterial fermentation collectively contributing to cancer prevention and improved gastrointestinal health

14.5 Therapeutic Potential of Prebiotics

14.5.1 Prebiotics in Cancer

14.5.1.1 Prebiotics in Cancer Growth, Immunity, and Therapy

Prebiotics impact on cancer has recently become a prominent subject of study and discourse. A growing body of research suggests that eating specific prebiotics, including FOSs or fructans, may help lower the risk of colon cancer. Prebiotics such as soybean oligosaccharides, oligofructose, FOS, GOS, and resistant starches may have a beneficial effect on the absorption and metabolism of minerals. Prebiotic mineral absorption and their anticancer efficacy are correlated in a complex way that is impacted by several variables.

Prebiotics produce SCFAs, which improve the solubility of minerals in the gastrointestinal system. As a result, this may lead to the formation of SCFA–salt complexes, which improve mineral absorption. Prebiotics also increase the amount of surface area that is available for the absorption of minerals, promote the synthesis of proteins that bind calcium, and break down the phytic acid–mineral complex, which releases bound minerals. Furthermore, prebiotics can synthesize compounds that regulate bone, like phytoestrogens found in particular foods. They further support a healthy gut microbiota by aiding in the stabilization of mucus and gut bacteria (Davis and Miller 2023).

Raftilose Synergy, a prebiotic mostly composed of inulin with oligofructose added, was used in a study carried out by Femia in 2002. Fischer 344 rats were used as the experimental model in this investigation. This study looked at how prebiotics affected the expression of the genes for cyclooxygenase 2 (COX2) and glutathione S-transferase Pi (GST-P) in tumor-induced and healthy rats. As compared to rats who did not receive any prebiotics, the data showed that rats given prebiotics expressed less GST-P. Regarding tumors, a consistent outcome was documented. They also discovered that prebiotics significantly reduced the amount of inducible nitric oxide synthase (i-NOS) produced in the tumors of rats. The results presented above suggest that prebiotics have a preventive effect on the body of the host and can lessen the development of cancer. Over the course of several weeks, prebiotics progressively alter the gut flora. Still, there is much to learn about the precise mechanism by which prebiotics work. The medication is nevertheless beneficial for medical conditions, even though the differences could go away after a short while when a regular diet is resumed.

Patients with gynecological cancer were the subjects of a recent clinical investigation. Prebiotics usually take 1–2 weeks to start working; however, even though they only last temporarily, they have a big impact. Immune checkpoint inhibitors are the driving force behind cancer immunotherapy, a highly significant development in the field of cancer treatment. Immune checkpoint inhibitors prevent T cells, B cells, and certain cancer cells from producing checkpoint proteins. These inhibitors can help T cells that tumor cells have deactivated and become functional again, allowing them to recognize and react to antigens. Inulin, a well-known prebiotic supplement, has demonstrated a decrease in transplantable melanoma growth in a model that depends on T cell activity (Nguyen and Lee 2023). When rat models were fed a diet rich in inulin and oligofructose, tumor growth was observed to abruptly stop, resulting in a considerable improvement in survival.

Prebiotics can improve the immune system of the host by altering the gut microbiota, which is essential for producing immune responses to fight tumor development. The gut microbiota makeup is significantly influenced by the food. By impeding tumor cell metabolism and enhancing the efficacy of anticancer T cells, both innate and adaptive immunity are essential in preventing the development of cancer. This is a basic component in all forms of cancer. CD8+ T cells play a pivotal role in the modulation of cancer and considerably aid in the inhibition of malignant proliferation. The main target of cancer treatment is T cells (Wang and Zhang 2023).

Therefore, nutritional intervention specifically, the use of prebiotics that modifies the characteristics of the gut microbiota serves as a treatment facilitator by preventing the development of malignancies. Studies have demonstrated the positive impact of dietary inulin doses on gut microbiomes, which results in a notable antitumor effect. An investigation employing an inulin-mediated diet revealed the beginning of immune responses against malignancies and the activation of specific T cells in epithelial tissues. Certain signals that are triggered by microbial metabolites are circulated throughout the body to induce immune responses. This controls how quickly tumors grow and how the immune system responds to them. The immune response is constantly influenced by the gut microbiota composition, which results in changes to both innate and adaptive immunity on several levels. To every antigen, the innate immune system reacts defensively.

Immune cells can be stimulated by prebiotics through several signaling pathways, which can cause pro- or anti-inflammatory responses. Prebiotics such as inulin, *Platycodon grandiflorus*, and short-chain/long-chain fructooligosaccharides (scGOS/lcFOS) have been demonstrated to inhibit the immune system by inducing the expression of the FOXP3 gene and releasing interleukin 10 (IL-10). However, some prebiotics, such as FOS and 3'-Sialyllactose, can prevent the expression of cytokines that promote inflammation, creating an anti-inflammatory milieu. To gain a thorough grasp of the signaling pathways and direct interactions between prebiotics and immune cells, more research is necessary.

14.5.1.2 The Relationship Between Colon Cancer and Prebiotics

Prebiotics, when administered alone or in combination with adopted probiotics and synbiotics, have the potential to prevent tumor formation. These have been proved through various animal and human studies. The in vivo rodent model has demonstrated the antiproliferative effects of prebiotics by inhibiting the formation of aberrant crypt foci (ACF) and colonic polyps along with changes in their gene expression, such as decreased antiapoptotic Bcl-2 and Survivin expression and increased pro-apoptotic Bax, Caspase 3, and Caspase 9 in vitro in cells. Prebiotics not only show positive advantages in avoiding colorectal cancer development but also reduce symptoms of the disease (Zhao and Ma 2023). Research and development are crucial to developing the prebiotic as an optimal tool in the battle against colorectal cancer.

The first visible signs of aberrant cell formation are called aberrant crypt foci (ACF), and they appear as big intestinal lesions. Chromoscopic colonoscopy with high magnification is employed to detect ACF lesions, which typically manifest as intestinal mucosa. The crypts are elevated on the upper surface of the typical mucosa. The structure is composed of highly compacted epithelial cells and has noticeably restricted luminal apertures compared to the adjacent crypts. Multiple genetic and metabolic changes accumulate during the course of ACF, leading to polyp, adenoma, and adenocarcinoma. There is a chance that even trace amounts of ACF can lead to colon cancer. The specific crypts responsible for tumour development remain unidentified. Numerous studies provide evidence that ACF is associated with the beginning of colon cancer. Research by Perrin et al. indicates that certain fibres facilitate the development of a stable colonic environment that produces butyrate. This particular intestinal environment lowered the incidence of ACF. Therefore, it is clear that the risk of developing colon cancer is considerably lower in a colonic environment that consistently produces butyrate.

14.5.1.3 The Use of Prebiotics in Mitigating the Adverse Effects of Chemotherapy

There are an estimated 39 trillion microorganisms in the human body, including various kinds of bacteria, viruses, and fungi. The development of illness and the preservation of health are significantly impacted by the human microbiome. Age, lifestyle, genetic makeup, and other factors are among the many that influence it. Any imbalance in the microbiota is called dysbiosis, and it can have serious consequences for human health. It may lead to fatal conditions like cancer and inflammatory bowel disease (IBD), and it may also affect vital physiological functions like homeostasis—maintaining internal balance—and host immunity, the body's defense against disease (Tan and Chan 2023).

The most common cancer treatment, chemotherapy, uses cytotoxic medications to kill the cancer cells as they divide quickly. However, because of their cytotoxic nature, they can have long-term impacts on the human body as well as equally harmful effects on normal cells. Some side effects can be serious and even fatal, while others might be mild and easily handled. Many adverse effects, including nausea, vomiting, mouth ulcers, alopecia, abnormalities of the nails and skin, fatigue, fever, and gastrointestinal problems like constipation and diarrhea, can be brought on by chemotherapy. According to recent studies, chemotherapy-induced gastrointestinal toxicity (CIGT) may also be exacerbated by chemotherapeutic medications due to their disruption of the normal gut microbiota composition and increase in intestinal permeability (Martinez and Ruiz 2023).

Eighty percent of patients receiving chemotherapy experience gastrointestinal toxicity brought on by the drug. The gut microbiota diversity was found to have significantly decreased by Montassier et al. (2015) during a 5-day study including high-dose chemotherapy. Bacteroidetes and Proteobacteria populations increased; however, Firmicutes and Actinobacteria numbers were drastically decreased. The term "gut microbiota dysbiosis" describes the disturbance of the gut's microbial ecology, which results in changes to the gut's composition and functionality. The body may produce more proinflammatory cytokines as a result of gut dysbiosis. Leaky gut syndrome is a condition brought on by cytokines' capacity to modify the gut lining. This disease allows pathogens and toxins to cross the intestinal epithelial layer and enter the bloodstream.

Among the substances that fall under this category are mannan-oligosaccharides (MOS), β-1,4-mannobiose (MNB), sialyl-oligosaccharides, xylooligosaccharides (XOS), epilactose, arabinoxylan oligosaccharides (AXOS), and chitooligosaccharides (COS). Aliyah and As'ad (2022) studied the influence of prebiotic supplements on chemotherapy-induced mucositis (CIM), a frequently reported adverse effect of chemotherapy. During their treatment, cancer patients may become more prone to gastrointestinal (GI) problems as a result of increased inflammation. In addition to causing physical symptoms, chemotherapy-induced gut dysbiosis causes anxiety, depression, and post-traumatic stress disorder, among other psychological and cognitive side effects in patients with cancers. During their observation, gastrointestinal problems such as nausea, vomiting, constipation, and sore throat significantly improved. SCFAs are produced when prebiotics are broken down by colonic bacteria.

The maintenance of intestinal health and optimal function and structure are dependent on the SCFAs, specifically acetic, propionic, and butyric acid. In particular, colon cells use the chemical butyrate as an energy source. It also has an effect on reducing the colon's pH level. By fortifying the gut's immune system and expanding the surface area that is accessible for absorption, lactate contributes to improved gut health. The mucin layer is essential for maintaining the integrity of the intestinal barrier (Yang and Wang 2023). Mucins coat the epithelial surface in a coating of mucus that prevents germs from penetrating the mucosa and growing excessively. Mucins act as a physical and chemical barrier, giving the epithelium more defense against luminal substances such as bacterial toxins found in the stomach. Mucins produce a wide variety of potential attachment sites for bacteria, including pathogenic and mutualistic microbes. Studies show that prebiotics work by interacting with their end sugars, which bind to the receptors that encourage bacterial adhesion and prevent bacteria from adhering to the mucus layer in the intestine.

Furthermore, it has been shown by Sinclair et al. that prebiotics have the ability to prevent cholera toxin from attaching to its cellular receptor, suggesting that this barrier is not exclusive to bacterial cells. Studies have linked the activation of mucin production to FOS and GOS (Clark and Green 2023). This increase in mucin synthesis lowers the risk of pathogenic infection by preventing the spread of germs from one location to another. Prebiotics can also lessen the negative effects of chemotherapy on the gastrointestinal tract by modifying mucin, which preserves the integrity of the mucosal layer. To fully comprehend how prebiotics affect mucin levels in the various intestinal regions impacted by mucositis, more research is required (Rodriguez and Garcia 2023).

14.5.2 Prebiotics in Other Aspects of Diseases

14.5.2.1 Effect of Prebiotics on the Composition of Gut Microflora

Prebiotic dietary fibers have a beneficial effect on the microorganisms in the intestines. Prebiotic fibers have a distinct role in cholesterol metabolism, particularly in the fermentation of substances and the regulation of microorganisms in the gut. Studies have shown that the nondigestible dietary items known as probiotics and prebiotics, which stimulate the growth of colonic bacteria, can prevent the formation of tumors and preneoplastic lesions in the colons of animals that have received treatment with carcinogenic chemicals. The metabolic activity of the complex mixture of oligosaccharides in human breast milk provides evidence for the presence of beneficial bacteria in the gastrointestinal tracts of neonates who are fed with this type of milk. The formula milk contains a more developed gut microbiome. Infants are administered the adult formulation to establish a specific type of gut bacteria. Dysbiosis of the gut microbiota can lead to autoimmune diseases, infections, and allergy reactions in older individuals (Lopez and Martinez 2023).

14.5.2.2 Prebiotics Modulation of Immune Function

Our diet has the ability to alter immunological activities, which in turn affects the host's resistance in multiple ways. Nonessential nutrients, such as undigested carbs, have a major effect on controlling immunological responses in addition to essential meal components, namely in the stomach's lymphoid tissues. Bodera conducted a comprehensive analysis of past scientific literature on the impact of prebiotics on host immunity. According to the findings, there is strong evidence that prebiotics, more especially, inulin have a function in regulating immune activity. In regard to the above study, prebiotics can be used to modify immune parameters in peripheral circulation, secondary lymphoid tissues, and gut-associated lymphoid tissues. As the first line of defense, the innate immune system gets rid of invasive infections and stops the spread of infectious agents. Physical barriers such as skin and mucous membranes are included in the concept, in addition to blood cells and tissues like phagocytes and natural killer cells, as well as soluble mediators like complement proteins and cytokines (Patel and Shah 2023).

14.5.2.3 Prebiotics as Dietary Agents That Affect the Gut Microbiota in Obese People

There is growing evidence that the gut microbiota is important for the development of obesity and related diseases. Therefore, targeting the gut microbiota could be a potential strategy for addressing obesity. It is suggested that reestablishing a mutually beneficial interaction between the gut microbiota and the host has tremendous potential, as obesity is thought to be associated with microbial dysbiosis. Prebiotics are ingested in order to alter the gut microbiome (Gonzalez and Perez 2023). Prebiotics are indigestible by host enzymes, but they are transported to the large intestine where they undergo fermentation in the gastrointestinal system. Beneficial bacteria, specifically *Bifidobacterium* and *Lactobacillus*, are encouraged to flourish during this fermentation process and have been linked to a number of health benefits.

14.5.2.4 Prebiotics Assist in the Management of Cognition

Schizophrenia is the most common cause of disability worldwide, resulting in a 10–20-year reduction in life expectancy. Pharmacotherapy and improper treatment methods are acknowledged as major factors contributing to cognitive impairments and weight gain. Consequently, it is imperative to discover remedies that can effectively alleviate one or both of these conditions. Utilizing gut microbiome derived from dietary sources such as prebiotics as a potential intervention for the treatment of various diseases could be a viable approach. Moreover, prebiotics affect how drugs are metabolized. When it comes to obesity, they raise the levels of anorexigenic gut hormones such as peptide tyrosine, glucagon-like 1, and leptin while lowering those of the orexigenic hormone ghrelin (Clarke et al. 2020).

14.5.2.5 Prebiotics to Increase Phosphate Absorption and Enzyme Activity

Prebiotics increase the activity of enzymes and make it easier for the colon to absorb phosphate. Researchers looked into how prebiotics added to baby formula affected the amount of enzyme activity and phosphate absorption in the small intestines of Sprague Dawley (SD) rats. At random, three sets of 48- and 15-day-old SD rats with identical weights were assigned: SD rats fed standard baby formula without prebiotic therapy were used to generate a control group. Group B rats received the standard newborn formula enriched with oligosaccharides, while group C rats received polysaccharides. After 28 days, the results were verified. In contrast to group A, groups B and C exhibited the following characteristics: (1) Group C SD rats had higher levels of lactase and sucrose activity in their small intestines. (2) In group C of SD rats, the lactase gene was expressed 1.68 and 2.26 times more in the anterior and posterior sections of the small intestine, respectively (Chong et al. 2018)

In addition, the posterior region of the small intestine showed a 0.99 improvement in the relative expression of the maltase-glucoamylase gene. The anterior and posterior segments of the small intestine of group C of SD rats showed increases in the relative expressions of the sodium-dependent phosphate transport protein 2B gene of 1.85 and 2.28, respectively. According to the research, prebiotics added to baby formula may reduce intestinal damage and boost the function of small intestinal enzymes by upregulating the expression of enzyme genes and the Na/Pi-IIb gene. When it came to health benefits, polysaccharides outperformed oligosaccharides.

14.5.2.6 The Relationship Between Prebiotics and the Bioavailability of Trace Elements and Minerals

Mineral and trace element absorption is positively impacted by prebiotic carbohydrates. There is strong evidence that prebiotic carbohydrates, like lactulose and nondigestible oligosaccharides, improve the absorption of minerals and trace elements. Nevertheless, the preponderance of research is centered on rodents because human studies are scarce. Prebiotics may have a potent osmotic effect, lowering pH and facilitating proton exchange, according to one theory. This leads to a rise in the concentrations of butyrate and calbindin. Improving the absorption of minerals and trace elements can be achieved by increasing the surface area of the colon (Martínez et al. 2009).

14.5.2.7 Prebiotics in the Treatment of Cardiovascular and Diabetes

Chronic inflammation is linked to prediabetes and is associated with a higher risk of type 2 diabetes and cardiovascular issues. An imbalance in the natural microbiota is linked to elevated lipopolysaccharide levels and has been linked to the emergence of type 2 diabetes and cardiovascular diseases. Prebiotics, including inulin, are involved in how certain natural microbes in the bodywork. They lessen intestinal permeability, lower endotoxin levels, and aid in preventing metabolic dysfunction in animals. There is still uncertain about how prebiotics affect cardiovascular health in people who are at risk of type 2 diabetes. Prebiotic supplements combined with inulin may be used as a prophylactic strategy to reduce cardiovascular disease risk in those who are susceptible to type 2 diabetes. Establishing and promoting dietary guidelines for prebiotics throughout the scientific and medical communities is one way that this strategy can impact therapeutic procedures (Slavin 2013).

14.5.2.8 Prebiotics and Renal Profile

Prebiotic, probiotic, and symbiotic supplementation have been linked to improved renal function; this is supported by an array of research. A meta-analysis of a renal profile that was acquired from a clinical trial was studied by Firouzi and Haghighatdoost (2018). The intervention group showed a higher quantity of creatinine and a lower glomerular filtration rate when compared to the placebo group, according to the researchers. However, there was no statistically significant difference between these two groups. The overall effect on blood urea nitrogen levels showed lower values than in the placebo group. In comparison to the placebo group, the uric acid levels were greater in the intervention group. Before well-designed, large-scale randomized controlled trials show the effectiveness and safety of prebiotic, probiotic, and symbiotic supplements in enhancing renal function, their use should be limited to people with impaired renal function.

14.5.2.9 Anxiolytic Effect of Prebiotics

Prolonged administration of prebiotics, specifically a blend of fructooligosaccharides and galactooligosaccharides, has been shown to exhibit antidepressant characteristics and mitigate feelings of anxiety. It has been found that the combination of fructooligosaccharide–galactooligosaccharide (FOGLG) and GOS lowers the corticosterone concentration in the body by primarily reducing its release. The activation of many genes associated with the levels of SCFAs in the hypothalamus and hippocampus is influenced by prebiotics (Louis and Flint 2017). After receiving prebiotics, the concentration of butyrate dropped and the levels of propionate and acetate rose. These changes have a strong correlation with the favorable effects on behavior. Moreover, the administration of fructooligosaccharides and galactooligosaccharides lowered the levels of corticosterone and proinflammatory cytokines resulting from extended stress. It also restored usual anxiety and depression-like behavior and mitigated the deleterious effects of stress on the microbiota.

14.6 Limitations and Future Directions

The limitations of prebiotics in cancer therapy include the need for more rigorous, large-scale clinical trials to confirm efficacy and safety, the complexity of individual microbiome variability affecting treatment outcomes, and the challenge of identifying reliable biomarkers for monitoring responses to prebiotic interventions. Future research should focus on conducting large-scale, randomized controlled trials to confirm the efficacy and safety of specific prebiotics in cancer prevention and treatment across diverse populations. It is crucial to delve deeper into the molecular mechanisms by which prebiotics influence cancer pathways, particularly their interactions with gut microbiota, immune modulation, and metabolic alterations. Personalized prebiotic therapies tailored to individual microbiome profiles and genetic backgrounds should be developed to enhance therapeutic benefits and minimize adverse effects. Exploring the synergistic effects of combining prebiotics with other therapeutic modalities, such as probiotics, chemotherapy, immunotherapy, and targeted therapies, could improve overall treatment efficacy. Additionally, identifying reliable biomarkers for predicting response to prebiotic interventions and monitoring their effects on cancer progression and patient outcomes is essential. Long-term studies assessing the impact of prebiotic consumption on cancer recurrence, survival rates, and quality of life in cancer patients will further inform clinical practice and integration of prebiotics into cancer care.

14.7 Conclusion

In conclusion, prebiotics represent an adjuvant strategy in the prevention and therapy of cancer by modulating gut microbiome and creating subsequent impact on systemic health. The evidence discussed in this chapter underscores the potent benefits of prebiotics in reducing cancer risk by improving the efficacy of conventional cancer treatments. However, several challenges persist, such as unresolved complex interactions among prebiotics, microbiomes, and cancer biology. More rigorous and well-designed clinical trials are necessary to establish standardized guidelines for the use of prebiotics in oncology. In addition, personalized approaches considering individual microbiome profiles may enhance the efficacy and safety of prebiotic interventions in cancer care.

References

Aliyah R, As'ad S (2022) Prebiotics for Chemotherapy-Induced Lower Intestinal Mucositis (CIM) With Moderate Protein Energy Malnutrition: A Case Report. Curr Dev Nutr 6(Suppl 1):734. https://doi.org/10.1093/cdn/nzac062.003. PMCID: PMC9193497

Chong CYL, Bloomfield FH, O'Sullivan JM (2018) Factors affecting gastrointestinal microbiome development in neonates. Nutrients 10(3):274

Clark B, Green S (2023) The impact of prebiotics on gut health and cancer outcomes. Gut Microbes 15(1):213–225

Clarke G, Sandhu KV, Dinan TG, Cryan JF (2020) Probiotics and mental health: a review of mechanisms of action and clinical effects. Neurosci Biobehav Rev 113:132–145

Davani-Davari D, Negahdaripour M, Karimzadeh I, Seifan M, Mohkam M, Masoumi SJ, Berenjian A, Ghasemi Y (2019) Prebiotics: definition, types, sources, mechanisms, and clinical applications. Food (Basel, Switzerland) 8(3):92. https://doi.org/10.3390/foods8030092

Davis R, Miller P (2023) The potential of prebiotics in enhancing cancer treatment outcomes. J Cancer Res Clin Oncol 149(7):1973–1985

Firouzi S, Haghighatdoost F (2018) The effects of prebiotic, probiotic, and synbiotic supplementation on blood parameters of renal function: A systematic review and meta-analysis of clinical trials. Nutrition 51–52:104–113. https://doi.org/10.1016/j.nut.2018.01.007. Epub 2018 Feb 6. PMID: 29626749

Garcia M, Lopez A (2023) Influence of prebiotics on immune response in cancer patients. Clin Nutr 42(2):345–355

Gonzalez L, Perez M (2023) Modulating gut microbiota with prebiotics in cancer therapy. Cancer Treat Rev 111:102532

Huang J, Tang Y (2023) Prebiotics and their role in cancer therapy: a review. Food Chem 399:134013

Kherade M, Solanke S, Tawar M, Wankhede S (2021) Fructooligosaccharides: a comprehensive review. J Ayurved Herb Med 7:193–200. https://doi.org/10.31254/jahm.2021.7305

Kim H, Park S (2023) Prebiotic fibers and cancer prevention: current evidence and future perspectives. Cancer Res Treat 55(2):345–356

Li X, Zhou L (2023) Mechanisms of prebiotics in cancer prevention. Nutrients 15(4):856

Lopez F, Martinez J (2023) Prebiotics in cancer management: clinical evidence and future directions. Int J Mol Sci 24(8):7845

Louis P, Flint HJ (2017) Formation of propionate and butyrate by the human colonic microbiota. Environ Microbiol 19(1):29–41

Martinez E, Ruiz C (2023) The emerging role of prebiotics in cancer therapy. Semin Cancer Biol 87:146–155

Martínez I, Wallace G, Zhang C, Legge R, Benson AK, Carr TP (2009) Dietary modulation of gut microbial composition and function in humans and animal models. Br J Nutr 101(2):148–161

Montassier E, Gastinne T, Vangay P, Al-Ghalith GA, Bruley des Varannes S, Massart S, Moreau P, Potel G, de La Cochetière MF, Batard E, Knights D (2015) Chemotherapy-driven dysbiosis in the intestinal microbiome. Aliment Pharmacol Ther 42(5):515–28. https://doi.org/10.10.1111/apt.13302. Epub 2015 Jul 6. PMID: 26147207

Nguyen T, Lee J (2023) Prebiotics as adjuvant therapy in cancer: benefits and mechanisms. Crit Rev Food Sci Nutr 63(6):854–868

de Oliveira DP, Todorov SD, Fabi JP (2024) Exploring the prebiotic potentials of hydrolyzed pectins: mechanisms of action and gut microbiota modulation. Nutr 16(21):3689. https://doi.org/10.3390/nu16213689

Patel K, Shah R (2023) Prebiotics and their role in cancer prevention. J Med Food 26(3):315–328

Pujari R, Banerjee G (2021) Impact of prebiotics on immune response: from the bench to the clinic. Immunol cell biolog, 99(3):255–273. https://doi.org/10.1111/imcb.12409

Rodriguez A, Garcia D (2023) Recent advances in prebiotics and cancer therapy. J Oncol 2023:5793124

Roy S, Dhaneshwar S (2023) Role of prebiotics, probiotics, and synbiotics in management of inflammatory bowel disease: current perspectives. World J Gastroenterol, 29(14):2078–2100. https://doi.org/10.3748/wjg.v29.i14.2078

Slavin J (2013) Fiber and prebiotics: mechanisms and health benefits. Nutrients 5(4):1417–1435

Sung H, Ferlay J, Siegel RL, Laversanne M, Soerjomataram I, Jemal A, Bray F (2021) Global cancer statistics 2020: GLOBOCAN estimates of incidence and mortality worldwide for 36 cancers in 185 countries. CA Cancer J Clin 71(3):209–249. https://doi.org/10.3322/caac.21660

Tan S, Chan K (2023) Prebiotic compounds in cancer prevention and therapy. Trends Cancer 9(4):234–245

Wang L, Zhang H (2023) Prebiotics and cancer: an overview of current research. J Funct Foods 100:105297

Yang Z, Wang P (2023) Efficacy of prebiotics in enhancing the immune response in cancer. Cancer Immunol Res 11(3):563–574

Zhao X, Ma L (2023) The role of prebiotics in modulating gut microbiota and enhancing cancer immunotherapy. Oncotarget 14(2):198–213

Priming the Nutrition-Gut-Immuno-Oncology Axis

15

Anupama Saha, Naman Kapoor, Navkiran Kaur, Sudeep Bose, Minu Kanchan Kesheri, and Shivani Sharda

Abstract

Nutrition, considered the silhouette of health and development, is essential throughout the bio ecosystems. The exquisite details of the prognosis and development of cancer have heralded into patronizing the quintessential role of microbiome. Gut architecture devises the outcomes of health and disease, which is now conclusively accepted in both communicable and non-communicable diseases. This understanding of complex diseases now encompasses not only genetic and immunological factors, but also the critical role of the microbial community in their development and progression. The molecular and cellular signatories relying on the physiological pulse of the host additionally concur a powerful modulation of health or disease based on the extraneous features of diet and nutrition of the individuals. This chapter aggregates the aspects of the nutrition-directed gut–immunology–oncology axes weathering the therapeutic outcomes. Priming the nutrition-gut-immuno-oncology axis entwines the strategic clinico-molecular aspects of leveraging the nutritional interventions to rebound gut parameters towards healthy states. This results in tuning the immune system functions, thereby enhancing the treatment efficacies and positive outcomes of nutrition science, microbiome research, oncology studies, and theragnostics.

Anupama Saha and Naman Kapoor contributed equally with all other contributors.

A. Saha · N. Kapoor · N. Kaur · S. Sharda (✉)
Amity Institute of Biotechnology, Amity University, Noida, Uttar Pradesh, India
e-mail: ssharda@amity.edu

S. Bose · M. K. Kesheri
Marshall University, Huntington, WV, USA

S. Pathak et al. (eds.), *Microbiota and Dietary Mediators in Colon Cancer Prevention and Treatment*, https://doi.org/10.1007/978-981-96-0297-1_15

Keywords

Gut–immunology axis · Nutrition–oncology axis · Gut microbiome

Chapter Learning

- Nutrition-gut axis: Gut health and dietary influences on gut microbiome composition and function
- Immuno-modulation and immunotherapy responses
- Priming: Specific dietary interventions, probiotics, prebiotics, and other nutritional strategies to create a more favourable environment for treatment.
- Nutrition-gut-immuno-oncology axis: Harnessing nutrition triggered impact on gut health and immune function, improving cancer treatment outcomes and overall patient well-being

15.1 Human Cancer Paradigm

The paradigm hypothesizes that cancer originates following a sequence of events that include a pathogenic stimulus, which can either be biological or chemical, followed by chronic inflammation, which ensues a development of fibrosis, leading to a pre-cancerous niche (Zhou et al. 2021). The pre-cancerous niche activates the deployment of a chronic stress escape strategy, but when this case fails to resolve, it could lead to the transition of a normal cell to a cancer cell. Most of the discoveries so far in cancer genetics reported are either late events or epiphenomena that occur after the appearance of the pre-cancerous niche (Brücher and Jamall 2014).

15.1.1 The Patterns and Profiling of the Paradigm

The theory that somatic mutations are the primary trigger for carcinogenesis, proposed early in the twentieth century as one of the predominant paradigms, is highly relevant to gastrointestinal (GI) tract cancers. It is crucial to understand the role of somatic mutations in driver genes in various tissues, including the GI tract. Mutations that result from damaged DNA and mis-repair play a significant role in the ageing process and tumorigenesis, including in the GI tract. The somatic mutation distributions in cancer genomes, including genes like TP53 and CDKN2a, are of particular importance in upper gastrointestinal tract (UGIT) cancers, such as oesophageal, esophagogastric junction, stomach, and duodenal cancers. The somatic mutations then contribute to the pathology, aetiology, and geographic distribution disparities observed in UGIT cancers. Additionally, understanding the interplay between somatic mutations and environmental tumour-promoting agents is crucial for assessing cancer risk, including GI tract cancers (Balmain 2020). Recently, research has demonstrated how the gut microbiome influences cancer development, acting as both promoters and inhibitors.

15.1.2 The Types of Cancers: Focussing on Gastrointestinal Tract (GI) Cancer

Cancer as a disease is caused by an interplay of a majority of genetic and environmental factors, approximately 20% of cancers are affected by microorganisms, such as *Helicobacter pylori, hepatitis* B and C viruses (HBV and HCV), including the *human papillomavirus* (HPV) that galvanizes susceptibility of humans towards a multitude of cancers. Individual and inter-individual microbiome differences are considered the underlying causes of various diseases and health issues, particularly in GI tract cancers. The gut microbiome plays a crucial role in GI cancer, as digestive system cancers contribute to about one-third of cancer-related deaths. Approximately, 15–20% of cancers are caused by infectious agents, while 20–30% are linked to tobacco use and 30–35% are associated with diet, physical inactivity, and obesity (Ott et al. 2010; Warren et al. 2014; Key et al. 2020). The dysregulation of gut microbiota can impact cancer development through various mechanisms (Ng et al. 2019). Shotgun metagenomics sequencing through random sequencing established that the bacterial microbiome of the human gut is dominated by five main phyla in the taxonomic classification stated at 60–80% *Firmicutes,* 15–25% *Bacteroidetes,* 2.5–5% *Actinobacteria,* 1–10% *Proteobacteria,* and 0.1–2.2% *Verrucomicrobia* phyla, therefore, *Firmicutes* and *Bacteroidetes* represent 90% of the gut microbiota (Yin et al. 2024) (Fig. 15.1).

The gut microbiota, which consists of a divergent community of microorganisms residing in the GI tract, plays an important role in regulating the host physiology, including immune functions, metabolism, and maintaining the gut barrier integrity. When this gut barrier integrity balance is disrupted, it leads to various detrimental effects, majorly in an increased risk of cancer development. Alterations in the composition and abundance of microbiomes have been associated with various GI cancers, including gastric, hepatic, and oesophageal cancers (ESCA) (Ağagündüz et al. 2023).

Dysbiosis, or the imbalance of gut microbiota, can lead to chronic inflammation in the gut, playing a crucial role in the development of gastric cancer (stomach cancer), and certain bacterial species, namely, *Helicobacter pylori*, are particularly relevant (Pereira-Marques et al. 2019). *Helicobacter pylori* infection is a commonly known risk factor for causing gastric cancer (Bakhti and Latifi-Navid 2021). The liver is closely connected to the gut through the portal vein, allowing gut-derived substances (including microbial metabolites) to reach the liver. Dysbiosis can lead to increased production of pro-inflammatory molecules, affecting liver health. Chronic liver inflammation (in conditions like non-alcoholic fatty liver disease [NAFLD]) is associated with an altered gut microbiome. Additionally, liver cirrhosis, a precursor to hepatocellular carcinoma (HCC) (common type of liver cancer), is influenced by gut microbial changes. The inflammatory response can also create an environment that is cohesive to tumour initiation and progression (Jiang and Schnabl 2020). Oesophageal adenocarcinoma (EAC) (a subtype of oesophageal cancer) has been linked to gastroesophageal reflux disease (GERD) (Hasan et al. 2020; Gillespie et al. 2021). The gut microbiome may influence GERD by

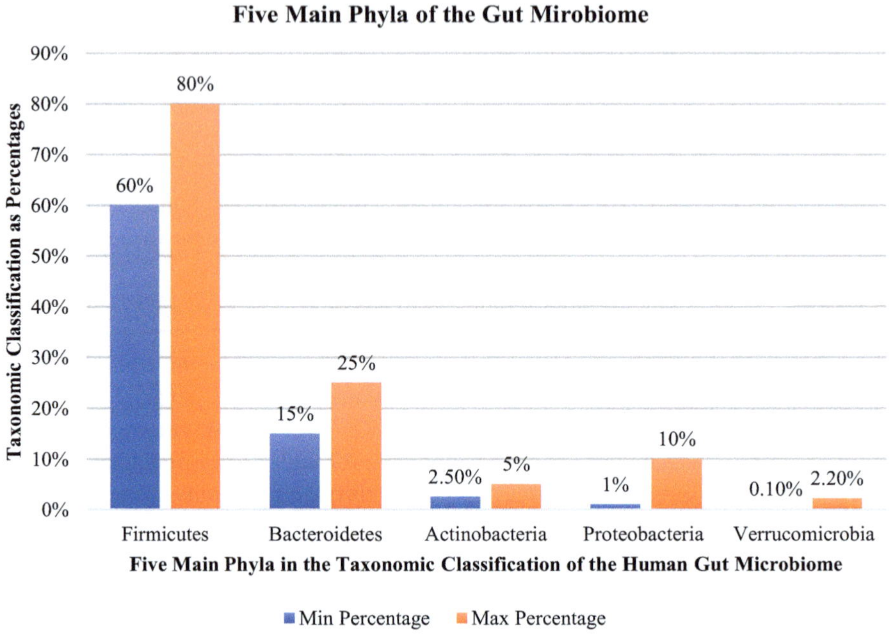

Fig. 15.1 Five main phyla of the gut microbiome along with approximate percentage composition

impacting inflammation and altering oesophageal motility. An imbalance in the microbiome, or dysbiosis, can lead to chronic inflammation of the oesophagus, potentially promoting cancer development.

In colorectal cancer (CRC), the gut microbiome is intensively involved in CRC development (Rebersek 2021). Diet, as a key modulator of gut microbiota, can rapidly alter the composition of the gut microbiome and its associated metabolites. In metabolite production, such as short-chain fatty acids (SCFAs), the gut microbiota can metabolize dietary components and produce various metabolites, some of which can have carcinogenic properties. Dietary fibre promotes the growth of beneficial bacteria in the gut. These bacteria ferment dietary fibre, producing SCFAs like butyrate. Butyrate has anti-inflammatory properties and supports gut health. For example, certain metabolites produced by gut bacteria, such as secondary bile acids, can promote carcinogenesis in the gut (Duygu Ağagündüz et al. 2023).

Regarding immunology or immune modulation, the gut microbiota plays a critical role in educating the immune system and maintaining immune homeostasis. Dysregulation of the gut microbiota can lead to immune dysregulation, contributing to an increased risk of cancer development. The human gut microbiota can also influence cancer immunotherapy, making a substantial contribution to human health and illness (Park et al. 2022). The link between gut microbiota and cancer immunotherapy is becoming widely recognized, indicating the microbiome's potential effect on precision medicine (Cheuk et al. 2023). According to research, microbial components such as lipopolysaccharides can trigger immunological responses, implying that the gut microbiome composition might influence immune system activity and consequently affect cancer progression (Li et al. 2024). Furthermore, gut bacteria have been shown to activate the immune system, enhancing those immunological responses to tumours and consequently altering cancer prognosis depending on tumour-associated gut microbiota (Hou et al. 2022). The interaction between the gut microbiota and cancer is of tremendous relevance in cancer treatment.

The gut microbiota has been shown to impact the effectiveness of the immune checkpoint inhibitors (ICIs) and impact the overall cancer immunotherapy outcomes. Studies have shown that the gut microbiota can modulate antitumor immunity and affect the efficacy of cancer immunotherapies. The mechanisms underlying the gut microbiota's impact on antitumor immunity reveal that it can reprogram the tumour microenvironment's (TME) immunity and influence the efficacy of immune checkpoint inhibitors by producing metabolites that can impact local and systemic antitumor immune responses and improve immune checkpoint inhibitor efficiency (Park et al. 2022). These findings highlight the crucial role of the gut microbiome in modulating immune responses and affecting cancer immunotherapy, as well as potential therapeutic implications for tumour progression and cancer treatment via precision medicine.

Understanding the functional role of the gut microbiota in modulating immune responses can be promising for the development of microbiome-based therapeutic techniques, perhaps leading to advancements in cancer therapeutics and effective treatments in the future (Cheuk et al. 2023), This has direct implication of the gut microbiota playing a critical role in modulating tumour growth and immune responses, and it bears potential as an area of study for predicting cancer prognosis and improving cancer treatment results. Extensive research has demonstrated that the gut microbiota also modulates the host signalling pathways, which can influence various host signalling pathways, including those involved in cell proliferation, apoptosis, and differentiation, which are key processes in cancer development and progression. The intricate interactions between the gut microbiota and host signalling pathways can either promote or inhibit cancer (Xie and Liu 2024). In this context, dysbiosis in the gut microbial population has been related to abnormal modulation of signalling cascades, which contributes to the deregulation of cellular processes that can lead to cancer progression. Understanding these complex interactions provides valuable insights into the potential mechanisms by which the gut microbiota influences cancer development, opening up a promising avenue for investigating targeted interventions that could manipulate these signalling pathways to effectively prevent or treat cancer (Luu et al. 2023).

Recently, the two-sample Mendelian randomization (MR) studies, such as those conducted by the MiBioGen consortium and the Dutch Microbiome Project, have provided valuable insights into the potential cause and effect link between gut microbiota and GI cancers (Su et al. 2023). These studies revealed specific microbial taxa implicated in these associations, shedding light on their potential impact on different GI cancers. For instance, the phylum *Verrucomicrobia* showed a strong negative association with intrahepatic cholangiocarcinoma, indicating a potential protective role. Similarly, the order *Bacillales* demonstrated a positive association with pancreatic cancer, suggesting a potential risk factor. These findings underscore the intricate relationship between gut microbiota and the development of GI cancers, providing valuable insights for further research and potential interventions. Therefore, understanding the role of specific gut bacteria in GI cancer may provide novel avenues for prevention and treatment and targeting the gut microbiome could potentially enhance our ability to manage GI cancers effectively (Long et al. 2023).

15.1.3 The Prevalence and Disease Fatality and Mortality Rates

When considering the highest to lowest prevalence of cancers and assessing disease mortality and fatality rates, it's important to note that mortality rates of cancer have declined over the last three decades in most high-income countries, reflecting improvements in cancer prevention, diagnosis, and management. However, substantial differences in mortality, incidence, and survival still exist worldwide, which requires a comprehensive understanding of global cancer trends. By 2040, the global estimation of cancer is projected to grow significantly, with approximately 21 out of 100 men and 18 out of 100 women globally expected to develop cancer by the age of 75 (Fig. 15.2).

Efforts to reduce global cancer disparities begin with an understanding of geographic patterns in cancer incidence, mortality, and prevalence. Such understanding is important for public health and to prevent the sudden surge of cancer in the future, providing vital insights to address the challenges posed by varying cancer prevalence and mortality rates globally (Cheuk et al. 2023). According to world statistics, stomach cancer (gastric cancer) has an incidence rate of 7.0 new cases per 100,000 individuals per year, with a mortality rate of 2.7 per 100,000 individuals per year. The 5-year relative survival rate for stomach cancer is 37% for localized cases, 14% for regional spread, and 4% for distant metastasis (Park et al. 2022). In 2019, India

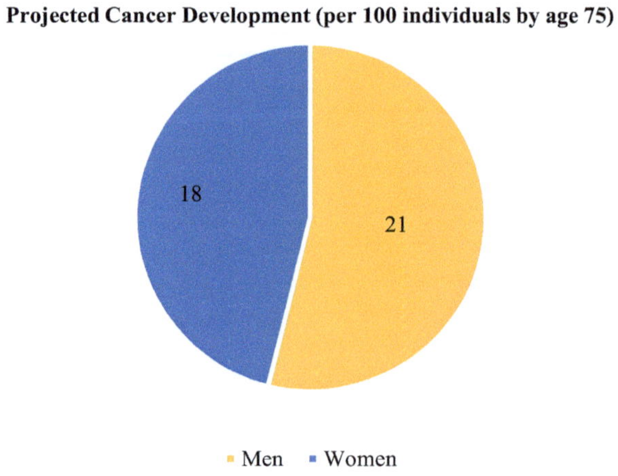

Fig. 15.2 Projected cancer development rates by 2040: Men 21 and women 18 per 100 individuals by age of 75

reported a total of 81,770 deaths (95% UI 70,139–94,684) due to stomach cancer, reflecting a substantial mortality rate. Research, therefore, suggests a critical role of the gut microbiome in affecting cancer outcomes, leading to an emerging area of research linking microbial imbalances with increased cancer risks and severity (Higashi and Kurokawa 2024). India's high fatality and mortality rates emphasize the importance of extensive studies into how gut microbiota alterations may contribute to carcinogenesis and cancer progression. These findings align with global observations that variations in the gut microbiome can significantly impact cancer incidence and outcomes, highlighting the need to explore microbiome-targeted therapies.

Colorectal cancer (CRC) is one of the most common cancers worldwide, contributing to substantial morbidity and mortality. From a global perspective, factors like geography, race, sex, and diet impact the composition of the microbiome, which is extremely important for screening, early detection and predicting treatment outcomes. CRC has an incidence rate of 36.5 new cases per 100,000 individuals per year, with a mortality rate of 12.9 per 100,000 individuals per year (Higashi and Kurokawa 2024). The survival rates vary by stage, but early detection and surgical removal improve outcomes. For localized cases, the 5-year survival rate is higher. In India, colorectal cancer is the fourth most common cancer type in males, following oral, lung, and stomach cancer (gastric cancer). The mortality rates for colorectal cancer are a major issue, specifically in India, as the research suggests that cases are increasing in countries with traditionally lower rates. In India, it has been noted that the changes in the gut microbiome have either a causative or associative relationship with colorectal cancer. The study disclosed that Indian patients with CRC demonstrated universal CRC-associated patterns, and the microbiome collection had the closest correlation with other Indian cohorts, emphasizing the need to uncover important local patterns in India. Findings related to cancers caused by the human gut microbiome in India: a study has amalgamated and sequenced microbial samples from a larger cohort of 46 Indian patients with CRC and 43 healthy volunteers. These samples resemble other Asian samples, demonstrating relatively high levels of the bacterium *Prevotella* and showing a correlation in predicting cancer status between groups. It was also stated that there are regional variations and that there is more similarity between location-matched comparisons (Bose et al. 2023; Ajab et al. 2024).

Hepatocellular carcinoma (HCC—liver cancer) accounted for 830,180 related deaths worldwide in 2020, making it the fourth leading cause of cancer-related death globally. The age-standardized mortality rate for HCC worldwide is 8.9 per 100,000 persons (Feng et al. 2022). The 5-year survival rate for advanced-stage HCC is about 18% (Gabbia and De Martin 2024). Hepatocellular carcinoma (HCC) primarily develops in patients with chronic liver diseases, including chronic viral hepatitis (HBV, HCV), NAFLD, and alcohol-related fatty liver disease. Dysbiosis has been associated with chronic liver diseases and HCC development. Dysbiosis contributes to inflammation, DNA damage, and tumour progression in HCC. Although the liver is not directly in contact with the microbiota, anatomic links exist between the liver and the gut. Gut-derived metabolites and products (microbiota-associated molecular patterns) impact liver health. In India, HCC is a significant health concern, and understanding the gut microbiome's role is crucial.

Recent research has focused on Indian CRC microbiomes, revealing both universal CRC-associated patterns and location-specific variations. Despite differences in sequencing technologies, concordance exists between Indian CRC cohorts, emphasizing the global nature of the CRC microbiome (Marin-Acevedo et al. 2018).

Oesophageal adenocarcinoma has an incidence rate of about 22,370 new cases diagnosed annually in the United States, with approximately 16,130 deaths occurring from oesophageal cancer each year in the United States (more common in men than women). The 5-year relative survival rate varies by stage, with localized cases having better outcomes. The survival rates are estimates based on large populations and may not predict individual outcomes. Early detection, treatment, and risk factor management are crucial in improving prognosis (Rodriguez et al. 2023; Hur et al. 2012). Oesophageal adenocarcinoma (EAC) constitutes approximately 20% of all oesophageal cancer (ESCA) cases and primarily affects individuals of Caucasian descent in developed nations, often arising in the distal oesophagus or gastroesophageal junction. Its incidence has surged by nearly 60% in recent decades, making it the fastest-growing malignancy. Dysbiosis has been associated with EAC development. Although the liver is not directly in contact with the microbiota, anatomic links exist between the liver and the gut. Gut-derived metabolites and products impact liver health, potentially influencing EAC progression. India's diverse population and unique dietary habits make its gut microbiome interesting to study (Wang and Geng 2023). The "Landscape of Gut Microbiome-Pan India Exploration" (LogMPIE) study identified *Prevotella copri* and *Faecalibacterium prausnitzii* as predominant organisms in Indians (Dubey et al. 2018). Investigating the gut microbiome's role in EAC could offer novel insights for diagnosis and treatment. Collaborative studies focusing on Indian populations might uncover unique associations that differ from those observed in Western populations. (Table 15.1).

15.1.4 Cancer Profiling and the Gut Microbiome: A Nexus of Health and Disease

Cancer profiling, also known as tumour profiling, is the systematic analysis of molecular, genetic, and clinical features of cancerous tumour cells and their microenvironment. The profiling aims to understand the underlying mechanisms

Table 15.1 Impact of gut microbiome on different cancer types and key bacteria involved

Cancer type	Gut microbiome impact	Key bacteria involved
Stomach cancer (gastric cancer)	Affects outcomes and severity	*Helicobacter pylori*
Colorectal cancer (CRC)	Changes in microbiome linked to CRC patterns	*Prevotella, Fusobacterium nucleatum, Bacteroides fragilis*
Hepatocellular carcinoma (HCC)	Dysbiosis contributes to HCC development	*Escherichia coli, Bacteroides*
Oesophageal adenocarcinoma (EAC)	Dysbiosis linked to EAC progression	*Prevotella copri, Faecalibacterium prausnitzii, Streptococcus, Veillonella*

navigating the cancer initiation, progression, and response to cancer therapy. Profiling encompasses various dimensions, including genomics, transcriptomics, proteomics, metabolomics, and cytomics to identify the specific genetic mutations, gene expression patterns, protein profiling, and other molecular and cell system features that contribute to the overall development and progression of cancer (Malone et al. 2020).

The primary goal of cancer profiling is to pinpoint the potential targets for personalized treatment and to predict the tumour's response to cancer therapies. This approach is vital for several reasons. First, it facilitates the selection of the most effective treatment options for individual patients. By understanding the specific molecular characteristics of a tumour, oncologists can identify targeted therapies that are more likely to be effective while working on minimizing the potential side effects. Second, cancer profiling can help predict a patient's prognosis, providing valuable information about the aggressiveness of the cancer and the likelihood of resurgence. Third, cancer profiling advances precision medicine by enabling the development of new targeted therapies and immunotherapies based on the distinct molecular features of different types of cancer. In recent years, an emerging player has entered this arena i.e., the gut microbiome. Advanced profiling techniques next-generation sequencing, gene expression profiling and proteomic analysis are a few of the techniques which can provide a comprehensive overview of the tumour's distinct molecular composition, identifying specific modifications that have functional implications for tumour biology. This comprehensive understanding of the tumour's molecular composition serves as the fundamental principle for devising personalized treatment plans, including targeted therapies, immunotherapies, and precision oncology approaches. In addition to the gut microbiome, dietary factors have been identified as key influencers of cancer immunotherapy. Interactions between diet and gut microbiota can significantly influence cancer development and treatment outcomes. The synthesis and regulation of bioactive compounds, metabolic regulation, and immune regulation within the host are influenced by dietary and gut microbial interactions, further emphasizing their relevance in cancer profiling (Wang and Geng 2023). Collectively, the combined impact of the gut microbiome and nutrition on cancer profiling is substantial, presenting a novel dimension to the understanding of cancer and its treatment strategies. As such, an in-depth analysis of a cancer patient's gut microbiome, dietary patterns, and their interplay with immune responses can provide valuable insights for personalized cancer profiling and therapeutic interventions. This approach aligns with the growing emphasis on precision medicine, highlighting the significance of considering individual variations in the nutrition-gut-immuno-oncology axis to optimize cancer treatment outcomes (Li et al. 2024). In conclusion, cancer profiling encompasses the intricate analysis of the characteristics and vulnerabilities of an individual's cancer to devise tailored treatment approaches. The emerging understanding of the role of the gut microbiome and nutrition in influencing immune responses and cancer treatment efficacy has added a pivotal dimension to this field. By considering the interplay between the gut microbiome, dietary factors, and immune responses, personalized cancer profiling and treatment strategies can be refined to improve patient outcomes. The process of cancer profiling and the gut microbiome entails several key steps:

First, the researcher should conduct an extensive literature review to understand the associations between the gut microbiome and GI tract cancers, exploring studies, reviews, and meta-analyses. Subsequently, the researcher needs to collect faecal or tissue samples from patients with GI tract cancers and process these samples to extract microbial DNA or RNA (Liu et al. 2022). The next step involves utilizing high-throughput sequencing techniques, such as 16S rRNA gene sequencing or shotgun metagenomics, to profile the gut microbiome, followed by the analysis of taxonomic composition and functional potential (Malone et al. 2020). For example, 16S rRNA gene sequencing identifies bacterial species by targeting a universal gene, while shotgun metagenomics provide a comprehensive view by sequencing the entire microbial genome. Once the microbiome profiling is complete, the raw sequencing data should be processed using bioinformatics tools to identify specific microbial taxa associated with cancer or health. Bioinformatic analysis should also involve exploring diversity indices, community structure, and functional pathways. Subsequently, statistical tests, including differential abundance analysis and machine learning, should be employed to identify significant associations and potential microbial biomarkers related to cancer risk, prognosis, or treatment response. Moreover, to gain functional insights, the impact of microbial metabolites on cancer should be explored through metabolomics data. The profiling process should also enfold correlating microbiome profiles with clinical parameters, such as tumour stage, patient demographics, and treatment outcomes, to assess whether specific microbial signatures predict cancer progression or therapeutic response. Mechanistic studies, including in vitro or animal model experiments, should also be conducted to understand how gut microbes influence cancer through immune modulation, inflammation, and metabolite production. Furthermore, potential interventions, such as probiotics, prebiotics, and dietary modifications, should be investigated to modulate the gut microbiome and evaluate their impact on cancer prevention or treatment (Wang and Geng 2023). Additionally, the researcher should explore how gut microbiota affects immunotherapy responses in patients with cancer and consider personalized approaches based on microbiome profiles (Dadgar et al. 2023). In parallel, microbial carcinogens have been associated with various types of cancer, with specific viruses such as HPV, HBV, HCV, EBV, and HTLV-1 playing crucial roles in carcinogenesis. Each of these viruses has been linked to specific types of cancer, underscoring the significant role of microbial carcinogens in cancer pathogenesis. Moreover, the elucidation of cancer-associated single-nucleotide polymorphisms (SNPs) in bacteria and the link between dysbiosis and GI cancers highlight the need for further research into microbial signatures and their impact on cancer development. Ultimately, these comprehensive profiling efforts are essential to advance our understanding of the nutrition-gut-immuno-oncology axis in GI tract cancers and to pave the way for innovative approaches to cancer prevention and treatment (Fig. 15.3).

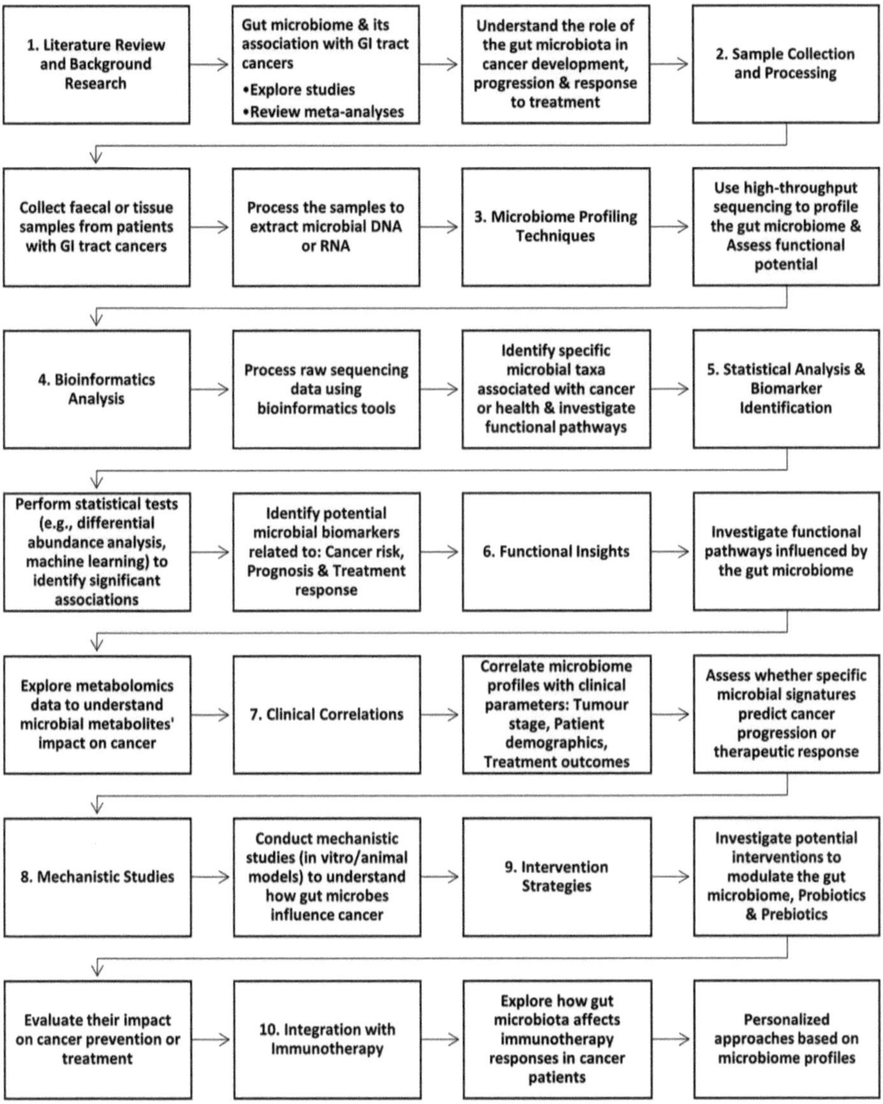

Fig. 15.3 Roadmap showing the steps involved in cancer profiling and gut microbiome analysis

15.1.5 Microbial Signatures Linked to Cancer: Microbial Carcinogens

The gut microbiome, composed of trillions of microorganisms in the GI tract, is vital for maintaining its health and modulating the immune system. Dysbiosis, or microbial imbalance, can contribute to stomach cancer by leading to chronic inflammation, producing carcinogenic metabolites, impairing immune surveillance, and altering the gastric environment. For instance, *Helicobacter pylori*, a bacterium associated with stomach cancer, produces toxins like CagA and VacA that damage DNA and promote cancer development. Dysbiosis can also affect the pH and mucosal environment of the stomach, influencing cancer cell proliferation. Furthermore, the gut microbiome's interaction with dietary factors, such as high salt intake, can exacerbate the harmful effects of *Helicobacter pylori* and lead to the formation of carcinogenic compounds. Understanding these mechanisms highlights the potential for microbiome-targeted therapies to reduce stomach cancer fatality rates. Insights into the association between the gut microbiome and cancer have shed light on the role of microbial carcinogens in oncogenesis (Sgamato et al. 2024) (Table 15.2).

Microbial carcinogens are microorganisms that possess the capability to instigate oncogenic processes within the host. These carcinogens can induce chronic inflammation, DNA damage and alterations in cellular behaviour that promote tumorigenesis. In gut-related cancers, certain microorganisms have been identified as potential contributors to cancer development. For example, *Fusobacterium nucleatum*, a common gut commensal, has been linked to colorectal cancer (CRC) progression through its pro-inflammatory properties and ability to modulate the tumour microenvironment. Similarly, *Bacteroides fragilis* and *Escherichia coli* have been associated with CRC pathogenesis, highlighting the intricate interplay between gut microbes and cancer development. *Helicobacter pylori*, a bacterial species that colonizes the stomach, is one of the most well-known microbial carcinogens, particularly in the context of gastric cancer. *Helicobacter pylori* infection is a major risk factor for the development of gastric malignancies, including gastric adenocarcinoma and gastric mucosa-associated lymphoid tissue (MALT) lymphoma. The bacterium's virulence factors, such as CagA and VacA, contribute to gastric inflammation, epithelial cell damage, and the disruption of host immune responses,

Table 15.2 Microbial signatures and their contributions in various cancers

Cancer type	Microbial signature	Pathology & aetiology contribution
Gastric cancer	*Helicobacter pylori*	Associated with an increased risk of developing a significant role in the onset
Colorectal cancer	*Fusobacterium nucleatum, Bacteroides fragilis*	Stimulation of tumour growth, activation of cancer pathways
Cervical, anogenital, oropharyngeal	*Human papillomavirus (HPV)*	Major etiological factors in these types of cancers
Liver cancer	*Hepatitis B Virus (HBV), Hepatitis C Virus (HCV)*	Implicated in carcinogenesis, inflammation
Lymphomas, nasopharyngeal carcinoma	*Epstein–Barr Virus (EBV)*	Associated with lymphomas and nasopharyngeal carcinoma
Adult T-cell leukaemia/ lymphoma	*Human T-cell lymphotropic virus type 1 (HTLV-1)*	Linked to adult T-cell leukaemia/ lymphoma
Oesophageal Cancer	*Fusobacterium, H. pylori*	Relevant in oesophageal development
Pancreatic Cancer	*Porphyromonas gingivalis, Clostridium species*	Potential role in pancreatic cancer

ultimately fostering a microenvironment conducive to cancer initiation and progression. In individuals infected with *Helicobacter pylori*, chronic gastritis often ensues, leading to the evolution of pre-neoplastic lesions such as gastric atrophy, intestinal metaplasia, and dysplasia (Butel 2000).

Over time, these histological changes can culminate in the development of gastric adenocarcinoma, one of the leading causes of cancer-related mortality worldwide. The pathogenic mechanisms employed by *Helicobacter pylori*, including the induction of oxidative stress, DNA damage, and chronic inflammation, contribute to the oncogenic transformation of gastric epithelial cells, underscoring the bacterium's role as a key player in gastric carcinogenesis. Exploring the microbial signatures linked to cancer, including microbial carcinogens like *Helicobacter pylori*, provides valuable insights into cancer pathogenesis and opportunities for innovative therapeutic interventions. Understanding the molecular mechanisms underpinning the interactions between microbial agents and host cells can pave the way for targeted therapies aimed at disrupting oncogenic processes fuelled by microbial carcinogens. Moreover, elucidating the specific microbial signatures associated with different cancer types can aid in the development of microbiome-based diagnostic tools, prognostic markers, and personalized treatment strategies for cancer patients. Lastly, the intricate relationship between microbial carcinogens and cancer highlights the multifaceted nature of oncogenic processes influenced by the gut microbiome. By unravelling the molecular intricacies of microbial signatures linked to cancer, researchers can gain valuable insights into cancer pathogenesis and identify novel avenues for therapeutic intervention, with a specific focus on well-known microbial carcinogens like *Helicobacter pylori* in the context of gastric cancer (Hatta et al. 2021; DiMaio et al. 2021).

Cancer is a multifactorial disease involving the interactions among host, microbial, and environmental factors. Microbial carcinogens, in particular, have been associated with various types of cancer. While *Helicobacter pylori* is recognized as a crucial risk factor for stomach cancer, several other microbial signatures and carcinogens have been identified as risk factors for different types of cancers. Recent evidence has shown that several specific microbial signatures are linked to the development of cancer. These microbial signatures play a critical role in carcinogenesis by promoting tumour growth, inducing chronic inflammation, and affecting the immune response. Some of the microbial carcinogens that have been implicated in cancer development include HPV, hepatitis B and C viruses (HBV and HCV), Epstein–Barr virus (EBV), and human T-cell lymphotropic virus type 1 (HTLV-1). Each of these viruses has been linked to specific types of cancer, further highlighting the significant role of microbial carcinogens in cancer pathogenesis. HPV, for instance, is recognized as the primary cause of cervical cancer, with specific genotypes, such as HPV-16 and HPV-18, identified as the most prevalent high-risk strains (Butel 2000; Liu et al. 2017; Fan et al. 2022). Moreover, HPV has also been linked to other anogenital cancers, as well as oropharyngeal cancers. HPV-associated cancers can be prevented through vaccination against the virus, highlighting the importance of understanding the microbial signatures and implementing preventive measures. In addition to HPV, the hepatitis B and C viruses (HBV and HCV) have a well-established link to liver cancer. Chronic infection with HBV or HCV can lead to liver cirrhosis, subsequently increasing the risk of hepatocellular carcinoma, one of the leading causes of cancer-related deaths worldwide. Targeted antiviral therapies have been developed to prevent and manage HBV and HCV infection to reduce the associated cancer burden. Another notable microbial carcinogen is the Epstein–Barr virus (EBV), which is associated with several types of lymphomas and nasopharyngeal carcinoma (Hau et al. 2020; Yu et al. 2020). EBV has been shown to induce the transformation of infected cells, leading to the development of these malignancies. Furthermore, human T-cell lymphotropic virus type 1 (HTLV-1) has been implicated in the development of adult T-cell leukaemia/lymphoma as well as other disorders (Liu et al. 2022). The identification of cancer-associated SNPs in bacteria, including *Helicobacter pylori*, has advanced our understanding of the aetiology and pathogenesis of cancer (Hartman-Houstman et al. 2024).

Several polymorphisms in the EPIYA-B phosphorylation motif of the injected effector protein CagA have been shown to affect cancer risk. Understanding the role of SNPs in *Helicobacter pylori* and other bacterial pathogens elucidates how genetic variations in these microorganisms can contribute to an individual's susceptibility to cancer (Rebersek 2021) (Fig. 15.4). Furthermore, studies have also demonstrated that microbiome dysbiosis, characterized by alterations in the composition and function of the gut microbiota, is associated with various gastrointestinal cancers, including gastric cancer.

The dysbiosis of the gut microbiome has been shown to disrupt host–microbiota interactions and promote inflammation, which could contribute to the development of cancerous conditions in the GI tract. Additionally, microbial dysbiosis has been implicated as a potential risk factor for colorectal cancer, highlighting the broader

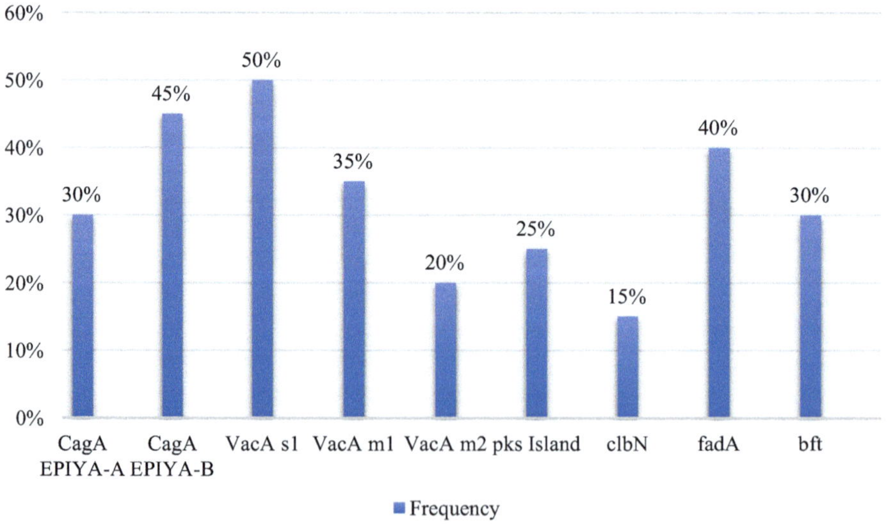

Fig. 15.4 Bacterial carcinogens and associated cancer risks: Frequency of specific SNPs and relative cancer risk levels

impact of microbial signatures on cancer development. These findings emphasize the need for further research into the associations between microbial signatures and cancer development. Understanding the complex interactions between host genetics, environmental factors, and microbial carcinogens is essential for developing effective strategies for cancer prevention, early detection, and treatment. In conclusion, microbial signatures play a significant role in the development of cancer, and several microbial carcinogens have been identified as risk factors for various types of cancers. Preventive measures, such as vaccination against specific viruses and targeted antiviral therapies, are critical for reducing the burden of virus-associated cancers. Moreover, the identification of cancer-associated SNPs in bacteria, as well as the link between dysbiosis and GI cancers, underscores the need for further investigation into microbial carcinogens and their impact on cancer development. Ultimately, advances in understanding microbial signatures linked to cancer will pave the way for the development of innovative approaches to cancer prevention and treatment.

In conclusion, the paradigm for the origin of most cancers has evolved beyond the primary trigger of somatic mutations, proposing a sequence of events involving chronic inflammation, fibrosis, and the development of a pre-cancerous niche. Personalized approaches, such as understanding an individual's microbiome profile, could lead to personalized cancer prevention and treatment strategies. Faecal microbiota transplantation (FMT) and probiotics are being explored as potential interventions (Mark et al. 2021). Understanding cancer prevalence, mortality, and patterns across different regions is crucial to addressing the global burden of cancer effectively, providing insights that can aid in developing and implementing comprehensive public health strategies for cancer control and prevention. This comprehensive understanding is essential to promote global public health and enhance the management and prevention of cancer on a global scale, thereby reducing the burden it poses to individuals and societies worldwide. Consequently, integrating microbiome research with traditional cancer prevention and treatment strategies could potentially enhance patient outcomes and reduce the high mortality rates associated with stomach cancer in India.

15.2 Immuno-oncology Axis in Cancer Studies

The immuno-oncology axis represents a transformational approach to cancer therapy, aiming to discover the inherent potential of the immune system's innate ability to recognize and eradicate cancer cells in the human body. The axis encompasses a comprehensive understanding of the interplay between the immune checkpoints, T cells, antigen-presenting cells (APCs) and the intricacies of the tumour microenvironment (TME). By integrating these components, immuno-oncology aims to overcome cancer's immune evasion strategies. The key advancements in this field include the development of checkpoint inhibitors, CAR-T cell therapy, cancer vaccines targeting neoantigens and the investigation of combination therapies. Additionally, biomarkers and prognostic factors are important in guiding patient selection and treatment planning, while emerging novel approaches, such as oncolytic viruses and bispecific antibodies, offer promising avenues for further research. However, as with any cutting-edge field, immuno-oncology also presents challenges, such as developing strategies to overcome resistance mechanisms and improving personalized treatment approaches (Pan et al. 2020).

The recognition of immune checkpoints, such as the PD-1/PD-L1 pathway and CTLA-4, as critical regulators of immune responses has revolutionized cancer therapy. The PD-1/PD-L1 interaction causes T-cell depletion, allowing cancer cells to evade immune surveillance or monitoring by the immune system of the human body. Checkpoint inhibitors, such as pembrolizumab and nivolumab, effectively block and disrupt this interaction, reinvigorating T-cell responses and enhancing anti-tumour immunity. Similarly, inhibitors targeting CTLA-4, such as ipilimumab, have shown promise in enhancing T-cell activation by blocking the inhibitory signals, thereby highlighting the potential of checkpoint inhibitors in overcoming immune evasion. Beyond checkpoint inhibitors, CAR-T-cell therapy has emerged as a groundbreaking approach in immuno-oncology. This technique is based on engineered T cells expressing tumour-specific receptors, known as chimeric antigen receptors (CARs), which form the basis of this therapeutic technique. CAR-T therapy, as found in products like axicabtagene and ciloleucel, has demonstrated impressive efficacy in haematological malignancies or cancers. This therapeutic approach highlights the possibility of genetically modifying and manipulating T cells to enhance their ability to target and eradicate cancer cells, thus bringing forth a new path in cancer immunotherapy (Marin-Acevedo et al. 2018).

Personalized cancer vaccines targeting neoantigens, which are specific to cancer cells, offer another route of administration in immuno-oncology. These vaccines aim to teach the immune system to recognize, attack, and further eradicate specific tumour-associated antigens, resulting in focused, targeted, and personalized immune responses against cancer. The discovery and development of mRNA-based vaccines, such as Moderna's mRNA-4157, demonstrates the ongoing innovation in this field of research, implying promising prospects for the integration of cancer vaccines into the immuno-oncology axis. The tumour microenvironment is a critical determinant of immune responses against cancer. Its impact on the complex interaction between cancer cells and immune cells contributes significantly to immunotherapy efficacy. The presence of immunosuppressive factors, such as regulatory T cells and myeloid-derived suppressor cells, within the TME significantly impairs anti-tumour immunity. Understanding the TME's impact on immune responses provides critical insights into improving immuno-oncology techniques and establishing the creation of a microenvironment favourable to powerful anti-tumour immune responses (Binnewies et al. 2018) (Fig. 15.5).

Combination therapies form a cornerstone in immuno-oncology, aiming to improve response rates and overcome plausible resistance mechanisms. Integrating checkpoint inhibitors with other therapies or treatments, such as chemotherapy, radiation, and targeted therapies, has shown the potential to synergistically enhance the immune response against cancer. This approach provides a critical framework for the rational design of combination regimens that benefit the understanding of

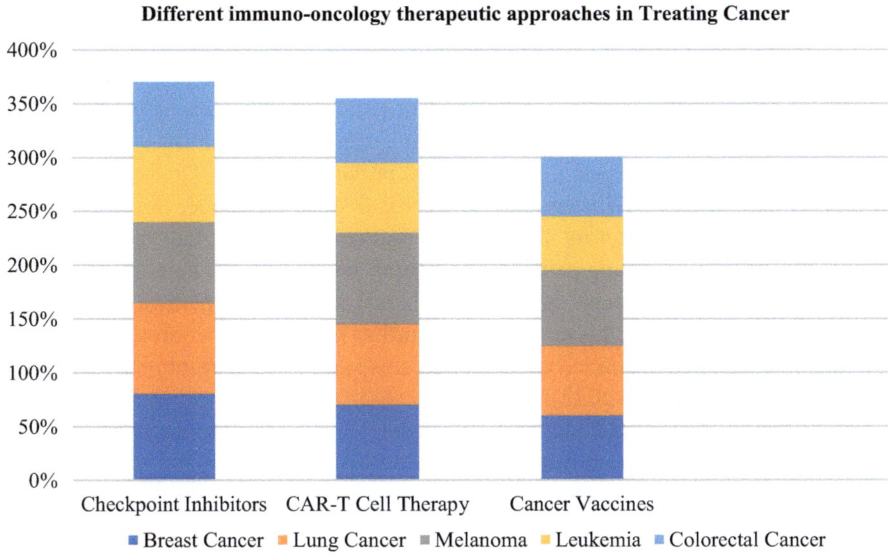

Fig. 15.5 Relative efficacy of immuno-oncology therapeutic approaches for specific types of cancer

tumour–immune interactions, demonstrating the promise of multiple therapeutic approaches within the immuno-oncology axis. In immuno-oncology, biomarkers and predictive factors play an important role, aiding in patient classification and therapy response prediction. The recognition of PD-L1 expression as a predictor of checkpoint inhibitor response, as well as tumour mutational burden (TMB) as a correlate of immunotherapy efficacy, highlights the increasing emphasis on personalized, precision medicine in immuno-oncology. These biomarkers and prediction tools and methods are critical in identifying patients most likely to benefit from immunotherapy, therefore contributing to treatment optimization and the minimization of adverse events (Ribas and Wolchok 2018).

Emerging techniques within the immuno-oncology axis offer intriguing opportunities for further research. Oncolytic viruses, which selectively infect and lyse cancer cells, triggering immune responses, constitute a novel and innovative immunotherapeutic approach. Similarly, bispecific antibodies, such as blinatumomab, can engage T cells and tumour cells, thereby enhancing immune-mediated anti-tumour activity. These emerging techniques signify the ongoing evolution and diversification of immuno-oncology, offering new opportunities for enhancing the immune response against cancer. However, immuno-oncology also brings forth challenges that require immediate attention. Understanding and overcoming resistance mechanisms, including the loss of antigen presentation and innate or acquired alterations in the TME, forms a critical aspect of ongoing research endeavours (Chen and Mellman 2017).

15.2.1 Immune Modulation in Cancer

The GI tract plays a critical role in immune modulation in cancer, especially in colorectal cancer. The interaction between epithelial cells, immune cells and the gut microbiota regulates immune system mediators and affects the intestinal barrier, making the GI tract the largest immunological organ in the body. Understanding the immune modulation in GI tract cancer is critical since GI cancer development is influenced by multiple factors such as diet, infection, environment, and genetics. Although activating immune pathways and components during cancer is critical for the host's survival, malignant cancerous cells can target those pathways to escape and survive. Thus, research is conducted to further investigate the gut microbiome–immune interactions, the underlying mechanisms and how they minimize the risk of GI cancer (Al-Ishaq et al. 2022).

Additionally, it highlights the underlying mechanisms of immune dysregulation caused by gut microbiota on four major cancerous pathways, including inflammation, cellular proliferation, apoptosis, and metastasis (Bose and Mukherjee 2019). The human gastrointestinal tract has a diverse microbial population, including bacteria, viruses, and fungi, and this organ plays an important role in modulating local mucosal and systemic immune responses via complicated cross-talks with immune cells (Cho et al. 2024). The gut microbiome can modulate the immune system and affect the efficacy of immunotherapies in cancer patients. Gut microbes can modulate the host immune regulation, thereby indirectly affecting the efficacy of immune checkpoint inhibitors (ICIs) in patients with cancer. ICIs enhance the immunological cytotoxicity of T cells by targeting coinhibitory molecules PD-1/PD-L1, which strengthens the natural immune response of the host and prevents tumour cells from evading immunological surveillance of the human body (Kang et al. 2024). Notably, specific gut microbiota may be transformed during various pathological conditions and produce several toxic compounds and oncoproteins, which may cause an inflammatory response and carcinogenesis (Rezasoltani et al. 2020). Therefore, the gut microbiome can have a substantial impact on how the immune system responds to cancer and its modulation.

The gut microbiome regulates the development and function of the immune system and plays an important role in maintaining immunological homeostasis, particularly in the context of GI cancer (Bose and Mukherjee 2019). The gut microbiome–immune interactions are critical in understanding the aetiology and progression of GI cancers, especially colorectal cancer. These interactions have a major effect on immune cells, such as T cells, B cells and dendritic cells. They can either enhance or suppress immunological responses in the context of cancer (Al-Ishaq et al. 2022). The underlying mechanisms of immunological dysregulation by gut microbiota can directly influence cancerous pathways, such as inflammation, cellular proliferation, apoptosis and metastasis, thereby affecting the progression and treatment of gastrointestinal malignancies or GI cancers. For example, gut dysbiosis can cause a low level of short-chain fatty acids, thereby triggering inflammatory pathways, cytokine and chemokine synthesis and activating STAT3 and NF–kB signalling pathways. This imbalance can contribute to the development and progression of GI cancers by altering the local and systemic immune responses and establishing a favourable environment for cancer cell proliferation. Furthermore, specific gut bacteria have been shown to modulate the efficacy of currently used anti-cancer drugs, highlighting the importance of the gut microbiome in cancer therapy (Bose and Mukherjee 2019).

Furthermore, the gut microbiome has been shown to influence cancer metastasis by targeting cellular interactions, response to chemotherapy, and other factors that contribute to the spread of cancer. The modulation of the gut microbiome could potentially impact the effectiveness of anti-cancer therapies and the overall prognosis of GI patients with cancer (Bose and Mukherjee 2019). This highlights the importance of understanding the profound influence of the gut microbiome on immune modulation in GI tract cancer and its potential implications for personalized cancer therapy. In conclusion, the gut microbiome plays an important role in modulating the local and systemic immune responses in the context of malignancies such as gastrointestinal (GI) cancers, specifically colorectal cancer. The complex interactions between the gut microbiome and the immune system influence the aetiology, progression, and treatment of GI cancers. Understanding the complex interaction between the gut microbiome and immunological responses is essential for developing novel strategies for cancer therapy and improving patient outcomes. Further research is needed to gain a more detailed understanding of the underlying mechanisms of immune modulation by the gut microbiome in the context of GI cancers and to explore potential therapeutic interventions based on these interactions. The insights gained from these studies have the potential to revolutionize immune-based therapies for GI cancers and pave the way for personalized and targeted treatment approaches.

15.2.2 Factors Affecting the Immuno-oncology Axis

The immuno-oncology-microbiome (IOM) axis plays a critical role in gastrointestinal (GI) tract cancers, and several factors influence this axis, impacting the immune responses to GI tract cancer (Lin et al. 2023a, b; Hecht et al. 2023). The aim is to comprehensively analyse the various factors affecting the immuno-oncology axis in GI tract cancer, providing insights into the complex mechanisms between the gut microbiome, immune responses and cancer development within the GI tract. The factors include the tumour microbiome, immune checkpoint inhibitors, gut microbiota, microbial metabolites and immunotherapy in GI malignancies, focusing on colorectal cancer, as it represents the second leading cause of cancer-related death worldwide (Zaborowski et al. 2021). The tumour microbiome, which consists of the collective genomes of the microorganisms in the tumour microenvironment, has emerged as an important factor affecting the immuno-oncology axis in GI tract cancers. The microbial composition within the tumour microenvironment has been demonstrated to impact tumour immunity, response to treatment and overall prognosis in various GI malignancies, most importantly in GI tract cancers (Lin et al. 2023a, b). Understanding the dynamic interactions between the tumour microbiome and the host immune system is critical for identifying novel therapeutic targets and enhancing the efficacy of immuno-oncology therapies in GI tract cancers.

Immune checkpoint inhibitors (ICIs) have transformed cancer treatment, but their efficacy in GI malignancies, such as microsatellite-stable colorectal cancer and pancreatic adenocarcinoma, has been limited. The factors influencing the response to ICIs include the tumour mutational burden, PD-L1 expression and the unique tumour microenvironment, which can pose significant challenges in achieving durable responses in GI tract cancers (Hecht et al. 2023). Understanding these factors is crucial for optimizing the use of ICIs and improving patient outcomes in GI tract cancers. The gut microbiota is critical in modulating the immune responses and cancer development within the GI tract, including colorectal cancer. The composition and diversity of the gut microbiota influence the efficacy of cancer immunotherapy and can impact the efficacy of GI tract cancers in immune-based therapies. Investigating the complex relationship between the gut microbiota and the immuno-oncology axis is crucial for developing precision medicine approaches tailored to the unique microbiome profiles of individual patients with GI tract cancers. Microbial metabolites, such as short-chain fatty acids and bile acids, have been identified as key modulators of immune responses and inflammation within the GI tract. These metabolites exert profound effects on the tumour microenvironment and the systemic immune system, influencing the progression of GI tract cancers and the efficacy of immunotherapy (Zaborowski et al. 2021). Elucidating the impact of microbial metabolites on the immuno-oncology axis is essential for developing innovative approaches to harness the immunomodulatory properties of these metabolites for therapeutic purposes in GI malignancies.

Advances in immunotherapy have resulted in promising outcomes for specific subgroups of patients with GI tract cancers, highlighting the need to identify predictive biomarkers and explore novel treatment techniques. Research into novel immunotherapeutic approaches, including vaccines, cellular therapy and combination strategies, may provide new opportunities for enhancing the immuno-oncology axis in GI tract cancers (Rao et al. 2019). Understanding the factors affecting immunotherapy in GI malignancies is critical for optimizing treatment strategies and improving patient outcomes. In conclusion, the immuno-oncology axis in gastrointestinal tract cancer is influenced by a multitude of factors, including the tumour microbiome, immune checkpoint inhibitors, gut microbiota, microbial metabolites, and immunotherapy approaches. Research elucidates the complex interactions between these factors and their impact on immune responses, cancer progression and treatment outcomes in GI tract malignancies.

15.3 Gut Microbiome in Malignancy

15.3.1 Microbial Infection and Cancer

Microbial infection shows a complex relationship in the development of cancer and has gained important attention in recent years by influencing the nutrition-gut-immuno-oncology axis. This interplay link between the gut microbiome and microbial infections leads to dysbiosis, which has been in association with distinct types of cancer development, progression, and treatment by understanding the importance of associated microbial communities. For example, dietary habits and patterns could influence the opportunistic and potentially carcinogenic bacteria colonization (*Helicobacter pylori)* in the gut, consequently elevating the probability of certain types of cancer (gastric cancer). The microbial infection may trigger dysbiosis, which can induce colonic epithelial tumorigenesis and show an immuno-permissive atmosphere for tumour growth (Poveshchenko et al. 2023; Ray et al. 2023; Asseri et al. 2023).

Several microorganisms, like bacteria, viruses, and fungi, have been connected to the pathophysiology of various cancers. *Helicobacter pylori* is linked with gastric cancer and causes chronic inflammation in the lining of the stomach, which triggers a cascade of molecular events that contribute to cellular damage, and many genetic abnormalities subsequently cancer (Reyes 2023). *HPV* develops the proteins that will hinder tumour suppressor pathways such retinoblastoma (Rb) and p53 leading to the triggering of cervical cancer and anal cancer (Haręża et al. 2022; Idres et al. 2022); moreover, the *Epstein–Barr virus (EBV)* directly affects the B lymphocytes and leads them to become immortalized cells thus contributing to the development of nasopharyngeal carcinoma and lymphomas (Perri et al. 2023; Ahmed et al. 2022). *Fusobacterium nucleatum*, *Escherichia coli,* and *Bacteroides fragilis* elevate the tumorigenesis by activating many inflammatory pathways and immuno-modulation responses by producing the genotoxins like colibactin, thereby leading to colorectal cancer (Shariati et al. 2021; Clay et al. 2022). *Hepatitis B and C viruses (HBV, HCV)* can directly infect liver cells and cause the transformation of hepatic cells also integrate into the host genome leading to genetic instability and triggering the hepatocellular carcinoma (Wang and Chen 2022). *Candida* spp. may provoke inflammation and tissue injury in the oesophagus therby, accelerating the progression of oesophagus cancer (Talapko et al. 2023) (Table 15.3).

It outlines (Table 15.3) the specific metabolites or genotoxins produced by specific microbial species and their corresponding impact on cellular pathways, DNA integrity, and immune modulation. Each offers some insights into the molecular mechanisms by which these microbial factors influence the pathogenesis and progression of cancer. From FadA protein produced by *Fusobacterium nucleatum* that activates the β-catenin pathway to the HBx protein from Hepatitis B virus (HBV) that modulates the Wnt/β-catenin signalling, these details shed light on the diverse strategies employed by microorganisms to potentially contribute to cancer development. The provided information can aid researchers and clinicians in understanding the complex interplay between microbial factors and cancer and may guide the development of targeted interventions.

The heatmap (Fig. 15.6) shows the prevalence of various microbial infections across different cancer types using a colour gradient to denote prevalence levels (Low: 1, Moderate: 2, High: 3). Key findings include a high prevalence of *Helicobacter pylori* in gastric cancer, highlighting its role in chronic inflammation and gastric neoplasia. Cervical and anal cancers show a high prevalence of human papillomavirus (HPV), emphasizing its inhibition of tumour suppressor pathways and cellular transformation. Lymphomas and nasopharyngeal cancers show a high

Table 15.3 Microbial genotoxins and metabolites affect cellular pathways, DNA integrity, and immune modulation in the context of infectious agents associated with cancer

Microbial species	Type	Genotoxins/ metabolites	Impact on cellular pathways	Impact on DNA damage	Impact on immune modulation
Fusobacterium nucleatum	Bacterium	FadA protein	Activates β-catenin pathway	Induces DNA damage	Modulates immune response
Escherichia coli	Bacterium	Colibactin	Disrupts cell cycle regulation	Induces DNA double-strand breaks	Suppresses immune response
Bacteroides fragilis	Bacterium	B. fragilis toxin (BFT)	Activates STAT3 signalling	Promotes DNA damage and chromosomal instability	Suppresses anti-tumour immunity
Human papillomavirus (HPV)	Virus	E6 and E7 oncoproteins	Inhibits p53 and Rb tumour suppressor pathways	Induces genomic instability	Evades immune surveillance
Candida spp.	Fungi	Candidalysin	Disrupts epithelial barrier integrity	Triggers DNA damage	Modulates inflammatory response
Helicobacter pylori	Bacterium	VacA, CagA	Alters PI3K signalling pathway	Triggers DNA damage	Modulates immune response
Epstein–Barr virus (EBV)	Virus	EBNA-1, LMP1	Activates NF–κB pathway	Induces chromosomal aberrations	Evades immune surveillance
Hepatitis C virus (HCV)	Virus	Core protein	Alters MAPK signalling pathway	Induces oxidative DNA damage	Suppresses antiviral immune response
Hepatitis B virus (HBV)	Virus	HBx protein	Modulates Wnt/β-catenin signalling	Induces DNA double-strand breaks	Impairs immune response

prevalence of *Epstein–Barr virus (EBV),* known for transforming B lymphocytes. Colorectal cancer shows a high prevalence of *Fusobacterium nucleatum, Bacteroides fragilis, and Escherichia coli*, linked to inflammatory pathways and carcinogenic metabolites contributing to tumorigenesis.

The events and mechanisms by which microbial infections and dysbiosis cause malignancy have been more understood by the preclinical model's use for studying the complex association between cancer and microbial infection. Additionally, promising clinical trials have shown the potential of dealing with microbes in cancer treatment by elucidating the influence of dietary alterations and microbial modulation on cancer outcomes.

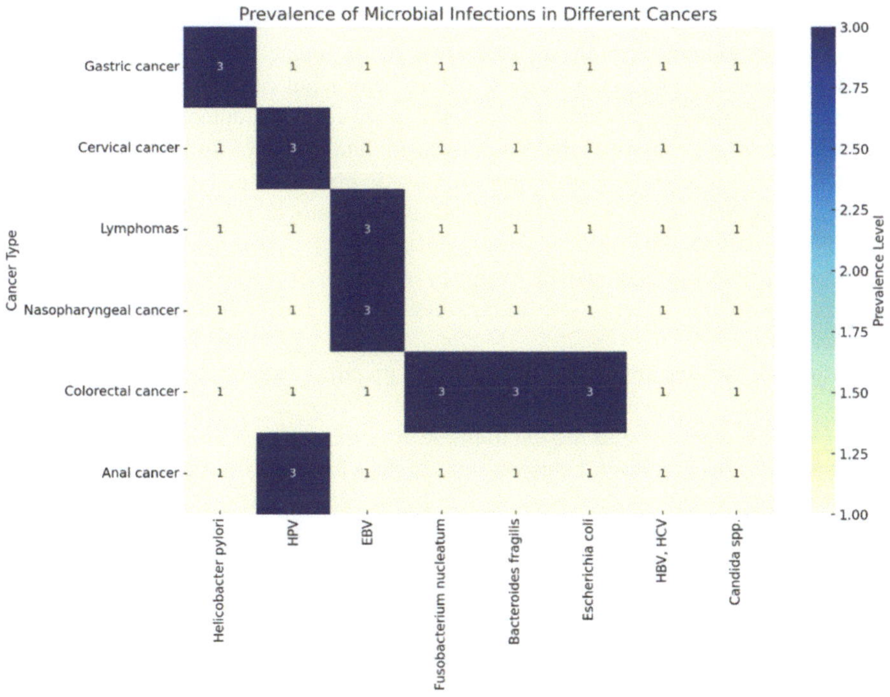

Fig. 15.6 This heatmap illustrates the prevalence levels of specific microbial infections across different cancer types. Prevalence levels are categorized as low (1), moderate (2), and high (3)

15.3.2 Microbiome Modulation and Cancer

As the human microbiome or microbiota forms a very complex, diverse, and dynamic population of microorganisms present in the various parts of the body and playing a significant role in modulating disease rates such as cancers. The microbiome activity and nature might affect the cancer's spread, development, and reaction to therapy. It also acts as a key factor in the interconnection between the health of the gut, immune function, nutrition, etc. Investigation has proven the potential of modulating the microbiota to cure or prevent cancer.

Modulation or alteration of microbiota involves changing the nature and functioning of the microbial population to reach the beneficial effects of the host. A variety of mechanisms like the formation of microbial metabolites, metabolism of dietary components, administration of immune tolerance and surveillance, and modulation of inflammatory responses have a profound impact on the development of cancer. The cumulative effect of metabolites, dietary elements and gut microbiota transforms the tumour microenvironment, functioning of immune cells and systemic inflammation, which promotes the onset and progression of cancer (Chai et al. 2023). For example, in the case of colorectal cancer, diets high in red meat and low in fibre are linked with increased chances of this cancer. Heme from red meat is metabolized by the gut microbiome into pro-inflammatory compounds, i.e. N-nitroso compounds, which are carcinogenic in nature (Rebersek 2021). Thus, these compounds damage the DNA of colonic epithelial cells and cause mutations and malignancy. Also, *F. nucleatum* is commonly found in colorectal cancer individuals, which can elevate the inflammation and enhance tumour growth. As *F. nucleatum* binds to the E-cadherin on colonic epithelial cells, this activates the β-catenin signalling and produces pro-inflammatory responses (McIlvanna et al. 2021).

Many numbers of techniques are employed to alter or modulate the microbiome, including probiotics, prebiotics, dietary alterations, and FMT (faecal microbiota transplantation).

Probiotics are live microorganisms, when given in sufficient amounts, provide health benefits by elevating the gut barrier function, suppressing the growth of pathogenic bacteria and altering the immune system such as strains of *Lactobacillus* and *Bifidobacterium* inhibit the growth of tumours and improve the effectiveness of cancer therapies and preventing the risk of colorectal cancer. On the other hand, indigestible food components, i.e. prebiotics can stimulate the functioning and activity of good microorganisms in the gut like fructooligosaccharides (FOS), a prebiotic which improve the growth of health-elevating bacteria such as *Lactobacillus* and *Bifidobacterium*. Hence, reducing the chances of cancer (Al-Yami et al. 2022; Yoo et al. 2024). Synbiotics can also elevate the survival and implantation of live microbial dietary supplements in the gut by the synergistic use of probiotics and prebiotics together. Hence, promoting the host's health, such as the symbiotic combination of *Bifidobacterium* and inulin, is associated with reducing systemic inflammation and improving immune function (Gourbeyre et al. 2011). The microbiome is greatly influenced by the diet, as omega-3 fatty acids and polyphenols are associated with a lower risk of colon cancer and enhance healthy microbiota (Gheorghe et al. 2022). FMT (faecal microbiota transplantation) incorporates the transfer of stool into the recipient's GI tract taken from a healthy donor. This approach has shown potential in managing diseases like IBD, which are linked to increased risk of cancer and help in restoring a healthy balance of microbes (Maroun Bou Zerdan et al. 2022). Another great approach, i.e. precision microbiome editing for precisely and targeted altering of the genetic composition of the microbiome through the use of CRISPR-cas9 technology to drop harmful ones and enhance beneficial effects, such as to target the unique strains of *Clostridium difficile* for lowering their virulence and prevalence the CRISPR-Cas9 Technology is employed (Muhammad Akram Tariq 2023). Finally, changing lifestyle factors like sleep, exercise, and stress management can also modulate the function and composition of the gut microbiome, a good and healthy lifestyle routine can enhance a balanced microbiome and decrease inflammation.

The above chart (Fig. 15.7) shows the prevalence or usage level of various microbiome modulation strategies, including probiotics, prebiotics, synbiotics, dietary interventions, faecal microbiota transplantation (FMT), precision microbiome editing, and lifestyle modification, across various types of cancer such as the oesophageal, stomach, colorectal, liver, pancreatic, gallbladder, small intestine, and anal

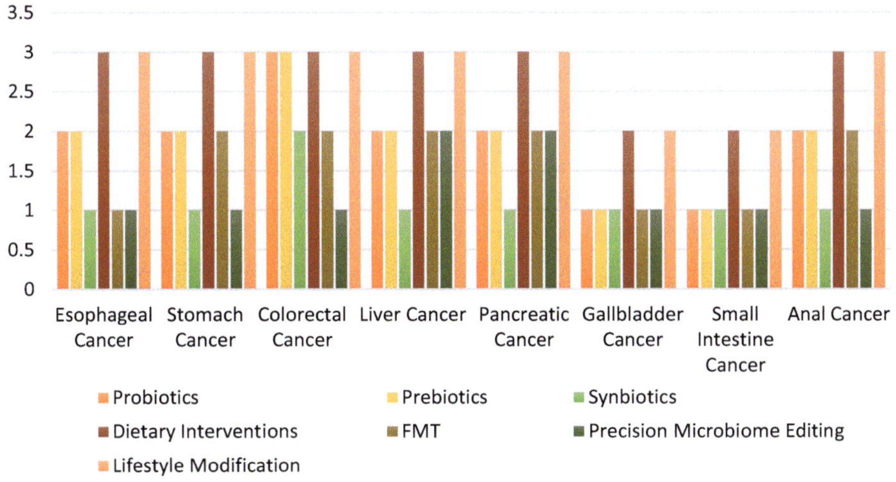

Fig. 15.7 Prevalence of microbiome modulation strategies across different cancer types

cancers. Where each bar stands for the level of focus or application of the strategy in each cancer type, offering essential knowledge about the relative importance of different methods in the treatment of cancer.

15.4 Priming the Nutrition: Modulation of the Immune-oncology Axis

15.4.1 Nutritional Oncology

Nutritional oncology, a burgeoning and multifaceted approach that explores the complex association with nutrition science and cancer treatment. Also, this approach combines with conventional oncological interventions by the employment of nutritional principles to boost the body's immune response towards cancer. The dietary factors and better nutrition can have a profound impact and influence on cancer progression, development, treatment response and quality of life. Within the context of the nutrition-gut-immuno-oncology axis, the function of nutrition in shaping the immune microenvironment and regulating the immune response against cancer has attracted more consideration in recent years (Parsons et al. 2023; Soldati et al. 2018).

Epidemiological research has shown a strong association between unique dietary patterns and cancer risk like plant-based diets are high in vegetables, fruits, legumes, whole grains, and lean proteins have shown potential in reducing the risk of developing several types of cancers due to the presence of phytoconstituents (Polyphenols) mainly found in fruits and vegetables such as curcumin and sulforaphane modulate immune responses by reducing pro-inflammatory cytokines such as IL-6 and TNF-alpha. Omega-3 fatty acids, found in fish and walnuts, further enhance anti-inflammatory pathways by promoting T regulatory cells. Thus, eventually preventing and overseeing the cancer risk (Wang et al. 2020; Molina-Montes et al. 2020). On the other hand, diets that are rich in processed and red meats, saturated fats, and sugars have a relatively high chance of causing cancer. These components increase cancer risk through various mechanisms such as inflammation, hormonal activity, and alteration of oxidative stress (World Cancer Research Fund 2018).

Figure 15. 8a focuses on dietary patterns linked to increased cancer risk, showing the relative contributions of different dietary habits in shaping this risk profile. The segments include diets "high in red and processed meats" (30%), known to significantly contribute to cancer risk, mainly colorectal cancer, attributed to factors like Heme iron and nitrates found in processed meats. Additionally, diets "high in saturated fats" (25%) are linked with an increased risk of certain cancers, such as breast and prostate cancer, owing to the role of fats in hormone regulation and inflammation. Furthermore, diets "high in sugars" (20%) are shown as potential contributors to cancer risk, as they can lead to obesity, which is a risk factor for various cancers like breast, liver, and pancreatic cancers. For instance, in glioblastoma, ketogenic diets have shown promise by reducing glucose availability, which tumors rely on for energy. Finally, the "other unhealthy dietary patterns segment" (25%) encompasses more habits, such as low intake of fibre and high consumption of processed foods and sugary beverages, adding to the overall cancer risk profile. In contrast, (Fig. 15.8b) chart emphasizes dietary patterns linked to decreased cancer risk, showcasing the beneficial impact of a healthy diet on cancer prevention. The segments include diets "high in fruits and vegetables" (35%), rich in vitamins, minerals, and antioxidants that help protect against cancer by reducing oxidative stress and inflammation. Additionally, diets "high in whole grains" (25%) are highlighted for their high fibre content, supporting a healthy gut microbiota and reducing the risk of colorectal cancer. Furthermore, diets "high in lean proteins" (20%), such as those found in fish, poultry, and legumes, are identified as sources of essential nutrients without the added risks associated with red and processed meats. Finally, the "other healthy dietary patterns" segment (20%) encompasses additional healthy habits, such as moderate consumption of dairy, nuts, and seeds, contributing to overall cancer prevention, while these hypothetical datasets provide valuable insights, individual variation, and evolving research may lead to ongoing refinement and updates in our understanding of the interplay between dietary patterns and cancer risk.

To elevate the immune function and its activity, several nutritional interventions are used, including micronutrients, phytochemicals, dietary fibre, omega-3 fatty acids, and calorie intake. Micronutrients like vitamins (A, C, D, and E) and minerals (selenium and zinc) possess anti-cancer properties and increase immune function; they also help regulate cell growth and differentiation, modulate immune response, and protect the cell membranes (Montégut et al. 2022; Diet during Cancer Treatment 2024). Phytochemicals including polyphenols, present in green tea, berries, etc., show anti-inflammatory, antioxidant properties that enhance the immune function and may have the potential to dysregulate the cancer cell proliferation (Imran et al. 2023).

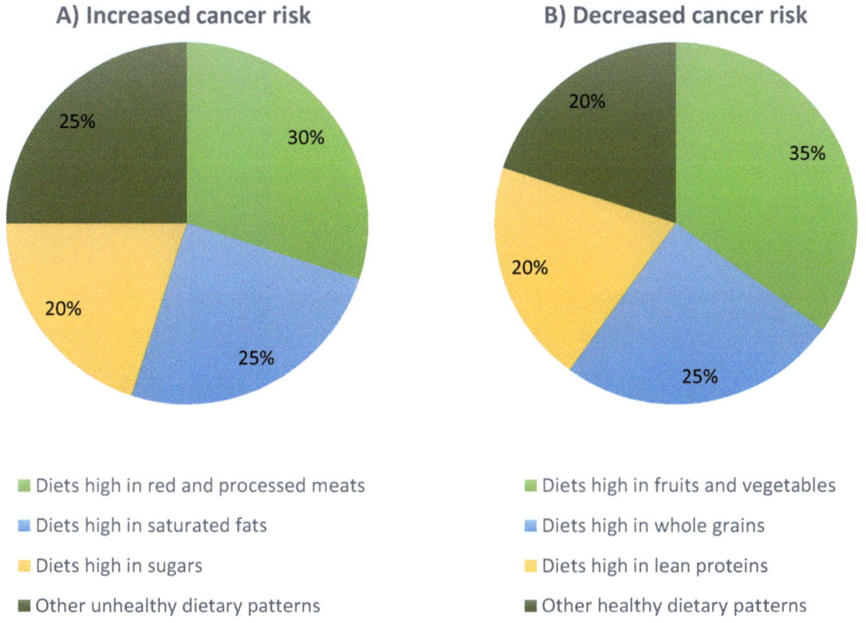

Fig. 15.8 Impact of dietary patterns on cancer risk, this chart shows the correlation between various dietary patterns and cancer risk, (**a**) highlighting the increased risk associated with unhealthy diets (left) and (**b**) the decreased risk associated with healthy diets (right)

As already mentioned above, dietary fibre, such as vegetables and fruits, plays a significant role in managing and maintaining a healthy gut (Koh et al. 2016).

Omega-3 fatty acids are majorly present in fatty fish, walnuts, and flaxseeds, which provide strong anti-inflammatory effects and immune stimulating activity. Also, these can alter the tumour microenvironment, creating an unfavourable environment for cancer cell growth, development, and metastasis (Ishihara et al. 2019). Ensuring a suitable calorie intake is vital for keeping immune system well-being and combating malnutrition. Excessive calorie intake can contribute to tumour growth and inflammation (Dung et al. 2023).

15.4.2 Faecal Microbiota Transplantation

Faecal microbiota transplantation (FMT) is a groundbreaking therapeutic approach that makes use of intricate interactions involving the gut microbiota, immune system, and oncological results. It involves providing faecal microbiota from a healthy donor to a recipient with an altered gut microbiome, with the aim to recalibrate the microbial equilibrium within the recipient's gut (Zuppi et al. 2024; Yadegar et al. 2023). The goal of this intervention is to reestablish, restore, or reshape a balanced and healthy gut microbiome and also to treat various GI disorders, which can have profound implications for the immune system and oncology outcomes. FMT has grabbed a lot of attention due to its ability to modulate the nutrition-gut-immune-oncology axis, which can strengthen the immune function, improve responses to cancer therapies, and potentially inhibit tumour growth (Mahmoudi and Hadi 2022). Through the use of microbial communities found in the gastrointestinal system, FMT presents a distinctive approach to regulating immune responses and promoting an immunological environment that is helpful in combating cancer (Burrello et al. 2019) (Fig. 15.9).

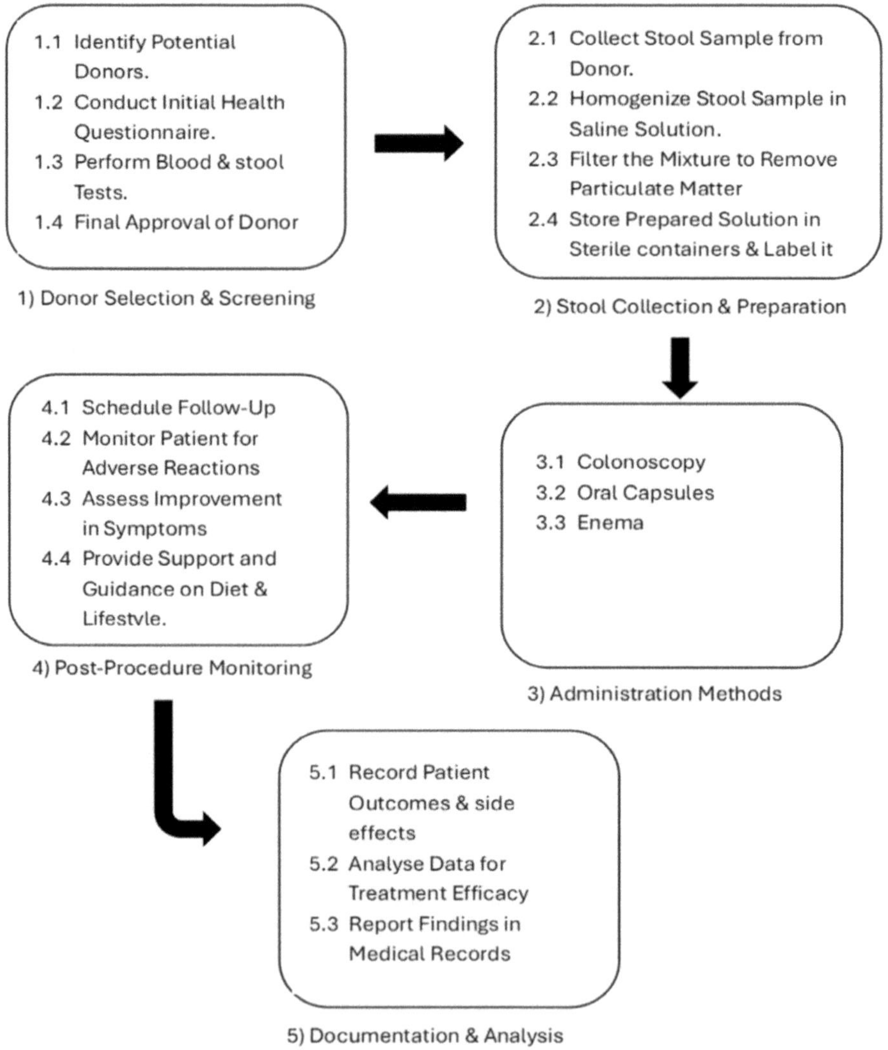

Fig. 15.9 This flowchart comprehensively outlines the FMT procedure, providing detailed steps for each stage of the process

Several mechanisms have been attributed to the beneficial effects of FMT in oncology:

1. *Restoration of Microbiota Diversity*

FMT can replenish a wide array of gut microbiota, which is often depleted in cancer individuals due to factors like the use of antibiotics, chemotherapy, and radiation therapy. Improved microbial diversity is connected with improved immune function and stronger responses to cancer therapies (Gopalakrishnan et al. 2018a, b).

2. *Enhancement of Immune Checkpoint Inhibitors*

Recent research implies that the gut microbiome can influence the efficacy of immune checkpoint inhibitors, a class of cancer immunotherapies. FMT from responders to non-responders has shown to promote the response to checkpoint inhibitors, highlighting that certain microbial populations can modulate anti-tumour immunity (Routy et al. 2018) (Fig. 15.10).

3. *Reduction of Treatment-Related Toxicities*

FMT may help mitigate the gastrointestinal toxicities imposed with cancer treatments, such as chemotherapy-induced mucositis and radiation-induced enteritis. FMT can lower inflammation and improve the gut's resilience by reestablishing a healthy gut flora (van Nood et al. 2013).

The data suggests that FMT can significantly enhance the activation of these immune cells, which are crucial for antitumor immunity. Before FMT, the baseline levels of immune cell activation were relatively low, with T cells at 50% and NK

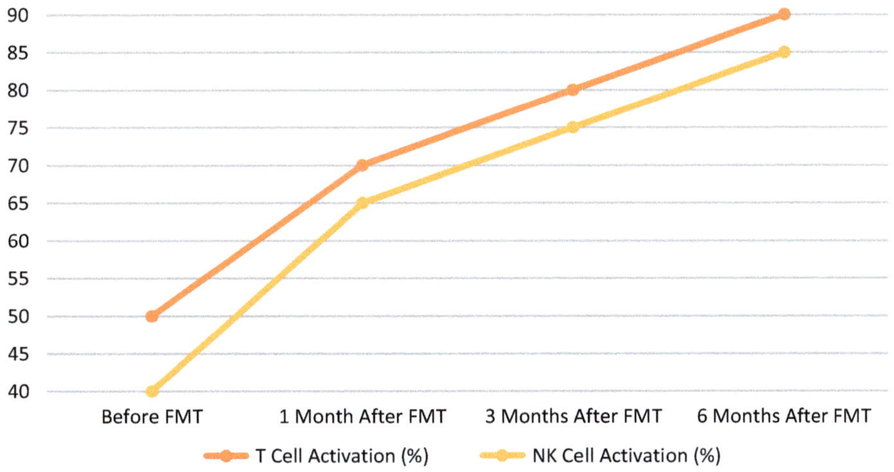

Fig. 15.10 Enhanced activation of T cells and NK cells post-FMT treatment over time

cells at 40%. One month after FMT, there was a noticeable increase in activation, with T cells rising to 70% and NK cells to 65%. The activation levels continued to increase three months after FMT, reaching 80% for T cells and 75% for NK cells. Six months after FMT, the activation levels peaked, with T cells at 90% and NK cells at 85%. This data illustrates the potential for FMT to stimulate immune responses and enhance antitumor immunity by significantly increasing the activation levels of T cells and NK cells over time. This enhancement can contribute to a more robust immune defence against cancer. It is important to note that the table and the data presented are hypothetical and can be changed accordingly based on further research and clinical findings.

Although FMT has promising results, there are also a few challenges associated with the clinical application of FMT. This will include potential pathogen transmission, the development of a standardized procedure for FMT preparation till monitoring for proper consistency and repeatability of outcomes. Another identification of acceptable or suitable donors with a wide range of healthy gut microbiota is vital for the efficacy of FMT. Also, deciding the suitability and effectiveness of FMT requires an awareness of its long-term impact on immune function and gut flora (Almeida et al. 2022).

15.4.3 Human Microbiome Context in Traditional Medicine: Ayurveda, Chinese, German

The intricate interactions among the immune system, human microbiome, and dietary habits are an integral part that can greatly influence oncological outcomes. Traditional medical systems like Ayurveda, Traditional Chinese Medicine (TCM), and German naturopathy have long recognized the significance of the gut and its inhabitants in overall health. This system offers a comprehensive view of preserving health and its impact on several physiological processes such as immunity and disease susceptibility and preventing illness involves balancing the microbiome with dietary habits, herbal remedies, and mindful lifestyle behaviours.

Ayurveda is an ancient system of medicine, originating over 3000 years ago in India, offers a holistic approach to health and wellness and provides essential insights and practical approaches to modulating the microbiome (Britannica 2024). Ayurveda is based on the concept of Doshas, i.e. Vata, Pitta, and Kapha, a balanced state of these doshas generally called “Tridoshic Balance” essential for overall health, as each doshas is associated with unique physiological functions, impacting dietary patterns and lifestyle practices, while imbalance leads to many disorders. The three doshas show different combinations of five fundamental elements, i.e. air, space, fire, water, and earth. Balance the doshas and support a healthy digestive system; Ayurveda prescribes personalized dietary plans and options based on individual-specific dosha. Another concept of Ayurveda is Agni or Digestive fire; Agni handles the digestion, absorption, and assimilation of food. Agni is like a digestive power to digest the food, a strong Agni is believed to contribute a healthy gut microbiota (Mishra 2023). Some Ayurvedic practices for modulating the microbiome are, offering fermented foods like buttermilk, yogurts, and pickles to introduce essential and beneficial bacteria into the gut, supporting a balanced microbiome. Providing food components that are rich in fibre, such as fruits, whole grains, and vegetables, promotes digestion and nourishes beneficial gut bacteria. "Triphala" formulation including three fruits, i.e. Amalaki, Bibhitaki, and Haritaki has prebiotic effects that enhance the growth of beneficial gut bacteria and digestive health (Makki et al. 2018; Peterson et al. 2017). In Ayurveda, a wide range of herbs are known for their digestive and immune-boosting effects such as Ashwagandha popularly known for its adaptogenic properties, in reducing stress, boosting immune function and curing many ailments (Dubey et al. 2023). For detoxification, rejuvenating the tissues and restoring balance, a set of five therapeutic procedures is used in Ayurveda named "Panchakarma", which includes Vamana (Therapeutic Emesis), Virechana (Therapeutic Purgation), Basti (Therapeutic Enema), Nasya (Nasal Administration), and Raktamokshana (Blood Letting). Yoga and meditation are more practices that are recommended in Ayurveda to boost-up mental well-being, reduce stress, and positively affect gut health.

Traditional Chinese Medicine (TCM), another ancient medical system, originated 2500 years ago and offers a holistic approach to health and provides meaningful insights and practices for altering the microbiome to promote immune-oncology outcomes (Zhang et al. 2023; Yu et al. 2023). TCM is based on the concept of the body's vital energy or Qi, which flows by pathways called meridians and balancing the vital energy is important for supporting a healthy microbiome (Feng et al. 2021). In TCM, Yin-Yang and the five elements theory explain the influence on the digestive system and the gut microbiome. Yin and Yang are two complementary forces where Yin shows aspects of cold, darkness, rest, and substance, while the Yang shows warmth, light, function, and activity, in context with the digestive system. Yin (nutritive substances, fluids) and Yang (digestive function, warmth) are significant for optimal digestion, absorption, and overall digestive health. A good health is when there is a dynamic equilibrium between Yin and Yang within the body disturbances in Ying and Yang balance leads to disruption in the digestive environment potentially influencing the gut microbiome (Li et al. 2023; Umeda 2019). Fire, earth, wood, metal, and water are five elements that explain the interconnection of several physiological and natural phenomena. Each element represents different organs and systems in the body, including the digestive system, which is crucial for supporting a healthy microbiome (Bao et al. 2022; Huang et al. 2018). For example, the earth element is associated with primary organs, i.e. stomach and spleen. Disturbances in earth elements show symptoms such as bloating, loose stool, and poor digestion which reflect disruption in gut microbiota. Yin-Yang and five-element approaches are employed by TCM principles to restore the balance of microbiota. Herbal medicine, acupuncture, mind–body practice and dietary therapy are several practices used in TCM. Herbal medicine formulas like Shen Ling Bai Zhu San improve the spleen and elevate digestion, and Huang Qi (Astragalus) is mainly known for immune-boosting properties and improving the body's defence mechanisms (Xu et al. 2024). Acupuncture methods to modulate gut motility and microbiome composition by inserting fine needles into particular points on the body to stimulate Qi flow and balance the body's energy such as SP6 (Sanyinjiao) and ST36 (Zusanli) are acupuncture points mainly used to regulate the spleen, liver, and kidney and help maintain digestive function and body's overall energy balance (Park et al. 2023). Mind–body practices like Tai Chi and Qigong are regular practices and exercises to promote the flow of Qi and support both mental and physical health (Osypiuk et al. 2018; Deledda et al. 2021). Thus contributing to a balanced microbiota. Dietary therapy focuses on balancing the thermal properties of foods such as cold, hot, warm, and neutral on the body's internal environment, probiotic foods (like kimchi and miso) and functional foods (like ginger and garlic) are used for their antimicrobial and anti-inflammatory properties which helps in harmonize the gut and support immune function (Cullin et al. 2021).

German medicines are known are for their rigorous scientific approach with the integration of holistic practices, representing its roots in traditional healing practices and scientific advancements, provides a unique perspective on modulating the microbiome to enhance immune responses against cancer (Kuhlmann 2017). In a holistic approach, German medicine describes the body as a networked system in which environment, physical, and mental factors contribute to overall health (Geissler et al. 2023). In the scientific approach, scientific research and evidence-based medicine are highly valued, fusing innovative medical discoveries with traditional procedures. Their approaches are patient-focused emphasizing disease prevention and its underlying causes as opposed to solely treating its symptoms (Zakharov 2022). In German medicine, humoral medicine theory explains the concept of balancing four bodily humours, i.e. blood, yellow bile, phlegm, and black bile. By this theory, the health is on an overall maintained by these humours, while in imbalance will result into to disorders. As each of the four humours is connected with particular qualities and bodily functions where blood is associated with air, yellow bile to fire, phlegm to water and black bile to earth and autumn (Nam 2014; Green et al. 2020). Probiotics and prebiotics, phytotherapy (herbal medicine), hydrotherapy, and dietary use are several practices used for modulating the microbiome in German medicine. Both probiotic (*Lactobacillus and Bifidobacterium*) and prebiotics (asparagus and garlic) support and encourage gut health, growth of beneficial gut bacteria and restore a healthy gut microbiome (Babich et al. 2020). German medicine uses herbs and other plant-based treatments (phytotherapy) to promote the immune system by reducing side effects including Echinacea and Chamomile are generally known for their anti-inflammatory, soothing properties and immune-boosting properties to prevent infection and maintain a balanced microbiome (Mahlovanyy et al. 2019). Another therapy, i.e. hydrotherapy involves the use of water in a therapeutic manner. It highlights the use of water in various forms, including baths, compresses, and affusions, in addition to other components like exercise, diet, herbal medicine, and lifestyle advice, which will impact the circulation system by promoting blood flow and lymphatic systems, also influencing the gut health and immune function and shoe potential in reducing the stress and boosting in relaxation. Hence, contributing to a more favourable gut microenvironment (Al Dweikat 2023; Eroğlu and Sanlier 2023). A dietary use including fermented foods (Sauerkraut and kefir) and an anti-inflammatory diet (nuts and

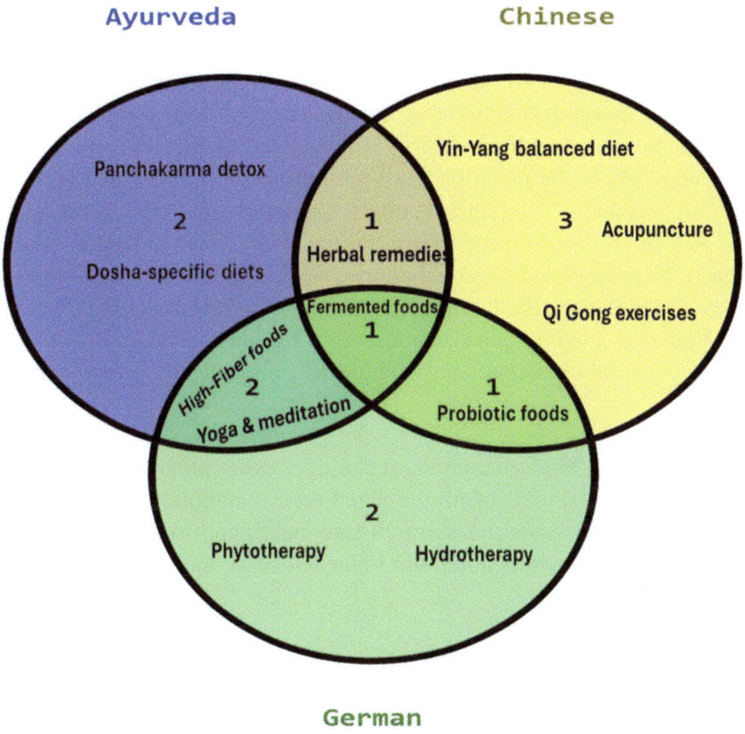

Fig. 15.11 Common and unique practices in microbiome modulation: Ayurveda, Chinese medicine, and German medicine

omega-3 fatty acids) are recommended to lower inflammation, improve gut flora and enhance immune function (Addissouky et al. 2024).

The diagram (Fig. 15.11) illustrates the common and unique practices of Ayurveda, Traditional Chinese Medicine (TCM), and German medicine in modulating and altering the human microbiome. Where each circle stands for one of the medical systems, showcasing their specific approaches to health and wellness. In Ayurveda, the emphasis lies on herbal remedies, Panchakarma detox, dosha-specific diets, yoga, meditation, and high-fibre foods to balance the doshas and support digestive health. Traditional Chinese Medicine (TCM) emphasizes acupuncture, herbal medicine, Qi Gong exercises, a Yin-Yang balanced diet, and probiotic foods to harmonize the body's vital energy or Qi. German Medicine integrates homoeopathy, hydrotherapy, modern pharmacology, probiotics, prebiotics, and an anti-inflammatory diet, combining holistic practices with scientific advancements. These systems have shared common practices such as the use of fermented foods, stress management techniques, natural products, and probiotics, showcasing their collective focus on supporting and handling a healthy gut microbiota to enhance immune function and support overall health.

15.5 Theragnostics and Stand-Offs

Theragnostics, an amalgamation of therapeutics and diagnostics, involves the merging of diagnostic imaging and targeted therapy. The rapidly developing field of theragnostics stands for a paradigm shift in personalized medicine, especially when it comes to oncology. In the nutrition-gut-immuno-oncology axis, theragnostics provides the opportunity to identify certain patient subgroups, forecast therapy response, and track disease progression (Feng et al. 2020; Behrouzi et al. 2019). Additionally, this integrated approach allows for the precise localization of cancerous tissues, tracking the effectiveness of treatments in real-time and the adaptation of therapeutic strategies based on the tumour's response (Vandamme et al. 2015). Nevertheless, several challenges, or "standoffs", prevent this promise from being fully realized.

Diagnostic Imaging Modern imaging techniques essential to theragnostics by providing detailed visualizations of tumours and their metabolic activity. These imaging modalities not only make it easier to locate malignant tissues precisely, but they also provide important details about the traits and behaviour of the tumour.

1. *Positron Emission Tomography (PET)*

PET scans use radioactive tracers that are created in cancer cells due to their increased metabolic activity. This allows for the visualization of metabolic processes in real time. Small tumour detection, metastasis detection, and treatment efficacy assessment are the areas of special utility. For example, fluorodeoxyglucose (FDG) PET is widely used to illustrate regions of elevated glucose metabolism, a typical characteristic of cancer cells (McRobbie et al. 2017).

2. *Magnetic Resonance Imaging (MRI)*

MRI uses powerful magnetic fields and radio waves to create finely detailed images of the body's soft tissues. It offers high-resolution images and has the ability to distinguish between various tissue types. It is beneficial for visualizing the anatomical structure of tumours, identifying tumours in tissues with intricate architecture, such as the brain and spinal cord, and viewing the tumour margins (Alqahtani 2022).

3. *Multiparametric MRI (mp-MRI)*

Multiple MRI sequences (such as T1-weighted, T2-weighted, DWI, and DCE) are incorporated in mp-MRI to provide detailed insight into tumour morphology, cellularity, vascularity, and functional characteristics. It facilitates a thorough evaluation of tumours, supporting diagnosis, staging, planning, and tracking of therapeutic outcomes (Nguyen et al. 2016).

4. *Single-Photon Emission Computed Tomography (SPECT)*

SPECT imaging employs the use of gamma-emitting radioisotopes to produce 3D images of internal organs. It gathers useful data about tissues by detecting the radiation that the isotopes emit. It is often used in combination with other imaging techniques to provide comprehensive diagnostic information. It is particularly effective for analysing bone metastases and measuring blood flow to organs (Lin et al. 2023a, b).

5. *Computed Tomography (CT) Perfusion Imaging*

CT perfusion imaging involves evaluating the blood flow and tissue perfusion in tumours by monitoring the track of contrast agents in the tissues. This method can help diagnose tumours and evaluate therapy response by offering important details regarding tumour vascularity, microcirculation, and tissue viability (Kawashima 2014).

Targeted Therapy Theragnostics use targeted therapies that target cancer cells directly while causing the least amount of harm to healthy tissues. This accuracy reduces side effects and enhances the therapy efficacy. The selection of targeted therapy is guided by the diagnostic findings, guaranteeing an individualized strategy for the treatment of cancer.

1. *Radiolabelled Antibodies*

These are radioactive isotopes conjugated with antibodies. The antibodies specifically bind to antigens expressed on cancer cells, delivering targeted radiation that kills the cancer cells. When treating non-Hodgkin lymphoma, radiolabelled antibodies like Ibritumomab tiuxetan (Zevalin) deliver targeted radiation to cancer cells while protecting healthy tissue (Wang et al. 2016).

2. *Nanoparticles*

Nanoparticles can be engineered to carry therapeutic agents, like drugs, proteins, or nucleic acids, directly to cancer cells. They can be made to increase the drug delivery, improve bioavailability, and lowering the toxicity. Chemotherapeutic chemicals are delivered by means of nanoparticles, including polymeric, dendrimer, and liposome nanoparticles. For example, the doxorubicin liposomal formulation Doxil is used to cure ovarian cancer and Kaposi's sarcoma (Gupta et al. 2019).

3. *Small Molecule Inhibitors*

These are compounds that selectively inhibit the action of enzymes or proteins necessary for cancer cell growth and survival. They focus on specific molecular pathways that are essential for tumour progression. In chronic myeloid leukaemia, small molecule inhibitors such as imatinib (Gleevec) target the BCR–ABL tyrosine kinase, effectively preventing cancer cell proliferation and triggering apoptosis (Yang and Wang 2023).

4. *Chimeric Antigen Receptor (CAR) T-Cell Therapy*

CAR T-cell therapy involves the genetic engineering of a patient's T cells to express chimeric antigen receptors targeting specific antigens on cancer cells. These modified T cells can identify and eliminate cancer cells when they are reinfused into the patient. Certain forms of leukaemia and lymphoma have been remarkably successfully treated by therapies like Kymriah and Yescarta, which offer a customized and focused immunotherapeutic approach (Geenen et al. 2021).

5. *Peptide Receptor Radionuclide Therapy (PRRT)*

PRRT involves the use of radiolabelled peptides that specifically bind to receptors overexpressed on the surface of cancer cells. The radiolabelled peptides attach to the cancer cells and release precise radiation. For the treatment of neuroendocrine tumours, PRRT, using radiolabelled peptides like Lutetium-177 DOTATATE, has been increasingly utilized, demonstrating favourable outcomes in patients with advanced or metastatic disease (Patel et al. 2022).

Theragnostics plays a critical role in forming a multimodal approach to cancer treatment by interacting with the interrelated dynamics of the nutrition-gut-immune-oncology axis. Leveraging this integrated framework, new approaches have the potential to completely transform individualized cancer care. One avenue of innovation lies in microbiome-targeted therapies, where theragnostic methods harness microbiome analysis to discern the impact of gut bacteria on the efficacy of cancer treatments. This method opens the door to customized dietary and probiotic therapies to maximize treatment outcomes by identifying bacterial profiles that either support or obstruct therapeutic responses. By delving into the complex interactions between the microbiome and cancer therapy, theragnostics enables healthcare practitioners to modify treatment regimens in a manner that accounts for the individualized microbial landscape (Kamel et al. 2024; de Castilhos et al. 2024).

Furthermore, the combination of nutritional evaluation and cutting-edge imaging methods opens up new possibilities for tailored cancer treatment. Healthcare providers can use nutritional imaging to learn more about the nutritional status and body composition of cancer patients. This information lays the groundwork for individualized nutritional therapies that are intended to strengthen immune system function and improve the body's reaction to therapy. Beyond the confines of conventional oncology, this holistic approach highlights the promise of theragnostics to highlight the symbiotic relationship between immune function, nutrition, and optimal treatment outcomes. Furthermore, the application of molecular imaging to immune surveillance is a challenging new area for therapeutic innovation. By monitoring immune cell infiltration and activity within tumours, molecular imaging enables real-time assessment of the immune response to dietary and microbiome-modulating interventions. This dynamic approach equips healthcare professionals with the tools to adjust therapeutic strategies in response to the evolving immune landscape, ultimately striving to enhance treatment efficacy through the orchestration of personalized, immune-centric care pathways (Dessale et al. 2022).

Although theragnostics holds great promise for the field of oncology, there are a few major obstacles that need to be resolved before its full potential can be reached. Complexity and expense are introduced by the integration of cutting-edge diagnostic and therapeutic technology, creating obstacles that prevent many patients from receiving these therapies. To mitigate this difficulty, methods that aim to reduce costs and streamline processes must be developed in order to ensure wider implementation and accessibility. Additionally, the dual nature of theragnostics, combining diagnostics and therapeutics, gives rise to regulatory hurdles that demand clear guidelines and streamlined approval processes for the development and clinical use of theragnostic agents. It is imperative that these regulatory obstacles be addressed in order to promote the development and application of theragnostics in clinical settings (Warenius 2009; Dubey et al. 2024).

Moreover, while theragnostics inherently embodies personalization, achieving optimal personalization requires extensive biomarker discovery and validation. Sustained investigation is necessary to find trustworthy biomarkers that can effectively direct the creation of customized theragnostic approaches, resolving the customization impasse. Furthermore, the imperative to ensure the safety and efficacy of theragnostic agents cannot be understated. Rigorous preclinical and clinical testing is essential to fully assess potential side effects and therapeutic advantages. The field of theragnostics can be ready to reach its full potential and have a big impact on the oncology care landscape by successfully overcoming these obstacles. The compilation presented in the chapter is towards a compendious journey of the Nutrition-Gut-Immuno-Oncology Axis that could significantly contribute to the growing knowledge in this fight against cancer.

References

Addissouky TA, El Sayed IET, Ali MMA (2024) Precision medicine and immunotherapy advances transforming colorectal cancer treatment. ResearchGate

Ağagündüz D, Cocozza E, Cemali Ö, Bayazıt AD, Nanì MF, Cerqua I, Morgillo F, Saygılı SK, Canani RB, Amero P, Capasso R (2023) Understanding the role of the gut microbiome in gastrointestinal cancer: a review. Front Pharmacol 14. https://doi.org/10.3389/fphar.2023.1130562

Ahmed N, Abusalah MAHA, Farzand A, Absar M, Yusof NY, Rabaan AA, AlSaihati H, Alshengeti A, Alwarthan S, Alsuwailem HS, Alrumaih ZA, Alsayyah A, Yean CY (2022) Updates on Epstein–Barr Virus (EBV)-associated nasopharyngeal carcinoma: emphasis on the latent gene products of EBV. Medicina. https://doi.org/10.3390/medicina59010002

Ajab SM, Zoughbor SH, Labania LA, Östlundh LM, Orsud HS, Olanda MA, Alkaabi O, Alkuwaiti SH, Alnuaimi SM, Al Rasbi Z (2024) Microbiota composition effect on immunotherapy outcomes in colorectal cancer patients: a systematic review. PLoS One 19(7):e0307639–e0307639. https://doi.org/10.1371/journal.pone.0307639

Al Dweikat T (2023) Complementary and integrative modalities. In: Core curriculum for oncology nursing-E-Book

Al-Ishaq RK, Koklesova L, Kubatka P, Büsselberg D (2022) Immunomodulation by gut microbiome on gastrointestinal cancers: focusing on colorectal cancer. Cancers 14(9):2140–2140. https://doi.org/10.3390/cancers14092140

Almeida C, Oliveira R, Baylina P, Fernandes R, Teixeira FG, Barata P (2022) Current trends and challenges of fecal microbiota transplantation—an easy method that works for all? Biomedicines 10:2742. https://doi.org/10.3390/biomedicines10112742

Alqahtani S (2022) Developing nomograms and predictive models based on multiparametric MRI characterised localised prostate cancer. University of Dundee

Al-Yami AM, Al-Mousa AT, Al-Otaibi SA, Khalifa AY (2022) Lactobacillus species as probiotic: isolation sources and health benefits. J Pure Appl Microbiol 16(4):2270–2291. https://doi.org/10.22207/JPAM.16.4.19

Asseri AH, Bakhsh TM, Samah S, Sajadi A, Ahmad A, Rather IA (2023) The gut dysbiosis-cancer axis: illuminating novel insights and implications for clinical practice. Front Pharmacol. https://doi.org/10.3389/fphar.2023.1208044

Babich O, Sukhikh S, Prosekov A, Asyakina L (2020) Medicinal plants to strengthen immunity during a pandemic. Pharmaceuticals 13(10):313

Bakhti SZ, Latifi-Navid S (2021) Interplay and cooperation of Helicobacter pylori and gut microbiota in gastric carcinogenesis. BMC Microbiol 21(1). https://doi.org/10.1186/s12866-021-02315-x

Balmain A (2020) The critical roles of somatic mutations and environmental tumor-promoting agents in cancer risk. Nat Genet 52(11):1139–1143. https://doi.org/10.1038/s41588-020-00727-5

Bao Y, Han X, Liu D, Tan Z, Deng Y (2022) Gut microbiota: the key to the treatment of metabolic syndrome in traditional Chinese medicine–a case study of diabetes and nonalcoholic fatty liver disease. Front Immunol 13:1072376

Behrouzi A, Nafari AH, Siadat SD (2019) The significance of microbiome in personalized medicine. Clin Transl Med 8(1):16

Binnewies M, Roberts EW, Kersten K, Chan V, Fearon DF, Merad M, Coussens LM, Gabrilovich DI, Ostrand-Rosenberg S, Hedrick CC, Vonderheide RH, Pittet MJ, Jain RK, Zou W, Kevin Howcroft T, Woodhouse EC, Weinberg RA, Krummel MF (2018) Understanding the tumor immune microenvironment (TIME) for effective therapy. Nat Med 24(5):541–550. https://doi.org/10.1038/s41591-018-0014-x

Bose M, Mukherjee P (2019) Role of microbiome in modulating immune responses in cancer. Mediat Inflamm 2019:1–7. https://doi.org/10.1155/2019/4107917

Bose M, Wood HM, Young C, Van Nang P, Van Doi M, Vaccaro C, Piñero TA, Arguero J, Melendez LC, Valladares CT, Quirke P, Seshadri RA (2023) Analysis of an Indian colorectal cancer faecal microbiome collection demonstrates universal colorectal cancer-associated patterns, but closest correlation with other Indian cohorts. BMC Microbiol 23(1). https://doi.org/10.1186/s12866-023-02805-0

Bou Zerdan M, Niforatos S, Nasr S, Nasr D, Ombada M, John S, Dutta D, Lim SH (2022) Fecal microbiota transplant for hematologic and oncologic diseases: principle and practice. Cancers 14:691. https://doi.org/10.3390/cancers14030691

Britannica T (2024, June 23) Editors of Encyclopaedia. Ayurveda. Encyclopedia Britannica. https://www.britannica.com/science/Ayurveda

Brücher BL, Jamall IS (2014) Epistemology of the origin of cancer: a new paradigm. BMC Cancer 14(1). https://doi.org/10.1186/1471-2407-14-331

Burrello C, Giuffrè MR, Macandog AD, Diaz-Basabe A, Cribiù FM, Lopez G, Borgo F, Nezi L, Caprioli F, Vecchi M, Facciotti F (2019) Fecal microbiota transplantation controls murine chronic intestinal inflammation by modulating immune cell functions and gut microbiota composition. Cells. https://doi.org/10.3390/CELLS8060517

Butel JS (2000) Viral carcinogenesis: revelation of molecular mechanisms and etiology of human disease. Carcinogenesis 21(3):405–426. https://doi.org/10.1093/carcin/21.3.405

Chai Y, Zheng-Hong H, Xuqiu S, Tianyu L, Yiyin Z, Xu F, Qijiang M, Yuelong L (2023) Microbiota regulates pancreatic cancer carcinogenesis through altered immune response. Microorganisms. https://doi.org/10.3390/microorganisms11051240
Chen DS, Mellman I (2017) Elements of cancer immunity and the cancer–immune set point. Nature 541(7637):321–330. https://doi.org/10.1038/nature21349
Cheuk H, Zhang X, Yu J (2023) Gut microbiota and immune alteration in cancer development: implication for immunotherapy. EGastroenterology 1(1):e100007–e100007. https://doi.org/10.1136/egastro-2023-100007
Cho YS, Han K, Xu J, Moon JJ (2024) Novel strategies for modulating the gut microbiome for cancer therapy. Adv Drug Deliv Rev 210:115332–115332. https://doi.org/10.1016/j.addr.2024.115332
Clay SL et al (2022) Colorectal cancer: the facts in the case of the microbiota. J Clin Invest 132(4):e155101. https://doi.org/10.1172/JCI155101
Cullin N, Azevedo Antunes C, Straussman R, Stein-Thoeringer CK, Elinav E (2021) Microbiome and cancer. Cancer Cell 39(10):1317–1341. https://doi.org/10.1016/j.ccell.2021.08.006
Dadgar N, Keshava VE, Raj MS, Wagner PL (2023) The influence of the microbiome on immunotherapy for gastroesophageal cancer. Cancers 15(18):4426–4426. https://doi.org/10.3390/cancers15184426
Das S, Das MK (2021) Technological challenges of theranostics in oncology. In: Multifunctional theranostic nanomedicines in cancer. Elsevier, pp 225–246
de Castilhos J, Tillmanns K, Blessing J, Laraño A (2024) Microbiome and pancreatic cancer: time to think about chemotherapy. Gut Microbes 15(2):2374596. https://doi.org/10.1080/19490976.2024.2374596
Deledda A, Annunziata G, Tenore GC, Palmas V, Manzin A, Velluzzi F (2021) Diet-derived antioxidants and their role in inflammation. Obes Gut Microbiota Modul Antioxid 10:708. https://doi.org/10.3390/antiox10050708
Deshpande N, Mokashi M (2024) Concept of Agni and its importance in Ayurveda: a review. Quest J Res Pharm Sci 10(5):01–06
Dessale M, Mengistu G, Mengist HM (2022) Nanotechnology: a promising approach for cancer diagnosis, therapeutics and theragnosis. Int J Nanomedicine 17:3735
Diet During Cancer Treatment (2024) In: Barnard ND (ed) Nutrition guide for clinicians, 3rd edn. Physicians Committee for Responsible Medicine. https://nutritionguide.pcrm.org/nutritionguide/view/Nutrition_Guide_for_Clinicians/1342067/all/Diet_during_Cancer_Treatment
DiMaio D, Emu B, Goodman AL, Mothes W, Justice A (2021) Cancer microbiology. J Natl Cancer Inst 114(5):651–663. https://doi.org/10.1093/jnci/djab212
Dizdar Ö, Kılıçkap S (2019) Global epidemiology of gastrointestinal cancers, vol 1–12. Springer EBooks. https://doi.org/10.1007/978-3-030-18890-0_1
Dubey AK, Uppadhyaya N, Nilawe P, Chauhan N, Kumar S, Gupta UA, Bhaduri A (2018) LogMPIE, pan-India profiling of the human gut microbiome using 16S rRNA sequencing. Sci Data 5(1). https://doi.org/10.1038/sdata.2018.232
Dubey SK, Verma KK, Qureshi A, Behar N (2023) Assessment of antimicrobial activity of commercially available Withania Somnifera (Ashwagandha) powder against pathogenic microorganisms. NewBioWorld J Alumni Assoc Biotechnol. https://doi.org/10.52228/nbw-jaab.2022-4-2-5
Dubey AK, Kaur I, Madaan R, Raheja S (2024) Unlocking the potential of oncology biomarkers: advancements in clinical theranostics. Drug Metabol Personal Ther 39:5
Dung NT, Susukida T, Ucche S, He K, Sasaki S-I, Hayashi R, Hayakawa Y (2023) Calorie restriction impairs anti-tumor immune responses in an immunogenic preclinical cancer model. Nutrients 15:3638. https://doi.org/10.3390/nu15163638
Eroğlu FE, Sanlier N (2023) Effect of fermented foods on some neurological diseases, microbiota, behaviors: mini review. Crit Rev Food Sci Nutr 63(3):455–470
Fan P, Zhang Z, Lu L, Guo X, Hao Z, Wang X, Ye Y (2022) Association of single nucleotide polymorphisms (SNPs) with gastric cancer susceptibility and prognosis in population in Wuwei, Gansu, China. World J Surg Oncol 20(1). https://doi.org/10.1186/s12957-022-02663-6

Feng W, Ao H, Peng C, Yan D (2019) Gut microbiota: a new frontier to understand traditional Chinese medicines. Pharmacol Res 142:176–191

Feng W, Liu J, Ao H, Yue S, Peng C (2020) Targeting gut microbiota for precision medicine: focusing on the efficacy and toxicity of drugs. Theranostics 10(20):8976–9004

Feng Y, Wang J, Jiang P, Li Z, Liu J (2021) Regulation of intestinal flora by external treatment of traditional Chinese medicine. MEDS Publ Health Preven Med 1:71

Gabbia D, De Martin S (2024) Insights into Hepatocellular Carcinoma: from pathophysiology to novel therapies. Int J Mol Sci 25(8):4188–4188. https://doi.org/10.3390/ijms25084188

Geenen L, Nonnekens J, Konijnenberg M, De Jong M (2021) Overcoming nephrotoxicity in peptide receptor radionuclide therapy using Lu-DOTA-TATE for the treatment of neuroendocrine tumors. Nucl Med Biol 98–99:29–35

Geissler J, Makaroff LE, Söhlke B (2023) Precision oncology medicines and the need for real world evidence acceptance in health technology assessment: importance of patient involvement in Germany. Eur J Cancer, Elsevier 193:113323

Gheorghe AS, Negru ŞM, Preda M, Mihăilă RI, Komporaly IA, Dumitrescu EA, Lungulescu CV, Kajanto LA, Georgescu B, Radu EA et al (2022) Biochemical and metabolical pathways associated with microbiota-derived butyrate in colorectal cancer and Omega-3 fatty acids implications: a narrative review. Nutrients 14:1152. https://doi.org/10.3390/nu14061152

Gillespie MR, Rai V, Agrawal S, Nandipati KC (2021) The role of microbiota in the pathogenesis of Esophageal Adenocarcinoma. Biology 10(8):697–697. https://doi.org/10.3390/biology10080697

Gopalakrishnan V, Helmink BA, Spencer CN, Reuben A, Wargo JA (2018a) The influence of the gut microbiome on cancer, immunity, and cancer immunotherapy. Cancer Cell 33(4):570–580. https://doi.org/10.1016/j.ccell.2018.03.015

Gopalakrishnan V, Spencer CN, Nezi L et al (2018b) Gut microbiome modulates response to anti-PD-1 immunotherapy in melanoma patients. Science 359(6371):97–103. https://doi.org/10.1126/science.aan4236

Gourbeyre P, Denery S, Bodinier M (2011) Probiotics, prebiotics, and synbiotics: impact on the gut immune system and allergic reactions. J Leukoc Biol 89(5):685–695. https://doi.org/10.1189/jlb.1109753

Green M, Arora K, Prakash S (2020) Microbial medicine: prebiotic and probiotic functional foods to target obesity and metabolic syndrome. Int J Mol Sci 21(8):2890

Gupta SK, Singh P, Ali V, Verma M (2019) Therapeutic approaches to enhance the BCR-ABL tyrosine kinase inhibitors efficacy in chronic myeloid leukemia. Asian Pac J Health Sci 6:58

Haręża DA, Wilczyński JR, Paradowska E (2022) Human papillomaviruses as infectious agents in gynecological cancers. Oncogenic properties of viral proteins. Int J Mol Sci. https://doi.org/10.3390/ijms23031818

Hartman-Houstman H, Swenson S, Minea RO, Sinha UK, Chiang M-F, Chen TC, Schönthal AH (2024) Activation of Epstein–Barr virus' lytic cycle in nasopharyngeal carcinoma cells by NEO212, a conjugate of Perillyl alcohol and Temozolomide. Cancers 16(5):936–936. https://doi.org/10.3390/cancers16050936

Hasan A, Hasan LK, Schnabl B, Greytak M, Yadlapati R (2020) Microbiome of the Aerodigestive tract in health and Esophageal disease. Dig Dis Sci 66(1):12–18. https://doi.org/10.1007/s10620-020-06720-6

Hatta A, Mohamad A, Chin S-F, Neoh H-m (2021) Pathogens and carcinogenesis: a review. Biology 10(6):533–533. https://doi.org/10.3390/biology10060533

Hau PM, Lung HL, Wu M, Tsang CM, Wong K-L, Mak NK, Lo KW (2020) Targeting Epstein-Barr virus in Nasopharyngeal Carcinoma. Front Oncol 10. https://doi.org/10.3389/fonc.2020.00600

Hecht JR, Mitchell J, Morelli MP, Anandappa G, Yang JC (2023) Next-generation approaches to immuno-oncology in GI cancers | American Society of Clinical Oncology educational book. American Society of Clinical Oncology Educational Book. https://ascopubs.org/doi/10.1200/EDBK_389072

Higashi T, Kurokawa Y (2024) Incidence, mortality, survival, and treatment statistics of cancers in digestive organs—Japanese cancer statistics 2024. Ann Gastroenterol Surg. https://doi.org/10.1002/ags3.12835

Hou X, Zheng Z, Wei J, Zhao L (2022) Effects of gut microbiota on immune responses and immunotherapy in colorectal cancer. Front Immunol 13. https://doi.org/10.3389/fimmu.2022.1030745

Huang Y et al (2018) The Yin-Yang property of Chinese Medicinal Herbs relates to chemical composition but not anti-oxidative activity: an illustration using Spleen-Meridian Herbs. Front Pharmacol 9. www.frontiersin.org/journals/pharmacology/articles/10.3389/fphar.2018.01304/full, https://doi.org/10.3389/fphar.2018.01304. Accessed 29 July 2024

Hur C, Miller M, Kong CY, Dowling EC, Nattinger KJ, Dunn M, Feuer EJ (2012) Trends in esophageal adenocarcinoma incidence and mortality. Cancer 119(6):1149–1158. https://doi.org/10.1002/cncr.27834

Idres YM, Nigel AJ, McMillan, Adi I (2022) Hyperactivating p53 in human Papillomavirus-driven cancers: a potential therapeutic intervention. Mol Diagn Ther. https://doi.org/10.1007/s40291-022-00583-5

Imran M, Insaf A, Hasan N, Sugandhi VV, Shrestha D, Paudel KR, Jha SK, Hansbro PM, Dua K, Devkota HP, Mohammed Y (2023) Exploring the remarkable chemotherapeutic potential of polyphenolic antioxidants in battling various forms of cancer. Molecules. https://doi.org/10.3390/molecules28083475

Ishihara T, Yoshida M, Arita M (2019) Omega-3 fatty acid-derived mediators that control inflammation and tissue homeostasis. Int Immunol 31(9):559–567. https://doi.org/10.1093/intimm/dxz001

Ji B, Nielsen J (2015) New insight into the gut microbiome through metagenomics. Adv Genom Genet. https://doi.org/10.2147//AGG.S57215

Jiang L, Schnabl B (2020) Gut microbiota in liver disease: what do we know and what do we not know? Physiology 35(4):261–274. https://doi.org/10.1152/physiol.00005.2020

Kamel M, Aleya S, Alsubih M, Aleya L (2024) Microbiome dynamics: a paradigm shift in combatting infectious diseases. J Personal Med 14(2):217. https://doi.org/10.3390/jpm14020217

Kang X, Lau HC-H, Yu J (2024) Modulating gut microbiome in cancer immunotherapy: harnessing microbes to enhance treatment efficacy. Cell Rep Med 5(4):101478–101478. https://doi.org/10.1016/j.xcrm.2024.101478

Kawashima H (2014) Radioimmunotherapy: a specific treatment protocol for cancer by cytotoxic radioisotopes conjugated to antibodies. Sci World J 14:492061

Key TJ, Bradbury KE, Perez-Cornago A, Sinha R, Tsilidis KK, Tsugane S (2020) Diet, nutrition, and cancer risk: what do we know and what is the way forward? BMJ:m511–m511. https://doi.org/10.1136/bmj.m511

Koh A et al (2016) The gut microbiota in health and disease. Nat Rev Microbiol 14(1):20–34

Kuhlmann K (2017) Health services research on people with mental disorders in the integrated care system

Li L, Cao Y, Zhan J, Wu C (2023) Perspective on the modern interpretation of the property theory of mild-natured and sweet-flavored traditional Chinese medicine via gut microbiota. Med Nephrol Artif Organs 1(2):122–130

Li Z, Xiong W, Liang Z, Wang J, Zeng Z, Kołat D, Li X, Zhou D, Xu X, Zhao L (2024) Critical role of the gut microbiota in immune responses and cancer immunotherapy. J Hematol Oncol 17(1). https://doi.org/10.1186/s13045-024-01541-w

Lin Q, Guan S-W, Yu H-B (2023a) Immuno-oncology-microbiome axis of gastrointestinal malignancy. World J Gastrointest Oncol 15(5):757–775. https://doi.org/10.4251/wjgo.v15.i5.757

Lin Q, Choyke PL, Sato N (2023b) Visualizing vasculature and its response to therapy in the tumor microenvironment. Theranostics 13:5223

Liu S, Liu Y, Zhang Q, Wu J, Liang J, Yu S, Wei G-H, White KP, Wang X (2017) Systematic identification of regulatory variants associated with cancer risk. Genome Biol 18(1). https://doi.org/10.1186/s13059-017-1322-z

Liu C, Ng S-K, Ding Y, Lin Y, Liu W, Wong SH, Sung JJ-Y, Yu J (2022) Meta-analysis of mucosal microbiota reveals universal microbial signatures and dysbiosis in gastric carcinogenesis. Oncogene 41(28):3599–3610. https://doi.org/10.1038/s41388-022-02377-9

Long Y, Tang L, Zhou Y, Zhao S, Zhu H (2023) Causal relationship between gut microbiota and cancers: a two-sample Mendelian randomisation study. BMC Med 21(1). https://doi.org/10.1186/s12916-023-02761-6

Luu M, Schütz B, Lauth M, Visekruna A (2023) The impact of gut microbiota-derived metabolites on the tumor immune microenvironment. Cancers 15(5):1588–1588. https://doi.org/10.3390/cancers15051588

Mahlovanyy AV, Hrynovets VS, Kuninets OB (2019) Bases of physical rehabilitation in medicine. Meduniv Lviv

Mahmoudi H, Hadi Hossainpour (2022) Application and development of fecal microbiota transplantation in the treatment of gastrointestinal and metabolic diseases: a review. Saudi J Gastroenterol. https://doi.org/10.4103/sjg.sjg_131_22

Makki K, Deehan EC, Walter J, Bäckhed F (2018) The impact of dietary fiber on gut microbiota in host health and disease. Cell Host Microbe 23(6):705–715. https://doi.org/10.1016/j.chom.2018.05.012

Malone ER, Oliva M, Sabatini PJB, Stockley TL, Siu LL (2020) Molecular profiling for precision cancer therapies. Genome Med 12(1). https://doi.org/10.1186/s13073-019-0703-1

Marin-Acevedo JA, Dholaria B, Soyano AE, Knutson KL, Chumsri S, Lou Y (2018) Next generation of immune checkpoint therapy in cancer: new developments and challenges. J Hematol Oncol 11(1). https://doi.org/10.1186/s13045-018-0582-8

Mark S, Terveer EM, Dahlerup JF, Erikstrup C, Arkkila P, Vehreschild MJ, Laniro G, Gasbarrini A, Sokol H, Kump PK, Satokari R, De Looze D, Vermeire S, Nakov R, Brezina J, Helms M, Kjeldsen J, Rode AA, Kousgaard SJ, Alric L (2021) The use of Faecal Microbiota Transplantation (FMT) in Europe: a Europe-wide survey. Lancet Region Health Europe 9:100181–100181. https://doi.org/10.1016/j.lanepe.2021.100181

McIlvanna E, Linden GJ, Craig SG et al (2021) Fusobacterium nucleatum and oral cancer: a critical review. BMC Cancer 21:1212. https://doi.org/10.1186/s12885-021-08903-4

McRobbie DW, Moore EA, Graves MJ, Prince MR (2017) MRI from picture to proton. Cambridge University Press

Mishra PK (2023) Concept of dietetics and its importance in Ayurveda. J Ayurveda Integ Med Sci. https://doi.org/10.21760/jaims.8.5.13

Molina-Montes E, Elena S-F, Elena S-F, Belén G-V, María J, Sánchez. (2020) The impact of plant-based dietary patterns on cancer-related outcomes: a rapid review and meta-analysis. Nutrients. https://doi.org/10.3390/NU12072010

Montégut L, de Cabo R, Zitvogel L, Kroemer G (2022) Science-driven nutritional interventions for the prevention and treatment of cancer. Cancer Discov 12(10):2258–2279. https://doi.org/10.1158/2159-8290.CD-22-0504

Nam JK (2014) Medieval European medicine and Asian spices. Korean J Med Hist 23(2):319–342

Ng R, Sutradhar R, Yao Z, Wodchis WP, Rosella LC (2019) Smoking, drinking, diet and physical activity—modifiable lifestyle risk factors and their associations with age to first chronic disease. Int J Epidemiol 49(1):113–130. https://doi.org/10.1093/ije/dyz078

Nguyen LT, Chang EI, Hanasono MM (2016) Evaluation of bone metastases by single-photon emission computed tomography (SPECT). J Surg Oncol 113(7):825–829

Osypiuk K, Thompson E, Wayne PM (2018) Can Tai Chi and Qigong postures shape our mood? Toward an embodied cognition framework for mind-body research. Front Hum Neurosci 12. https://doi.org/10.3389/fnhum.2018.00174

Ott JJ, Ullrich A, Mascarenhas M, Stevens GA (2010) Global cancer incidence and mortality caused by behavior and infection. J Public Health 33(2):223–233. https://doi.org/10.1093/pubmed/fdq076

Pan C, Liu H, Robins E, Song W, Liu D, Li Z, Zheng L (2020) Next-generation immuno-oncology agents: current momentum shifts in cancer immunotherapy. J Hematol Oncol 13(1). https://doi.org/10.1186/s13045-020-00862-w

Park EM, Chelvanambi M, Bhutiani N, Kroemer G, Zitvogel L, Wargo JA (2022) Targeting the gut and tumor microbiota in cancer. Nat Med 28(4):690–703. https://doi.org/10.1038/s41591-022-01779-2

Park M, Song R, Ju K, Shin JC, Seo J, Fan X, Gao X, Ryu A, Li Y (2023) Effects of Tai Chi and Qigong on cognitive and physical functions in older adults: systematic review, meta-analysis, and meta-regression of randomized clinical trials. BMC Geriatr 23(1). https://doi.org/10.1186/s12877-023-04070-2

Parsons HM, Forte ML, Abdi HI, Brandt S, Claussen AM, Wilt T, Klein M, Ester E, Landsteiner A, Shaukut A, Sibley SS, Slavin J, Sowerby C, Ng W, Butler M (2023) Nutrition as prevention for improved cancer health outcomes: a systematic literature review, JNCI cancer. Spectrum 7(3):pkad035. https://doi.org/10.1093/jncics/pkad035

Patel M, McAllister M, Nagaraju R (2022) The intestinal microbiota in colorectal cancer metastasis–passive observer or key player? Crit Rev Oncol Hematol 169:103599. https://doi.org/10.1016/j.critrevonc.2022.103599

Pereira-Marques J, Ferreira RM, Pinto-Ribeiro I, Figueiredo C (2019) Helicobacter pylori infection, the gastric microbiome and gastric cancer. Adv Exp Med Biol 195–210. https://doi.org/10.1007/5584_2019_366

Perri F, Sabbatino F, Ottaiano A, Fusco R, Caraglia M, Cascella M, Longo F, Rega RA, Salzano G, Pontone M, Marciano ML, Piccirillo A, Montano M, Fasano M, Ciardiello F, Scarpati GDV, Ionna F (2023) Impact of Epstein Barr virus infection on treatment opportunities in patients with nasopharyngeal cancer. Cancers. https://doi.org/10.3390/cancers15051626

Peterson CT, Denniston K, Chopra D (2017) Therapeutic uses of Triphala in Ayurvedic medicine. J Altern Complement Med (New York, NY) 23(8):607–614. https://doi.org/10.1089/acm.2017.0083

Poveshchenko AF, Kabakov AV, Kazakov OV (2023) Gut microbiota and carcinogenesis: actual aspects. J Microbiol Epidemiol. https://doi.org/10.36233/0372-9311-356

Rao D, Parakrama R, Augustine T, Liu Q, Goel S, Maitra R (2019) Immunotherapeutic advances in gastrointestinal malignancies. Npj Precis Oncol 3(1). https://doi.org/10.1038/s41698-018-0076-8

Ray A, Rajashree P, Adam D, Borsch BD (2023) An overview of selected bacterial infections in cancer, their virulence factors, and some aspects of infection management. Biology. https://doi.org/10.3390/biology12070963

Rebersek M (2021) Gut microbiome and its role in colorectal cancer. BMC Cancer 21(1). https://doi.org/10.1186/s12885-021-09054-2

Reyes VE (2023) Helicobacter pylori and its role in gastric cancer. Microorganisms. https://doi.org/10.3390/microorganisms11051312

Rezasoltani S, Yadegar A, Aghdaei HA, Zali MR (2020) Modulatory effects of gut microbiome in cancer immunotherapy: a novel paradigm for blockade of immune checkpoint inhibitors. Cancer Med 10(3):1141–1154. https://doi.org/10.1002/cam4.3694

Ribas A, Wolchok JD (2018) Cancer immunotherapy using checkpoint blockade. Science 359(6382):1350–1355. https://doi.org/10.1126/science.aar4060

Rodriguez GM, DePuy D, Aljehani M, Bien J, Jerry W, Lin AY (2023) Trends in epidemiology of Esophageal cancer in the US, 1975-2018. JAMA Netw Open 6(8):e2329497–e2329497. https://doi.org/10.1001/jamanetworkopen.2023.29497

Routy B, Le Chatelier E, Derosa L et al (2018) Gut microbiome influences efficacy of PD-1–based immunotherapy against epithelial tumors. Science 359(6371):91–97. https://doi.org/10.1126/science.aan3706

Sgamato C, Rocco A, Compare D, Priadko K, Romano M, Nardone G (2024) Exploring the link between Helicobacter pylori, gastric microbiota and gastric cancer. Antibiotics 13(6):484–484. https://doi.org/10.3390/antibiotics13060484

Shariati A, Razavi S, Ghaznavi-Rad E et al (2021) Association between colorectal cancer and Fusobacterium nucleatum and Bacteroides fragilis bacteria in Iranian patients: a preliminary study. Infect Agents Cancer 16:41. https://doi.org/10.1186/s13027-021-00381-4

Snehal S (2023) Ayurveda for cancer therapy - a review. J Ayurveda Integ Med Sci. https://doi.org/10.21760/jaims.8.5.32
Soldati L, Di Renzo L, Jirillo E et al (2018) The influence of diet on anti-cancer immune responsiveness. J Transl Med 16:75. https://doi.org/10.1186/s12967-018-1448-0
Su Q, Jin C, Bo Z, Yang Y, Wang J, Wang J, Zhou J, Chen Y, Zeng H, Chen G, Wang Y (2023) Association between gut microbiota and gastrointestinal cancer: a two-sample bi-directional Mendelian randomization study. Front Microbiol 14. https://doi.org/10.3389/fmicb.2023.1181328
Talapko J, Meštrović T, Dmitrović B, Juzbašić M, Matijević T, Bekić S, Erić S, Flam J, Belić D, Petek Erić A et al (2023) A putative role of Candida albicans in promoting cancer development: a current state of evidence and proposed mechanisms. Microorganisms 11:1476. https://doi.org/10.3390/microorganisms11061476
Tariq MA (2023) CRISPR-Cas9: the future of precision genome engineering. Futur Biotechnol. https://doi.org/10.54393/fbt.v3i01.35
Umeda N (2019) Gut flora "the second brain" connects Eastern and Western medicine: intestinal hyper-permeability or Qi deficiency can affect brain, mind, and whole body. Longhua Chinese Med 2(1):14–25
van Nood E, Vrieze A, Nieuwdorp M et al (2013) Duodenal infusion of donor feces for recurrent Clostridium difficile. N Engl J Med 368(5):407–415. https://doi.org/10.1056/NEJMoa1205037
Vandamme TF, Gilissen LJ, Postnov AA (2015) Positron emission tomography in drug discovery and drug monitoring. Trends Pharmacol Sci 36(1):3–12
Wang G, Chen Z (2022) HBV genomic integration and hepatocellular carcinoma. Adv Gut Microb Res. https://doi.org/10.1155/2022/2140886
Wang X, Geng S (2023) Diet-gut microbial interactions influence cancer immunotherapy. Front Oncol 13. https://doi.org/10.3389/fonc.2023.1138362
Wang X, Yang L, Chen ZG, Shin DM (2016) Application of nanotechnology in cancer therapy and imaging. CA Cancer J Clin 58(2):97–110
Wang L-S, Mo YY, Huang Y-W, Echeveste CE, Wang H-T, Chen J, Oshima K, Yearsley M, Simal-Gandaraf J, Battino M, Xiao J, Chen J, Sun C, Yu J, Bai W (2020) Effects of dietary interventions on gut microbiota in humans and the possible impacts of foods on Patients' responses to cancer immunotherapy. eFood. https://doi.org/10.2991/EFOOD.K.200824.002
Warren GW, Sobus S, Gritz ER (2014) The biological and clinical effects of smoking by patients with cancer and strategies to implement evidence-based tobacco cessation support. Lancet Oncol 15(12):e568–e580. https://doi.org/10.1016/s1470-2045(14)70266-9
World Cancer Research Fund/American Institute for Cancer Research (2018) Diet, nutrition, physical activity and cancer: a global perspective. continuous update project expert report 2018
Xie Y, Liu F (2024) The role of the gut microbiota in tumor, immunity, and immunotherapy. Front Immunol 15. https://doi.org/10.3389/fimmu.2024.1410928
Xu H, Luo Y, Li Q, Zhu H (2024) Acupuncture influences multiple diseases by regulating gut microbiota. Front Cell Infect Microbiol 14. https://doi.org/10.3389/fcimb.2024.1371543
Yadegar A, Pakpoor S, Ibrahim FF, Nabavi-Rad A, Cook L, Walter J et al (2023) Beneficial effects of fecal microbiota transplantation in recurrent Clostridioides difficile infection. Cell Host Microbe 31(5):695–711
Yang Z, Wang Y (2023) Clinical development of chimeric antigen receptor-T cell therapy for hematological malignancies. Chin Med J 136:2285
Yin P-K, Xiao H, Yang Z-B, Yang D-S, Yang Y-H (2024) Shotgun metagenomics reveals the gut microbial diversity and functions in Vespa mandarinia (Hymenoptera: Vespidae) at multiple life stages. Front Microbiol 15. https://doi.org/10.3389/fmicb.2024.1288051
Yoo S, Jung S-C, Kwak K, Kim J-S (2024) The role of prebiotics in modulating gut microbiota: implications for human health. Int J Mol Sci 25:4834. https://doi.org/10.3390/ijms25094834
Yu F, Lu Y, Li Y, Uchio Y, Pangnguriseng UA, Kartika AV, Iizasa H, Yoshiyama H, Loh KS (2020) Epstein–Barr virus infection of pseudostratified nasopharyngeal epithelium disrupts epithelial integrity. Cancers 12(9):2722–2722. https://doi.org/10.3390/cancers12092722

Yu YX, Wang S, Liu ZN, Zhang X, Hu ZX, Dong HJ, Lu XY, Zheng JB, Cui HJ (2023) Traditional Chinese medicine in the era of immune checkpoint inhibitor: theory, development, and future directions. Chin Med 18(1):59. https://doi.org/10.1186/s13020-023-00751-7

Zaborowski AM, Winter DC, Lynch L (2021) The therapeutic and prognostic implications of immunobiology in colorectal cancer: a review. Br J Cancer 125(10):1341–1349. https://doi.org/10.1038/s41416-021-01475-x

Zakharov D (2022) Enlightenment medicine. ResearchGate

Zhang M, Liu J, Xia Q (2023) Role of gut microbiome in cancer immunotherapy: from predictive biomarker to therapeutic target. Exp Hematol Oncol 12:84. https://doi.org/10.1186/s40164-023-00442-x

Zhou C-B, Zhou Y-L, Fang J-Y (2021) Gut microbiota in cancer immune response and immunotherapy. Trends Cancer 7(7):647–660. https://doi.org/10.1016/j.trecan.2021.01.010

Zuppi M, Vatanen T, Wilson BC et al (2024) Fecal microbiota transplantation alters gut phage communities in a clinical trial for obesity. Microbiome 12:122. https://doi.org/10.1186/s40168-024-01833-w

Therapeutic Role of Probiotics and Prebiotics in Irritable Bowel Syndrome (IBS)

16

Swati Chandrakant Hosvakkal, Kanthesh M. Basalingappa, and K. Sumana

Abstract

Irritable bowel syndrome (IBS), a common condition affecting the gastrointestinal tract, causes bloating, problems with bowel movements, and recurring abdominal pain. Recent data suggest a role for probiotics and prebiotics in altering the gut flora, providing a possible avenue to treat IBS. Probiotics (live microorganisms that produce health benefits) have demonstrated a range of positive effects on individuals with IBS. These mechanisms include restoring the intestinal barrier integrity, normalization of gastrointestinal motility and visceral hypersensitivity, amelioration of gut–brain interaction, and modulation of the composition of gut microbiota. Although the efficacy of various probiotic strains is debated, some may be helpful in easing symptoms such as bloating and abdominal pain characterizing IBS, including *Bacillus coagulans*, *Bifidobacterium bifidum*, and *Lactobacillus plantarum*. Prebiotics, or indigestible carbohydrates that selectively promote the growth of beneficial gut bacteria, are less straightforward in how they relate to IBS. But prebiotics may likewise help to control the gut microbiota and mitigate symptoms, according to some research. Synbiotics are an investigational method, consisting of both probiotics and prebiotics. In conclusion, there is probable efficacy and sufficient safety to support the use of these probiotics as an adjunct/alternative treatment for people with IBS (and possibly prebiotics, too) amidst their otherwise galaxy of symptoms. Further research is needed to identify the specific pathways and products involved in IBS pathogenesis, as well as what may be the most effective probiotic/prebiotic ther-

S. C. Hosvakkal · K. Sumana (✉)
Department of Microbiology, School of Life Sciences, JSS Academy of Higher Education and Research, Mysore, Karnataka, India
e-mail: sumana.k@jssuni.edu.in

K. M. Basalingappa
Division of Molecular Biology, School of Life Sciences, JSS Academy of Higher Education and Research, Mysore, Karnataka, India

S. Pathak et al. (eds.), *Microbiota and Dietary Mediators in Colon Cancer Prevention and Treatment*, https://doi.org/10.1007/978-981-96-0297-1_16

apies for managing this enigmatic syndrome. In this chapter, we discuss the intricate relationship between gut microbiota and irritable bowel syndrome (IBS) focusing on treatment possibilities using probiotics and prebiotics. It reviews the complex relationship existing between gut microbiota and irritable bowel syndrome, focusing on the possible uses of probiotics and prebiotics as treatments. This chapter underlines the mechanisms through which probiotics load the function of the gastrointestinal barrier, modulate the gut microbiota, regulate the immune system, and alter stomach–brain axis communication in order to alleviate irritable bowel syndrome symptoms.

Keywords

Probiotics · IBS · Prebiotics · Gut microbiome · Synbiotics

16.1 Introduction

The intestinal functional disorder known as irritable bowel syndrome (IBS) is categorized as a non-life-threatening illness. It lowers one's capacity to contribute to society and results in a decline in life quality. Up to 10% of people globally are thought to be afflicted by this illness; 45% is the even higher prevalence estimated by certain sources (Chlebicz-Wójcik and Śliżewska 2021). Prevalence rates differ between countries; it is estimated to affect 1 in 10 individuals globally. Some developing countries adopting a Western diet and lifestyle might see a consistent increase in the IBS prevalence rate (Black and Ford 2020). The prevalence of IBS in India varies about 0.4–4.2% (Ghoshal et al. 2023). IBS is more common in younger age groups (26 to 55 years old) and in women. Women who are menstruating seem to experience recurrent and severe disease symptoms because of reduced ovarian hormone levels, which are linked to lower sensory edges for rectal swelling. Moreover, indications differ by gender, with males reporting added symptoms linked to diarrhea (IBS-D) and women reporting more symptoms of constipation (IBS-C). Older people also have constipation symptoms more frequently, most likely as a result of their increased comorbidities and decreased mobility (Ferreira et al. 2020).

Patients with IBS are classified into four categories based on the prevailing symptoms and bowel habits: constipation-predominant IBS (IBS-C), diarrhea-predominant IBS (IBS-D), mixed IBS (IBS-M), and unclassified patients with IBS. Disordered gastrointestinal motility, aberrant intestinal secretion, visceral hypersensitivity, altered gut–brain axis, and intestinal permeability are all linked to the etiology of IBS, and the gut microbial ecology can have an impact on these conditions. The gut microbiota's part in the pathophysiology of IBS has been steadily revealed, it is still unknown if microbial dysbiosis causes IBS (Xiao et al. 2021). Compared to healthy people, patients with IBS are also more likely to have minor intestine bacterial overgrowth.

The most often prescribed medications for IBS are antispasmodics, antidiarrheal medications, and laxatives; however, prolonged use of these medications can cause serious side effects (Chlebicz-Wójcik and Śliżewska 2021). In addition, treatments that aim to modify the microbiota, such as probiotics, prebiotics, and antibiotics, have shown promise in treating IBS (Wu et al. 2022). Numerous studies have demonstrated that an abnormal composition or metabolic rate of intestinal flora in patients with IBS suggests that intestinal microbes play a significant role in the illness. One potential treatment for IBS is the use of probiotics to regulate the gut microbiota. The microbial strains found in the intestines are referred to as the gut microbiota, gut microbiome, or gut microflora in common usage. The gastrointestinal tract is the primary location where healthy persons' natural bacteria survive. The gastrointestinal tract is the primary location where healthy persons' natural bacteria thrive. Multiple strains of bacteria that are not pathogenic represent a healthy, balanced microbiome in the gut. Its diversity and composition typically vary with the stages of a person's life, and it typically varies from people in good health to those with various diseases. It also typically varies throughout and following chemotherapy when prescribed oral antibiotics are used (Chen et al. 2021).

Because the gut microbiota regulates host physiology, it has a significant impact on an individual's overall health. An unbalanced gut flora can cause harm and raise the risk of gastrointestinal problems, including diarrhea. Unchecked dysbiosis can weaken the immune system and cause inflammation in certain areas of the gut, which can result in cancer of the colon, irritable bowel disorder, and inflammatory bowel disease (Dahiya and Nigam 2023). Probiotics have the ability to enhance intestinal flora and reduce the growth of harmful microorganisms. When consumed in sufficient quantities, prebiotics and probiotics are substances that can alter the microbiome and metabolome's makeup, which has significant effects on both the preservation of health and the treatment of illness. Prebiotics and probiotics have several main therapeutic benefits, such as regulating the immune and nervous systems of the host, enhancing the function of the intestinal barrier and improving nutritional absorption, competing with pathogens for minerals and adherence to the gut mucosa, and producing antimicrobial substances. Although the precise process by which they achieve these effects is still not well understood, recent research indicates that the generation of metabolites that modulate immunity is a major factor (Rau et al. 2024).

16.2 Understanding Irritable Bowel Syndrome

IBS is an enduring bowel disorder that disturbs the compromised intestinal function and causes pain in the abdomen associated with it. It also reduces work productivity and superiority of life of the patient (Camilleri 2021). The analysis of IBS is critical for reasons like variations in the symptoms, it can mimic other disorders, i.e. lactose and fructose intolerance, there are no exact biomarkers for IBS diagnosis, patients may order tests to identify the cause but routine testing results generally will be normal. As a result, doctors and researchers have relied on a number of standards that have been developed over time, i.e., Manning, Kruis, and Rome (Lacy and Patel 2017). William Oster was the first to describe the condition and coined it as "Mucous colitis" in 1892. Diagnostic progress was seen when Adrian Manning elaborated his diagnosis criteria, i.e., Manning criteria in Bristol in 1978 (Jones 2004). The Manning Criteria was the initial diagnostic measure to be presented and studied more broadly. The primary symptoms included more frequent bowel movements resulting in the initiation of discomfort, alleviation of stomach pain with a bowel movement, looser stools as soon as the pain started, and distension of the abdomen. Kurtis et al. first reported a comparable collection of symptoms to account for IBS in 1984.; bloating, intestinal pain, and altered gut function. Kurtis Criteria, in contrast to Manning Criteria, emphasizes the need to take into account warning signs, or "red flags." It also emphasizes the duration of symptoms, suggesting a two-year period, and further emphasizes the need to rule off organic illness using a typical physical exam in conjunction with basic laboratory testing (CBC and ESR) (Naaz Kazi et al. 2020).

In order to categorize functional gastrointestinal diseases (FGIDs) utilizing a symptom-based classification scheme, a panel of worldwide experts convened in Rome. This resulted in the 1992 publication of the Rome Criteria, which became known as Rome I (Naaz Kazi et al. 2020). Rome I criteria were shown to have an 85% sensitivity and 71% specificity for diagnostic accuracy. The Rome committee updated the Rome I standards after a few years, and the updated Rome II criteria were released in 1999. The most significant modification, which classified IBS into subtypes, was included in the 2006 introduction of the Rome III criteria. Following the publication of the Rome III criteria, creative studies in the basic and clinical sciences discovered novel IBS causes and expanded our understanding of the intricate pathophysiology of IBS (Corazziari 2004). These developments in knowledge caused several changes in the Rome III criteria to improve its clinical efficacy. The Rome IV criteria were realized in 2016 with new definitions and new rationale (Table 16.1).

16.3 Symptoms and Diagnostic Criteria

Some of the common symptoms of IBS are abdominal pain, changes in bowel habits (diarrhea, constipation, and mixed bowel habits), bloating, gas, mucus in stool, fatigue, anxiety, or depression and nausea. Symptom-based diagnosis particularly cannot be used for diagnosis of IBS as limited testing is required to reject conditions that mimic IBS (Camilleri 2021).

16.4 Pathophysiology of Irritable Bowel Syndrome

Conventionally, IBS is defined as a condition of GI motor abnormalities and visceral hypersensitivity, which causes diarrhea or constipation and abdominal pain or discomfort, respectively. However, our understanding of the pathophysiology of IBS is still incomplete. This illness may be exacerbated by pathogenic variables such as innate immunity, dysbiosis, gut–brain axis dysfunction, hereditary

Table 16.1 Rome criteria in detail

Rome criteria	Symptoms	Sub-types
I	Three months or longer of persistent or recurring symptoms, such as steady or connected with a shift in the frequency or texture of feces; discomfort or soreness in the abdomen that is eased by bowel motions; additionally. At least on a quarter of the events or days, at least two of the following: altered amount of bowel movements (for research purposes, "altered" can mean over three daily intestinal movements or less than three weekly stool movements); alterations in the stool's composition (lumpy, hard, loose, watery); straining, urgency, or the feeling of an incomplete evacuation in relation to stool passage; mucus passage; bloating; or a sensation of a bloated abdomen (Peyton and Greene 2014).	–
II	Stomach pain or discomfort for a minimum of twelve weeks (it doesn't have to be consecutive), accompanied by more than one of the following symptoms: alleviated upon defecation; or onset associated with a shift in the amount of stools; or onset associated with a modification in the appearance (shape) of stools (Saha 2014).	–
III	Development of symptoms greater than 6 months before diagnosis, with the last 3 months seeing the fulfillment of the following requirements: At least two of the following requirements must be satisfied: (a) Less than three bowel motions per week. (b) More than 25% of the time, manual defecation aids are required. (c) More than 25% of the time, hard or lumpy stools, (d) Feelings like an incomplete evacuation. (e) Feels like an anorectal obstruction, (f) Requires straining during feces more than 25% of the time (Moayyedi et al. 2017).	–
IV	During the previous 3 months, recurring stomach pain, on usual, at minimum one daytime each week, discomfort brought on by two or more of the following: Affecting the act of defecating. b. Linked to a variation in the frequency of stools. c. Correlated with a modification in the structure or appearance of stools, before a diagnosis, symptoms had to have begun at least six months earlier (Camilleri 2020).	Mixed IBS (IBS-M), diarrhea-predominant (IBS-D), and constipation-predominant (IBS-C), IBS-U, or unclassified IBS.

vulnerability, and dietary intolerance. IBS symptoms vary greatly amongst individuals; thus, it is still uncertain which of these pathogenic factors can cause or exacerbate the condition. Increased intestinal permeability, or "leaky gut," has been observed in a subdivision of patients with IBS, potentially allowing for the infiltration of luminal antigens and the subsequent stimulation of the mucosal immune structure (Tang et al. 2021). This chronic inferior inflammation can contribute to the characteristic symptoms of IBS, such as altered bowel function and abdominal pain (Ng et al. 2018). The disruption of the intestinal epithelial barrier, which helps as a critical interface between the host and the external atmosphere, is one important aspect in the pathogenesis of IBS. The pathophysiology of IBS has also been linked to disruption of the gut–brain axis in addition to barrier failure (Vancamelbeke and Vermeire 2017). Changes in the bidirectional relationship between the digestive system and the brain's nervous system can affect visceral understanding, movement of the gut, and the interpretation of pain signals. These effects can be caused by a range of natural and psychological variables (Gwak and Chang 2021).

16.5 The Gut Microbiome and Irritable Bowel Syndrome

The collection of bacteria that have colonized our bodies since childhood and are dispersed throughout them is known as the human microbiome. Because they supply necessary amino acids, vitamins, and fatty acids with short chains, as well as promoting the proper growth and stimulation of the intestinal immune system, these microbes are needed for appropriate gut development. The gut is the most studied organ and has the highest bacteria count. Millions of bacteria are found in a healthy human's saliva, which we swallow every day along with our food (Gilbert et al. 2018). However, a number of variables, such as the stomach's acidity and the formation of bile acids, limit the microbes' ability to remain in the gut. Numerous other chemical and biological factors such as pH, oxygen concentrations, redox potential, also production of mucus, bile, antibodies, and physical factors such as gut structure, all have an impact on the further downstream microbial colonization. As a result, a gradient of microbial concentration is present throughout the small intestine. This is because, although having fairly identical microbial taxa, the microbial abundance in duodenal aspirates was found to be 1000 times lower than that of oral samples (de Vos et al. 2022).

The initial colonization of the GI tract depends on the means by which the neonate is born. Because the newborn is exposed to the mothers' vaginal microbes during vaginal delivery, their early composition tends to imitate this section. These microorganisms belong to the *Lactobacillus, Prevotella*, and *Sneathia* genera, more precisely. On the other hand, those born by caesarean delivery typically have less diversified microbiotas and are heavily influenced by the skin flora of their mothers (Houghteling and Walker 2015). As a result, they include larger concentrations of skin-dwelling bacteria, such as species of *Staphylococcus, Corynebacterium*, and *Propionibacterium*. It has also been demonstrated that the administration of intrapartum antibiotics during caesarean sections alters the gut flora of the newborn. Overall, compared to kids delivered via C-section, there is a higher level of gut microbiota similarity between moms and newborns delivered vaginally. The type of milk consumed early in life also affects the gut microbiota, *Enterobacteriaceae* species are more common in formula-fed infants, but *Bifidobacterium* species dominate in the microbiome of breastfed infants. The diet also plays a vital role in gut microbiota.

Probably one of the most remarkable features of the gut microbiome in IBS patients is low abundance or reduced diversity. Healthy individuals have a rich and diverse microbial community that helps them in gut homeostasis and resilience. On the other hand, a patient with IBS shows lesser diversity of the microbiome, which gives way to lesser functional capacity and a higher risk of perturbations. Specific taxonomic changes in the gut microbiota in IBS have been reported by many studies (Huang et al. 2022). For instance, a decrease in positive bacteria, such as *Lactobacillus and Bifidobacterium* species, has been reported in many patients with IBS. These bacteria have anti-inflammatory properties and act on maintaining gut barrier integrity; hence, their reduction can be attributed to the low-quality inflammation and increased intestinal absorptivity seen in IBS (Lobionda et al. 2019).

On the contrary, levels of potentially harmful bacteria are increased in patients with IBS. For example, a few studies found a higher level of *Enterobacteriaceae* pathogenic species. Overgrowth of sulfate-reducing bacteria has also been reported and may contribute to visceral hypersensitivity through the production of hydrogen sulfide. The ratio of *Firmicutes* to *Bacteroidetes,* two major bacterial phyla in the gut, is often changed in IBS patients. However, the direction of change may be different; some of the studies have reported an increase and others a decrease, underpinning that changes in the microbiota are heterogeneous in IBS (Menees and Chey 2018).

Moreover, the gut microbiome interacts with the host immune system. In that case, dysbiosis of patients with IBS could be responsible for the low-grade chronic inflammation characteristic of this syndrome. Thus, shifts in the composition of the microbiome might raise the levels of proinflammatory cytokines and modulate the T-regulatory cell balance that underestimated IBS symptoms (Yoo et al. 2020).

16.6 Probiotics: Therapeutic Potential in Irritable Bowel Syndrome

When taken in sufficient quantities, probiotics—live microorganisms—benefit the host through colonizing the body. Probiotics have the ability to adapt to the anatomy of the human intestine and stop pathogen growth. By strengthening the intestinal mucosa's protective layer, boosting the intestinal barrier effect, and boosting immunity, they benefit the body (You et al. 2022).

16.7 Modulation of Gut Microbiota

In the fight for gut nutrients, minerals, and receptors that trigger mucosal immunity, probiotics outcompete harmful bacteria. When dangerous bacteria adhere to and colonize human mucosa, infections may ensue. As a result, attachment to intestinal epithelial cells is the initial step for *Lactobacillus* strains in their colonization and subsequent contact with the gut, which is essential for preventing harmful bacteria and boosting immune system performance. A further example is *Bifidobacterium adolescentis* S2-1, which suppresses the growth of *Porphyromonas gingivalis* and competes with it for vitamin K. Probiotics can also change the physical environment in a way that makes it impossible for harmful bacteria to flourish Their capacity to generate compounds with antibacterial properties allows them to prevent gastrointestinal illnesses (Zhang 2022). By strengthening the functions of the intestine mucosa barrier and producing intestinal defensive metabolites, including glutamine, arginine, short-chain lipids, and conjugation linoleic acids, these chemicals prevent the proliferation of dangerous germs. Probiotics generate a variety of antipathogenic substances, including peptides, ethanol, diacetyl, acetaldehydes, organic acids, and bacteriocins. Target cells' membrane permeability is increased by peptides and bacteriocins, which alter the membrane and eventually cause cell death. Antibiotics are also produced by probiotic microbes; one such antibiotic is rutin, which has inhibitory effects on viruses, yeast, fungi, bacteria, and bacterial species that are both Gram-positive and Gram-negative (Latif et al. 2023).

16.8 Improvement of Gut Barrier Function

As the initial line of defense against invasive infections, the immune system of the mouth is crucial. It is made up of chemical and biological compounds that work together to keep microorganisms out. By releasing the bile salts, probiotics can improve mucus output and attract water to the colon. Secretion antibody A (SIgA), which serves a crucial protective role for the mucosal and can stop pathogen invasion, is secreted by the mucosa of the intestine. Probiotics also stimulate the gut defense pathways by promoting the manufacture of defensing cationic antimicrobial peptides generated by a range of types of cells, particularly Paneth cells, which are located in the tiny intestine crypts and intestine epithelial tissues. These peptides, also known as low-molecular-weight proteins, influence how the intestinal barrier operates and are effective against bacteria, fungi, and viruses (Vancamelbeke and Vermeire 2017).

16.9 Immunomodulation

The immune system is regulated by the microbial flora in the gut, which produces chemicals that have anti-inflammatory and immunological properties. Interactions between probiotic bacteria and dendrites, macrophages, monocytes, and epithelial cells lymphocytes, humans have evolved a highly sophisticated mucosal and systemic immune system. Mucosal immunity is recognized as the initial line of protection that lowers the need for systemic immunity (Aleman and Yadav 2023). Both the mucosal and systemic immune systems can be modified by probiotics. The initial line of defense for natural killer (NK) cells is innate immunity, which is the primary cell implicated in the spontaneous recognition and death of invasion targets (tumor cells, cells infected with viruses, bone marrow cell lines, and embryonic tissues). Fundamental immune responses are crucial because they serve as the body's initial line of defense and a trigger for the emergence of immunological responses. Several strains of lactobacilli promote the growth of cells called dendritic cells (DCs), which improves the immune system's innate and acquired functions. As a result, numerous findings show that lactic acid bacteria effectively enhance both innate and acquired immunity, preventing allergies of the stomach mucosa and protecting against infections by intestinal pathogens (Aleman and Yadav 2023).

Probiotics can boost immunity by engulfing antigens from deceased cells. When compared to the ingestion of live probiotic bacteria, the consumption of nonliving cells dramatically reduces the generation of cytokines. Dead bacteria are swiftly removed from the intestinal lumen after engaging in several encounters with immune cells (Mazziotta et al. 2023).

Probiotics increase immunological responses by inducing T cells, which are employed to produce cytokines, increase Th1 responses, and decrease Th2 responses. Probiotics have the power to increase the production of interferon-γ, IgG, and IgM antibodies, as well as to repair deficient NK cell function. Regarding pathogen phagocytosis, probiotics have the power to improve and restore macrophage activity. Phagocytic drugs inhibit lytic digestive enzymes, reactive oxygen intermediary molecules, and dangerous chemicals throughout various stages of inflammation. Probiotics increase the development of mucosal antibodies, especially IgA, and secrete more immunoglobulin, which can decrease the amount of pathogenic microbes in the colon while enhancing the microbial flora. Probiotics alter the microbiota of the gut by preserving balance and preventing the proliferation of potentially hazardous microorganisms (Zhang et al. 2022).

Probiotics also change the microbiota in the gut as they have the ability to boost immunological responses and T-regulatory cell development. Irritable bowel syndrome (IBS), autoimmune disorders, and allergies are examples of inflammatory conditions that probiotics can help cure. The Th1/Th2 cell ratio is crucial for the innate defense mechanism. Moreover, a particular category of Th1 cells within the immune response produces IFN-γ, and these types of cells are critical for promoting inflammatory responses. Th1 reduces immunity in cells by ensnaring intracellular bacteria and viruses, whereas Th2 reacts to humoral defense and IL-4 production by secreting IFN-γ cytokines (Aleman and Yadav 2023) (Fig. 16.1).

16.10 Prebiotics: Therapeutic Potential in Irritable Bowel Syndrome

The prebiotics concept was first introduced by Glenn Gibson & Marcel Roberfroid in 1995. "An indigestible food ingredient which selectively stimulates the development and/or function of a few kinds of microbes in the colon, thus enhancing host health," according to the definition of prebiotics. It is vital for human health to have prebiotics. Many food items, including rye, soybean, bananas, wheat, tomato plants, the asparagus plant, sugar beets, cloves of garlic, chicory, onion, Palestine artichokes, grain, honey, and beans and peas, and, lately, seaweeds and microalgae, are naturally rich in them. Prebiotics come in a variety of varieties. Most of them are oligosaccharide carbohydrates, which are a subset of groups of carbohydrates (OSCs) (Davani-Davari et al. 2019). Fructooligosaccharides (FOS),

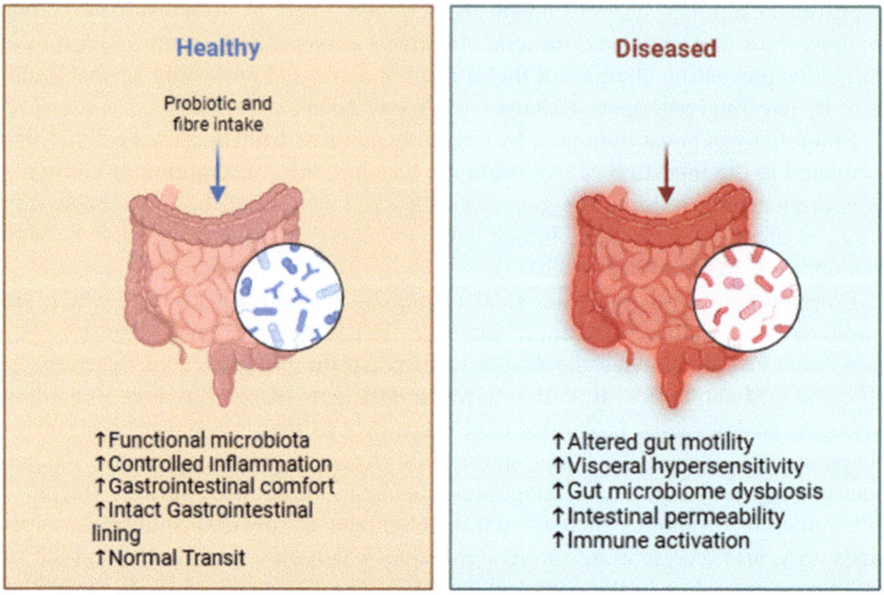

Fig. 16.1 Role of probiotics in modulation of the gut

galactooligosaccharide (GOS), and trans-galactooligosaccharides (TOS) are the three most common forms of prebiotics. Short-chain fatty acids (SCFAs), which are created when the gut bacteria ferment prebiotics, include lactic acid, butyric acid, and propionic acid. These could have a number of effects on the body. The composition of the prebiotics and the variety of bacteria in the gut dictate the fermentation products. Human health advantages from prebiotics are transmitted by the byproducts of bacteria that break them down (Davani-Davari et al. 2019).

The possible use of prebiotics to manage the condition known as IBS has been investigated in a number of studies; however, the exact processes through which prebiotics may affect IBS remain unclear.

Modulation of the Gut Microbiota The popular prebiotic research focuses on the improvement of *Bifidobacterium* and *Lactobacillus* strains, which ferment prebiotics to create lactate and acetate and improve host health. Prebiotics may have a wider impact on the gut community, where bacterial cooperation and competition are important, according to the growing understanding of the gut microbiota. Prebiotic foods include those that can be fermented by the gut and increase the generation of health-promoting compounds in the intestine (Bedu-Ferrari et al. 2022). These complex carbohydrates provide a substantial portion of the 20–60 g of carbohydrates per day that enter the colon and act as fermenting substrates for the gut bacteria. They include resistant starch and plant cell-wall polysaccharides. Fermentation in the gut creates short-chain lipids (SCFAs), mainly butyrate, acetate, and propionate. They offer the host a number of health benefits, including energizing colorectal tissues, supporting cellular functions that preserve tissue integrity, enhancing immunity, and maybe having anti-inflammatory qualities. Dietary fiber fermentation results in the production of formate, lactate, and succinate, which lower gut pH and inhibit the growth of pathogenic microorganisms. Therefore, the selective preponderance of the bacteria that produce these metabolites enhances the prebiotic qualities of the complex carbohydrates (Umu et al. 2017).

Improvement of Gut Barrier Function By altering intestinal tight junctions, several prebiotics support the function of the intestinal barrier. Occludin, claudin-3, and ZO-1 are among the TJ genes that are significantly upregulated when inulin fermentation products are supplemented in vitro. Certain proteins, such as ZO-1 and occludin, are redistributed to the area around the TJs as a result of supplementation with fructooligosaccharides and butyrate, a bacterial metabolite generated during prebiotic fermentation. Studies conducted both in vivo and in vivo showed that pre-treatment with galactooligosaccharide causes ZO-1, occludin, and claudin-1 gene expression to be upregulated in mice (Rose et al. 2021).

Immunomodulation Prebiotics have a variety of effects on gut immunity. As the first step in any pathogenic invasion, pathogens adhere to intestinal epithelial cells and subsequently colonize them. Several prebiotics, such as lactulose, raffinose, inulin, GOS, FOS, and pectin oligosaccharides, have been reported to react to this effect by acting insoluble decoy receptors that mimic the pathogen-binding site, making it simpler for the infectious agent to bind and be removed from the gut (Vieira et al. 2013).

Mechanistic investigations have demonstrated the significance that supports the integrity of the epithelium barrier and immune system maintenance provided by fiber from the diet, probiotics, and their fermentation byproducts (SCFAs). DCs play a crucial role in regulating the immune system's response to both foreign and self-antigens since they are antigen-presenting cells. These receptors can be contacted by prebiotics or their fermentation byproducts, which then initiate the maturation process by secreting cytokines & up- or reducing membrane molecules such as CD83, CD86, HLA-DR, and DC-SIGN. The stimulation of DCs through certain pattern recognition receptors determines the tendency of cytotoxic T cells toward the Th1, Th2, Th17, or Treg phenotype (Shah et al. 2020).

16.11 Synbiotics: Combining Probiotics and Prebiotics

The word "symbiotic" usually refers to a probiotic and prebiotic combination that is beneficial to the host and functions well together. In May 2019, the definition of synbiotics was revised by the International Scientific Association for Probiotics and Prebiotics (ISAPP). Synbiotics are described as "a combination comprising living organisms and substrate(s) specifically utilized by host microbes that confers a health benefit on the host" by a panel of specialists in the subject, which included microbiologists, physiologists, and nutritionists (Simon et al. 2021).

Cappello et al. conducted the first investigation of Probinul® and its effects on people with IBS in 2013. The synbiotic contained strains of *Bifidobacterium* spp., *Lactobacillus* spp., and Streptococcus thermophilus, as well as inulin and tapioca-resistant starch. Probinul® did not reduce bloating and gas to the degree that trial participants found satisfactory (Chlebicz-Wójcik and Śliżewska 2021). Synbiotic products can be used to enhance the intestinal or extraintestinal microbial communities of animals and humans by feeding them as foods, nonfoods, nutritional supplements, or feed additives. Probiotics and synbiotics have been shown to have beneficial benefits on chronic diseases linked to oxidative stress because of their antioxidant qualities, which also help to reduce inflammation, boost immunity, and lessen oxidative stress in organs and DNA damage (Mounir et al. 2022). The host's digestive enzymatic activities can be improved by dietary synbiotic delivery, which enables the host to break down more nutrients. This dietary strategy speeds up the process of digestion, which probably improves feed efficiency and/or weight gain rate. Feed supplements containing both probiotics and prebiotics work together to protect well-organized fermentation while also increasing the survival of beneficial organisms. In the end, this provides the host with a fitting response. Via the GIT, probiotic and prebiotic dietary supplements (MOS, FOS, and inulin) strengthen the host's immune systems. Given its ability to resist infections and regulate dangerous bacteria, a confrontation examination can be used to assess the efficacy of a synbiotic treatment in settings of defense against infectious agents (Srirengaraj et al. 2023).

16.12 Conclusion

With the investigation of probiotics and prebiotics as therapeutic agents, our understanding is greatly advanced with regard to this most complex and heterogeneous condition, the syndrome called irritable bowel syndrome. As previously reviewed in this chapter, bloating, irregular bowel patterns, stomach pain, and other associated symptoms define IBS and cause noteworthy impairment in the superiority of life in these patients. As gut microbial dysbiosis emerged as one of the major factors in the pathophysiology of IBS, new avenues for treatment have opened up. Of these, several, including the use of probiotics and prebiotics, are particularly promising. The disturbance of gut microbiota has been implicated in a variety of gastrointestinal disorders, such as IBS. It was proven that the gut microbiome played a very important role in maintaining intestinal homeostasis. Probiotics can decrease IBS symptoms through several mechanisms. Probiotics are live microorganisms that, when administered in adequate amounts, convey health benefits. These changes include events such as alterations to the gut microbiota composition, enhancement of the function of the intestinal barrier, modulation of immune responses, and influence on the gut–brain axis. Variable efficacy in controlling symptoms of IBS was reportedly indicated through probiotic strains in clinical studies. Notably, some of the strains have been overly impressive in reducing stomach discomfort and bloating.

Application of personalized medicine techniques in probiotic and prebiotic therapies is possible, which can be based on their unique microbiome profile, thereafter tailoring the same as it suits each patient.

Other IBS treatment options being emphasized are prebiotics, which are food-indigestible ingredients that stimulate the growth or activity of good gut flora. Prebiotics may help in improving gut health or alleviating symptoms associated with IBS through activities linked to good bacteria and stimulation of good short-chain fatty acid production. While prebiotics have fewer studies backing up their efficacy in the treatment of IBS as compared with probiotics, they can still be helpful, especially if given as synbiotics in association with probiotics. The challenges and limitations in this area must not be ignored. It is hard to draw firm conclusions and direct comparisons due to IBS heterogeneity, variations in study methods, and the probiotic strains and prebiotic chemicals used by different studies. Supplementary studies are warranted to establish the long-standing efficacy and safety of such treatments.

References

Aleman RS, Yadav A (2023) Systematic review of probiotics and their potential for developing functional nondairy foods. Appl Microbiol 4(1):47–69

Bedu-Ferrari C, Biscarrat P, Langella P, Cherbuy C (2022) Prebiotics and the human gut microbiota: from breakdown mechanisms to the impact on metabolic health. Nutrients 14(10):2096

Black CJ, Ford AC (2020) Global burden of irritable bowel syndrome: trends, predictions and risk factors. Nat Rev Gastroenterol Hepatol 17(8):473–486

Camilleri M (2020) Irritable bowel syndrome: straightening the road from the Rome criteria. Neurogastroenterol Motil 32(11):e13957

Camilleri M (2021) Diagnosis and treatment of irritable bowel syndrome. JAMA 325(9):865

Chen Y, Zhou J, Wang L (2021) Role and mechanism of gut microbiota in human disease. Front Cell Infect Microbiol 17:11

Chlebicz-Wójcik A, Śliżewska K (2021) Probiotics, prebiotics, and synbiotics in the irritable bowel syndrome treatment: a review. Biomol Ther 11(8):1154

Corazziari E (2004) The Rome criteria for functional gastrointestinal disorders: a critical reappraisal. J Pediatr Gastroenterol Nutr 39(Supplement 3):S754–S755

Dahiya D, Nigam PS (2023) Biotherapy using probiotics as therapeutic agents to restore the gut microbiota to relieve gastrointestinal tract inflammation, IBD, IBS and prevent induction of cancer. Int J Mol Sci 24(6):5748

Davani-Davari D, Negahdaripour M, Karimzadeh I, Seifan M, Mohkam M, Masoumi S et al (2019) Prebiotics: definition, types, sources, mechanisms, and clinical applications. Food Secur 8(3):92

Ferreira AI, Garrido M, Castro-Poças F (2020) Irritable bowel syndrome: news from an old disorder. GE Port J Gastroenterol 27(4):255–268

Ghoshal UC, Sachdeva S, Pratap N, Karyampudi A, Mustafa U, Abraham P et al (2023) Indian consensus statements on irritable bowel syndrome in adults: a guideline by the Indian Neurogastroenterology and Motility Association and jointly supported by the Indian Society of Gastroenterology. Indian J Gastroenterol 42(2):249–273

Gilbert JA, Blaser MJ, Caporaso JG, Jansson JK, Lynch SV, Knight R (2018) Current understanding of the human microbiome. Nat Med 24(4):392–400

Gwak MG, Chang SY (2021) Gut-brain connection: microbiome, gut barrier, and environmental sensors. Immune Netw 21(3):e20

Houghteling PD, Walker WA (2015 Mar) Why is initial bacterial colonization of the intestine important to infants' and children's health? J Pediatr Gastroenterol Nutr 60(3):294–307

Huang Z, Liu K, Ma W, Li D, Mo T, Liu Q (2022) The gut microbiome in human health and disease—where are we and where are we going? A bibliometric analysis. Front Microbiol 15:13

Jones R (2004) Irritable bowel syndrome: management of expectations and disease. Br J Gen Pract 54(504):490–491

Lacy BE, Patel NK (2017) Rome criteria and a diagnostic approach to irritable bowel syndrome. J Clin Med 6(11):99

Latif A, Shehzad A, Niazi S, Zahid A, Ashraf W, Iqbal MW et al (2023) Probiotics: mechanism of action, health benefits and their application in food industries. Front Microbiol 17:14

Lobionda S, Sittipo P, Kwon HY, Lee YK (2019) The role of gut microbiota in intestinal inflammation with respect to diet and extrinsic stressors. Microorganisms 7(8):271

Mazziotta C, Tognon M, Martini F, Torreggiani E, Rotondo JC (2023) Probiotics mechanism of action on immune cells and beneficial effects on human health. Cells 12(1):184

Menees S, Chey W (2018) The gut microbiome and irritable bowel syndrome. F1000Res 7:1029

Moayyedi P, Mearin F, Azpiroz F, Andresen V, Barbara G, Corsetti M et al (2017 Oct) Irritable bowel syndrome diagnosis and management: a simplified algorithm for clinical practice. United European Gastroenterol J 5(6):773–788

Mounir M, Ibijbijen A, Farih K, Rabetafika HN, Razafindralambo HL (2022) Synbiotics and their antioxidant properties, mechanisms, and benefits on human and animal health: a narrative review. Biomol Ther 12(10):1443

Naaz Kazi F, Agarwal A, Kanni P (2020) Comparison of Kruis, Manning and Rome IV criteria in irritable bowel syndrome. Int J Gastroenterol 4(2):34

Ng QX, Sen SAY, Loke W, Lim DY, Yeo WS (2018) The role of inflammation in irritable bowel syndrome (IBS). J Inflamm Res 11:345–349

Peyton L, Greene J (2014) Irritable bowel syndrome: current and emerging treatment options. P T 39(8):567–578

Rau S, Gregg A, Yaceczko S, Limketkai B (2024) Prebiotics and probiotics for gastrointestinal disorders. Nutrients 16(6):778

Rose EC, Odle J, Blikslager AT, Ziegler AL (2021) Probiotics, prebiotics and epithelial tight junctions: a promising approach to modulate intestinal barrier function. Int J Mol Sci 22(13):6729

Saha L (2014) Irritable bowel syndrome: pathogenesis, diagnosis, treatment, and evidence-based medicine. World J Gastroenterol 20(22):6759–6773

Shah BR, Li B, Al Sabbah H, Xu W, Mráz J (2020) Effects of prebiotic dietary fibers and probiotics on human health: with special focus on recent advancement in their encapsulated formulations. Trends Food Sci Technol 102:178–192

Simon E, Călinoiu LF, Mitrea L, Vodnar DC (2021) Probiotics, prebiotics, and synbiotics: implications and beneficial effects against irritable bowel syndrome. Nutrients 13(6):2112

Srirengaraj V, Razafindralambo HL, Rabetafika HN, Nguyen HT, Sun YZ (2023) Synbiotic agents and their active components for sustainable aquaculture: concepts, action mechanisms, and applications. Biology (Basel) 12(12):1498

Tang HY, Jiang AJ, Wang XY, Wang H, Guan YY, Li F et al (2021) Uncovering the pathophysiology of irritable bowel syndrome by exploring the gut-brain axis: a narrative review. Ann Transl Med 9(14):1187–1187

Umu ÖCO, Rudi K, Diep DB (2017) Modulation of the gut microbiota by prebiotic fibres and bacteriocins. Microb Ecol Health Dis 28(1):1348886

Vancamelbeke M, Vermeire S (2017) The intestinal barrier: a fundamental role in health and disease. Expert Rev Gastroenterol Hepatol 11(9):821–834

Vieira AT, Teixeira MM, Martins FS (2013) The role of probiotics and prebiotics in inducing gut immunity. Front Immunol 4:445

de Vos WM, Tilg H, Van Hul M, Cani PD (2022) Gut microbiome and health: mechanistic insights. Gut 71(5):1020–1032

Wu J, Lv L, Wang C (2022) Efficacy of fecal microbiota transplantation in irritable bowel syndrome: a meta-analysis of randomized controlled trials. Front Cell Infect Microbiol 28:12

Xiao L, Liu Q, Luo M, Xiong L (2021) Gut microbiota-derived metabolites in irritable bowel syndrome. Front Cell Infect Microbiol 23:11

Yoo J, Groer M, Dutra S, Sarkar A, McSkimming D (2020) Gut microbiota and immune system interactions. Microorganisms 8(10):1587

You S, Ma Y, Yan B, Pei W, Wu Q, Ding C et al (2022) The promotion mechanism of prebiotics for probiotics: a review. Front Nutr 5:9

Zhang P (2022) Influence of foods and nutrition on the gut microbiome and implications for intestinal health. Int J Mol Sci 23(17):9588

Zhang T, Zhang C, Zhang J, Sun F, Duan L (2022) Efficacy of probiotics for irritable bowel syndrome: a systematic review and network meta-analysis. Front Cell Infect Microbiol 1:12

MIX
Papier aus verantwortungsvollen Quellen
Paper from responsible sources
FSC® C105338

If you have any concerns about our products,
you can contact us on
ProductSafety@springernature.com

In case Publisher is established outside the EU,
the EU authorized representative is:
Springer Nature Customer Service Center GmbH
Europaplatz 3, 69115 Heidelberg, Germany

Printed by Libri Plureos GmbH
in Hamburg, Germany